Melatonin: Signal Transduction Mechanisms and Defense Networks in Plants

Rahul Kumar Tiwari
Milan Kumar Lal
Ravinder Kumar • Muhammad Ahsan Altaf
Editors

Melatonin: Signal Transduction Mechanisms and Defense Networks in Plants

Springer

Editors
Rahul Kumar Tiwari
Division of Crop Protection
ICAR-Indian Institute of Sugarcane Research
Lucknow, Uttar Pradesh, India

Milan Kumar Lal
Division of Crop Physiology and Biochemistry
ICAR-National Rice Research Institute
Cuttack, Odisha, India

Ravinder Kumar
Division of Plant Pathology
ICAR-Indian Agricultural Research Institute
New Delhi, Delhi, India

Muhammad Ahsan Altaf
School of Breeding and Multiplication (Sanya Institute of Breeding and Multiplication)
Hainan University
Sanya, Hainan, China

ISBN 978-981-96-8868-5 ISBN 978-981-96-8869-2 (eBook)
https://doi.org/10.1007/978-981-96-8869-2

This Springer imprint is published by the registered company Springer Nature Singapore Pte Ltd.
The registered company address is: 152 Beach Road, #21-01/04 Gateway East, Singapore 189721, Singapore

Preface

In the twenty-first century, the emerging role of melatonin, a multifaceted phytoprotectant associated with animal biology, has extended extensively into the domain of plant science. In the plant kingdom, phytomelatonin has emerged as a master regulator that modulates growth and development and enhances resistance against diverse abiotic and biotic stressors. The experimental evidences of prominent role of melatonin in plants has motivated researchers worldwide to explore its physiological, molecular, and biochemical functions, which will contribute to crop improvement, food preservation, and sustainable agriculture.

This book, *Melatonin: Signal Transduction Mechanisms and Defense Networks in Plants*, is a collaborative work aimed at delivering a comprehensive understanding of functional roles of phytomelatonin in plants. This book comprises 17 crafted chapters that document fruitful evidences, omics-driven mechanistic insights, and extensive studies highlighting how melatonin modulates responses to various abiotic and biotic stresses, regulates signaling pathways and crosstalk with phytohormones, and maximizes post-harvest life of perishable food products.

The contributors of this book have disciplinary expertise and diverse geographical representation. The chapters have been written by subject matter specialists, experienced researchers, and scholars working in the area of plant signaling and defense networks. The authors represent India, China, Egypt, Iran, Malaysia, and several other countries. This suggests the increasing momentum of global research on phytomelatonin in managing agricultural challenges and adversities arising due to climate change, soil degradation, and emergence and re-emergence of crop pathogens.

The need for this book is both timely and urgent. In the era of sustainable agriculture and eco-friendly plant health management practices, the investigation on natural and eco-friendly plant growth regulators like melatonin becomes essential. This book is an essential resource for agronomists, horticulturists, plant protection scientists, molecular biologists, research scholars, and industry stakeholders looking for sustainable solutions for crop production and protection.

We sincerely acknowledge all the contributors for their scholarly insights, meticulous scientific work and the reviewers who offered valuable, positive, and constructive insights to maintain the reputed standards of this volume. We are grateful to the editorial and production team at Springer for their valuable support and professionalism throughout the publication process.

Lucknow, Uttar Pradesh, India — Rahul Kumar Tiwari
Cuttack, Odisha, India — Milan Kumar Lal
New Delhi, Delhi, India — Ravinder Kumar
Sanya, Hainan, China — Muhammad Ahsan Altaf

Contents

Editors and Contributors

About the Editors

Rahul Kumar Tiwari, PhD in Plant Pathology, is currently researching melatonin-mediated mitigation of soil and tuber-borne diseases and physiological disorders in crop plants. He has also worked on the signaling mechanisms of pathogens, pathogenomics, and the management of hemibiotrophic fungi. Additionally, he has explored the genetic diversity of pathogenic microbes in field and horticultural crops. His research interests also include various aspects of pathophysiological alterations in food crops under the influence of diverse biotic and abiotic stresses. He has published more than 60 articles in nationally and internationally reputed journals.

Milan Kumar Lal, PhD in Plant Physiology, focuses on abiotic stress and the nutritional aspects of potatoes and other starchy crops. He is an expert in the field of abiotic stress, such as heat, drought, salinity, and heavy metals. Additionally, he is researching the effects of biotic stresses such as fungi, viruses, and bacteria on plant physiological, biochemical, and molecular responses. He also specializes in the nutritional and quality aspects of starchy crops, including resistant starch, glycemic index, phytonutrients, functional fermented foods and beverages, bioactive compounds, and various processing techniques to enhance these components in starchy crop-based food products. His work has resulted in more than 90 publications in international peer-reviewed journals.

Ravinder Kumar, PhD, a Senior Scientist in Plant Pathology, has over 15 years of research experience in managing biotic and abiotic stresses in plants, particularly in potato biotechnology. His work includes the formulation of dsRNA for late blight, the development of transgenic lines with ToLCNDV resistance, and potato genetic resource management. He has developed several diagnostic tools, including uniplex/multiplex RT-PCR, real-time RT-PCR, LAMP, and RT-RPA protocols for the detection of potato pathogens, as well as the molecular characterization and genome sequencing of plant pathogens. He has published over 120 research papers, review articles in national and international peer-reviewed journals, training manuals, and book chapters. He has also edited institutional publications such as newsletters and annual reports. He has received several awards, including the IPA-Kaushalya Sikka Memorial Award, Chandra Prabha Singh Young Scientist Award, Young Scientist Associate Award, and best oral/poster awards from various scientific and professional societies.

Muhammad Ahsan Altaf completed his PhD at the School of Life Sciences at Hainan University, China. He is currently employed as a postdoctoral researcher in the College of Horticulture at Hainan University, China. He has published around 30 research articles in leading journals worldwide, many with high impact factors. His research interests focus on the physiological, biochemical, and molecular aspects of horticultural plants, especially solanaceous vegetable crops. He is actively engaged in investigating the role of melatonin in photosynthetic efficiency and mineral nutrient uptake from root to shoot under abiotic stress.

Contributors

H. M. Akshay Kumar Division of Crop Protection, ICAR-Indian Agricultural Research Institute, Gogamukh, Assam, India

Abeer Ali Division of Fruit Science, ICAR-Indian Agricultural Research Institute, New Delhi, India

Ansheef Ali School of Crop Improvement, ICAR-Indian Agricultural Research Institute, Gogamukh, Dhemaji, Assam, India

Muhammad Ahsan Altaf Key Laboratory for Quality Regulation of Tropical Horticultural Crops of Hainan Province, School of Breeding and Multiplication (Sanya Institute of Breeding and Multiplication), Center of Nanfan and High-Efficiency Tropical Agriculture, Hainan University, Sanya, China

Key Laboratory for Quality Regulation of Tropical Horticultural Crops of Hainan Province, School of Tropical Agriculture and Forestry, Hainan University, Danzhou, China

Hainan University, Haikou, China

Nawal Asif Genetics and Breeding Center for Blunt Snout Bream, Ministry of Agriculture, Shanghai, China

Key Laboratory of Freshwater Aquatic Genetic Resources, Ministry of Agriculture, Shanghai, China

National Demonstration Center for Experimental Fisheries Science Education, Shanghai Ocean University, Shanghai, China

Ahsan Ayyaz Institute of Crop Science, Ministry of Agriculture and Rural Affairs Key Laboratory of Spectroscopy Sensing, Zhejiang University, Hangzhou, China

Elamin Baillo Agricultural Research Corporation (ARC), Ministry of Agriculture, Gezira, Sudan

Iram Batool Institute of Crop Science, Ministry of Agriculture and Rural Affairs Key Laboratory of Spectroscopy Sensing, Zhejiang University, Hangzhou, China

Tanmaya Kumar Bhoi Forest Protection Division, ICFRE-Arid Forest Research Institute (ICFRE-AFRI), Jodhpur, Rajasthan, India

Ali Chenari Bouket East Azarbaijan Agricultural and Natural Resources Research and Education Centre, Plant Protection Research Department, Agricultural Research, Education and Extension Organization (AREEO), Tabriz, Iran

Karandeep Singh Chahal Department of Plant, Soil and Microbial Sciences, Michigan State University, East Lansing, MI, USA

Moxian Chen National Key Laboratory of Green Pesticide, Key Laboratory of Green Pesticide and Agricultural Bioengineering, Ministry of Education, Center for R&D of Fine Chemicals of Guizhou University, Guiyang, China

K. Darshan Forest Protection Division, ICFRE-Tropical Forest Research Institute (TFRI), Jabalpur, Madhya Pradesh, India

Tariq Dildar STU-UMT Joint Shellfish Research Laboratory, Shantou University, Shantou, China

Guangdong Provincial Key Laboratory of Marine Biotechnology, Shantou University, Shantou, China

International Joint Research Center for the Development and Utilization of Important Mariculture Varieties Surrounding the South China Sea Region, Shantou University, Shantou, China

Abolghassem Emamverdian Co-Innovation Center for Sustainable Forestry in Southern China, Nanjing Forestry University, Nanjing, China

Muhammad Ahsan Farooq Zhejiang Provincial Key Laboratory for Water Environment and Marine Biological Resources Protection, College of Life and Environmental Science, Wenzhou University, Wenzhou, China

Nikita Gambhir FMC Stine Research Center, Newark, DE, USA

Mohammad Sani Gambo Department of Plant Pathology, CCS Haryana Agricultural University, Hisar, Haryana, India

Abazar Ghorbani National Key Laboratory of Green Pesticide, Key Laboratory of Green Pesticide and Agricultural Bioengineering, Ministry of Education, Center for R&D of Fine Chemicals of Guizhou University, Guiyang, China

James Godwin Donald Danforth Plant Science Center, St. Louis, MO, USA

Arumoy Halder Department of Soil Science, Gazipur Agricultural University, Gazipur, Bangladesh

Maha M. Hamada Department of Agronomy, Faculty of Agriculture, Ain Shams University, Cairo, Egypt

Muhammad Sajid Hanif Centre for Agricultural Resources Research, Institute of Genetics and Developmental Biology, Chinese Academy of Sciences, Shijiazhuang, China

University of Chinese Academy of Sciences, Beijing, China

Fakhir Hannan Institute of Crop Science, Ministry of Agriculture and Rural Affairs Key Laboratory of Spectroscopy Sensing, Zhejiang University, Hangzhou, China

Akbar Hossain Soil Science Division, Bangladesh Wheat and Maize Research Institute, Dinajpur, Bangladesh

Mohamed F. M. Ibrahim Department of Agricultural Botany, Faculty of Agriculture, Ain Shams University, Cairo, Egypt

Ariful Islam Jute Agriculture Experimental Station, Bangladesh Jute Research Institute (BJRI), Manikgonj, Bangladesh

Mohammad Shah Jahan Department of Horticulture, Sher-e-Bangla Agricultural University, Dhaka, Bangladesh

Shimi Jose Hawkesbury Institute for the Environment, Western Sydney University, Richmond, VIC, Australia

Md. Parvez Kabir Soil Science Division, Bangladesh Wheat and Maize Research Institute, Dinajpur, Bangladesh

Malini Muthu Karpagam Smith College, Northampton, MA, USA

Malavika Ram Amanthra Keloth Hawkesbury Institute for the Environment, Western Sydney University, Richmond, VIC, Australia

Hafiz Muhammad Kashif Khalil Department of Plant Breeding and Genetics/CAS-AFS, University of Agriculture, Faisalabad, Pakistan

J. Komal Basic Seed Multiplication and Training Centre, Central Silk Board, Kharsawan, Jharkhand, India

Manish Kumar Dr. Rajendra Prasad Central Agricultural University, Pusa, Samastipur, Bihar, India

Rajeev Kumar Division of Plant Physiology and Biochemistry, ICAR-Indian Institute of Sugarcane Research, Lucknow, Uttar Pradesh, India

Ravinder Kumar Division of Plant Pathology, ICAR-Indian Agricultural Research Institute, New Delhi, India

Sumit Kumar Krishi Vigyan Kendra, Mau, Uttar Pradesh, India

Acharya Narendra Deva University of Agriculture and Technology, Ayodhya, Uttar Pradesh, India

Arti Kumari Nalanda College of Horticulture, Noorsarai, Nalanda, Bihar, India

Jyoti Kumari Nalanda College of Horticulture, Noorsarai, Nalanda, Bihar, India

Mala Kumari Nalanda College of Horticulture, Noorsarai, Nalanda, Bihar, India

Milan Kumar Lal Division of Crop Physiology and Biochemistry, ICAR-National Rice Research Institute, Cuttack, Odisha, India

Deepak Kumar Mahanta Forest Entomology Discipline, Forest Protection Division, Indian Council of Forestry Research and Education (ICFRE)-Forest Research Institute (ICFRE-FRI), Dehradun, Uttarakhand, India

Debashis Mahapatra Department of Plant Physiology, College of Agriculture, Odisha University of Agriculture and Technology, Bhubaneswar, Odisha, India

Sayanti G. Majumdar ICAR-Indian Institute of Sugarcane Research, Lucknow, Uttar Pradesh, India

Azade Masoumi Department of Biology, Faculty of Sciences, Mashhad Branch, Islamic Azad University, Mashhad, Iran

Sucharita Mohapatra Department of Plant Pathology, College of Agriculture, Odisha University of Agriculture and Technology, Bhubaneswar, Odisha, India

Naveed Mushtaq Key Laboratory for Quality Regulation of Tropical Horticultural Crops of Hainan Province, School of Breeding and Multiplication (Sanya Institute of Breeding And Multiplication), Center of Nanfan and High-Efficiency Tropical Agriculture, Hainan University, Sanya, China

Key Laboratory for Quality Regulation of Tropical Horticultural Crops of Hainan Province, School of Tropical Agriculture and Forestry, Hainan University, Danzhou, China

Mir Muhammad Nizamani Institute of Marine Sciences, Guangdong Provincial Key Laboratory of Marine Disaster Prediction and Prevention, Shantou University, Shantou, China

Asharani Patel Division of Plant Pathology, ICAR-Indian Agricultural Research Institute, Hazaribag, Jharkhand, India

H. R. Raghavendra School of Crop Improvement, ICAR-Indian Agricultural Research Institute, Gogamukh, Dhemaji, Assam, India

Mehjebin Rahman Department of Plant Pathology, College of Agriculture, Assam Agricultural University, Jorhat, Assam, India

S. R. Ramseena College of Agriculture, Kerala Agricultural University, Thiruvananthapuram, Kerala, India

Komal Rani ICFRE-Tropical Forest Research Institute (TFRI), Jabalpur, Madhya Pradesh, India

E. R. Rashmi ICAR-CITH, Central Institute of Temperate Horticulture, Regional Station, Mukteshwar, Uttarakhand, India

Lellapalli Rithesh Department of Plant Pathology, College of Agriculture, Kerala Agricultural University, Thiruvananthapuram, Kerala, India

Nasim Roodbari Department of Biology, Kahnooj Branch, Islamic Azad University, Kahnooj, Iran

Kuleshwar Parsad Sahu College of Agriculture, Rani Lakshmi Bai Central Agricultural University, Jhansi, India

Ipsita Samal ICAR-National Research Centre on Litchi, Mushahari, Ramna, Muzaffarpur, Bihar, India

S. Shreedevasena Directorate of Medicinal and Aromatic Plants Research (DMAPR), Anand, Gujarat, India

Jagmohan Singh Department of Plant Pathology, CCS Haryana Agricultural University, Hisar, Haryana, India

Shiv Singla Department of Plant and Soil Sciences, University of Delaware, Newark, DE, USA

V. Soundararajan ICFRE-Tropical Forest Research Institute (TFRI), Jabalpur, Madhya Pradesh, India

Vanapalli Lohitha Sai Sree Division of Plant Pathology, ICAR-Indian Agricultural Research Institute, New Delhi, India

B. K. Surendra Shree Banshagopal Secondary School (Technical School), Makwanpur, Bagmati, Nepal

Kirtika Swaraj ICAR-Indian Institute of Sugarcane Research, Lucknow, Uttar Pradesh, India

Aditi Tailor ICFRE-Arid Forest Research Institute, Jodhpur, Rajasthan, India

Muhammad Tayyab Institute of Marine Sciences and Guangdong Provincial Key Laboratory of Marine Biotechnology, Shantou University, Shantou, China

Swagata Thakur Division of Plant Pathology, ICAR-Indian Agricultural Research Institute, New Delhi, India

Rahul Kumar Tiwari Division of Crop Protection, ICAR-Indian Institute of Sugarcane Research, Lucknow, Uttar Pradesh, India

Md. Nasir Uddin Sustainable Agricultural Systems, International Maize and Wheat Improvement Center, Dhaka, Bangladesh

Xin-Xin Zhang National Key Laboratory of Green Pesticide, Key Laboratory of Green Pesticide and Agricultural Bioengineering, Ministry of Education, Center for R&D of Fine Chemicals of Guizhou University, Guiyang, China

Weijun Zhou Institute of Crop Science, Ministry of Agriculture and Rural Affairs Key Laboratory of Spectroscopy Sensing, Zhejiang University, Hangzhou, China

Prominent Role of Melatonin in Plant: Experimental Evidence and Advancements in Phytomelatonin Research

1

K. Darshan, Aditi Tailor, E. R. Rashmi, Komal Rani, S. Shreedevasena, and V. Soundararajan

Abstract

Phytomelatonin is a powerful biomolecule that plays a key role in regulating various physiological and biochemical processes in plants. It is present in various plant tissues, such as roots, leaves, and seeds, and helps in plant growth regulation, seed germination, root development, flowering, and fruit development. The multifaceted melatonin has a recent interest for its role in plant growth and development and protection from biotic and abiotic stresses. Due to the elevated harmful effects of pesticide and global climate change, melatonin has led to great interest in plant applications. The omics advancements have unraveled the most complex and in-depth molecular processes behind melatonin-induced changes in plants. The role of melatonin in maintaining cellular homeostasis via post translational changes and signaling cascades is well studied with the proteomic

K. Darshan, Aditi Tailor, E. R. Rashmi, Komal Rani, S. Shreedevasena and V. Soundararajan contributed equally with all other contributors.

K. Darshan (✉)
Forest Protection Division, ICFRE-Tropical Forest Research Institute (TFRI), Jabalpur, Madhya Pradesh, India

A. Tailor
ICFRE-Arid Forest Research Institute, Jodhpur, Rajasthan, India

E. R. Rashmi
ICAR-CITH, Central Institute of Temperate Horticulture, Regional Station, Mukteshwar, Uttarakhand, India

K. Rani · V. Soundararajan
ICFRE-Tropical Forest Research Institute (TFRI), Jabalpur, Madhya Pradesh, India

S. Shreedevasena
Directorate of Medicinal and Aromatic Plants Research (DMAPR), Anand, Gujarat, India

R. K. Tiwari et al. (eds.), *Melatonin: Signal Transduction Mechanisms and Defense Networks in Plants*, https://doi.org/10.1007/978-981-96-8869-2_1

approach. Melatonin increases plant resistance to biotic and abiotic stressors by regulating gene expression, protein function, and metabolic pathways. Experimental evidence for its multifunction, plant growth, development, and stress adaptation using various techniques is needed for its field usage agriculture, horticulture, and forestry ecosystems. The recent experimental evidence and discovery of melatonin effects on plant growth and development, regulation of phytohormone, and biotic and abiotic stress adaptation have been summarized in this chapter, and also avenues for future work. Gaining insight into the recent advancements in phytomelatonin research will pave the way for sustainable agricultural practices and improved global food security.

Keywords

Antioxidant defense · Stress adaptation · Hormonal cross-talk · Abiotic stress · Growth

1.1 Introduction

Melatonin (*N*-acetyl-5-methoxytryptamine) is a ubiquitous hormone secreted by the pineal gland in the brain that plays an important role in various biological processes. It is a small, lipophilic molecule derived from a precursor amino acid tryptophan, which is converted into serotonin and subsequently acetylated and methylated to form melatonin (Chen and Arnao 2022). This molecule was discovered in 1958 and has since been extensively studied for its role in various physiological processes. Its production is induced by darkness and repressed by light, so melatonin levels peak at night and decrease during the day. The length of melatonin secretion acts as a cue of changing photoperiods and allows photoperiodic species to control their seasonal physiological activities. Melatonin's diverse biological roles have led it to be a component in numerous aging, cancer, cardiovascular diseases, and immune disorder studies (Sun et al. 2020). There is evidence that melatonin can also be synthesized in other sites of the body, acting in an autocrine or paracrine manner: skin, gastrointestinal tract, retina, bone marrow, and placenta.

Melatonin found in plants is known as "phytomelatonin" and it plays an important role in regulating various physiological and biochemical processes in plants (Altaf et al. 2022; Tiwari et al. 2020, 2021a, b). Its presence in various plant tissues, such as roots, leaves, and seeds, underlines its importance in plant physiology. Specifically in plant growth regulation, seed germination, root development, flowering, and fruit development (Arnao and Hernández-Ruiz 2019; Zhang et al. 2017a). Phytomelatonin is found to be involved in giving tolerance to biotic and abiotic stress through crosstalk with the hormones such as auxins, abscisic acid, and ethylene (Zhang et al. 2015; Zhao et al. 2021). Phytomelatonin is involved in regulating abiotic stresses such as drought, salinity, extreme temperatures (High and low), and heavy metal toxicity. Melatonin in plants acts as a powerful antioxidant, growth regulator, and stress mitigator (Sharma et al. 2020), unlike its role in animals, where it mainly regulates circadian rhythms (Cajochen et al. 2003). Tolerance is achieved by scavenging reactive oxygen species (ROS) and boosting the activity of enzymes involved in antioxidation (Arnao and Hernández-Ruiz 2015; Tan et al. 2012; Shi et al. 2015). Melatonin improves photosynthesis efficiency, senescence regulation, and enhancement of crop yield as stated by Tan et al. (2012). Under sustainable crop production, melatonin has become the most useful molecule in building crop resilience to various stresses and climate change impacts (Cai et al. 2025). Melatonin helps in improving seed germination, provides better shoot and root growth, improves the photosynthesis efficiency, regulates senescence, tolerance to biotic and abiotic stresses, including drought, salinity, and extreme temperatures by modulating the oxidative damage and enhancing the antioxidant activity.

The advanced molecular biology tools, such as transcriptomics, proteomics, and gene editing tools in melatonin research, help to discover the genes and pathways of melatonin signaling in plants, providing resilience to plants under stress conditions. This chapter highlights the importance of melatonin in sustainable crop production by addressing the challenges posed by climate change. The three important genes involved in *Arabidopsis thaliana* melatonin biosynthesis are TDC (tryptophan decarboxylase), ASMT (acetylserotonin methyltransferase), and SNAT (serotonin N-acetyltransferase) (Lee et al. 2019). The OsTDC and OsSNAT genes for stress tolerance in rice are modified by using CRISPR/cas9 technology (Park et al. 2013). The advanced imaging techniques like fluorescence microscopy, confocal microscopy, mass spectrometry, and autoradiography are used to visualize the melatonin at the subcellular level. The high-resolution image and three-dimensional (3D) images at the cellular and subcellular levels can be visualized by confocal microscopy (Liu et al. 2019b) and fluorescence microscopy (Wang et al. 2017). The movement of melatonin within the plant tissues and the spatial distribution of melatonin can be visualized by mass spectrometry imaging (MSI) (Wang et al. 2020). Autoradiography is an advanced imaging technique that uses radioactive isotopes bound to the molecules of melatonin, and whenever the photographic emulsion is exposed, it creates an image of melatonin localization within the plant tissues (Jin et al. 2014). The harmful effects of pesticides can be substituted by the usage of biopesticides and bio stimulants (Archana et al. 2022; Krishnappa et al. 2023). The melatonin effect on various physiological activities, metabolic processes, gene expression, and protein activity at the system levels sheds light on the melatonin effect on improving crop resilience and productivity. In brief, this chapter highlights the experimental evidence and advanced research on melatonin in plants, emphasizing their usage in sustainable agriculture.

1.2 Multi-functional Roles of Melatonin in Plants

Melatonin improves growth attributes and reduces the stress impact by altering the physiological and biochemical pathways, and also enhances the photosynthetic rates, improves the root development, regulates the hormone levels, leading to increased crop productivity (Fig. 1.1). Melatonin has a positive interaction with other phytohormones like salicyclic acid, auxin, zeatin, abscisic acid, and

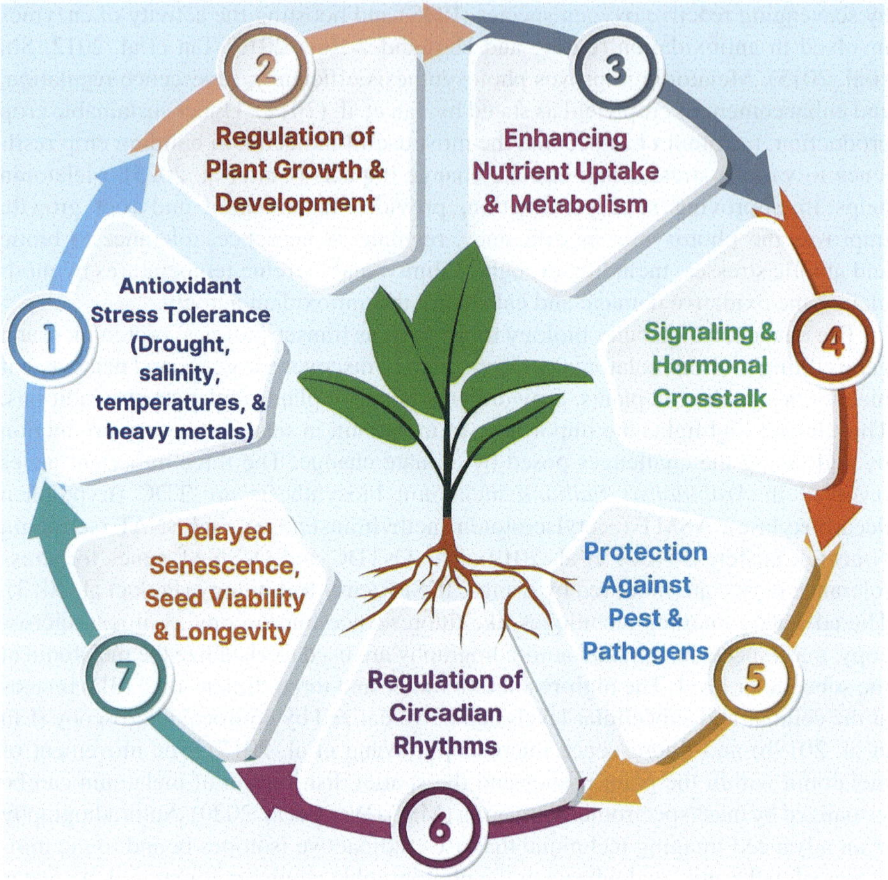

Fig. 1.1 Multifaceted role of melatonin in regulating plant growth, development, stress responses, and metabolic pathways

gibberellin, which further improves growth and hormonal homeostasis (Shaffique et al. 2024; Shi et al. 2024). This hormonal regulation by melatonin is one of the key factors in the absorption of nutrients and overall root robustness by the formation of lateral roots and root elongation (Khan et al. 2024). According to Hussain et al. (2024), melatonin interacts with nitric oxide (NO) to produce N-nitrosomelatonin (NOMela), a compound that is involved in plant development and stress responses by working in conjunction with other phytohormones and signaling molecules to activate defense pathways against biotic stressors. It increases biomass accumulation and overall productivity by enhancing photosynthesis (Shi et al. 2024; Khan et al. 2024). Yan et al. (2024) have shown that foliar melatonin application in rice significantly increases the number of spikelets and viable grains, which in turn boost rice productivity. According to Hussain et al. (2024), it plays a major role in regulating plants' circadian cycle, has an impact on modulating seed germination, flowering time, and leaf senescence. Melatonin has been shown to boost fertilizer usage efficiency by as much as 20%, which raises biomass and yield (Shahani et al. 2023).

The beneficial roles of melatonin in the regulation of plant defense have been identified in a vast number of species. Melatonin plays multifaceted roles in boosting the stress resistance, such as abiotic and biotic stress factors in plants. It enhances the abiotic stress tolerance such as low and high temperature, drought, cold, heat stress, tolerance to salt, and heavy metal toxicity by activating stress-responsive genes and modulating signaling pathways by scavenging reactive oxygen species (ROS) and enhancing antioxidant systems by increasing the activities of enzymes such as superoxide dismutase (SOD), catalase (CAT), and ascorbate peroxidase (APX) (Himanshu et al. 2024; Khan et al. 2024). The abiotic stress poses a threat to plant growth due to the excess production and accumulation of reactive oxygen species (ROS), which leads to cellular and metabolic processes. Melatonin, a phytohormone, has gained great interest in research owing to its wide applications in plant systems. To cope with most of the abiotic stresses and pathogen infections, the reactive oxygen species scavenging and activation of antioxidant defense responses are the two major mechanisms mediated through melatonin. By modifying stress-responsive genes and antioxidant defenses, melatonin increases a plant's resistance to abiotic conditions, including drought and salt (Chaurasia et al., 2023; Muhammad et al. 2024). Melatonin improved the micronutrients and antioxidant concentrations in citrus plants under various abiotic stress conditions, such as drought, waterlogging, and salinity (Abbas et al. 2023).

Melatonin application also improves seed germination in a diverse range of abiotic stresses, which encompass water deficiency, high temperature, chilling, salt stress (Zhang et al. 2013), and also chelates heavy metal (Yu et al. 2021). The presence of high doses of these heavy metals within the plant body adversely affects many critical aspects of growth by impairing the normal metabolism, eventually resulting in the manifestation of toxicity in the form of symptoms at multiple levels, including morphology, physiology, and biochemistry (Ghuge et al. 2023). Some major symptoms associated with metal toxicity include chlorosis, necrosis, and stunted growth (Ghuge et al. 2023). They also interfere with the uptake and assimilation of other vital mineral nutrients (Naeem et al. 2019). Evidence has been gathered through many investigations that melatonin may also exert a positive effect and relieve metal toxicity in plants (Hoque et al. 2021). Exogenous application of melatonin has also been demonstrated to help alleviate metal toxicity caused by a range of heavy metals and enhance multiple growth attributes by positively modulating plant metabolism (Hoque et al. 2021; Darshan et al. 2024). High NaCl levels in soils negatively affect plants as they reduce water availability and cause toxicity in plants as Na^+ and Cl^- ions over accumulate in their body over time (Shahid et al. 2020). The accumulation of Na^+ and Cl^- ions in sensitive plant tissues triggers salt stress (Flowers et al. 2010). Melatonin exerts beneficial effects on plants growing under stressful conditions by strengthening the antioxidative capacity and mitigating salt-induced damage symptoms (Li et al. 2019a; Zhan et al. 2019). It is reported that melatonin treatment could significantly reduce Na^+ accumulation and K^+ content under salt stress, which might regulate the expression of ion channel genes (MdNHX1 and MdAKT1) to maintain ion homeostasis and thus improve plants' salt resistance (Meloni et al. 2003). It has been demonstrated that applying melatonin to legumes under salt stress enhances germination, seedling biomass, and total yield (Chaurasia et al. 2023). In drought conditions, exogenous melatonin application improves agronomic traits by boosting antioxidant enzyme activities and enhancing the interactions with other plant hormones and modulating the signaling molecules (Khan et al. 2024; He et al. 2021). This helps in improving the turgor pressure, recovery of chlorophyll, and maintenance of cell and photosynthetic machinery stability under drought stress conditions (Shaffique et al. 2024). Melatonin increases water usage efficiency and decreases transpiration rates in cereal crops, hence improving drought tolerance (Muhammad et al. 2024). Melatonin's significance in stress reduction is demonstrated by the enhanced water usage efficiency and decreased transpiration rates seen in cereal crops such as sorghum and maize after treatment (Muhammad et al. 2024). Similarly, under heat stress, exogenous melatonin application can enhance antioxidant defense systems via control of ROS accumulation and enhancement of proline metabolism (Zeng et al. 2022). Melatonin application is known to trigger the phenylpropanoid pathway by enhancing the activities of PAL, 4-coumarate-coenzyme A ligase (4CL), cinnamate-4-hydroxylase (4CH), and POD, which may be accompanied by the accumulation of higher contents of total phenols and lignin, which may improve temperature tolerance (Qu et al. 2022).

Melatonin acts as a master regulator of host defense against insects, fungi, bacteria, viruses, protozoa, nematodes, and phanerogamic plant parasites and environmental challenges by acting as a multifaceted signaling molecule by activating defense signaling networks (Darshan et al. 2024, 2025). It acts as a nontoxic biomolecule and regulates the growth and development of plants by providing them tolerance against biotic and abiotic stresses. It has piqued the curiosity of researchers because of its numerous benefits in plant systems in maintaining cellular integrity and promoting survival under stress conditions through melatonin-mediated reactive oxygen species scavenging and activation of antioxidative defense responses during pathogen attacks (Moustafa Farag et al. 2019; Tiwari et al. 2021b; Khan et al. 2024). By forming complex regulatory networks with other signaling molecules like salicylic acid (SA) and nitric oxide (NO), melatonin helps plants fight off fungal infections (Eke et al. 2024; Guo et al. 2024). It promotes stomatal immunity and hormonal crosstalk to enhance immunity, like plant pattern-triggered immunity (PTI) and effector-triggered immunity (ETI) against attacking pathogens (Khalil et al. 2023). Research suggests that melatonin enhances the efficacy of fungicides and reduces pathogen damage, thereby increasing resistance to a variety of pathogens, such as fungi like *Phytophthora* infestans, bacteria like *Xanthomonas* species, and viruses (Rithesh et al. 2024; Hernández-Ruiz et al. 2023). The exogenous application of melatonin to plants has been demonstrated to decrease the infection of viral diseases, including tobacco mosaic virus by the induction of systemic acquired resistance, upregulates defense hormones, and activates pathogenesis-related genes

(Hernández-Ruiz et al. 2023). Melatonin protects against another Tobamovirus infection, Pepper mild mottle virus (PMMoV), by specifically upregulating the cysteine-rich secretory protein (defense-related gene CRISP1), implying that melatonin has a regulatory function in plant pathogen defense systems (Yang et al. 2023). Melatonin regulates immunological responses, reduces bacterial, fungal, and viral infections, forms a complex loop with reactive oxygen and nitrogen species, and interacts with phytohormones such as salicylic acid, jasmonic acid, auxin, and abscisic acid to protect plants and strengthen their defense (Zeng et al. 2022).

Melatonin concentrations of 1.5 mg/L boosted adventitious root total phenolics and flavonoids, antioxidant activity, and secondary metabolite synthesis (Ali et al. 2023). Melatonin treatment considerably boosted flavonoids, phenolics, caffeic acid, ferulic acid, rosmarinic acid, and hyperoside accumulation in *Prunella vulgaris*, particularly at a 100 μM dose. Improved photosynthetic and root absorption lead to increased plant yield and medicinal component content. Melatonin-treated citrus crops had higher total flavonoid and total phenolic content. Melatonin (100 μM) generated the largest levels of essential oil components, including limonene. Melatonin increases plant secondary metabolite synthesis (Jafari and Shahsavar 2021). Melatonin influences various metabolic processes by upregulating the genes involved in the phenylpropanoid pathway, increasing resveratrol contents and their derivatives in treated grape seeds (Shi et al. 2024; Shahani et al. 2023). It also altered the secondary metabolite profile of grape berry skin by boosting ethylene production via the VvMYB14-mediated pathway, which affected the accumulation of flavonoids, phenolic acids, and stilbenes (Ma et al. 2021). Melatonin treatment in *Lithocarpus litseifolius* and *Stevia rebaudiana* considerably increased dihydrochalcone, stevioside, phenolics, and total flavonoid content by regulating antioxidant enzyme activities and hormonal pathways (Zhang et al. 2024; Naz et al. 2024). Melatonin boosts secondary metabolite production in broccoli hairy roots. In the transcriptomics research, 500 μM treatment with melatonin led to a 1.22-fold increase in glucoraphanin (GRA) production at 12 hours compared to the control. It also promoted secondary metabolite production by upregulating critical glucosinolate biosynthesis pathway genes, which were among 13,234 differentially expressed genes (Tian et al. 2021).

1.3 Experimental Evidence Supporting Melatonin Functions in Plants

1.3.1 Phytomelatonin-Mediated Mitigation of ROS and Oxidative Stress

The ability to synthesize melatonin is believed to have been acquired almost more than 2.5–3.5 billion years ago in organisms that evolved with the capacity for aerobic metabolism as a counteracting measure to mitigate the oxidative stress arising due to the production of reactive oxygen species (ROS) (Mannino et al. 2021). The presence of melatonin as an evolutionary feature is a clear indication of the immense potential as an antioxidative molecule that it is believed to have played throughout the evolution of organisms. The first evidence of the ability of phytomelatonin to act as a scavenger of free radicals ($^{\cdot}OH$ and $O_2^{\cdot-}$) came as early as the 1990s, and since then, its role has been described to be fundamental in the management of oxidative stress triggered by environmental cues (Tan et al. 1993). In tomato, 100 μM pre-treatment of melatonin induces resilience against drought stress by restoring root architecture, chlorophyll content, and gas exchange attributes. It also upregulated gene expression of antioxidant enzymes, improving seedling growth and development (Altaf et al. 2022).

In general, the redox network in plants comprises a set of factors that interact to maintain cellular redox homeostasis, including metabolites, enzymes, genes, and regulatory elements (Lal et al. 2022; Nayak et al. 2023). Various ROS, such as superoxide anion ($O_2^{\cdot-}$), hydroxyl radical ($^{\cdot}OH$), or hydrogen peroxide (H_2O_2), are produced even during the normal physiological function in a plant body as a consequence of aerobic metabolism (Ahammed et al. 2024). The plant experiences oxidative stress due to unavoidable environmental stress resulting in the production of excessive ROS. Antioxidants are essential for plants to maintain redox equilibrium and combat the accumulation of reactive oxygen species (ROS). This antioxidant apparatus functions in conjunction with both enzymatic and nonenzymatic mechanisms. Superoxide dismutase (SOD), ascorbate peroxidase (APX), catalase (CAT), guaiacol peroxidase (POD), glutathione peroxidase (GPX), glutathione reductase (GR), and peroxiredoxins (PRX) are enzymes in the enzymatic antioxidant pathway that scavenge reactive oxygen species (ROS). It employs enzymatic and nonenzymatic mechanisms to mitigate oxidative damage and reactive oxygen species (ROS). It also protects plants from stress by interacting with other signaling channels like NO, Ca^{2+}, and phytohormones at the plant redox hub (Khanna et al. 2023). Mounting evidence hints toward melatonin-mediated protection against environmental stressors, including salinity, drought, heat, cold, and heavy metals, by stimulating the ROS scavenging system (Li et al. 2022; Muhammad et al. 2022, 2023; Ali et al. 2024; Altaf et al. 2024; Awan et al. 2024; Munir et al. 2024; Zhu et al. 2024). The antioxidant capacity of melatonin originates from its ability to act as a reducing agent and donate electrons, thereby disarming ROS, like the hydroxyl radical ($^{\cdot}OH$), singlet oxygen ($^1O_2$), hydrogen peroxide (H_2O_2), other free radicals, along with reactive nitrogen species (RNS), like peroxynitrite ($ONOO^-$) (Galano and Reiter 2018). Thus, it can directly contribute to the scavenging of ROS and prevent oxidative damage. Additionally, it has been shown to regulate the expression of stress-responsive genes and defense compounds (nonenzymatic antioxidants like ascorbate (AsA), reduced glutathione (GSH), osmolytes, proline), and upregulates genes of antioxidants like CAT, SOD, GPX, and APX (Siddiqui et al. 2019; Sharma et al. 2020; Ali et al. 2024; Awan et al. 2024; Kumar et al. 2025).

1.3.2 Phytomelatonin-Mediated Regulation of Plant Growth and Development

The physiological functions of melatonin in plant growth and development are primarily attributed to the similarity of its molecular structure to that of auxin, indole-3-acetic acid (IAA). Owing to this molecular similarity, melatonin behaves as an auxin-like molecule with functional similarity to the action of auxin (Arnao and Hernández-Ruiz 2017). The auxin-like activity of melatonin has been experimentally confirmed through a wide range of scientific explorations in the past, mainly involving exogenous melatonin application. Several investigations have demonstrated its role, albeit dose-dependent and species-specific, in the stimulation of elongation growth in above-ground parts, root architecture and morphogenesis, development of lateral roots, flowering, fruit ripening, seed germination, pigment development, circadian rhythms, and aging process or senescence in plants (Mannino et al. 2021; Shi et al. 2024a; Fig. 1.2). Phytomelatonin regulates stomatal conductance by modulating the stomatal movements. Particularly, in response to stress, receptor-mediated phytomelatonin signaling facilitates stomatal closure by activating the NADPH oxidase-mediated H_2O_2 generation, which further stimulates Ca^{2+} influx and K^+ efflux (Wei et al. 2018). It is also involved in regulating nutrient uptake and use efficiency through the regulation of the expression and/or activities

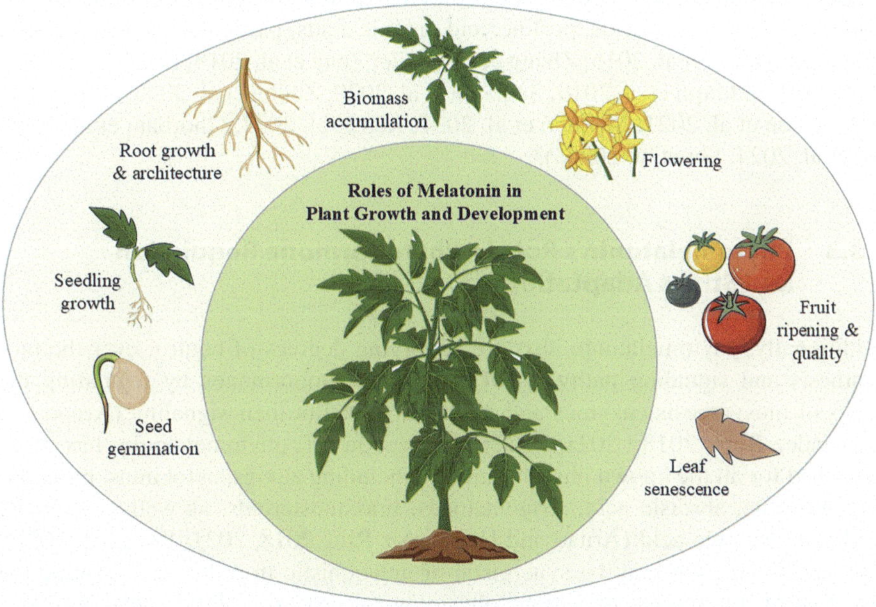

Fig. 1.2 Illustration depicting the major growth and development attributes regulated by phytomelatonin in plants

of ion transporters, H^+-ATPase, and the ROS–Ca^{2+} signaling (Huang et al. 2022; Sun et al. 2022; Ghorbani et al. 2023). Apart from regulating defense-related pathways, it also improves the nutritional value of crops, improving their productivity. In soybean, exogenous application of 100 and 300 μmol/L melatonin significantly increased the length of sprouts and germination percentage by 53% and 12%. Also, the isoflavone content was increased to 5.101 g/kg on the 5th day of sprouting (Wu et al. 2023).

Phytomelatonin has established roles in enhancing growth parameters in plants, also known to improve energy status and provide protection to the photosynthetic machinery under stressed conditions, including chlorophyll, and the photosystems, thereby contributing to the improvement of the physiological attributes, like net photosynthesis, transpiration rate, and intracellular CO_2 (Arnao and Hernández-Ruiz 2009; Tal et al. 2011; Shi et al. 2015; Ye et al. 2016; Sharif et al. 2018; Sharma and Zheng 2019; Ahmad et al. 2020; Sharma et al. 2020; Wang et al. 2020; Chen et al. 2021; Azizi et al. 2022; Yang et al. 2022; Ali et al. 2024; Alomran et al. 2024; Feng et al. 2024; Kumar et al. 2025). It has been shown to influence the expression genes and/or activities of the enzymes involved in carbon metabolism, including the enzymes of the Calvin cycle (RUBISCO), glycolysis, tricarboxylic acid (TCA) cycle, and fatty acid pathway (Shi et al. 2015; Zhao et al. 2015; Siddiqui et al. 2019; Arnao et al. 2021; Teng et al. 2022). It also regulates the biosynthesis, and thereby accumulation, of metabolites of nitrogen metabolism, sugar metabolism, and secondary metabolism, like total sugars, osmolytes (polyamines (PAs), γ-aminobutyric acid (GABA)), amino acids, proline, and organic acids, particularly under stressed conditions (Zhao et al. 2015; Zhang et al. 2017c; Zeng et al. 2018; Erdal 2019; Qiao et al. 2019; Siddiqui et al. 2019; Samanta et al. 2020; Zhong et al. 2020; Arnao et al. 2021; Chen et al. 2021; Madebo et al. 2021; Ren et al. 2022; Ghorbani et al. 2023; Ali et al. 2024; Shi et al. 2024a).

1.3.3 Phytomelatonin's Role in Phytohormone Regulation and Stress Adaptation

Additionally, phytomelatonin also exerts varying degrees of control over the biosynthesis and signaling pathways of multiple phytohormones by regulating the genes of biosynthesis enzymes and factors involved in their signaling (Arnao and Hernández-Ruiz 2018, 2021). Such interaction of phytomelatonin has been described for all the known phytohormones, including auxins, cytokinins, gibberellins, ethylene, abscisic acid, strigolactones, brassinosteroids, as well as salicylic acid and jasmonic acid (Arnao and Hernández-Ruiz 2018, 2021; Liu et al. 2022). The said interaction may be synergistic or antagonistic in order to determine the direction of progression of various physiological processes. Phytomelatonin, in a manner mimicking that of the auxin IAA, inhibits the primary root (Chen et al. 2009). However, this inhibition imposed on root growth is often ascribed to the downstream increase in auxin levels as reported in mustard (*Brassica juncea* (L.) Czern.) wherein the elevated melatonin concentration subsequently led to enhanced IAA levels (Chen et al. 2009). In lower doses, phytomelatonin has also been found to stimulate root growth, possibly by regulating related genes (Yang et al. 2021a, b). Similar to IAA, it also promotes lateral root development by inducing a greater number of lateral root primordia, by affecting the expression of a variety of genes and fine-tuning auxin transport (Arnao and Hernández-Ruiz 2007; Zhang et al. 2013, 2014b; Ren et al. 2019; Yang et al. 2021a, b). Phytomelatonin-mediated root growth stimulation under salt stress is also correlated with improved antioxidants and osmolyte contents (Zeng et al. 2018). As already mentioned, it ensures intracellular redox homeostasis and mitigates oxidative stress by reinforcing the ROS scavenging system by acting as a scavenger itself and by activating other antioxidants. Additionally, melatonin also stimulates the production of H_2O_2 via the activity of polyamine oxidase (PAO), which is associated with improved lateral root development (Chen et al. 2019). It positively influences the process of seed germination, regulating the GA_3 and ABA balance to improve the GA/ABA ratio (Zhang et al. 2013, 2023; Xiao et al. 2019; Chen et al. 2021). Such improvement in seed germination by melatonin is also reported under stress where it regulates hormonal balance, and glucose metabolism by promoting the related proteins and enzymes while also increasing energy generation (Zhang et al. 2017b; Wang et al. 2022a). However, its inhibitory role on seed germination and seedling growth has also been brought to light, which is ascribed to distorted phytohormone (GA, ABA, and auxin) levels and signaling (Pan et al. 2021; Yin et al. 2022).

It acts to delay senescence and aging by alleviating damage caused to chlorophyll through the induction of cytokinin, IAA, and gibberellic acid biosynthesis, activating antioxidant defense, and downregulation of senescence-associated genes and ABA biosynthesis and signaling pathway genes (Zhang et al. 2017a; Tan et al. 2019a; Jing et al. 2022). More recently, a role for H_2O_2-dependent calcium signaling has also been deduced for the melatonin-mediated delay in senescence (Guo et al. 2023). It regulates fruit ripening by augmenting anthocyanin and phenolic contents, antioxidant machinery, and controlling the contents of ABA, NO, H_2O_2, and ethylene (Xu et al. 2018; Mansouri et al. 2021; Aghdam et al. 2023; Corpas et al. 2022). Its role in controlling flowering has been proposed in Arabidopsis, wherein its biosynthesis mutants *comt1* and *snat1* display early flowering phenotype as compared to the wild-type Col-0 (Zhang et al. 2019). In contrast, it also delays flowering by regulating hormonal interactions, ROS, and flowering gene regulation networks, probably in a PMTR1-dependent pathway (Chen et al. 2022; Yin et al. 2022). It plays a significant role in improving germination, especially under high temperatures; in rice plants, it enhances the antioxidant properties and boosts the tolerance of rice to germinate under high temperatures (Yu et al. 2022). This implies its critical role under changing climate scenarios where strengthening food security under increasing temperature and climatic events.

1.4 Molecular Mechanisms Underlying Melatonin Actions in Plants

To understand the molecular mechanism underlying phytomelatonin-mediated responses in plants, it is important to develop an understanding of phytomelatonin perception and signaling, leading up to the regulation of phytomelatonin response genes. The discovery of the first phytomelatonin receptor CAND2/PMTR1 (referred to as PMTR1 henceforth) came following investigations in *Arabidopsis thaliana*, wherein the *cand2* mutants reportedly lacking the melatonin-stimulated H_2O_2 synthesis, failed to elicit Ca^{2+} influx required for stomatal closure (Wei et al. 2018). AtPMTR1 (Phytomelatonin Receptor 1) was shown to exhibit melatonin binding activity under in vitro conditions. It is a candidate GPCR-like (G protein-coupled receptor-like) transmembrane protein located in the plasma membrane capable of directly interacting with the G protein α subunit (GPA1), and thus triggering receptor-mediated phytomelatonin signaling (Wei et al. 2018). To trigger stomatal closure, AtPMTR1 activates the NADPH oxidase-mediated H_2O_2 generation and promotes Ca^{2+} influx along with K^+ efflux. Further downstream signaling may converge on distinct pathways like the G protein signaling pathway and mitogen-activated protein kinase (MAPK) signaling cascade to transmit the phytomelatonin signal, and may further involve Ca^{2+}, calmodulin (CaM), ROS, and protein kinases (e.g., calcium-dependent protein kinases (CDPK), SnRK1 (Snf1-related protein kinase 1)) (Lee and Back 2016, 2017, Lee et al. 2022; Chen et al. 2022; Chang et al. 2024; Fig. 1.3).

Since the discovery of the first and only phytomelatonin receptor, several homologs of AtPMTR1 have been identified in different plant species, like alfalfa, tobacco, maize, cassava, rice, and tomato (Kong et al. 2021; Yu et al. 2021; Bai et al. 2022; Wang et al. 2022b; Barman et al. 2023; Liu et al. 2025). Owing to the dedicated research and much progress made in the recent past, the roles of phytomelatonin receptor-mediated signaling have expanded beyond their involvement in stomatal closure. PMTR1-mediated signaling has also been implicated in the regulation of the circadian stomatal movement, stomatal and plant immunity, seedling growth, leaf senescence, wound-induced adventitious rooting, as well as drought and salt stress tolerance (Yang et al. 2021b; Bai et al. 2022; Jing et al. 2022; Liu et al. 2025). Upon sensing phytomelatonin signals, PMTR1 binds to the Gα subunit. It has been speculated that Gα remains inactive in the resting state through binding of an inhibitor harboring GTPase activity, hence it is rendered incapable of undergoing spontaneous GDP-GTP exchange (Liu et al. 2022). Following the perception of phytomelatonin by PMTR1, the activation of the kinase induces phosphorylation of the inhibitor, thereby facilitating its dissociation from Gα. Upon being freed from this suppression, Gα is able to self-activate and dissociate from the Gβγ dimer to transmit signaling to downstream effectors. Eventually, the signal relay results in the modulation of the expression of transcription factors (TFs) and the phytomelatonin response genes. Phytomelatonin controls the expression of an array of genes, and this molecular control over distinct categories of genes belonging to several pathways imparts phytomelatonin the status of a master regulator. Another level of

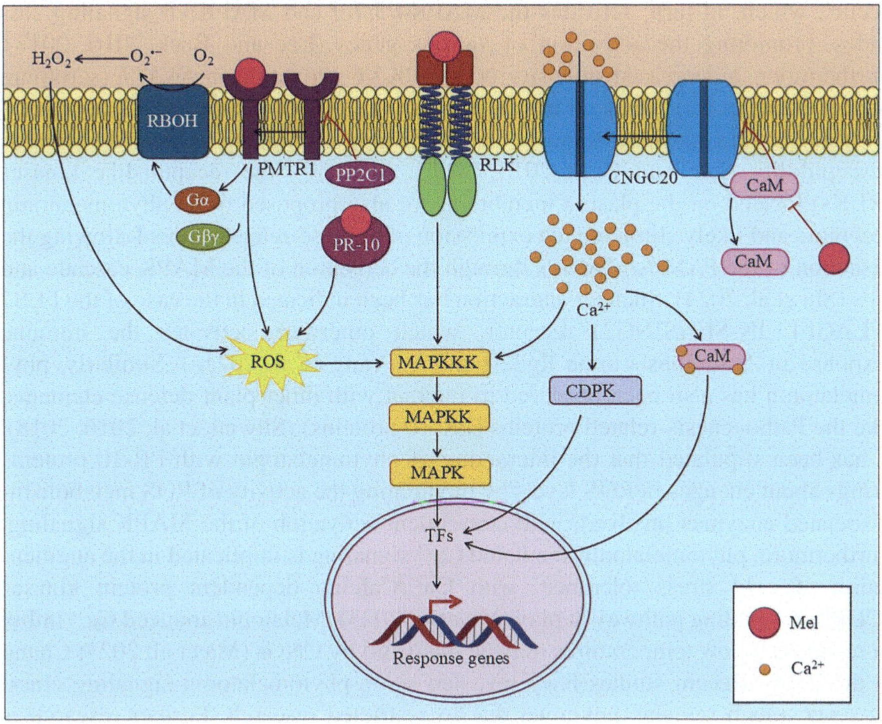

Fig. 1.3 Schematic representation of the phytomelatonin signaling pathway in plants. Abbreviations: CaM Calmodulin, CDPK Calcium-dependent protein kinase, CNGC20 Cyclic nucleotide gated channel 20, MAPK Mitogen-activated protein kinase, MAPKK Mitogen-activated protein kinase kinase, MAPKKK Mitogen-activated protein kinase kinase kinase, Mel Melatonin, PMTR1 Phytomelatonin receptor 1, PP2C1 Protein phosphatase 2C 1, PR-10 Pathogenesis related protein 10, RLK Receptor-like kinase, RBOH Respiratory burst oxidase homolog, ROS Reactive oxygen species

regulation of the receptor-mediated phytomelatonin signaling has been revealed in cassava, wherein PMTR1 was demonstrated to interact with PP2C1 (Protein Phosphatase 2C1) and become dephosphorylated at the Ser11 residue upon this interaction, resulting in a diminished ability to bind melatonin (Bai et al. 2022).

The other signaling pathway, the MAPK signaling cascade, is utilized in other phytomelatonin responses, like immunity to pathogen attack, activation of autophagy, and drought stress tolerance (Lee and Back 2016, 2017; Supriya et al. 2024). It has been suggested that the activation of the MAPK cascade by PMTR1 may operate directly or indirectly. Moreover, this phytomelatonin-induced activation of the MAPK signaling cascade is proposed to operate via the degradation of one or more intermediate inhibitors that keep MAPKKKs in the inactive state in the absence of stimulus (Liu et al. 2022). In plants witnessing a pathogen challenge, phytomelatonin mediates pathogen resistance by activating the MAPK cascade (Mansoor et al. 2024). In pathogen-afflicted plants, a transient increase in phytomelatonin levels occurs, which, in turn, activates the MAPK4/5/7/9 and MAPK3/6 signaling cascades, promoting the activation of various genes (Lee and Back 2016, 2017). Furthermore, a decreased intensity of MAPK3/6 activation in *SNAT1* (serotonin N-acetyltransferase) knockout mutants is associated with a diminished phytomelatonin-mediated resistance response to a pathogen, causing an enhanced pathogen susceptibility (Lee and Back 2016, 2017). Similarly, the receptor-like kinases (RLKs) located on the plasma membrane are also proposed to be phytomelatonin receptors and likely stimulate the expression of defense-related genes following the detection of the PAMPs/MAMPs through the activation of the MAPK cascade and TFs (Shi et al. 2024). Such an interaction has been indicated in the case of the FLS2 (FLAGELLIN-SENSING2) receptor, which otherwise activates the immune response in Arabidopsis upon flg22 binding (Yang et al. 2021c). Similarly, phytomelatonin has also been observed to interact with other plant defense elements, like the Pathogenesis-related proteins (PR-10 proteins) (Sliwiak et al. 2016, 2018). It has been stipulated that the interaction of phytomelatonin with PR-10 proteins brings about changes in ROS levels by modulating the activity of ROS metabolism-associated enzymes involved, with consequent activation of the MAPK signaling. Furthermore, phytomelatonin-mediated Ca^{2+} signaling is implicated in the augmentation of cold stress tolerance, with Ca^{2+}/Calcium-dependent protein kinases (CDPKs) signaling pathway at play (Ma et al. 2023). Melatonin-induced Ca^{2+} influx in response to low temperatures may be mediated by CNGs (Ma et al. 2023; Chang et al. 2024). Recent studies have revealed some phytomelatonin signaling clues, however, much remains unknown due to restricted research. Further research is needed to understand phytomelatonin-mediated reactions' molecular processes.

1.5 Cross-Talk Between Phytomelatonin and Phytohormones

Phytomelatonin controls plant growth, development, and defense. Such regulation puts phytomelatonin and phytohormones in conflict. Phytomelatonin controls plant physiological processes synergistically and antagonistically with phytohormones (Arnao and Hernández-Ruiz 2021; Liu et al. 2022; Shi et al. 2024a, Table 1.1). In recent years, numerous studies have illustrated this interaction. Phytomelatonin acts in a manner similar to auxin, functioning generally synergistically with auxin, and the perception of phytomelatonin signals has been shown to activate the auxin synthesis pathway and regulate the expression of auxin transporter genes *PIN1/3/7*, and auxin-responsive genes, like *Aux/IAAs*, *ARFs*, *GH3s*, and *SAUR* (Wen et al. 2016; Liang et al. 2017). Its interaction with auxin has been shown to regulate root growth and architecture, and the development of lateral and adventitious roots (Yang et al. 2021a). For example, it promotes primary root growth in an IAA-dependent pathway (Yang et al. 2021b). However, a counteracting role has also been reported at elevated phytomelatonin concentrations, where it inhibits the expression of genes of auxin biosynthesis enzymes (*YUC1*, *YUC2*, *YUC5*, *YUC6*, *TAA1*, and *TAR2*) and PIN transporter proteins (*PIN1*, *PIN3*, and *PIN7*) in response to certain

Table 1.1 Cross-talk between phytomelatonin and phytohormones in regulation of physiological processes

Plant hormone	Nature of interaction	Physiological effects	Responses underlying the effects	References
Auxin	Synergistic	Promotion of primary root growth	Modulation of expression and distribution of auxin transporters *PIN2/3* and *AUX1*	Yang et al. (2021b) and Wang et al. (2023)
	Synergistic	Promotion of lateral and adventitious roots	Regulation of auxin synthesis, transporter, signaling and response genes (*PINs*, *Aux/IAAs*, *YUCs*, *TAR2*, *SAURs*, *ARF*, *WOX11*)	Wen et al. (2016) and Liang et al. (2017)
	Synergistic	Stimulation of seed germination under stress		Golovatskaya et al. (2024), Yuan et al. (2024), and Zhang et al. (2024)
	Synergistic	Delayed senescence, extension of post-harvest shelf-life	Increased IAA and GA contents	Shi et al. (2024b)
	Synergistic	Promotion of seed germination under salt stress	Improved photosynthesis, reduced MDA content, increased proline content and the antioxidant enzymes	Zhang et al. (2024)
	Synergistic	Promotion of cold endurance	Regulation of *TOR*, *PIN1*, *and YUC4* genes, and antioxidant defense machinery	Pei et al. (2024)
Abscisic acid (ABA)	Synergistic	Inhibition of seed germination, promotion of dormancy		Lv et al. (2021)
	Synergistic	Stimulation of seed germination under salt stress	Increases ABA contents, upregulation of ABA signaling gene *ABF2*	Sun et al. (2024)
	Antagonistic	Delayed senescence	Inhibition of *ABFs*, chlorophyll catabolic enzyme genes	Tan et al. (2019a)
	Antagonistic	Drought tolerance	Regulation of stomatal movements, promotion of stomatal closure, downregulation of ABA synthesis genes (and upregulation of ABA catabolism genes (*ABFs*, *NCED1/2/3*, *CYP707A1/2*)	Wei et al. (2018), Arnao and Hernández-Ruiz (2021), and Jahan et al. (2024)
	Synergistic	High salt and cold endurance	Enhanced antioxidant capacity and improved photosynthesis parameters, higher ABA levels	Feng et al. (2024) and Yan et al. (2024)
		Delayed ripening and softening	Downregulation of *NCED*, *ACS* and *ACO*, lower ABA and ethylene levels, prevention of changes in pectin-modifying enzymes	Liu et al. (2020)

(continued)

Table 1.1 (continued)

Plant hormone	Nature of interaction	Physiological effects	Responses underlying the effects	References
Cytokinin (CK)	Synergistic	Improvement of growth under stress	Improvement of photochemical efficiency, chlorophyll content, and RWC, upregulation of cytokinin biosynthesis and signaling genes (*Type-A* and *Type-B ARRs*, *AHKs*, *AHPs*)	Zhang et al. (2017a), Ma et al. (2018), and Tan et al. (2019b)
	Antagonistic	Stimulation of primary root growth	Downregulation of *AHK4*, *AHP2/3/5*, and type-A *ARR15*, modulation of expression and distribution of auxin transporters *PIN2/3* and *AUX1*	Wang et al. (2023)
Gibberellin (GA)	Synergistic	Stimulation of seed germination under high salinity and fluorine, enhancement in cold tolerance	Upregulation of GA biosynthesis and GA signaling genes, suppression of DELLA proteins, improved morphological parameters, chlorophyll synthesis, proline content, antioxidant system, and methylglyoxal detoxification	Zhang et al. (2014a), Li et al. (2017), Banerjee and Roychoudhury (2019), Tan et al. (2019b), Khan et al. (2020), Siddiqui et al. (2020), and Sun et al. (2024)
	Antagonistic	Inhibition of seed germination		Lv et al. (2021)
	Synergistic	Induction of parthenocarpy	Stimulation of cell division, and expansion of mesocarp, upregulation of GA3 and GA4 biosynthesis genes	Liu et al. (2018)
	Antagonistic	Delayed flowering	Inhibition of *KS* and *FT* genes	Lee et al. (2019)
Ethylene (ETH)	Synergistic	Promotion of fruit ripening	Regulation of ethylene synthesis and signaling genes (*ACC*, *ACSs*, *EIN*, *EILs*, *ERFs*, *ETRs*)	Sun et al. (2015) and Shan et al. (2022)
Brassinosteroids (BRs)	Synergistic	Promotion of plant growth, seed size, delayed senescence, and stress tolerance	Upregulation of BR biosynthesis genes (*DWARF4 and BZR1*), enhanced BR levels	Tan et al. (2019b), Sharma et al. (2020), Yang et al. (2021a), and Lee and Back (2024)
Strigolactones (SLs)	Synergistic	Drought and heavy metal tolerance	Improvement in physico-biochemical attributes and antioxidant system	Raja et al. (2023) and Xie et al. (2024)
	Antagonistic	Delayed flowering	Inhibition of *FLC* expression	Zhang et al. (2019) and Arnao and Hernández-Ruiz (2020)

Jasmonic acid (JA)	Synergistic	Augmentation of abiotic and biotic stress tolerance	Upregulation of JA biosynthesis genes (e.g., *LOXs*, *AOS*, *AOC*, and *JMT*) and downregulation of JA signaling inhibitors *MYC2* and *JAZ1*, enhanced JA accumulation, higher total chlorophyll content, improvement of the Fv/Fm, improved photosynthesis, decline in EL and H_2O_2 content, activation of antioxidant defense, regulation of ethylene level, improved S and N metabolism	Liu et al. (2019a), Ding et al. (2022), Luo et al. (2023), Sehar et al. (2023), Aybeke (2024), Li et al. (2024), and Xu et al. (2024)
	Synergistic	Induction of adventitious rooting in response to wounding	Activation of *MYC2*, *PMTR1/2*, and *BRL1*, JA-mediated enhancement of phytomelatonin signals	Liu et al. (2025)
Salicylic acid (SA)	Synergistic	Augmentation of abiotic and biotic stress tolerance	Upregulation of SA biosynthesis gene *PAL*, improved biomass production, fresh and dry weights, photosynthetic pigments, PSII efficiency, proline and phytochelatins contents, and augmented antioxidant system, methylglyoxal detoxifcation, and nitrogen metabolism, and decreased H_2O_2, and MDA levels	Ni et al. (2018), Imran et al. (2021), Park et al. (2021), Kaya et al. (2022, 2023), Xing et al. (2023), Nasiri et al. (2024), and Yan et al. (2024)
	Synergistic	Delayed fruit senescence	Higher antioxidant enzyme activity, reduced ETH release, and cell wall degrading enzyme activity	Cheng et al. (2025)

environmental stress cues, like drought and unfavourable temperature (Wang et al. 2016). Phytomelatonin also works synergistically with auxin to stimulate seed germination, along with mitigating stresses (Golovatskaya et al. 2024; Yuan et al. 2024; Zhang et al. 2024).

Phytomelatonin has been shown to counteract the effects of ABA in regulating seed germination, stomatal movement, senescence, and abiotic stress tolerance (Zhang et al. 2017a, b; Tan et al. 2019a, b; Li et al. 2021a;Feng et al. 2024; Jahan et al. 2024; Yan et al. 2024). Phytomelatonin-mediated drought stress mitigation has been linked with its antagonistic effect on the synthesis of the stress hormone abscisic acid (ABA). For instance, the exogenous application of melatonin reduces ABA levels under water-scarce conditions by inhibiting ABA synthesis. During drought stress, it downregulates the expression of an ABA synthesis pathway gene *NCED3* (encodes epoxycarotenoid oxygenase) while concurrently inducing the catabolic enzymes, like CYP707 (ABA catabolism cytochrome P450 monooxygenase family 707), thereby bringing about reduction in ABA levels and delay in plant senescence (Arnao and Hernández-Ruiz 2021). It also triggers RBOHD-mediated H_2O_2 synthesis (respiratory burst oxidase homolog), which, in turn, enhances Ca^{2+} influx, thereby regulating ABA resistance and delaying senescence (Guo et al. 2023). By inhibiting the expression of ABF (ABA-responsive element binding factors) genes, which govern the expression of genes of ABA biosynthesis and chlorophyll catabolic pathways, melatonin also delays senescence (Tan et al. 2019a). Phytomelatonin also exerts an inhibitory effect on seed germination in a synergistic interaction with ABA, and the inhibitory effect of melatonin on seed germination is weakened in the ABA signaling mutants *abi3-8* and *abi5-7* (Lv et al. 2021). Melatonin, alone or in combined supplementation with ABA, enhances cold inducing by boosting the antioxidant capacity and improving the photosynthesis parameters (Feng et al. 2024). Likewise, melatonin also contributes to enhanced salinity tolerance under ABA-deficient conditions by relieving the inhibition on the antioxidant system and the process of photosynthesis (Yan et al. 2024). On the other hand, exogenous melatonin improves drought tolerance by suppressing ABA biosynthesis and concurrently enhancing sucrose synthesis (Jahan et al. 2024).

Melatonin has a synergistic interaction with the phytohormone cytokinin (CKs), Under stress conditions, melatonin upregulates the expression of cytokinin biosynthesis and signaling genes, Type-A and Type-B response regulators (ARRs), histidine kinases (AHKs), and histidine phosphotransfers (AHPs) and downregulates the ABA biosynthesis genes (Zhang et al. 2017a; Ma et al. 2018; Tan et al. 2019a). Under the drought conditions, this synergistic interaction of phytomelatonin with cytokinin improves the photochemical efficiency, chlorophyll content, and relative water content (RWC) (Ma et al. 2018). Melatonin also has a synergistic interaction with the gibberellins (GAs) phytohormone, which plays a role in seed germination and cell division (Zhang et al. 2014a; Banerjee and Roychoudhury 2019; Tan et al. 2019a). The exogenous application of melatonin upregulates the expression of GA biosynthesis genes and provides tolerance to cold (Li et al. 2017). Meanwhile, a combined melatonin and GA_3 supplementation has been demonstrated to augment the morphological parameters, chlorophyll synthesis, proline content, antioxidant system, and methylglyoxal detoxification (Khan et al. 2020; Siddiqui et al. 2020).

Phytomelatonin's interaction with ethylene (ETH) orchestrates the processes of fruit ripening, senescence, and tolerance to (a)biotic stresses. This relationship has been majorly ascertained by exogenous melatonin application, which has been described to regulate ethylene metabolism by modulating the expression of ethylene biosynthesis and signaling genes, like 1-aminocyclopropane-1-car carboxylic acid (ACC) synthase (*ACS*), *ACO*, *NR*, *ETRs*, *EIL1*s, *EINs*, and *ERF* (Sun et al. 2015, 2020; Shan et al. 2022). In tomato and apples, melatonin application activates ethylene biosynthesis and signaling genes (*ACS10*, *ACO*, *EIN3*, *ERF-A1*, and *ERT10*) to promote fruit maturation, while in mango and cabbage, it leads to lowering of ethylene levels to modulate ripening and softening and delay senescence via downregulation of *ACS*, *ACO*, and pectin-modifying enzyme genes, and suppressing ERF2/BrERF109-mediated inhibition on flavonoid synthesis (Liu et al. 2020; Sun et al. 2020; Shan et al. 2022; Verde et al. 2023; Yue et al. 2023). Cross-talk of phytomelatonin with other phytohormones, like brassinosteroids (BRs), strigolactones (SLs), jasmonic acid (JA), and salicylic acid (SA), has also been documented. It interacts with BR to improve plant growth, seed size, senescence, and stress tolerance (drought and salinity) by upregulating BR biosynthesis genes while impeding growth at high concentrations by inhibiting the expression of BR biosynthesis genes (Tan et al. 2019a; Sharma et al. 2020; Yang et al. 2021a; Lee and Back 2024). It has been suggested to act downstream of SLs in the regulation of flowering via the activation of *FLC* expression (Zhang et al. 2019; Arnao and Hernández-Ruiz 2020). Melatonin functions in concert with SLs to enhance tolerance to stresses like drought and chromium stress through improvement in physico-biochemical attributes and antioxidant system (Raja et al. 2023; Xie et al. 2024). In defense against pathogens, melatonin functions in inducing an immune response through its interplay with jasmonic acid (JA) and salicylic acid (SA). Melatonin assists in enhanced accumulation of JA to impart pathogen resistance to *Botrytis cinerea* in tomatoes and *Botryosphaeria dothidea* in pears, as well as salinity, cold, and drought tolerance (Liu et al. 2019a; Luo et al. 2023; Aybeke 2024; Li et al. 2024; Xu et al. 2024). JA biosynthesis genes such as LOXs, AOS, AOC, and JMT are upregulated, while JA signaling inhibitors MYC2 and JAZ1 are downregulated, resulting in an increase in endogenous JAs (Liu et al. 2019a; Luo et al. 2023; Li et al. 2024). On the other hand, it reduces JA production as reported in response to salt stress, thus allowing seedling recovery (Tan et al. 2019a). Their cooperative interaction is associated with improved cold stress endurance via improvement of the Fv/Fm with a concomitant decline in electrolyte leakage (EL) and H_2O_2 content (Ding et al. 2022). Furthermore, when provided in combined treatment, melatonin and JA restored the total chlorophyll content and antioxidant defense to normal levels in plants under salt stress (Aybeke 2024). In addition, phytomelatonin-induced adventitious rooting in response to wounding also involves crosstalk between JA along with auxin, as confirmed by the induction of fewer roots in *myc2* mutants, which fail to accumulate JA in response to injury (Liu et al. 2025).

With respect to SA, melatonin application has been described to induce its enhanced accumulation by increasing the expression of the SA biosynthesis gene *PAL2* during heat and salt stress and improving immunity against pathogens (Imran et al. 2021; Park et al. 2021; Nasiri et al. 2024). In contrast, a decline in SA content in *SNAT* mutants with reduced melatonin content is associated with enhanced susceptibility to pathogens (Lee et al. 2015; Arnao and Hernández-Ruiz 2019). A combined treatment of melatonin and SA has been shown to alleviate salt stress, drought, and heavy metal (Zn, and Cd) stress by way of improving biomass production, fresh and dry weights, photosynthetic pigments, PSII efficiency, proline and phytochelatins contents, and augmenting the antioxidant system, methylglyoxal detoxifcation, and nitrogen metabolism, and decreasing H_2O_2 and MDA levels (Ni et al. 2018; Kaya et al. 2022, 2023; Xing et al. 2023; Yan et al. 2024). Their synergistic relationship is also effective in delaying fruit senescence and decay by improving the fruit quality traits and antioxidant enzyme activity, and reducing ethylene release and cell wall degrading enzyme activity (Cheng et al. 2025).

1.6 Use of Omics Approaches for Studying Melatonin in Plants

The transcriptomics and proteomics offer valuable insights into the molecular networks of melatonin in plants. This approach helps in the identification of genes, proteins, and signaling pathways influenced by melatonin in various physiological conditions (Cao et al. 2022; Yang et al. 2022). RNA sequencing (RNA-seq) is a powerful transcriptomic technique that involves analyzing the complete set of RNA transcripts in a plant for understanding the gene expression patterns. The RNA-Seq analysis showed that melatonin regulates the genes involved in antioxidant defense, hormone signaling, and secondary metabolism (Ren et al. 2022; Zhao et al. 2021). The differential gene expression analysis identified melatonin-responsive transcription factors such as MYB and WRKY mediating stress adaptation and also altering the photosynthesis-related genes with improved plant growth under stress conditions. The expression, modifications, and interactions of proteins are discussed in detail in proteomics. The proteomic analyses have identified melatonin-induced proteins involved in ROS scavenging, stress adaptation, and hormone metabolism (Yang et al. 2021a, 2024; Wang et al. 2025). The post-translational modifications of phosphorylation and ubiquitination of melatonin-responsive proteins have been characterized, shedding light on melatonin's mechanisms of action. These tools have significantly advanced the understanding of biosynthesis and the functions of melatonin. The qPCR helped the identification of gene expression of key melatonin biosynthetic genes, while CRISPR-Cas9 gene editing provides functional validation of melatonin's role in stress responses (Duan et al. 2024). Additionally, omics approaches unraveled complex regulatory networks controlled by melatonin. Future research should focus on integrating these molecular tools to develop melatonin-enriched crops with improved stress resilience and enhanced productivity. Over the past few years, research has proven melatonin's multifaceted role in improving crop resilience and productivity. Understanding the complex pathways through which it operates requires a comprehensive approach integrating various omics technologies, and are listed in Table 1.2.

1.7 Imaging Techniques for Melatonin Localization in Plants

The spatial distribution of melatonin within plant tissues analyzed by various imaging techniques, including confocal microscopy, immunohistochemistry, mass spectrometry imaging (MSI), and advanced fluorescent probes, which localize the melatonin accumulation, movement, and function at the cellular and subcellular levels under normal and stress conditions. Confocal microscopy uses a laser to excite fluorophores and produces a high-resolution optical and spatial resolution. The melatonin-specific fluorescent probes like 2-hydroxybenzylamine (HBA)-based probes have been developed to enhance detection accuracy (Byeon et al. 2014; Yang et al. 2024). In *Arabidopsis thaliana*, confocal microscopy revealed that melatonin accumulates in the meristematic zones and vascular tissues of roots. This highlights the role of melatonin in the involvement in root development and stress adaptation (Wang et al. 2017; Bargunam et al. 2024). In *Oryza sativa*, melatonin fluorescence was detected in the chloroplasts, indicating a potential role in photosynthetic regulation. Confocal microscopy provides high spatial resolution and real-time imaging that co-localizes melatonin even in cellular components (Byeon and Back 2014). Though confocal microscopy has its own advantages, certain limitations include the requirement of specific fluorescent probes, and effect of photobleaching and their ability to detect melatonin at very low concentrations making them less usable. Immunohistochemistry (IHC), an antibody-based technique detects melatonin in plant tissues in situ. The use of primary and secondary antibodies conjugated with fluorescent probes provides both qualitative and semi quantitative data on melatonin localization at the tissue and cellular levels. In *Solanum lycopersicum*, immunohistochemical staining shows the accumulation of melatonin in guard cells during drought stress, regulating the stomatal closure (Colizzi 2023; Li et al. 2021a, b). In *Zea mays* (maize), melatonin was found to localize in phloem companion cells, suggesting its role in nutrient transport and stress signaling. Their ability to detect melatonin at different plant tissues, even in low concentrations, and also their high compatibility with fluorescence detection make them a relative advantage. The specific antibody requirements and cross-reactivity issues are their major limitations of using immunohistochemistry. The emerging new technology, the mass spectroscopy imaging (MSI) is highly used for direct visualization of melatonin at a cellular level without the need for probes or antibodies. The MSI uses laser desorption/ionization (MALDI) and desorption electroscopic ionization (DESI) to locate melatonin within plant tissues. MSI provides molecular-level specificity without the need of chemical labeling unlike other techniques (Delcourt et al. 2018). The MSI results showed higher accumulation of melatonin in salt stressed tomato plants compared to control, highlighting the role of melatonin in stress

Table 1.2 Summarizing the research on melatonin using omics approaches and molecular techniques

Omics approach	Molecular techniques	Plant studied	Research findings	References
Genomics	Real-time- quantitative PCR/genome sequencing	Rice	Abiotic stress, such as drought, downregulated the genes (tryptophan decarboxylase (TDC), tryptamine 5-hydroxylase (T5H), and serotonin N-acetyltransferase, SNAT) involved in melatonin biosynthesis in rice.	Ahn et al. (2021)
		Tomato	Melatonin enhances H_2O_2 levels by activating NADPH oxidase enzyme and enhances stress tolerance in tomato.	Gong et al. (2017)
		Carya cathayensis	Genes involved in chlorophyll metabolism, phenylpropanoid biosynthesis, and antioxidative defense were upregulated in *Carya cathayensis* after the exogenous application of melatonin.	Sharma et al. (2020)
		Catharanthus roseus	Nine reference genes (FBOX, CACS, TIP, RSP9, EXP, EXPR, SAND, F17M5, ACT) were evaluated for stability under melatonin supplementation and during abiotic stress. EXP and EXPR were the most stable.	Sheshadri et al. (2018)
		Rice	This study revealed that 2-hydroxynelatonin was less effective than melatonin in inducing defense-related signaling in rice plants.	Byeon et al. (2015)
	Transcriptome analysis/ RNA sequencing	Arabidopsis	Melatonin application in Arabidopsis upregulated genes involved in ABA, ET, SA, and jasmonic acid signaling, giving insights into its possible role in providing defense against stresses.	Weeda et al. (2014)
		Cucumber	Melatonin influences the lateral root development in cucumber, and during this process, peroxide-related genes aid this process.	Zhang et al. (2014b)
		Watermelon	Exogenous application of melatonin and genetic transformation with the SNAT gene in watermelon boosted the resistance against powdery mildew fungal pathogens, suggesting its role in defense pathways.	Mandal et al. (2018)
		Soybean	Synergistic effects of auxin and melatonin were observed in soybean, as melatonin upregulated genes involved in auxin metabolism, such as YUCCA, TIR, AFB3, AUX1, and PIN.	Wei et al. (2022)
		Arabidopsis	Functional characterization of genes involved in serotonin and melatonin biosynthesis was conducted. MiRNAs such as miR-6249a and miR-1846e also regulated their biosynthesis.	Bhowal et al. (2021)
		Tomato	DREB1α and IAA3 were key transcription factors that improved melatonin-induced sodic alkaline stress tolerance in tomato.	Yan et al. (2019)
		Zoysia japonica	Melatonin improves seed germination by enhancing the antioxidant potential of Zoysia seeds.	Dong et al. (2021)

Proteomics	Western blotting, knockout mutant analysis	Arabidopsis	Melatonin maintains chloroplast proteins by upregulating Clp and CpHSPs proteins after its exogenous application.	Lee and Back (2021)
	NMR and mass spectroscopy	*Melia toosendan*	Lignans extracted from the fruit of *Melia toosendan* showed antagonistic activity towards melatonin receptor type 1, which is crucial in therapy for psychiatric problems in humans.	Wang et al. (2015)
Integrative approaches	Transgenic plant development and enzymatic assays	Arabidopsis	Plant mitochondrial localized enzyme MzSNAT5 is involved in melatonin biosynthesis, enhancing drought resilience in Arabidopsis. It possesses high kinetic energy and serves from 5–75 °C.	Wang et al. (2017)
	Transgenic lines production and quantitative PCR analysis	Arabidopsis	Upregulation of genes involved in UV-B signaling pathways post melatonin application, suggesting its role as a robust antioxidant system.	Yao et al. (2021)
	Western blotting, quantitative PCR, and RNAi	Arabidopsis	Melatonin activates MAP3 and MAP6 in Arabidopsis and tobacco leaves, highlighting its role in providing innate immunity.	Lee and Back (2016)
	qRT-PCR and dot-ELISA	Tobacco	Exogenous application of melatonin reduced tobacco mosaic virus RNA levels and upregulated the expression of defense genes like PR1 and PR5.	Zhao et al. (2019)

adaptation (Mukherjee et al. 2014; Wang et al. 2016). In *Brassica napus* (canola), MSI revealed that melatonin deposits in endodermal cells regulate root architecture and nutrient absorption. The MSI has high specificity but requires expensive equipment and complex data analysis is its major limitations. In the advanced fluorescent probes detection, benzimidazole-based probes are used which provide rapid detection in live plant tissues. In Arabidopsis, these novel fluorescent probes track the real-time movement of melatonin between the root tip and shoot apex, showing their systemic movement of stress signaling (Liu et al. 2022). In *Nicotiana benthamiana* (tobacco), real-time imaging revealed that melatonin is rapidly transported via phloem, supporting its role as a mobile signaling molecule. Future research should focus on refining these methods to improve the detection sensitivity with enhanced real-time tracking imaging approaches.

1.8 Recent Advancements in Phytomelatonin Research

Melatonin is a key critical signaling molecule in plants involved in regulating growth, development, and stress responses. The advanced molecular tools, such as gene expression studies, CRISPR-cas9 gene editing, and omics approaches, are highly used to understand the underlying molecular mechanism of melatonin biosynthesis in plant physiology (Fan et al. 2018; Murch and Erland 2021). These advanced molecular methods analyze the gene expression and regulation, help to study the functional knockouts of targeted genes, and unravel the large-scale molecular networks influenced by melatonin (Wei et al. 2015). The qPCR reverses transcription of RNA into complementary DNA (cDNA) and amplifies the specific genes using fluorescent dyes such as SYBR green or Tag man probes which provide high sensitivity with precise quantifications (Zhang et al. 2014b; Weeda et al. 2014) The gene expression using quantitative PCR (Qpcr) helps to understand the expression levels of key genes of melatonin involved in melatonin synthesis in plants (Gu et al. 2017; Chen et al. 2020). In the melatonin biosynthesis, three important hormones, the tryptophan decarboxylase (TDC), serotonin N-acetyltransferase (SNAT) and acetyl serotonin methyltransferase (ASMT), convert tryptamine to melatonin (Mandal et al. 2018; Sheshadri et al. 2018). The differential expression of biosynthesis genes TDC and ASMT gene through qPCR reveals elevated levels of melatonin in extreme environmental stress like drought, salinity, and temperatures (Zhao et al. 2019; Wang et al. 2020). The precise gene alterations of plant genomes by CRISPR technology have been used to knock out the melatonin biosynthesis genes for studying the physiological and signaling role of melatonin (Khattak et al. 2023; Seth et al. 2023). The cas9 endonuclease protein and Guide RNA (g RNA) cause the double-stranded breaks at the target site and direct gene editing accurately. After the Cas9 induced cleavage, the natural DNA repair of plants induces mutations via nonhomologous end joining (NHEJ) and homology-directed repair (HDR), which results in gene knockout with specific gene changes (Kaya and Ugurlar 2023). The CRISPR-Cas9 technology in melatonin has been used extensively to knock out the TDC and ASMT genes, which shows the reduced melatonin levels and increases the susceptibility to the oxidative stress in edited plants (Zhang et al. 2023; Wang et al. 2022a). Through gene knockout, the plants with impaired biosynthesis exhibit reduced drought and salt tolerance, highlighting melatonin role in stress resistance. By introducing specific mutations in the desired regions, stress-resistant crops can be developed by targeting specific melatonin biosynthesis pathways. The powerful CRISPR-Cas9 has certain challenges such as off-target effects, gene redundancy, and low efficiency in some plant species (Ma et al. 2017; Chang et al. 2021; Wani et al. 2023). By improving the focus on delivery methods and developing accurate targeted modifications of melatonin biosynthesis genes, editing systems play an important role in developing resistant crop plants. The transcriptomics and proteomics offer valuable insights into the molecular networks of melatonin in plants. This approach helps in the identification of genes, proteins, and signaling pathways influenced by melatonin in various physiological conditions (Cao et al. 2022; Yang et al. 2022). The RNA sequencing (RNA-seq) is a powerful transcriptomic technique that involves analyzing the complete set of RNA transcripts in a plant for understanding the gene expression patterns. The RNA-Seq analysis showed that melatonin regulates the genes involved in antioxidant defense, hormone signaling, and secondary metabolism (Ren et al. 2022; Zhao et al. 2021). The differential gene expression analysis identified melatonin-responsive transcriptions factors such as MYB and WRKY mediating stress adaptation and also altering the photosynthesis-related genes with improved plant growth under stress conditions. The expression, modifications, and interactions of proteins are discussed in detail in proteomics. The proteomic analyses have identified melatonin-induced proteins involved in ROS scavenging, stress adaptation, and hormone metabolism (Yang et al. 2021a, b, c, 2024; Wang et al. 2025). The post-translational modifications of phosphorylation and ubiquitination of melatonin-responsive proteins have been characterized, shedding light on melatonin's mechanisms of action. These tools have significantly advanced the understanding of biosynthesis and the functions of melatonin. The qPCR helped the identification of gene expression of key melatonin biosynthetic genes, while CRISPR-Cas9 gene editing provides functional validation of melatonin's role in stress responses (Duan et al. 2024). Additionally, omics approaches unraveled complex regulatory networks controlled by melatonin. Future research need focusing on integrating these molecular tools to develop melatonin-enriched crops with improved stress resilience and enhanced productivity.

Melatonin (N-acetyl-5-methoxytrytamine) is an indoleamine that plays a role in animal physiology. In plants, phytomelatonin has been extensively studied (Kolář and Macháčková 2005). Melatonin is an important bioactive molecule involved in various physiological processes of the plant, including plant growth regulation, stress responses, and circadian rhythms (Arnao and Hernández-Ruiz 2006; Agathokleous et al. 2019; Wang et al. 2022a, b). The accurate quantification of melatonin in plants is crucial for understanding their physiological and biochemical roles. There are various advanced analytical techniques, including Liquid Chromatography-Tandem Mass Spectrometry (LC-MS/MS), Gas Chromatography-Mass Spectrometry (GC-MS), and High-Performance Liquid Chromatography (HPLC), employed to detect and quantify melatonin, offering higher sensitivity, specificity, and application to different plant matrices (Arnao and Hernández-Ruiz 2015; Tan et al. 2012). The melatonin physiological role in growth attributes and interaction during stress responses requires precise quantification. The chromatography studies are highly useful to study the melatonin interaction in plants for their high accuracy and sensitivity, even in complex matrices and low concentration samples (Albu and Radu 2018; Albu et al. 2020; Sezgin et al. 2024). HPLC quantified melatonin in Alfa alfa (*Medicago sativa*) seeds with high sensitivity and accuracy using ultraviolet and fluorescence detectors. The HPLC measurement reveals that melatonin plays an important role in seed germination of alfa alfa seeds (Martins and Mainardes 2017). Simultaneously, the Gas Chromatography-Mass Spectrometry (GC-MS) and Liquid Chromatography-Tandem Mass Spectrometry (LCMS) were also used to know the interaction of melatonin in plant systems. The GC-MS separates the melatonin in volatile compounds after derivatization from the fruit extract (Paik et al. 2010; Ma et al. 2006; Van Tassel et al. 2001). GC-MS due to their high sensitivity in metabolic profiling, it is widely used to study the interaction of melatonin in fruits, medicinal plants, and seeds (Hasan et al. 2024; Rajotia et al. 2023). LCMS integrates liquid chromatography with mass spectrometry that detects even trace amounts of melatonin in phloem saps and root exudates. It is highly useful in drought and stress conditions, due to their ultra-high sensitivity and specificity, even a minor change in melatonin can be identified (Hardeland et al. 2011).

1.9 Research Gaps, Future Directions, and Conclusion

Melatonin biosynthesis in plants is well characterized, but still few questions on its signaling and response of melatonin in diverse environmental conditions remain unaddressed, and also, field applications of melatonin need to be focused more. For example, it is unknown how stress regulates enzymes such as serotonin N-acetyltransferase (SNAT) and N-acetylserotonin methyltransferase (ASMT) (Tan et al. 2012). While melatonin has shown an opportunity in controlled situations to enhance plant growth and stress tolerance, its use in the field offers obstacles. Environmental variation, interactions with soil microorganisms, and the development of cost-effective delivery techniques for large-scale application all require more investigation (Arnao and Hernández-Ruiz 2019). In addition, a prolonged study must be undertaken to prove scientifically the ecological and economic viability of melatonin-based agricultural applications. It is imperative to identify those key genes and pathways that govern melatonin sensitivity in order to apply this substance to specific crops (Zhao et al. 2021). For the future, research on plants with melatonin should combine a systems biology approach for the unraveling of signaling and metabolic networks, mechanistically, thereby providing large-scale applicability to agriculture. Studies from future researchers must also emphasize the relevance of phytomelatonin in improving resistance to abiotic stresses.

References

Abbas T, Ahmad I, Nawaz R, Nazim M, Gatasheh MK, Alamri AM, Muneeb A (2023) Physiological responses and antioxidant properties of Citrus reticulata under different abiotic stresses mitigated by endogenous melatonin. Sci Hortic 322:112442

Agathokleous E, Kitao M, Calabrese EJ (2019) New insights into the role of melatonin in plants and animals. Chem Biol Interact 299:163–167

Aghdam MS, Mukherjee S, Flores FB, Arnao MB, Luo Z, Corpas FJ (2023) Functions of melatonin during postharvest of horticultural crops. Plant Cell Physiol 63(12):1764–1786. https://doi.org/10.1093/pcp/pcab175

Ahammed GJ, Li Z, Chen J, Dong Y, Qu K, Guo T, Wang F, Liu A, Chen S, Li X (2024) Reactive oxygen species signaling in melatonin-mediated plant stress response. Plant Physiol Biochem 207:108398. https://doi.org/10.1016/j.plaphy.2024.108398

Ahmad S, Su W, Kamran M, Ahmad I, Meng X, Wu X, Javed T, Han Q (2020) Foliar application of melatonin delay leaf senescence in maize by improving the antioxidant defense system and enhancing photosynthetic capacity under semi-arid regions. Protoplasma 257(4):1079–1092. https://doi.org/10.1007/s00709-020-01491-3

Ahn HR, Kim YJ, Lim YJ, Duan S, Eom SH, Jung KH (2021) Key genes in the melatonin biosynthesis pathway with circadian rhythm are associated with various abiotic stresses. Plants 10(1):129

Albu C, Radu GL (2018) Development and application of a HPLC-PDA-FL method for the determination of melatonin and its precursors in infant formulas. Food Anal Methods 11:951–958

Albu C, Radu LE, Radu GL (2020) Assessment of melatonin and its precursors content by a HPLC-MS/MS method from different Romanian wines. ACS Omega 5(42):27254–27260

Ali S, Nawaz A, Naz S, Ali M, Ejaz S, Azam M, Razzaq K (2023) Exogenous melatonin mitigates chilling injury in zucchini fruit by enhancing antioxidant system activity, promoting endogenous proline and GABA accumulation, and preserving cell wall stability. Postharvest Biology and Technology 204:112445

Ali M, Malik Z, Abbasi GH, Irfan M, Ahmad S, Ameen M, Ali A, Sohaib M, Rizwan M, Ali S (2024) Potential of melatonin in enhancing antioxidant defense system and yield of maize (*Zea mays* L.) hybrids under saline condition. Sci Hortic 325:112665. https://doi.org/10.1016/j.scienta.2023.112665

Alomran MM, Noman A, Khalid N, Hadayat N, Alqahtani FM, Alzuaibr FM, Akhter N, Hashem M, Habeeb T, Al-Zoubi OM, Aqeel M (2024) Harnessing melatonin protective efficacy in capsicum plants against nickel-contaminated soil. J Soil Sci Plant Nutr 24(1):408–420. https://doi.org/10.1007/s42729-023-01551-6

Altaf MA, Shahid R, Ren MX, Naz S, Altaf MM, Khan LU, Ahmad P (2022) Melatonin improves drought stress tolerance of tomato by modulating plant growth, root architecture, photosynthesis, and antioxidant defense system. Antioxidants 11(2):309

Altaf MA, Hao Y, Shu H, Jin W, Chen C, Li L, Zhang Y, Mumtaz MA, Fu H, Cheng S, Zhu G, Wang Z (2024) Melatonin mitigates cold-induced damage to pepper seedlings by promoting redox homeostasis and regulating antioxidant profiling. Hortic Plant J 10(2):532–544. https://doi.org/10.1016/j.hpj.2023.02.006

Archana HR, Darshan K, Lakshmi MA, Ghoshal T, Bashayal BM, Aggarwal R (2022) Biopesticides: a key player in agro-environmental sustainability. In: Trends of applied microbiology for sustainable economy. Academic Press, pp 613–653

Arnao MB, Hernández-Ruiz J (2006) The physiological function of melatonin in plants. Plant Signal Behav 1(3):89–95

Arnao MB, Hernández-Ruiz J (2007) Melatonin promotes adventitious- and lateral root regeneration in etiolated hypocotyls of *Lupinus albus* L. J Pineal Res 42(2):147–152. https://doi.org/10.1111/j.1600-079X.2006.00396.x

Arnao MB, Hernández-Ruiz J (2009) Protective effect of melatonin against chlorophyll degradation during the senescence of barley leaves. J Pineal Res 46(1):58–63. https://doi.org/10.1111/j.1600-079X.2008.00625.x

Arnao MB, Hernández-Ruiz J (2015) Functions of melatonin in plants: a review. J Pineal Res 59(2):133–150

Arnao MB, Hernández-Ruiz J (2017) Growth activity, rooting capacity, and tropism: three auxinic precepts fulfilled by melatonin. Acta Physiol Plant 39(6):127. https://doi.org/10.1007/s11738-017-2428-3

Arnao MB, Hernández-Ruiz J (2018) Melatonin and its relationship to plant hormones. Ann Bot 121(2):195–207. https://doi.org/10.1093/aob/mcx114

Arnao MB, Hernández-Ruiz J (2019) Melatonin: a new plant hormone and/or a plant master regulator? Trends Plant Sci 24(1):38–48. https://doi.org/10.1016/j.tplants.2018.10.010

Arnao MB, Hernández-Ruiz J (2020) Melatonin in flowering, fruit set and fruit ripening. Plant Reprod 33(2):77–87. https://doi.org/10.1007/s00497-020-00388-8

Arnao MB, Hernández-Ruiz J (2021) Melatonin as a regulatory hub of plant hormone levels and action in stress situations. Plant Biol J 23(S1):7–19. https://doi.org/10.1111/plb.13202

Arnao MB, Hernández-Ruiz J, Cano A, Reiter RJ (2021) Melatonin and carbohydrate metabolism in plant cells. Plants 10(9):1917. https://doi.org/10.3390/plants10091917

Awan SA, Khan I, Rizwan M, Brestic M, Wang X, Zhang X, Huang L (2024) Exogenous melatonin regulates the expression pattern of antioxidant-responsive genes, antioxidant enzyme activities, and physio-chemical traits in pearl millet under drought stress. J Plant Growth Regul 43(4):1061–1075. https://doi.org/10.1007/s00344-023-11159-w

Aybeke M (2024) Interplay of melatonin and jasmonic acid in modulating wheat (*Triticum aestivum* L.) physiology and molecular pathways under salt stress. J Plant Growth Regul. https://doi.org/10.1007/s00344-024-11567-6

Azizi F, Amiri H, Ismaili A (2022) Melatonin improves salinity stress tolerance of Phaseolus vulgaris L. cv. Pak by changing antioxidant enzymes and photosynthetic parameters. Acta Physiol Plant 44(4):40. https://doi.org/10.1007/s11738-022-03373-y

Bai Y, Wei Y, Yin H, Hu W, Cheng X, Guo J, Dong Y, Zheng L, Xie H, Zeng H, Reiter RJ, Shi H (2022) PP2C1 fine-tunes melatonin biosynthesis and phytomelatonin receptor PMTR1 binding to melatonin in cassava. J Pineal Res 73(1):e12804. https://doi.org/10.1111/jpi.12804

Banerjee A, Roychoudhury A (2019) Melatonin application reduces fluoride uptake and toxicity in rice seedlings by altering abscisic acid, gibberellin, auxin and antioxidant homeostasis. Plant Physiol Biochem 145:164–173. https://doi.org/10.1016/j.plaphy.2019.10.033

Bargunam S, Roy R, Amisha SH, Shetty D, Babu VS (2024) Deciphering regulatory role of melatonin in foliar nastic movements of *Portulaca oleracea*: visual insights from histology and anatomy. J Plant Growth Regul:1–15

Barman D, Kumar MN, Dalal M, Khan FN, Yadav J, Nagar S, Santosh Kumar VV, Singh MP, Sathee L, Gopala Krishnan S, Chinnusamy V (2023) Identification of rice melatonin receptor OsPMTR and its comparative in silico analysis with arabidopsis AtCAND2 receptor. S Afr J Bot 162:813–829. https://doi.org/10.1016/j.sajb.2023.09.054

Bhowal B, Bhattacharjee A, Goswami K, Sanan-Mishra N, Singla-Pareek SL, Kaur C, Sopory S (2021) Serotonin and melatonin biosynthesis in plants: genome-wide identification of the genes and their expression reveal a conserved role in stress and development. Int J Mol Sci 22(20):11034

Byeon Y, Back K (2014) Melatonin synthesis in rice seedlings in vivo is enhanced at high temperatures and under dark conditions due to increased serotonin N-acetyltransferase and N-acetylserotonin methyltransferase activities. J Pineal Res 56(2):189–195

Byeon Y, Lee HY, Lee K, Park S, Back K (2014) Cellular localization and kinetics of the rice melatonin biosynthetic enzymes SNAT and ASMT. J Pineal Res 56(1):107–114

Byeon Y, Lee HY, Hwang OJ, Lee HJ, Lee K, Back K (2015) Coordinated regulation of melatonin synthesis and degradation genes in rice leaves in response to cadmium treatment. J Pineal Res 58(4):470–478

Cai H, Li J, Li J, Teng H (2025) Melatonin—angel of plant growth regulation and protection, vol 4. Adv Agrochem, p 114

Cajochen C, Kräuchi K, Wirz-Justice A (2003) Role of melatonin in the regulation of human circadian rhythms and sleep. J Neuroendocrinol 15(4):432–437

Cao Y, Du P, Ji J, He X, Zhang J, Shang Y, Liang B (2022) Ionomic combined with transcriptomic and metabolomic analyses to explore the mechanism underlying the effect of melatonin in relieving nutrient stress in apple. Int J Mol Sci 23(17):9855

Chang J, Guo Y, Yan J, Zhang Z, Yuan L, Wei C, Li H (2021) The role of watermelon caffeic acid O-methyltransferase (ClCOMT1) in melatonin biosynthesis and abiotic stress tolerance. Hortic Res 8:210

Chang J, Guo Y, Li J, Liu L, Liu J, Yuan L, Wei C, Ma J, Zhang Y, Ahammed GJ, Luan F, Liu Y, Zhang X, Li H (2024) Cyclic nucleotide-gated ion channel 20 regulates melatonin-induced calcium signaling and cold tolerance in watermelon. Plant Physiol:630. https://doi.org/10.1093/plphys/kiae630

Chaurasia S, Sapna S, Padhy AK, Bhatia S (2023) Emerging roles of melatonin in mitigating salinity stress of legumes. South African Journal of Botany 163:181–190

Chen Q, Arnao MB (2022) Phytomelatonin: an emerging new hormone in plants. J Exp Bot 73:5773–5778. https://doi.org/10.1093/jxb/erac307

Chen Q, Qi W, Reiter RJ, Wei W, Wang B (2009) Exogenously applied melatonin stimulates root growth and raises endogenous indoleacetic acid in roots of etiolated seedlings of Brassica juncea. J Plant Physiol 166(3):324–328. https://doi.org/10.1016/j.jplph.2008.06.002

Chen J, Li H, Yang K, Wang Y, Yang L, Hu L, Liu R, Shi Z (2019) Melatonin facilitates lateral root development by coordinating PAO-derived hydrogen peroxide and Rboh-derived superoxide radical. Free Radic Biol Med 143:534–544. https://doi.org/10.1016/j.freeradbiomed.2019.09.011

Chen X, Laborda P, Dong Y, Liu F (2020) Evaluation of suitable reference genes for normalization of quantitative real-time PCR analysis in rice plants under Xanthomonas oryzae pv. oryzae--infection and melatonin supplementation. Food Prod Process Nutr 2:1–10

Chen Z, Cao X, Niu J (2021) Effects of melatonin on morphological characteristics, mineral nutrition, nitrogen metabolism, and energy status in alfalfa under high-nitrate stress. Front Plant Sci 12:694179. https://doi.org/10.3389/fpls.2021.694179

Chen Q, Hou S, Pu X, Li X, Li R, Yang Q, Wang X, Guan M, Rengel Z (2022) Dark secrets of phytomelatonin. J Exp Bot 73(17):5828–5839. https://doi.org/10.1093/jxb/erac168

Cheng A, Zhao P, Wang X, Luo S, Xu P, Ye Y, Yuan H, Ji Y, Ma H, Xu J, Zhang Y, Zhang H (2025) Relationships between melatonin and salicylic acid treatments in delaying the senescence of postharvest pear fruit. Postharvest Biol Technol 219:113288. https://doi.org/10.1016/j.postharvbio.2024.113288

Colizzi FS (2023) Immunohistochemical analysis of the pea aphid circadian clock. Doctoral dissertation. Bayerische Julius-Maximilians-Universitaet Wuerzburg (Germany)

Corpas FJ, Rodríguez-Ruiz M, Muñoz-Vargas MA, González-Gordo S, Reiter RJ, Palma JM (2022) Interactions of melatonin, reactive oxygen species, and nitric oxide during fruit ripening: an update and prospective view. J Exp Bot 73(17):5947–5960. https://doi.org/10.1093/jxb/erac128

Darshan K, Harshitha KN, Shreedevasena S, Tailor A, Bhoi TK, Nigam S, Kulkarni N (2024) Role of melatonin in management of stress tolerance of forest tree species. In: Melatonin in plants: a pleiotropic molecule for abiotic stresses and pathogen infection. Springer Nature Singapore, pp 177–194

Darshan K, Tailor A, Rani K, Chittaragi A, Loona D, Santhoshinii SN, Shreedevasena S (2025) Biotechnological approaches for combatting tree diseases to enhance the forest ecosystem sustainability. In: Tree biology and biotechnology. Springer, pp 261–285

Delcourt V, Franck J, Quanico J, Gimeno JP, Wisztorski M, Raffo-Romero A, Fournier I (2018) Spatially-resolved top-down proteomics bridged to MALDI MS imaging reveals the molecular physiome of brain regions. Mol Cell Proteomics 17(2):357–372

Ding F, Ren L, Xie F, Wang M, Zhang S (2022) Jasmonate and melatonin act synergistically to potentiate cold tolerance in tomato plants. Front Plant Sci 12:763284. https://doi.org/10.3389/fpls.2021.763284

Dong D, Wang M, Li Y, Liu Z, Li S, Chao Y, Han L (2021) Melatonin influences the early growth stage in *Zoysia japonica* Steud. by regulating plant oxidation and genes of hormones. Sci Rep 11(1):12381

Duan Y, Wang X, Jiao Y, Liu Y, Li Y, Song Y, Wang S (2024) Elucidating the role of exogenous melatonin in mitigating alkaline stress in soybeans across different growth stages: a transcriptomic and metabolomic approach. BMC Plant Biol 24(1):380

Erdal S (2019) Melatonin promotes plant growth by maintaining integration and coordination between carbon and nitrogen metabolisms. Plant Cell Rep 38(8):1001–1012. https://doi.org/10.1007/s00299-019-02423-z

Eke P, Mabou LCN, Tchonang DN, Kouokap LRK, Youmbi DY, Dinango VN, Fokom R (2024) Explicating the Role of Melatonin in the Mitigation of Fungal Diseases in Plants. In Melatonin in Plants: A Pleiotropic Molecule for Abiotic Stresses and Pathogen Infection (pp. 155–176). Singapore: Springer Nature Singapore.

Fan J, Xie Y, Zhang Z, Chen L (2018) Melatonin: a multifunctional factor in plants. Int J Mol Sci 19(5):1528

Feng Y, Liu C, Gong B, Ai X, Bi H (2024) Abscisic acid participates in melatonin-induced chilling tolerance of cucumber via regulating photosynthesis and the antioxidant system. Veg Res 4:e025. https://doi.org/10.48130/vegres-0024-0024

Flowers T, Galal H, Bromham L (2010) Evolution of halophytes: multiple origins of salt tolerance in land plants. Funct Plant Biol 37:604–612. https://doi.org/10.1071/FP09269

Galano A, Reiter RJ (2018) Melatonin and its metabolites vs oxidative stress: from individual actions to collective protection. J Pineal Res 65(1):e12514. https://doi.org/10.1111/jpi.12514

Ghorbani A, Pishkar L, Saravi KV, Chen M (2023) Melatonin-mediated endogenous nitric oxide coordinately boosts stability through proline and nitrogen metabolism, antioxidant capacity, and Na+/K+ transporters in tomato under NaCl stress. Front Plant Sci 14:1135943. https://doi.org/10.3389/fpls.2023.1135943

Ghuge SA, Nikalje GC, Kadam US et al (2023) Comprehensive mechanisms of heavy metal toxicity in plants, detoxification, and remediation. J Hazard Mater 450:131039. https://doi.org/10.1016/j.jhazmat.2023.131039

Golovatskaya IF, Kadyrbaev MK, Boyko EV (2024) Protective role of melatonin and IAA in the regulation of resistance of potato regenerants to cold stress. Potato Res 67(2):421–449. https://doi.org/10.1007/s11540-023-09642-8

Gong B, Yan Y, Wen D, Shi Q (2017) Hydrogen peroxide produced by NADPH oxidase: a novel downstream signaling pathway in melatonin-induced stress tolerance in *Solanum lycopersicum*. Physiol Plant 160(4):396–409

Gu Q, Chen Z, Yu X, Cui W, Pan J, Zhao G, Shen W (2017) Melatonin confers plant tolerance against cadmium stress via the decrease of cadmium accumulation and reestablishment of microRNA-mediated redox homeostasis. Plant Sci 261:28–37

Guo Y, Zhu J, Liu J, Xue Y, Chang J, Zhang Y, Ahammed GJ, Wei C, Ma J, Li P, Zhang X, Li H (2023) Melatonin delays ABA-induced leaf senescence via H2O2-dependent calcium signalling. Plant Cell Environ 46(1):171–184. https://doi.org/10.1111/pce.14482

Guo X, Xu J, Zhao Y, Wang J, Fu T, Richard ML, Peng S (2024) Melatonin alleviates heat stress-induced spermatogenesis dysfunction in male dairy goats by regulating arachidonic acid metabolism mediated by remodeling the gut microbiota. Microbiome 12(1):233

Hardeland R, Cardinali DP, Srinivasan V, Spence DW, Brown GM, Pandi-Perumal SR (2011) Melatonin—A pleiotropic, orchestrating regulator molecule. Progress in neurobiology 93(3):350–384

Hasan MK, Shopan J, Ahammed GJ, Zhou J (2024) Advances in tools and techniques to quantify melatonin in plants and foodstuff. In: Sharma A, Ahammed GJ (eds) Melatonin in plants: role in plant growth, development, and stress response. Plant life and environment dynamics. Springer, Singapore

He L, Norris C, Cui X, Li Z, Barkjohn KK, Teng Y, Zhang JJ (2021) Role of endogenous melatonin in pathophysiologic and oxidative stress responses to personal air pollutant exposures in asthmatic children. Science of The Total Environment 773:145709

Hernández-Ruiz J, Giraldo-Acosta M, El Mihyaoui A, Cano A, Arnao MB (2023) Melatonin as a possible natural anti-viral compound in plant biocontrol. Plants 12(4):781

Himanshu H, Sunny S, Vishal R, Ankit A, Vikanksha T, Arun K, Prachi A, Shivender T, Neha S (2024) Unlocking the sustainable role of melatonin in fruit production and stress tolerance: a review. CABI Agric Biosci 5(1). https://doi.org/10.1186/s43170-024-00309-z

Hoque MN, Tahjib-Ul-Arif M, Hannan A et al (2021) Melatonin modulates plant tolerance to heavy metal stress: morphological responses to molecular mechanisms. Int J Mol Sci 22:11445. https://doi.org/10.3390/ijms222111445

Huang X, Tanveer M, Min Y, Shabala S (2022) Melatonin as a regulator of plant ionic homeostasis: implications for abiotic stress tolerance. J Exp Bot 73(17):5886–5902. https://doi.org/10.1093/jxb/erac224

Hussain A, Brekhna Faheem HY, Jang D-S, Lee B-G, Mun N, Rolly K, Yun B-W (2024) Melatonin–nitric oxide crosstalk in plants and the prospects of NOMela as a nitric oxide donor. Int J Mol Sci 25(15):8535. https://doi.org/10.3390/ijms25158535

Imran M, Aaqil Khan M, Shahzad R, Bilal S, Khan M, Yun B-W, Khan AL, Lee I-J (2021) Melatonin ameliorates thermotolerance in soybean seedling through balancing redox homeostasis and modulating antioxidant defense, phytohormones and polyamines biosynthesis. Molecules 26(17):5116. https://doi.org/10.3390/molecules26175116

Jafari M, Shahsavar A (2021) The effect of foliar application of melatonin on changes in secondary metabolite contents in two citrus species under drought stress conditions. Front Plant Sci 12:692735

Jahan MS, Yang JY, Althaqafi MM, Alharbi BM, Wu HY, Zhou XB (2024) Melatonin mitigates drought stress by increasing sucrose synthesis and suppressing abscisic acid biosynthesis in tomato seedlings. Physiol Plant 176(4):e14457. https://doi.org/10.1111/ppl.14457

Jin Z, Zhang H, Li L (2014) Tracking radiolabeled melatonin in *Arabidopsis* using autoradiography. Plant J 79(6):1012–1023

Jing T, Liu K, Wang Y, Ai X, Bi H (2022) Melatonin positively regulates both dark- and age-induced leaf senescence by reducing ROS accumulation and modulating abscisic acid and auxin biosynthesis in cucumber plants. Int J Mol Sci 23(7):3576. https://doi.org/10.3390/ijms23073576

Kaya C, Ugurlar F (2023) Melatonin and stress tolerance in horticultural crops: insights into gene regulation, epigenetic modifications, and hormonal interplay. Sci Hortic 322:112432

Kaya C, Sarıoglu A, Ashraf M, Alyemeni MN, Ahmad P (2022) The combined supplementation of melatonin and salicylic acid effectively detoxifies arsenic toxicity by modulating phytochelatins and nitrogen metabolism in pepper plants. Environ Pollut 297:118727. https://doi.org/10.1016/j.envpol.2021.118727

Kaya C, Ugurlar F, Ashraf M, Alyemeni MN, Ahmad P (2023) Exploring the synergistic effects of melatonin and salicylic acid in enhancing drought stress tolerance in tomato plants through fine-tuning oxidative-nitrosative processes and methylglyoxal metabolism. Sci Hortic 321:112368. https://doi.org/10.1016/j.scienta.2023.112368

Khalil HM, Ahmed M, Kamel A, Mohamed D, Hesham Y, Mahmoud R, Mraweh I, Salama N, El-Sayed M (2023) Emerging roles of melatonin in mitigating pathogen stress. In: Plant in challenging environments. https://doi.org/10.1007/978-3-031-40173-2_12

Khan MN, Khan Z, Luo T, Liu J, Rizwan M, Zhang J, Xu Z, Wu H, Hu L (2020) Seed priming with gibberellic acid and melatonin in rapeseed: consequences for improving yield and seed quality under drought and non-stress conditions. Ind Crop Prod 156:112850. https://doi.org/10.1016/j.indcrop.2020.112850

Khan Z, Jan R, Asif S, Farooq M, Jang YH, Kim EG, Kim KM (2024) Exogenous melatonin induces salt and drought stress tolerance in rice by promoting plant growth and defense system. Scientific reports 14(1):1214

Khanna K, Bhardwaj R, Alam P, Reiter RJ, Ahmad P (2023) Phytomelatonin: a master regulator for plant oxidative stress management. Plant Physiol Biochem 196:260–269. https://doi.org/10.1016/j.plaphy.2023.01.035

Khattak WA, Sun J, Abbas A, Hameed R, Jalal A, Niaz N, Wang Y (2023) Melatonin alleviating drought stress in plants: a review. S Afr J Bot 161:192–201

Kolář J, Macháčková I (2005) Melatonin in higher plants: occurrence and possible functions. J Pineal Res 39(4):333–341

Kong M, Sheng T, Liang J, Ali Q, Gu Q, Wu H, Chen J, Liu J, Gao X (2021) Melatonin and its homologs induce immune responses via receptors trP47363-trP13076 in *Nicotiana benthamiana*. Front Plant Sci 12:691835. https://doi.org/10.3389/fpls.2021.691835

Krishnappa C, Kavya BS, Akshay Kumar HM, Reddy P, Rajeshwar Rao G, Darshan K (2023) Endophytic microbes and their role in plant health. In: Microbial symbionts and plant health: trends and applications for changing climate. Springer, Singapore, pp 301–328

Kumar S, Yu R, Liu Y, Liu Y, Khan MN, Liu Y, Wang M, Zhu G (2025) Exogenous melatonin enhances heat stress tolerance in sweetpotato by modulating antioxidant defense system, osmotic homeostasis and stomatal traits. Hortic Plant J 11(1):431–445. https://doi.org/10.1016/j.hpj.2023.12.006

Lal MK, Sharma N, Adavi SB, Sharma E, Altaf MA, Tiwari RK, Kumar R, Kumar A, Dey A, Paul V, Singh B (2022) From source to sink: mechanistic insight of photoassimilates synthesis and partitioning under high temperature and elevated [CO2]. Plant Mol Biol 110(4):305–324

Lee HY, Back K (2016) Mitogen-activated protein kinase pathways are required for melatonin-mediated defense responses in plants. J Pineal Res 60(3):327–335. https://doi.org/10.1111/jpi.12314

Lee HY, Back K (2017) Melatonin is required for H2O2- and NO-mediated defense signaling through MAPKKK3 and OXI1 in *Arabidopsis thaliana*. J Pineal Res 62(2):e12379. https://doi.org/10.1111/jpi.12379

Lee HY, Back K (2021) Melatonin regulates chloroplast protein quality control via a mitogen-activated protein kinase signaling pathway. Antioxidants 10(4):511

Lee K, Back K (2024) Melatonin-induced brassinosteroid biosynthesis enhances seed size, senescence tolerance, and salt stress susceptibility in transgenic rice plants overexpressing an archaeal serotonin N-acetyltransferase. Melatonin Res 7(3):249–259. https://doi.org/10.32794/mr112500180

Lee HY, Byeon Y, Tan D, Reiter RJ, Back K (2015) *Arabidopsis* serotonin N-acetyltransferase knockout mutant plants exhibit decreased melatonin and salicylic acid levels resulting in susceptibility to an avirulent pathogen. J Pineal Res 58(3):291–299. https://doi.org/10.1111/jpi.12214

Lee S, Byun JK, Kim NY, Jin J, Woo H, Choi YK, Park KG (2022) Melatonin inhibits glycolysis in hepatocellular carcinoma cells by downregulating mitochondrial respiration and mTORC1 activity. Bmb Reports 55(9):459

Lee HY, Back K, Kim SJ (2019) RNAi suppression of serotonin N-acetyltransferase in *Arabidopsis* reduces melatonin levels and increases sensitivity to oxidative stress. Antioxidants 8(9):338

Li H, Chang J, Zheng J, Dong Y, Liu Q, Yang X, Wei C, Zhang Y, Ma J, Zhang X (2017) Local melatonin application induces cold tolerance in distant organs of *Citrullus lanatus* L. via long-distance transport. Sci Rep 7(1):40858. https://doi.org/10.1038/srep40858

Li H, Chang J, Chen H, Wang Z, Gu X, Wei C, Zhang Y (2019a) Exogenous melatonin enhances the phenolic content and antioxidant capacity of grape berries under drought stress. Front Plant Sci 10:1426

Li S, Guo J, Wang T, Gong L, Liu F, Brestic M, Li X (2021a) Melatonin reduces nanoplastic uptake, translocation, and toxicity in wheat. J Pineal Res 71(3):e12761

Li Z, Su X, Chen Y, Fan X, He L, Guo J, Wang Y, Yang Q (2021b) Melatonin improves drought resistance in maize seedlings by enhancing the antioxidant system and regulating abscisic acid metabolism to maintain stomatal opening under PEG-induced drought. J Plant Biol 64(4):299–312. https://doi.org/10.1007/s12374-021-09297-3

Li J, Liu Y, Zhang M, Xu H, Ning K, Wang B, Chen M (2022) Melatonin increases growth and salt tolerance of *Limonium bicolor* by improving photosynthetic and antioxidant capacity. BMC Plant Biol 22(1):16. https://doi.org/10.1186/s12870-021-03402-x

Li S, Ma R, Xu J, Shen Z, Yu M (2024) Effect of melatonin treatment on improving jasmonates content and cell wall stability involved in enhanced chilling tolerance of peach fruit during cold storage. Postharvest Biol Technol 213:112957. https://doi.org/10.1016/j.postharvbio.2024.112957

Liang C, Li A, Yu H, Li W, Liang C, Guo S, Zhang R, Chu C (2017) Melatonin regulates root architecture by modulating auxin response in rice. Front Plant Sci 8. https://doi.org/10.3389/fpls.2017.00134

Liu J, Zhai R, Liu F, Zhao Y, Wang H, Liu L, Yang C, Wang Z, Ma F, Xu L (2018) Melatonin induces parthenocarpy by regulating genes in gibberellin pathways of *Pyrus communis* L. Front Plant Sci 9:946. https://doi.org/10.3389/fpls.2018.00946

Liu C, Chen L, Zhao R, Li R, Zhang S, Yu W, Sheng J, Shen L (2019a) Melatonin induces disease resistance to *Botrytis cinerea* in tomato fruit by activating the jasmonic acid signaling pathway. J Agric Food Chem 67(22):6116–6124. https://doi.org/10.1021/acs.jafc.9b00058

Liu J, Zhang Z, Zhang H (2019b) Confocal microscopy reveals melatonin's role in chloroplast and mitochondrial function in grapevine. J Exp Bot 70(12):3239–3249

Liu S, Huang H, Huber DJ, Pan Y, Shi X, Zhang Z (2020) Delay of ripening and softening in 'Guifei' mango fruit by postharvest application of melatonin. Postharvest Biol Technol 163:111136. https://doi.org/10.1016/j.postharvbio.2020.111136

Liu Y, Wang X, Lv H, Cao M, Li Y, Yuan X, Zhang X, Guo YD, Zhang N (2022) Anabolism and signaling pathways of phytomelatonin. J Exp Bot 73(17):5801–5817. https://doi.org/10.1093/jxb/erac158

Liu Y, Wang X, Jing S, Jia C, Li H, Li C, He Q, Zhang N, Guo Y (2025) Integration of phytomelatonin signaling with jasmonic acid in wound-induced adventitious root regeneration. Adv Sci:2413485. https://doi.org/10.1002/advs.202413485

Luo M, Wang D, Delaplace P, Pan Y, Zhou Y, Tang W, Chen K, Chen J, Xu Z, Ma Y, Chen M (2023) Melatonin enhances drought tolerance by affecting jasmonic acid and lignin biosynthesis in wheat (*Triticum aestivum* L.). Plant Physiol Biochem 202:107974. https://doi.org/10.1016/j.plaphy.2023.107974

Lv Y, Pan J, Wang H, Reiter RJ, Li X, Mou Z, Zhang J, Yao Z, Zhao D, Yu D (2021) Melatonin inhibits seed germination by crosstalk with abscisic acid, gibberellin, and auxin in *Arabidopsis*. J Pineal Res 70(4):e12736. https://doi.org/10.1111/jpi.12736

Ma X, Idle JR, Krausz KW, Tan DX, Ceraulo L, Gonzalez FJ (2006) Urinary metabolites and antioxidant products of exogenous melatonin in the mouse. J Pineal Res 40(4):343–349

Ma T, Tao J, Yang M, He C, Tian X, Zhang X, Liu G (2017) An AANAT/ASMT transgenic animal model constructed with CRISPR/Cas9 system serving as the mammary gland bioreactor to produce melatonin-enriched milk in sheep. J Pineal Res 63(1):e12406

Ma SJ, Li C, Yan C, Liu N, Jiang GY, Yang HR, Gao C (2023) Melatonin alleviates early brain injury by inhibiting the NRF2-mediated ferroptosis pathway after subarachnoid hemorrhage. Free Radical Biology and Medicine 208:555–570.

Ma X, Zhang J, Burgess P, Rossi S, Huang B (2018) Interactive effects of melatonin and cytokinin on alleviating drought-induced leaf senescence in creeping bentgrass (*Agrostis stolonifera*). Environ Exp Bot 145:1–11. https://doi.org/10.1016/j.envexpbot.2017.10.010

Ma W, Xu L, Gao S, Lyu X, Cao X, Yao Y (2021) Melatonin alters the secondary metabolite profile of grape berry skin by promoting VvMYB14-mediated ethylene biosynthesis. Hortic Res 8:43

Madebo MP, Luo S, Wang L, Zheng Y, Jin P (2021) Melatonin treatment induces chilling tolerance by regulating the contents of polyamine, γ-aminobutyric acid, and proline in cucumber fruit. J Integr Agric 20(11):3060–3074. https://doi.org/10.1016/S2095-3119(20)63485-2

Mandal MK, Suren H, Ward B, Boroujerdi A, Kousik C (2018) Differential roles of melatonin in plant-host resistance and pathogen suppression in cucurbits. J Pineal Res 65(3):e12505

Mannino G, Pernici C, Serio G, Gentile C, Bertea CM (2021) Melatonin and phytomelatonin: chemistry, biosynthesis, metabolism, distribution and bioactivity in plants and animals—an overview. Int J Mol Sci 22(18):9996. https://doi.org/10.3390/ijms22189996

Mansouri S, Sarikhani H, Sayyari M, Soleimani Aghdam M (2021) Melatonin accelerates strawberry fruit ripening by triggering GAMYB gene expression and promoting ABA accumulation. Sci Hortic 281:109919. https://doi.org/10.1016/j.scienta.2021.109919

Mansoor S, Farooq I, Wani OA, Ahmad P, Reiter RJ, Boo KH, Chung YS (2024) Melatonin as a modulator of MAPK cascade and ROS-RNS feedforward loop during plant pathogen interaction. Physiological and Molecular Plant Pathology 102367.

Martins LG, Mainardes RM (2017) Application of a validated HPLC-PDA method for the determination of melatonin content and its release from poly (lactic acid) nanoparticles. J Pharm Anal 7(6):388–393

Meloni DA, Oliva MA, Martinez CA, Cambraia J (2003) Photosynthesis and activity of superoxide dismutase, peroxidase and glutathione reductase in cotton under salt stress. Environ Exp Bot 49:69–76. https://doi.org/10.1016/S0098-8472(02)00058-8

Moustafa Farag M, Almoneafy A, Mahmoud A, Elkelish A, Arnao MB, Li L, Ai S (2019) Melatonin and its protective role against biotic stress impacts on plants. Biomolecules 10(1):54

Muhammad I, Yang L, Ahmad S, Mosaad ISM, Al-Ghamdi AA, Abbasi AM, Zhou X-B (2022) Melatonin application alleviates stress-induced photosynthetic inhibition and oxidative damage by regulating antioxidant defense system of maize: a meta-analysis. Antioxidants 11(3):512. https://doi.org/10.3390/antiox11030512

Muhammad I, Yang L, Ahmad S, Farooq S, Khan A, Muhammad N, Ullah S, Adnan M, Ali S, Liang QP, Zhou XB (2023) Melatonin-priming enhances maize seedling drought tolerance by regulating the antioxidant defense system. Plant Physiol 191(4):2301–2315. https://doi.org/10.1093/plphys/kiad027

Muhammad I, Khan A, Mustafa AEZM, Elshikh MS, Shen W (2024) Elucidating the modulatory effect of melatonin on enzyme activity and oxidative stress in wheat: a global meta-analysis. Physiologia plantarum 176(2):e14294

Mukherjee S, David A, Yadav S, Baluška F, Bhatla SC (2014) Salt stress-induced seedling growth inhibition coincides with differential distribution of serotonin and melatonin in sunflower seedling roots and cotyledons. Physiol Plant 152(4):714–728

Munir R, Yasin MU, Afzal M, Jan M, Muhammad S, Jan N, Nana C, Munir F, Iqbal H, Tawab F, Gan Y (2024) Melatonin alleviated cadmium accumulation and toxicity by modulating phytohormonal balance and antioxidant metabolism in rice. Chemosphere 346:140590. https://doi.org/10.1016/j.chemosphere.2023.140590

Murch SJ, Erland LA (2021) A systematic review of melatonin in plants: an example of evolution of literature. Front Plant Sci 12:683047

Naeem A, Zafar M, Khalid H, Zia-ur-Rehman M, Ahmad Z, Ayub MA, Qayyum MF (2019) Cadmium-induced imbalance in nutrient and water uptake by plants. In: Cadmium toxicity and tolerance in plants. Elsevier, London, pp 299–326

Nasiri S, Rezaei M, Mojerlou S (2024) Impacts of preharvest treatment with salicylic acid and melatonin in suppressing gray mold (*Botrytis cinerea* Pers.) in bell pepper. J Crop Health 76(5):981–994. https://doi.org/10.1007/s10343-024-01007-6

Nayak L, Lal MK, Tiwari RK, Kumar R, Lal P, Das R, Behera B, Thakur P, Altaf MA, Kumar A (2023) A balancing act: exploring the interplay between HSPs and osmoprotectants in temperature stress responses. S Afr J Bot 162:64–71

Naz B, Afzal A, Ali H, Ahmad N, Fazal H, Ullah R, Dilbar S (2024) Melatonin-induced stress enhanced biomass and production of high-value secondary cell products in submerged adventitious root cultures of *Stevia rebaudiana* (Bert.). ACS Omega 9(5):5548–5562

Ni J, Wang Q, Shah FA, Liu W, Wang D, Huang S, Wu L (2018) Exogenous melatonin confers cadmium tolerance by counterbalancing the hydrogen peroxide homeostasis in wheat seedlings. Molecules 23(4):799

Paik MJ, Nguyen DT, Kim YJ, Shin JY, Shim W, Cho EY, Ahn YH (2010) Simultaneous GC–MS analysis of melatonin and its precursors as ethoxycarbonyl/pentafluoropropionyl derivatives in rat urine. Chromatographia 72:1213–1217

Pan J, Wang H, Chen W, You Q, Li X, Yu D (2021) Phytomelatonin inhibits seed germination by regulating germination-related hormone signaling in Arabidopsis. Plant Signal Behav 16(12):1970447. https://doi.org/10.1080/15592324.2021.1970447

Park S, Byeon Y, Lee HY, Kim YS, Ahn T, Back K (2013) Elevated melatonin levels enhance stress tolerance and reduce leaf senescence in transgenic rice plants. J Pineal Res 54(3):258–263

Park HS, Kazerooni EA, Kang SM, Al-Sadi AM, Lee IJ (2021) Melatonin enhances the tolerance and recovery mechanisms in *Brassica juncea* (L.) Czern. Under saline conditions. Front Plant Sci 12:593717. https://doi.org/10.3389/fpls.2021.593717

Pei ZQ, Ma C, Dong CY, Xu TT, Chai CH, Zhu Q, Wang J, Zheng S, Zhang TG (2024) Target of rapamycin coordinates auxin are involved in exogenous melatonin regulated low temperature tolerance in cucumber seedlings. Plant Physiol Biochem 215:109055. https://doi.org/10.1016/j.plaphy.2024.109055

Qiao Y, Yin L, Wang B, Ke Q, Deng X, Wang S (2019) Melatonin promotes plant growth by increasing nitrogen uptake and assimilation under nitrogen deficient condition in winter wheat. Plant Physiol Biochem 139:342–349. https://doi.org/10.1016/j.plaphy.2019.03.037

Qu G, Wu W, Ba L, Ma C, Ji N, Cao S (2022) Melatonin enhances the postharvest disease resistance of blueberries fruit by modulating the jasmonic acid signaling pathway and phenylpropanoid metabolites. Front Chem 10:957581

Raja V, Qadir SU, Kumar N, Alsahli AA, Rinklebe J, Ahmad P (2023) Melatonin and strigolactone mitigate chromium toxicity through modulation of ascorbate-glutathione pathway and gene expression in tomato. Plant Physiol Biochem 201:107872. https://doi.org/10.1016/j.plaphy.2023.107872

Rajotia MS, Sharma A, Bhatt A, Mangal V, Sood S, Thakur AK, Verma LK (2023) Melatonin detection and quantification techniques. In: Melatonin in plants: a regulator for plant growth and development. Springer Nature Singapore, Singapore, pp 19–38

Ren S, Rutto L, Katuuramu D (2019) Melatonin acts synergistically with auxin to promote lateral root development through fine tuning auxin transport in *Arabidopsis thaliana*. PLoS One 14(8):e0221687. https://doi.org/10.1371/journal.pone.0221687

Ren W, Chen L, Xie ZM, Peng X (2022) Combined transcriptome and metabolome analysis revealed pathways involved in improved salt tolerance of *Gossypium hirsutum* L. seedlings in response to exogenous melatonin application. BMC Plant Biol 22(1):552

Rithesh L, Krishna GV, Rao SS, Shiva B (2024) Emerging Role of Melatonin in Integrated Management of Crop Pathogens. In Melatonin in Plants: A Pleiotropic Molecule for Abiotic Stresses and Pathogen Infection (pp. 195–221). Singapore: Springer Nature Singapore.

Samanta S, Singh A, Banerjee A, Roychoudhury A (2020) Exogenous supplementation of melatonin alters representative organic acids and enzymes of respiratory cycle as well as sugar metabolism during arsenic stress in two contrasting indica rice cultivars. J Biotechnol 324:220–232. https://doi.org/10.1016/j.jbiotec.2020.10.013

Sehar Z, Fatma M, Khan S, Mir IR, Abdi G, Khan NA (2023) Melatonin influences methyl jasmonate-induced protection of photosynthetic activity in wheat plants against heat stress by regulating ethylene-synthesis genes and antioxidant metabolism. Sci Rep 13(1):7468. https://doi.org/10.1038/s41598-023-34682-y

Seth T, Asija S, Khatoon S, Iqbal N, Umar S, Khan MIR (2023) A new perspective of melatonin in stress tolerance through regulation of nutrients. S Afr J Bot 163:316–329

Sezgin B, Soyseven M, Arli G (2024) Greenness assessment and comparison of the developed and validated green HPLC-PDA, HPLC-FLD, and HPLC-ELSD methods for the determination of melatonin in various products using analytical eco-scale, NEMI, GAPI, and AGREE greenness metric tools. Microchem J 205:111211

Shaffique S, Shah AA, Kang SM, Injamum-Ul-Hoque M, Shahzad R, Azzawi TNIA, Lee IJ (2024) Melatonin: dual players mitigating drought-induced stress in tomatoes via modulation of phytohormones and antioxidant signaling cascades. BMC Plant Biology 24(1):1101

Shahani AA, Abbas A, Siddiqui JA, Ullah S, Hameed R, Cheng T, Farooq TH (2023) Opportunities and challenges in mitigating nutrient stress in plants via melatonin-mediated signaling pathways. S Afr J Bot 161:607–616

Shahid MA, Sarkhosh A, Khan N et al (2020) Insights into the physiological and biochemical impacts of salt stress on plant growth and development. Agronomy 10:938. https://doi.org/10.3390/agronomy10070938

Shan S, Wang Z, Pu H, Duan W, Song H, Li J, Zhang Z, Xu X (2022) DNA methylation mediated by melatonin was involved in ethylene signal transmission and ripening of tomato fruit. Sci Hortic 291:110566. https://doi.org/10.1016/j.scienta.2021.110566

Sharif R, Xie C, Zhang H, Arnao MB, Ali M, Ali Q, Muhammad I, Shalmani A, Nawaz MA, Chen P, Li Y (2018) Melatonin and its effects on plant systems. Molecules 23(9):2352. https://doi.org/10.3390/molecules23092352

Sharma A, Zheng B (2019) Melatonin mediated regulation of drought stress: physiological and molecular aspects. Plants 8(7):190. https://doi.org/10.3390/plants8070190

Sharma A, Wang J, Xu D, Tao S, Chong S, Yan D, Li Z, Yuan H, Zheng B (2020) Melatonin regulates the functional components of photosynthesis, antioxidant system, gene expression, and metabolic pathways to induce drought resistance in grafted *Carya cathayensis* plants. Sci Total Environ 713:136675. https://doi.org/10.1016/j.scitotenv.2020.136675

Sheshadri SA, Nishanth MJ, Yamine V, Simon B (2018) Effect of Melatonin on the stability and expression of reference genes in *Catharanthus roseus*. Sci Rep 8(1):2222

Shi H, Jiang C, Ye T, Tan D, Reiter RJ, Zhang H, Liu R, Chan Z (2015) Comparative physiological, metabolomic, and transcriptomic analyses reveal mechanisms of improved abiotic stress resistance in bermudagrass [*Cynodon dactylon* (L). Pers.] by exogenous melatonin. J Exp Bot 66(3):681–694. https://doi.org/10.1093/jxb/eru373

Shi D, Zhao L, Zhang R, Song Q (2024) Regulation of plant metabolism by melatonin signaling. https://doi.org/10.20944/preprints202406.0113.v1

Shi D, Zhao L, Zhang R, Song Q (2024a) Regulation of plant growth and development by melatonin. Life 14(12):1606. https://doi.org/10.3390/life14121606

Shi L, Chen Y, Dong W, Li S, Chen W, Yang Z, Cao S (2024b) Melatonin delayed senescence by modulating the contents of plant signalling molecules in postharvest okras. Front Plant Sci 15:1304913. https://doi.org/10.3389/fpls.2024.1304913

Siddiqui MH, Alamri S, Al-Khaishany MY, Khan MN, Al-Amri A, Ali HM, Alaraidh IA, Alsahli AA (2019) Exogenous melatonin counteracts NaCl-induced damage by regulating the antioxidant system, proline and carbohydrates metabolism in tomato seedlings. Int J Mol Sci 20(2):353. https://doi.org/10.3390/ijms20020353

Siddiqui MH, Alamri S, Alsubaie QD, Ali HM (2020) Melatonin and gibberellic acid promote growth and chlorophyll biosynthesis by regulating antioxidant and methylglyoxal detoxification system in tomato seedlings under salinity. J Plant Growth Regul 39(4):1488–1502. https://doi.org/10.1007/s00344-020-10122-3

Sliwiak J, Dauter Z, Jaskolski M (2016) Crystal structure of Hyp-1, a Hypericum perforatum PR-10 protein, in complex with melatonin. Front Plant Sci 7. https://doi.org/10.3389/fpls.2016.00668

Sliwiak J, Sikorski M, Jaskolski M (2018) PR -10 proteins as potential mediators of melatonin-cytokinin cross-talk in plants: crystallographic studies of Ll PR -10.2B isoform from yellow lupine. FEBS J 285(10):1907–1922. https://doi.org/10.1111/febs.14455

Sun Q, Zhang N, Wang J, Zhang H, Li D, Shi J, Li R, Weeda S, Zhao B, Ren S, Guo YD (2015) Melatonin promotes ripening and improves quality of tomato fruit during postharvest life. J Exp Bot 66(3):657–668. https://doi.org/10.1093/jxb/eru332

Sun Q, Liu L, Zhang L, Lv H, He Q, Guo L, Zhang X, He H, Ren S, Zhang N, Zhao B, Guo YD (2020) Melatonin promotes carotenoid biosynthesis in an ethylene-dependent manner in tomato fruits. Plant Sci 298:110580. https://doi.org/10.1016/j.plantsci.2020.110580

Sun C, Sun N, Ou Y, Gong B, Jin C, Shi Q, Lin X (2022) Phytomelatonin and plant mineral nutrition. J Exp Bot 73(17):5903–5917. https://doi.org/10.1093/jxb/erac289

Sun Y, Chen J, Shi X, Li Z, Wan L, Yan H, Yang G (2024) Safety and efficacy of melatonin supplementation as an add-on treatment for infantile epileptic spasms syndrome: A randomized, placebocontrolled, double-blind trial. Journal of pineal research 76(1):e12922

Supriya L, Dake D, Muthamilarasan M, Padmaja G (2024) Melatonin-mediated regulation of autophagy is independent of ABA under drought stress in sensitive variety of *Gossypium hirsutum* L. Plant Physiol Biochem 207:108409. https://doi.org/10.1016/j.plaphy.2024.108409

Tal O, Haim A, Harel O, Gerchman Y (2011) Melatonin as an antioxidant and its semi-lunar rhythm in green macroalga Ulva sp. J Exp Bot 62(6):1903–1910. https://doi.org/10.1093/jxb/erq378

Tan DX, Chen L, Poeggeler B, Manchester L, Reiter R (1993) Melatonin: a potent hydroxyl radical scavenger. Endocr J 1:57–60

Tan DX, Hardeland R, Manchester LC, Korkmaz A, Ma S, Rosales-Corral S, Reiter RJ (2012) Functional roles of melatonin in plants, and perspectives in nutritional and agricultural science. J Exp Bot 63(2):577–597

Tan X, Fan Z, Kuang J, Lu W, Reiter RJ, Lakshmanan P, Su X, Zhou J, Chen J, Shan W (2019a) Melatonin delays leaf senescence of Chinese flowering cabbage by suppressing ABFs-mediated abscisic acid biosynthesis and chlorophyll degradation. J Pineal Res 67(1):e12570. https://doi.org/10.1111/jpi.12570

Tan X, Long W, Zeng L, Ding X, Cheng Y, Zhang X, Zou X (2019b) Melatonin-induced transcriptome variation of rapeseed seedlings under salt stress. Int J Mol Sci 20(21):5355. https://doi.org/10.3390/ijms2021535

Teng Z, Zheng W, Jiang S, Hong S-B, Zhu Z, Zang Y (2022) Role of melatonin in promoting plant growth by regulating carbon assimilation and ATP accumulation. Plant Sci 319:111276. https://doi.org/10.1016/j.plantsci.2022.111276

Tian P, Lu X, Bao J, Zhang X, Lu Y, Zhang X, Ma S (2021) Transcriptomics analysis of genes induced by melatonin related to glucosinolates synthesis in broccoli hairy roots. Plant Signal Behav 16(11):1952742

Tiwari RK, Lal MK, Naga KC, Kumar R, Chourasia KN, Kumar D, Sharma S (2020) Emerging roles of melatonin in mitigating abiotic and biotic stresses of horticultural crops. Sci Hortic 272:109592

Tiwari RK, Lal MK, Kumar R, Chourasia KN, Naga KC, Kumar D, Das SK, Zinta G (2021a) Mechanistic insights on melatonin-mediated drought stress mitigation in plants. Physiol Plant 172(2):1212–1226

Tiwari RK, Lal MK, Kumar R, Mangal V, Altaf MA, Sharma S, Kumar M (2021b) Insight into melatonin-mediated response and signaling in the regulation of plant defense under biotic stress. Plant Mol Biol 109:1–15

Van Tassel DL, Roberts N, Lewy A, O'Neill SD (2001) Melatonin in plant organs. J Pineal Res 31(1):8–15

Verde A, Míguez JM, Gallardo M (2023) Melatonin stimulates postharvest ripening of apples by up-regulating gene expression of ethylene synthesis enzymes. Postharvest Biol Technol 195:112133. https://doi.org/10.1016/j.postharvbio.2022.112133

Wang H, Geng CA, Xu HB, Huang XY, Ma YB, Yang CY, Chen JJ (2015) Lignans from the fruits of *Melia toosendan* and their agonistic activities on melatonin receptor MT1. Planta Med 81(10):847–854

Wang Q, An B, Wei Y, Reiter RJ, Shi H, Luo H, He C (2016) Melatonin regulates root meristem by repressing auxin synthesis and polar auxin transport in *Arabidopsis*. Front Plant Sci 07. https://doi.org/10.3389/fpls.2016.01882

Wang L, Feng C, Zheng X, Guo Y, Zhou F, Shan D, Kong J (2017) Plant mitochondria synthesize melatonin and enhance the tolerance of plants to drought stress. J Pineal Res 63(3):e12429

Wang M, Zhang S, Ding F (2020) Melatonin mitigates chilling-induced oxidative stress and photosynthesis inhibition in *tomato* plants. Antioxidants 9(3):218. https://doi.org/10.3390/antiox9030218

Wang K, Xing Q, Ahammed GJ, Zhou J (2022a) Functions and prospects of melatonin in plant growth, yield, and quality. J Exp Bot 73(17):5928–5946

Wang LF, Lu KK, Li TT, Zhang Y, Guo JX, Song RF, Liu WC (2022b) Maize *PHYTOMELATONIN RECEPTOR1* functions in plant tolerance to osmotic and drought stress. J Exp Bot 73(17):5961–5973. https://doi.org/10.1093/jxb/erab553

Wang Y, Li J, Yang L, Chan Z (2023) Melatonin antagonizes cytokinin responses to stimulate root growth in *Arabidopsis*. J Plant Growth Regul 42(3):1833–1845. https://doi.org/10.1007/s00344-022-10663-9

Wang H, Zhang Y, Jiang H, Ding Q, Wang Y, Wang M, Jia L (2025) Transcriptomic and metabolomic analysis reveals the molecular mechanism of exogenous melatonin improving salt tolerance in eggplants. Front Plant Sci 15:1523582

Wani AK, Akhtar N, Sharif Usman S, Uba AI, Rahayu F, Side THR, Murianingrum M (2023) Positive regulatory role of melatonin in conferring drought resistance to plants. In: Melatonin: role in plant signaling, growth and stress tolerance: phytomelatonin in normal and challenging environments. Springer International Publishing, Cham, pp 343–361

Weeda S, Zhang N, Zhao X, Ndip G, Guo Y, Buck GA, Ren S (2014) Arabidopsis transcriptome analysis reveals key roles of melatonin in plant defense systems. PLoS One 9(3):e93462

Wei W, Li QT, Chu YN, Reiter RJ, Yu XM, Zhu DH, Chen SY (2015) Melatonin enhances plant growth and abiotic stress tolerance in soybean plants. J Exp Bot 66(3):695–707

Wei J, Li D, Zhang J, Shan C, Rengel Z, Song Z, Chen Q (2018) Phytomelatonin receptor PMTR 1-mediated signaling regulates stomatal closure in *Arabidopsis thaliana*. J Pineal Res 65(2):e12500. https://doi.org/10.1111/jpi.12500

Wei W, Tao JJ, Yin CC, Chen SY, Zhang JS, Zhang WK (2022) Melatonin regulates gene expressions through activating auxin synthesis and signaling pathways. Front Plant Sci 13:1057993

Wen D, Gong B, Sun S, Liu S, Wang X, Wei M, Yang F, Li Y, Shi Q (2016) Promoting roles of melatonin in adventitious root development of *Solanum lycopersicum* L. by regulating auxin and nitric oxide signaling. Front Plant Sci 7:7. https://doi.org/10.3389/fpls.2016.00718

Wu SQ, Wang YX, Beta T, Wang SY, Mendez-Zamora G, Laborda P, Herrera-Balandrano DD (2023) Effect of exogenous melatonin on the isoflavone content and antioxidant properties of soybean sprouts. LWT 175:114498

Xiao S, Liu L, Wang H, Li D, Bai Z, Zhang Y, Sun H, Zhang K, Li C (2019) Exogenous melatonin accelerates seed germination in cotton (*Gossypium hirsutum* L.). PLoS One 14(6):e0216575. https://doi.org/10.1371/journal.pone.0216575

Xie F, Liu Y, Zhao Q, Liu X, Wang C, Wang Q, Wei Q, Zhao X, Jiang J, Liu R, Chen Y, Cui G, Han J (2024) Exogenous application of melatonin and strigolactone by regulating morpho-physiological responses and gene expression to improve drought resistance in fodder soybean seedlings. Agronomy 14(8):1803. https://doi.org/10.3390/agronomy14081803

Xing Q, Hasan MDK, Li Z, Yang T, Jin W, Qi Z, Yang P, Wang G, Ahammed GJ, Zhou J (2023) Melatonin-induced plant adaptation to cadmium stress involves enhanced phytochelatin synthesis and nutrient homeostasis in *Solanum lycopersicum* L.J. Hazard Mater 456:131670. https://doi.org/10.1016/j.jhazmat.2023.131670

Xu L, Yue Q, Xiang G, Bian F, Yao Y (2018) Melatonin promotes ripening of grape berry via increasing the levels of ABA, H2O2, and particularly ethylene. Hortic Res 5(1):41. https://doi.org/10.1038/s41438-018-0045-y

Xu H, Zhang S, Liang C, Li M, Wang R, Song J, Cui Z, Yang Y, Liu J, Li D (2024) Melatonin enhances resistance to Botryosphaeria dothidea in pear by promoting jasmonic acid and phlorizin biosynthesis. BMC Plant Biol 24(1):470. https://doi.org/10.1186/s12870-024-05187-1

Yan Y, Jing X, Tang H, Li X, Gong B, Shi Q (2019) Using transcriptome to discover a novel melatonin-induced sodic alkaline stress resistant pathway in *Solanum lycopersicum* L. Plant Cell Physiol 60(9):2051–2064

Yan F, Chen X, Wang Z, Xia Y, Zheng D, Xue S, Zhao H, Huang Z, Niu Y, Zhang G (2024) Melatonin alleviates abscisic acid deficiency inhibition on photosynthesis and antioxidant systems in rice under salt stress. Phyton 93(7):1421–1440

Yang L, Sun Q, Wang Y, Chan Z (2021a) Global transcriptomic network of melatonin-regulated root growth in *Arabidopsis*. Gene 764:145082. https://doi.org/10.1016/j.gene.2020.145082

Yang L, You J, Li J, Wang Y, Chan Z (2021b) Melatonin promotes *Arabidopsis* primary root growth in an IAA-dependent manner. J Exp Bot 72(15):5599–5611. https://doi.org/10.1093/jxb/erab196

Yang Q, Peng Z, Ma W, Zhang S, Hou S, Wei J, Dong S, Yu X, Song Y, Gao W, Rengel Z, Huang L, Cui X, Chen Q (2021c) Melatonin functions in priming of stomatal immunity in *Panax notoginseng* and *Arabidopsis thaliana*. Plant Physiol 187(4):2837–2851. https://doi.org/10.1093/plphys/kiab419

Yang N, Han M-H, Teng R-M, Yang Y-Z, Wang Y-H, Xiong A-S, Zhuang J (2022) Exogenous melatonin enhances photosynthetic capacity and related gene expression in a dose-dependent manner in the tea plant (*Camellia sinensis* (L.) Kuntze). Int J Mol Sci 23(12):6694. https://doi.org/10.3390/ijms23126694

Yang LL, Li QL, Han XY, Jiang XL, Wang H, Shi YJ, Shi Y (2023) A cysteine-rich secretory protein involves in phytohormone melatonin-mediated plant resistance to CGMMV. BMC Plant Biol 23(1):215

Yang X, Liu D, Liu C, Li M, Yan Z, Zhang Y, Feng G (2024) Possible melatonin-induced salt stress tolerance pathway in *Phaseolus vulgaris* L. using transcriptomic and metabolomic analyses. BMC Plant Biol 24(1):72

Yao JW, Ma Z, Ma YQ, Zhu Y, Lei MQ, Hao CY, Huang X (2021) Role of melatonin in UV-B signaling pathway and UV-B stress resistance in *Arabidopsis thaliana*. Plant Cell Environ 44(1):114–129

Ye J, Wang S, Deng X, Yin L, Xiong B, Wang X (2016) Melatonin increased maize (*Zea mays* L.) seedling drought tolerance by alleviating drought-induced photosynthetic inhibition and oxidative damage. Acta Physiol Plant 38(2):48. https://doi.org/10.1007/s11738-015-2045-y

Yin X, Bai Y, Gong C, Song W, Wu Y, Ye T, Feng Y (2022) The phytomelatonin receptor PMTR1 regulates seed development and germination by modulating abscisic acid homeostasis in *Arabidopsis thaliana*. J Pineal Res 72(4):e12797. https://doi.org/10.1111/jpi.12797

Yu R, Zuo T, Diao P, Fu J, Fan Y, Wang Y, Zhao Q, Ma X, Lu W, Li A, Wang R, Yan F, Pu L, Niu Y, Wuriyanghan H (2021) Melatonin enhances seed germination and seedling growth of *Medicago sativa* under salinity via a putative melatonin receptor MsPMTR1. Front Plant Sci 12:702875. https://doi.org/10.3389/fpls.2021.702875

Yu Y, Deng L, Zhou L, Chen G, Wang Y (2022) Exogenous melatonin activates antioxidant systems to increase the ability of rice seeds to germinate under high temperature conditions. Plants 11(7):886

Yuan X, Li J, Zhang X, Ai X, Bi H (2024) Auxin as a downstream signal positively participates in melatonin-mediated chilling tolerance of cucumber. Physiol Plant 176(5):e14526. https://doi.org/10.1111/ppl.14526

Yue L, Kang Y, Zhong M, Kang D, Zhao P, Chai X, Yang X (2023) Melatonin delays postharvest senescence through suppressing the inhibition of BrERF2/BrERF109 on flavonoid biosynthesis in flowering Chinese cabbage. Int J Mol Sci 24(3):2933. https://doi.org/10.3390/ijms24032933

Zeng L, Cai J, Li J, Lu G, Li C, Fu G, Zhang X, Ma H, Liu Q, Zou X, Cheng Y (2018) Exogenous application of a low concentration of melatonin enhances salt tolerance in rapeseed (*Brassica napus* L.) seedlings. J Integr Agric 17(2):328–335. https://doi.org/10.1016/S2095-3119(17)61757-X

Zeng W, Mostafa S, Lu Z, Jin B (2022) Melatonin-mediated abiotic stress tolerance in plants. Front Plant Sci 13. https://doi.org/10.3389/fpls.2022.847175

Zhan H, Nie X, Zhang T et al (2019) Melatonin: a small molecule but important for salt stress tolerance in plants. Int J Mol Sci 20:709. https://doi.org/10.3390/ijms20030709

Zhang N, Zhao B, Zhang H, Weeda S, Yang C, Yang Z, Ren S, Guo Y (2013) Melatonin promotes water-stress tolerance, lateral root formation, and seed germination in cucumber (*Cucumis sativus* L.). J Pineal Res 54(1):15–23. https://doi.org/10.1111/j.1600-079X.2012.01015.x

Zhang H, Zhang N, Yang R, Wang L, Sun Q, Li D, Cao Y, Weeda S, Zhao B, Ren S, Guo Y (2014a) Melatonin promotes seed germination under high salinity by regulating antioxidant systems, ABA and GA4 interaction in cucumber (*Cucumis sativus* L.). J Pineal Res 57(3):269–279. https://doi.org/10.1111/jpi.12167

Zhang N, Zhang H, Zhao B, Sun Q, Cao Y, Li R, Wu X, Weeda S, Li L, Ren S, Reiter RJ, Guo Y (2014b) The RNA-seq approach to discriminate gene expression profiles in response to melatonin on cucumber lateral root formation. J Pineal Res 56(1):39–50. https://doi.org/10.1111/jpi.12095

Zhang N, Sun Q, Zhang H, Cao Y, Weeda S, Ren S, Guo YD (2015) Roles of melatonin in abiotic stress resistance in plants. J Exp Bot 66(3):647–656

Zhang J, Shi Y, Zhang X, Du H, Xu B, Huang B (2017a) Melatonin suppression of heat-induced leaf senescence involves changes in abscisic acid and cytokinin biosynthesis and signaling pathways in perennial ryegrass (*Lolium perenne L.*). Environ Exp Bot 138:36–45. https://doi.org/10.1016/j.envexpbot.2017.02.012

Zhang N, Zhang HJ, Sun QQ, Cao YY, Li X, Zhao B, Wu P, Guo YD (2017b) Proteomic analysis reveals a role of melatonin in promoting cucumber seed germination under high salinity by regulating energy production. Sci Rep 7(1):503. https://doi.org/10.1038/s41598-017-00566-1

Zhang YP, Xu S, Yang SJ, Chen YY (2017c) Melatonin alleviates cold-induced oxidative damage by regulation of ascorbate–glutathione and proline metabolism in melon seedlings (*Cucumis melo* L.). J Hortic Sci 92(3):313–324. https://doi.org/10.1080/14620316.2016.1266915

Zhang Z, Hu Q, Liu Y, Cheng P, Cheng H, Liu W, Xing X, Guan Z, Fang W, Chen S, Jiang J, Chen F (2019) Strigolactone represses the synthesis of melatonin, thereby inducing floral transition in *Arabidopsis thaliana* in an FLC-dependent manner. J Pineal Res 67(2):e12582. https://doi.org/10.1111/jpi.12582

Zhang Z, Zhang X, Chen Y, Jiang W, Zhang J, Wang J, Zhang Y (2023) Understanding the mechanism of red light-induced melatonin biosynthesis facilitates the engineering of melatonin-enriched tomatoes. Nat Commun 14(1):5525

Zhang Y, Wang L, Li X, Wen H, Yu X, Wang Y (2024) Synergistic effects of exogenous IAA and melatonin on seed priming and physiological biochemistry of three desert plants in saline-alkali soil. Plant Signal Behav 19(1):2379695. https://doi.org/10.1080/15592324.2024.2379695

Zhao H, Su T, Huo L, Wei H, Jiang Y, Xu L, Ma F (2015) Unveiling the mechanism of melatonin impacts on maize seedling growth: sugar metabolism as a case. J Pineal Res 59(2):255–266. https://doi.org/10.1111/jpi.12258

Zhao L, Chen L, Gu P, Zhan X, Zhang Y, Hou C, Wang QC (2019) Exogenous application of melatonin improves plant resistance to virus infection. Plant Pathol 68(7):1287–1295

Zhao C, Yang M, Wu X, Wang Y, Zhang R (2021) Physiological and transcriptomic analyses of the effects of exogenous melatonin on drought tolerance in maize (*Zea mays* L.). Plant Physiol Biochem 168:128–142

Zhong L, Lin L, Yang L, Liao M, Wang X, Wang J, Lv X, Deng H, Liang D, Xia H, Tang Y (2020) Exogenous melatonin promotes growth and sucrose metabolism of grape seedlings. PLoS One 15(4):e0232033. https://doi.org/10.1371/journal.pone.0232033

Zhu J, Lou H, Yan C, Zhang W, Li Z (2024) Exogenous melatonin enhances cold tolerance by regulating the expression of photosynthetic performance, antioxidant system, and related genes in cotton. Plants 13(15):2010. https://doi.org/10.3390/plants13152010

Omics-Based Strategies to Unravel the Role of Melatonin in Biotic and Abiotic Stress Amelioration

2

Shimi Jose, Mehjebin Rahman, Malavika Ram Amanthra Keloth, Sucharita Mohapatra, Lellapalli Rithesh, Sumit Kumar, B. K. Surendra, S. R. Ramseena, and Ali Chenari Bouket

Abstract

Melatonin, a well-known indoleamine, has emerged as a critical regulator of plant immunity, modulating various physiological and molecular pathways to mitigate biotic and abiotic stresses. Recent advances in omics technologies have provided unprecedented insights into the intricate roles of melatonin in plant–pathogen interactions. Genomic studies have identified key genes involved in melatonin biosynthesis and signaling, while transcriptomic analyses have

S. Jose · M. R. A. Keloth
Hawkesbury Institute for the Environment, Western Sydney University, Richmond, VIC, Australia

M. Rahman
College of Agriculture, Assam Agricultural University, Jorhat, Assam, India

S. Mohapatra
Department of Plant Pathology, College of Agriculture, Odisha University of Agriculture and Technology, Bhubaneswar, Odisha, India

L. Rithesh · S. R. Ramseena
College of Agriculture, Kerala Agricultural University, Thiruvananthapuram, Kerala, India

S. Kumar
KVK, Mau, Acharya Narendra Deva University of Agriculture and Technology, Ayodhya, Uttar Pradesh, India

B. K. Surendra
Shree Banshagopal Secondary School (Technical School), Makwanpur, Bagmati, Nepal

A. C. Bouket (✉)
East Azarbaijan Agricultural and Natural Resources Research and Education Centre, Plant Protection Research Department, Agricultural Research, Education and Extension Organization (AREEO), Tabriz, Iran
e-mail: a.chenari@areeo.ac.ir

R. K. Tiwari et al. (eds.), *Melatonin: Signal Transduction Mechanisms and Defense Networks in Plants*, https://doi.org/10.1007/978-981-96-8869-2_2

elucidated its influence on defense-related gene expression networks. Proteomic approaches have revealed melatonin-induced changes in pathogenesis-related proteins and antioxidant enzymes, which contribute to enhanced disease resistance. Furthermore, metabolomic studies highlight the melatonin-mediated accumulation of secondary metabolites with antimicrobial properties, while epigenomic investigations uncover its role in immune memory through DNA methylation and histone modifications. This chapter integrates multi-omics strategies to decipher the molecular basis of melatonin-induced biotic and abiotic responses, emphasizing its potential applications in sustainable crop management.

Keywords
Melatonin · Plant immunity · Omics · Disease management · CRISPR-Cas

2.1 Introduction

The global human population is expanding rapidly, with predictions indicating that the Earth will have the capacity to support around 10 billion people by 2050, primarily due to elevated birth rates (Vollset et al. 2020; Becker and Fanzo 2023). To meet the food requirements of this growing population, it is necessary to increase the production of food and agricultural products by 70% from current levels. In this context, only crop plants are capable of directly providing 80% of protein and 90% of calorie requirements for humans (Nadathur et al. 2017). Unfortunately, plants encounter various abiotic and biotic stressors throughout their life cycle, from germination to maturity (Nayak et al. 2023; Devi et al. 2021). Abiotic stressors, such as drought, high and low temperatures, salinity, acidic soils, and nutrient starvation, are the primary factors limiting agricultural production and productivity (Kamanga and Mndala 2019; Chaudhry and Sidhu 2022; Lal et al. 2022). Among these, salinity and drought pose significant challenges to agricultural crop production, directly affecting plant metabolic systems and resulting in crop losses worth billions of dollars (Fita et al. 2015; Khondoker et al. 2023; Hossain and Li 2024). Biotic stressors, which include harmful plant pathogens such as fungi, bacteria, viruses, and nematodes, as well as weeds and herbivorous insects, present a significant threat to plant growth and development (Bhatla et al. 2018; Umar et al. 2021; Kumar et al. 2023a, b). Plant pathogens can cause severe and detrimental diseases in various crop plants, resulting in a range of symptoms (Buttar et al. 2023; Watpade et al. 2023).

Plant diseases pose a significant threat to global food security, as multiple reports indicate that they decrease yields of a number of economically important agricultural crops by 21–30% worldwide (Savary et al. 2012; Fones et al. 2020; Junaid and Gokce 2024). Important crops, such as wheat, rice, maize, peanuts, potatoes, tomatoes, and cucumbers, are affected by various phytopathogens, which diminish their nutritional quality and yield. Notable crop diseases, such as early and late blight of potato, rust of wheat, bacterial blight of rice, Karnal bunt of wheat, sheath blight of rice, leaf curl of tomato, and yellow vein of okra, are induced by fungal, bacterial, and viral pathogens (Savary and Willocquet 2020; Kumar et al. 2022). To mitigate yield loss due to crop diseases, there has been a significant dependency among farmers on the use of synthetic agrochemicals in recent years. There is no doubt that agrochemicals reduce pathogen infections in both fields and greenhouses, thereby increasing crop yields. However, if used improperly or for prolonged periods, these chemicals can have lasting negative effects on human and animal health and can adversely impact the populations of beneficial soil microbes (Kalia and Gosal 2011; Kalyabina et al. 2021; Rastogi et al. 2023). Thus, in light of these issues and in accordance with the 2030 Agenda for Sustainable Development Goals adopted by United Nations member states in 2015 (United Nations 2015), it is essential to discover innovative and environmentally friendly technologies that can achieve sustainable crop production without harming the environment. In this context, the application of biomolecules presents the most effective alternative to synthetic chemical methods, as they possess a wide range of beneficial mechanisms that combat plant diseases without compromising environmental integrity. Among various biomolecules, melatonin has recently garnered significant attention from the scientific community and researchers due to its multifaceted beneficial roles.

Melatonin, known as N-acetyl-5-methoxytryptamine, was first discovered by the dermatologist Lerner et al. in 1958 in the pineal gland of cattle as an animal hormone; it was later identified in plants by Hattori et al. in 1995. Melatonin functions as a multifaceted signaling molecule, with its biosynthesis in plants initiated by the enzymes tryptophan hydroxylase, tryptamine 5-hydroxylase, caffeic acid O-methyltransferase, N-acetylserotonin methyltransferase, tryptophan decarboxylase, and serotonin N-acetyltransferase (Kang et al. 2011). Plants typically synthesize melatonin in the chloroplasts and mitochondria of their roots and leaves, subsequently transporting it to the meristem, flowers, and fruits (Wang et al. 2016). While melatonin is widely recognized as a sleep-regulating hormone in animals, research has demonstrated that it also plays a significant role in enhancing plant immunity (Wei et al. 2023). Melatonin enhances plant defense through several mechanisms. It helps eliminate harmful reactive oxygen species (ROS), which can damage plant cells during stress (Altaf et al. 2022). Additionally, melatonin activates immune-related genes and proteins that strengthen the plant's defense system. It also interacts closely with other plant hormones, such as salicylic acid (SA) and jasmonic acid (JA), which are crucial for combating infections. Another unique feature of melatonin is its ability to help plants "remember" past infections, allowing them to respond more quickly and effectively to future attacks (Tiwari et al. 2020, 2021a, b).

With recent advancements in science, researchers are increasingly using omics technologies to study melatonin's role in plant disease resistance in greater depth. These approaches provide a detailed understanding of how melatonin influences plant defense mechanisms. Genomics helps identify the genes responsible for melatonin production, while transcriptomics examines which genes are activated when melatonin combats stress. Proteomics explores changes in plant proteins involved in resistance, and metabolomics investigates how melatonin influences the production of natural plant chemicals that help suppress various stresses. Additionally, epigenomics examines how melatonin can create long-term immunity by modifying

DNA and proteins in plant cells. By integrating these omics-based strategies, scientists are uncovering the full potential of melatonin in plant management. Understanding how melatonin strengthens plant immunity can lead to innovative methods for protecting crops naturally, reducing the reliance on chemicals, and promoting sustainable agriculture. In this chapter, we will explore how these advanced technologies are assisting researchers in revealing the critical role of melatonin in plant defense.

2.2 Role of Melatonin in Plant Stress Responses

2.2.1 Biochemical Role: Antioxidant Activity, ROS Scavenging, and Hormonal Regulation

Melatonin is an essential antioxidant that supports plant vigor under both biotic and abiotic challenges. More than 20 years ago, it was discovered that melatonin scavenges superoxide anions (O^{2-}) and free hydroxyl radicals (OH•) (Ianas et al. 1991). Research has shown that melatonin stimulates the endogenous accumulation and synthesis of nitric oxide (NO), directly neutralizes reactive oxygen species (ROS), and modulates plant defense mechanisms (Arnao and Hernández-Ruiz 2018; Okant and Kaya 2019). By regulating both enzymatic and nonenzymatic antioxidant defense mechanisms, melatonin enhances a plant's resistance to stress. Enzymatic antioxidants activated by exogenous melatonin include superoxide dismutase (SOD), catalase (CAT), glutathione peroxidase, and enzymes of the ascorbate–glutathione cycle, such as dehydroascorbate reductase, ascorbate peroxidase (APX), glutathione reductase, and monodehydroascorbate reductase. Furthermore, under stressful conditions, melatonin elevates the levels of nonenzymatic antioxidants such as carotenoids, reduced glutathione, tocopherols, ascorbate, and phenolics. Consequently, exogenously applied melatonin can enhance physiological activity and strengthen a plant's resistance to stress.

Due to its amphipathic nature, melatonin can easily cross cell membranes into the cytoplasm (Zhang and Zhang 2014). While small molecules are typically absorbed through the plasma membrane, melatonin's unique properties allow it to pass through effectively. As a potent antioxidant in plants, melatonin combats oxidative stress through both direct and indirect mechanisms. Its direct scavenging activity helps maintain cell integrity by reducing oxidative damage to vital cellular components and nucleic acids. In addition to its direct antioxidant properties, melatonin enhances the activity of key antioxidant enzymes such as APX, CAT, and SOD (Altaf et al. 2022). These enzymes are crucial for mitigating oxidative stress and detoxifying ROS. SOD converts superoxide radicals into hydrogen peroxide (H2O2), which is subsequently detoxified by CAT and APX, creating a synergistic enzymatic defense mechanism that is further enhanced by melatonin (Sharma et al. 2023a; Kumar et al. 2024a, b).

Additionally, melatonin stabilizes cell membranes and inhibits lipid peroxidation, which is vital for preserving cellular homeostasis under stress (Jafari et al.

2022). Melatonin signaling is critical for ROS-mediated activities, particularly in maintaining H_2O_2 equilibrium, and it functions as a direct antioxidant to keep intracellular ROS levels low. Interestingly, melatonin outperforms ascorbic acid (AsA) in reducing ROS levels. The total antioxidant capacity of melatonin is further enhanced by the notable antioxidant properties of several metabolites produced during melatonin biosynthesis, including cyclic 3-hydroxymelatonin (3-OHM), 2-hydroxymelatonin, and N1-acetyl-N2-formyl-5-methoxykynuramine (AFMK) (Tan et al. 2015). Apart from its direct antioxidant properties, melatonin mediates various antioxidant processes, including the glutathione–ascorbate cycle and the enzymatic activity of CAT, peroxidases, and SOD. Through these pathways, melatonin induces abiotic and biotic stress responses in plants and helps regulate oxidative stress (Debnath et al. 2019). These diverse functions underscore the crucial role of melatonin in maintaining cellular homeostasis and fostering stress tolerance.

Melatonin production is stimulated by stress-induced ROS released by mitochondria, chloroplasts, and respiratory burst oxidase homologs (RBOH). Melatonin and its metabolites, such as 5-methoxytryptamine (5-MT), N1-acetyl-5-methoxykynuramine (AMK), N1-acetyl-N2-formyl-5-methoxykynuramine (AFMK), 2-hydroxymelatonin (2-OHM), and 3-hydroxymelatonin (3-OHM), actively scavenge reactive nitrogen species (RNS) and ROS, aiding in cellular redox homeostasis. Notably, melatonin's activation of Mitogen-Activated Protein Kinase (MAPK) is independent of ROS, indicating its role downstream of ROS bursts, in contrast to stress-induced MAPK activation mediated by ROS (Lee and Back 2017). This suggests that specific receptors and related signal transduction pathways may mediate plant responses to melatonin. For instance, it has been identified that the phytomelatonin receptor CAND2/PMTR1 may serve as a putative melatonin receptor with dual functions in signaling (Lee and Back 2018).

2.2.2 Molecular Role: Interaction with Signaling Pathways and Gene Expression Modulation

By regulating the MAPK pathway, melatonin controls a variety of physiological functions in plants, including stress tolerance. It has been suggested that melatonin receptors may activate the upstream enzyme in the MAPK signaling cascade, MAPKKK (mitogen-activated protein kinase kinase kinase), thereby linking melatonin to the regulation of the stress response (Jagodzik et al. 2018). Furthermore, melatonin binding promotes stomatal closure through the generation of H_2O_2 and the consequent influx of Ca^{2+} signaling (Wei et al. 2018a).

The expression of genes involved in melatonin biosynthesis, including tryptophan decarboxylase (TDC), tryptamine 5-hydroxylase (T5H), serotonin N-acetyltransferase (SNAT), caffeic acid O-methyltransferase (COMT), and N-acetylserotonin O-methyltransferase (ASMT), is elevated under biotic and abiotic stresses, leading to increased melatonin production. This rise in melatonin activates stress response pathways by binding to receptors such as CAND2/PMTR1 and initiating antioxidant defenses.

As plants age or are exposed to abiotic stimuli, they tend to produce melatonin precursors such as serotonin, tryptophan, and tryptamine. In the biosynthetic pathway, serotonin is first O-methylated by COMT to form 5-methoxytryptamine (5-MT), which is then acetylated by SNAT to produce melatonin. Thus, serotonin, tryptamine, and 5-MT serve as intermediates in the transformation of tryptophan into melatonin (Byeon et al. 2015).

Moreover, melatonin-mediated systemic acquired resistance (SAR), which is crucial for managing plant diseases, involves cross-talk with salicylic acid (SA) and jasmonic acid (JA). Melatonin enhances plant defense systems by modulating key signaling pathways, including interactions with OXI1/MAPKKK3 (oxidative signal-inducible 1 kinases), MAPKK4/5/7/9, and MAPK3/6. Additionally, it has been demonstrated that melatonin signaling through its receptor activates Nonexpressor of Pathogenesis-Related Genes 1 (NPR1), which subsequently triggers the nuclear expression of important defense genes, including Pathogenesis-Related Protein 1 (PR1) and Isochorismate Synthase 1 (ICS1).

Further supporting melatonin's role in strengthening plant defense systems is the correlation between elevated endogenous melatonin levels and the overexpression of genes associated with effector-triggered immunity (ETI) and pathogen-associated molecular pattern (PAMP)-triggered immunity (PTI).

2.2.3 Pathogen Resistance: Evidence of Melatonin-Mediated Defense Against Bacterial, Fungal, and Viral Pathogens

Melatonin enhances resistance to pathogens by regulating the transcription of defense-related genes and influencing various physiological functions in plants. By promoting the expression of defense-associated genes, heat shock proteins, and genes related to fungicide sensitivity, melatonin acts upstream of mitogen-activated protein kinases (MAPKs), thereby aiding plants in defending themselves against pathogenic attacks. Additionally, exogenous application of melatonin alleviates endoplasmic reticulum (ER) stress by activating the basic leucine zipper (bZIP) transcription factor (bZIP60) signaling pathway in MAPK3/6, improving cellular homeostasis and stress resistance through the upregulation of key ER stress-relief components, such as heat shock protein 70 (Hsp70) family proteins BIP2, BIP3, and calnexin 1 (CNX1). These diverse functions underscore the significance of melatonin as a key modulator of plant defense and stress responses.

Significant antimicrobial effects have been observed with direct application of melatonin to pathogenic bacteria and fungi, including changes in spore morphology, destabilization of cellular ultrastructure, reduced biofilm formation, and increased pathogen mortality. When combined, these benefits alleviate stress in plants treated with melatonin, as detailed in Fig. 2.1.

One study demonstrated that plants with elevated melatonin levels exhibited enhanced resistance to soil-borne fungal pathogens of the oomycete group, such as *Phytophthora capsici* and *Podosphaera xanthii*, which cause powdery mildew. A transcriptome analysis of watermelon leaves treated with melatonin (1 mmol/L)

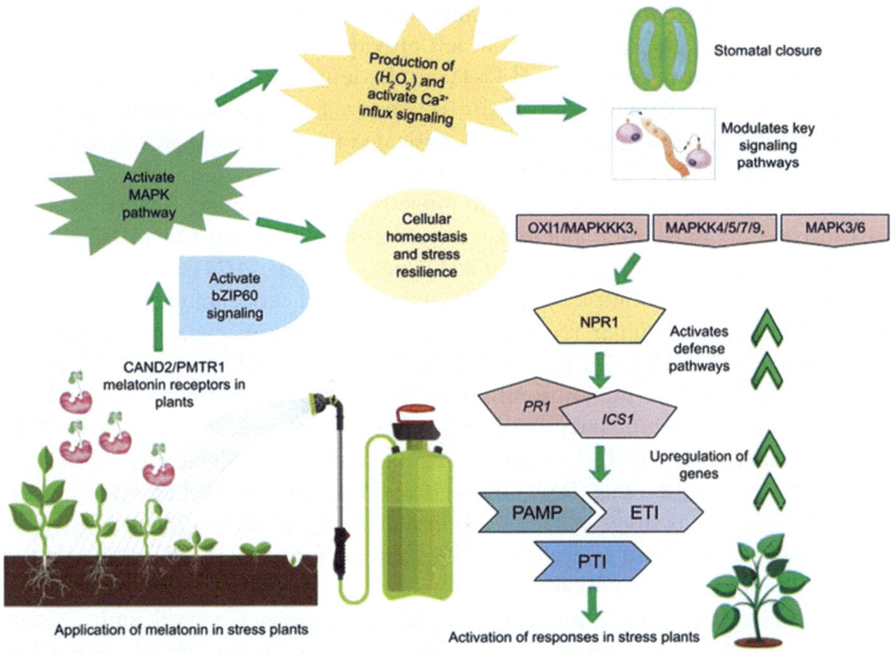

Fig. 2.1 Illustrative diagram describing the role of melatonin in inducing plant defense mechanisms

revealed that melatonin alters the expression of genes involved in both effector-triggered immunity (ETI) and pathogen-associated molecular pattern (PAMP)-triggered immunity. The study highlighted that 27 genes associated with constitutive defense mechanisms and early priming of the melatonin-induced resistance response were upregulated (Mandal et al. 2018).

In another investigation, melatonin's role in protecting *Arabidopsis thaliana* against the aggressive bacterial pathogen *Pseudomonas syringae* DC3000 was examined. Treatment with 10 μM melatonin induced several pathogenesis-related (PR) genes and various defense-related genes regulated by salicylic acid (SA) and ethylene (ET) in Arabidopsis and tobacco leaves. Melatonin-treated Arabidopsis exhibited a significant increase in resistance to the pathogen, reducing bacterial proliferation by nearly tenfold compared to mock-treated leaves (Lee et al. 2014). Furthermore, melatonin treatment of roots for the first time demonstrated improved resistance of tomato plants to *P. syringae* DC3000, with a systemic resistance effect leading to a 51% reduction in disease incidence. This treatment was also associated with the concurrent activation of SA-mediated and JA/ET-mediated defense mechanisms.

This was evidenced by increased expression of relevant PR2 and PR3 genes, along with a rise in the enzymatic activity of β-1,3-glucanase (PR2; 103%), a marker of the SA pathway, and chitinase (PR3; 44%), a marker of the JA/ET pathway. Furthermore, transgenic Arabidopsis lines showed a significant increase in melatonin-mediated expression of the chitinase gene CaChiIII2 following infection by *Colletotrichum gloeosporioides*, which provided resistance by reducing H_2O_2 generation and maintaining stable H_2O_2 levels. By temporarily increasing its expression in both control plants and CaChiIII2-silenced plants, melatonin treatment improved tolerance in pepper plants and mitigated the effects of CaChiIII2 knockdown. Studies utilizing CaChiIII2 knockdown and overexpression lines highlighted melatonin's crucial role in the innate immune response of pepper against anthracnose disease, demonstrating that melatonin effectively promotes CaChiIII2 expression along with other pathogenesis-related genes (Ali et al. 2021).

When melatonin and nitric oxide (NO) were applied exogenously to rice plants, it was observed that higher levels of melatonin and NO in melatonin-pretreated plants significantly reduced disease incidence. In contrast, only NO levels increased in plants treated with sodium nitroprusside (SNP), a reagent that releases NO. Both melatonin and NO significantly elevated the expression levels of the rice basic PR1 (OsPR1b) gene and OsWRKY45, indicating that melatonin enhances rice susceptibility to the rice stripe virus (RSV) through a NO-dependent mechanism (Lu et al. 2019). These findings suggest that melatonin, a naturally occurring plant compound, has the potential to serve as a cost-effective and environmentally friendly tool for managing biotic stress in tomato cultivation.

Exogenous application of melatonin has also been shown to significantly control infection with the Cucurbitaceae green mottle mosaic virus (CGMMV). A notable control effect was observed 3 days after root irrigation with a 50 μM melatonin concentration. In tobacco and cucumber plants, melatonin demonstrated early-stage prevention and treatment actions against CGMMV infection. RNA sequencing comparing expression patterns of tobacco leaves infected with CGMMV, melatonin+CGMMV, and mock-inoculated tobacco revealed that melatonin particularly increased the expression of the defense-related gene CRISP1, while SA had no effect. Although CRISP1 silencing did not influence the infection itself, it enhanced melatonin's ability to prevent CGMMV infection. Moreover, exogenous melatonin also exhibited protective effects against infection with the pepper mild mottle virus (PMMoV), another Tobamovirus (Yang et al. 2023).

2.3 Omics Approaches in Melatonin Research

2.3.1 Genomics

Melatonin biosynthesis in plants occurs through multiple pathways involving six key enzymes: tryptophan hydroxylase (TPH), tryptophan decarboxylase (TDC), tryptamine 5-hydroxylase (T5H), N-acetylserotonin O-methyltransferase (ASMT), caffeic acid O-methyltransferase (COMT), and serotonin N-acetyltransferase (SNAT). Genes encoding these enzymes have been identified, with the exception of the TPH enzyme. All pathways include serotonin as an intermediate, highlighting the compound's significance in melatonin synthesis.

Serotonin is synthesized from tryptophan through the combined action of TPH, T5H, and TDC, and is subsequently converted to melatonin by SNAT, ASMT, and COMT. The synthesis of serotonin takes place in either the endoplasmic reticulum or the cytoplasm of plant cells, while melatonin synthesis occurs in the chloroplasts or cytoplasm (Tan et al. 2015; Back et al. 2016; Arnao et al. 2022; Liu et al. 2022). A brief overview of the melatonin biosynthesis pathways is illustrated in Fig. 2.2.

The genes SNAT, ASMT, TDC, T5H, and COMT are involved in regulating responses to various abiotic stresses, including drought, salinity, heavy metals, high temperature, alkalinity, cold, and osmotic stress. They also play roles in managing biotic stresses, such as bacterial, fungal, nematode, and viral infections. Additionally, these genes influence important plant growth traits, including stem elongation, root development, seedling growth, and seed growth, as well as productive traits such as yield, flowering, leaf senescence, fruit quality, and ripening (Gao et al. 2023).

Melatonin is metabolized to 2-hydroxymelatonin by melatonin 2-hydroxylase (M2H), highlighting M2H's role in melatonin metabolism (Byeon and Back 2015). Overexpression of TDC isoforms in rice has resulted in increased melatonin production (Byeon and Back 2015). Serotonin and serotonin-derived metabolites contribute to pathogen resistance, as the T5H gene is induced by pathogen infection and chitin elicitors (Fujiwara et al. 2010). The overexpression of the PtoASMT gene in

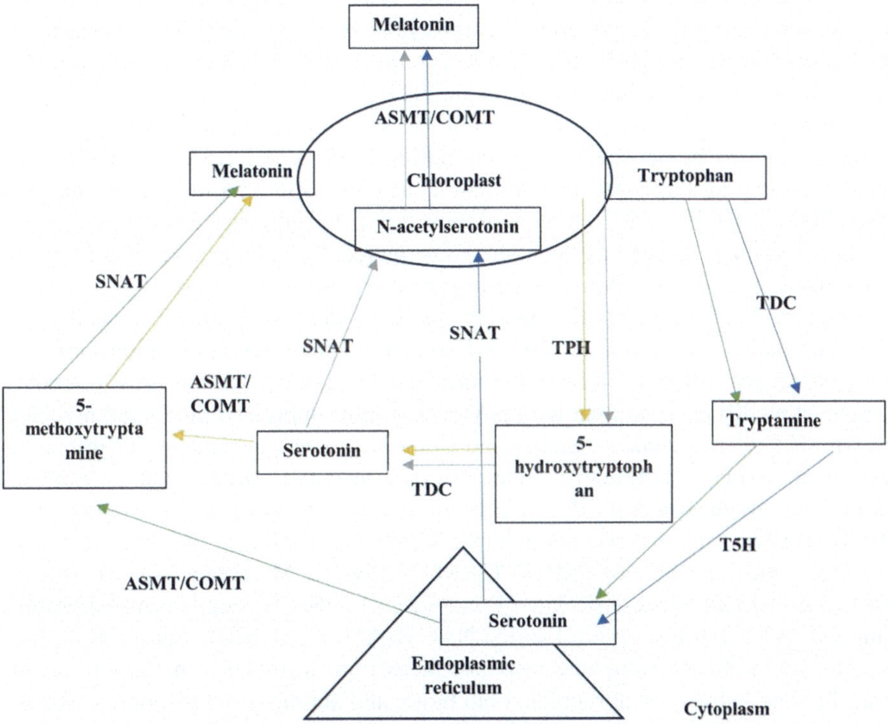

Fig. 2.2 Overview of melatonin pathways in a plant cell

poplar plants conferred resistance to the fungus *Dothiorella gregaria* through enhanced melatonin biosynthesis (Yu et al. 2025).

The enzymes SNAT, ASMT, hydroxyindole-O-methyltransferase (HIOMT), and TDC mediate redox homeostasis by scavenging reactive oxygen species (ROS) such as superoxide (O2−) and hydrogen peroxide (H2O2), as well as reactive nitrogen species (RNS). They also induce ion homeostasis and mitigate lipid peroxidation. These enzymes activate antioxidant enzymes such as catalase (CAT), peroxidase, superoxide dismutase (SOD), glutathione peroxidase (GPX), and ascorbate peroxidase (APX), while elevating levels of nonenzymatic components like glutathione (GSH), malondialdehyde (MDA), relative electrolyte leakage (REL), gamma-glutamylcysteine (GCH), phytochelatins (PCs), lignin, flavonoids, polyamines, and proline. This activation plays a significant role in regulating abiotic stress in plants (Gao et al. 2023).

Furthermore, SNAT, T5H, TDC, and COMT also influence phytohormone homeostasis, including brassinosteroids (BR), gibberellins (GA), abscisic acid (ABA), ethylene, indole-3-acetic acid (IAA), and jasmonic acid (JA) during abiotic stress. They activate JA signaling transcription factors such as MYC2 and genes involved in ABA signaling and metabolism pathways like AB1, MdNCED3, MdCYP707A1, and MdCYP707A2, as well as transcription factors responsive to ethylene like 1BERF1B and YUC, and IAA signaling genes like IAR3. Additionally, they can upregulate resistance-related genes such as ATP-binding cassette (ABC) transporters, SOD, CAT, pyrroline-5-carboxylate synthetase (P5CS), plasma membrane intrinsic proteins (PIP), heat shock proteins (HSP), and other stress-responsive genes and transcription factors (Gao et al. 2023).

A similar mechanism is observed in biotic stress, where melatonin biosynthetic genes alter redox homeostasis by accumulating ROS. SNAT, ASMT, and TDC can regulate signaling pathways for salicylic acid (SA) (PR1, NPR1), jasmonic acid (JA) (PDF1.2, COI1), ABA, IAA, cytokinins (CK), and ethylene. Silencing of melatonin biosynthetic genes has been shown to increase plant susceptibility to biotic stress due to the downregulation of pathogenesis-related proteins, WRKY transcription factors, MYC2 transcription factors, and plant defensin proteins, as well as JA signaling pathways (Gao et al. 2023). Exogenous melatonin application inhibited mycelial growth, altered cell ultrastructure, and decreased pathogen stress tolerance (Zhang et al. 2017). Treatment with phytomelatonin suppressed the fungal mycelial growth of *Colletotrichum acutatum* and *Colletotrichum gloeosporioides* in pepper (Ali et al. 2021). Additionally, melatonin demonstrated efficacy against *Botrytis cinerea* in tomato (Liu et al. 2019), *Podosphaera xanthii* in cucurbits (Mandal et al. 2018), and *Fusarium oxysporum* in banana (Wei et al. 2017a).

Gene families such as PMRT1/CAND2 (Plasma Membrane Transporter 1/ Chitinase and Pathogenesis-Related Protein 2), MAPKs (mitogen-activated protein kinases), WRKY transcription factors, NAC (NAM, ATAF, CUC transcription factor family), and ERF (ethylene response factor) are involved in melatonin signal transduction pathways that regulate both biotic and abiotic stress responses, as well as plant growth (Gao et al. 2023). Seven melatonin biosynthesis genes (one MeSNAT, one MeT5H, two MeTDCs, and three MeASMTs) enhanced endogenous melatonin synthesis in *Nicotiana benthamiana* leaves (Shi et al. 2016). When the pathogen *Xanthomonas axonopodis* pv. *manihotis* infects cassava, the transcripts MeWRKY79 and MeHsf20 are induced, which activate the expression of the MeASMT2 gene, leading to increased melatonin production and improved disease resistance (Wei et al. 2017b).

CRISPR-Cas9-based gene editing has been performed to target the binding motif of SlPIF4 (*Solanum lycopersicum* PIF4), a phytochrome-interacting factor in the promoter of SlCOMT2 (*S. lycopersicum* COMT2), to enhance melatonin production in tomato fruit (Zhang et al. 2023). CRISPR-Cas9-mediated mutations were carried out in the transcription factor elongated hypocotyl 5 (HY5) to investigate its effect on the expression of the AtSNAT6 gene in *Arabidopsis thaliana*. The knockout of HY5 inhibited the expression of AtSNAT6, thereby preventing melatonin production. However, in darkness, the HY5-mediated suppression of melatonin synthesis was reduced. Increased melatonin levels were also shown to suppress cotyledon opening, indicating the direct effect of HY5 in regulating cotyledon opening through melatonin production (Wang et al. 2022a).

An RNA interference (RNAi) approach was used to develop transgenic rice plants with downregulated BR and GA biosynthesis and signaling genes to investigate their effect on melatonin biosynthesis. Inhibition of a GA biosynthesis gene (GA3ox2) and a signaling gene (Gα) resulted in reduced melatonin synthesis, indicating their role as elicitors (Ok Jin and Back 2022a). Similar observations were made with the suppression of three BR genes, namely RAV-Like1 (RAVL1), DWARF11, and DWARF4, in rice plants, further confirming their significant role as elicitors (Ok Jin and Back 2022b). Additionally, RNAi was employed to suppress the expression of the CsSNAT (*Cucumis sativus* SNAT) gene to understand its role in regulating salt tolerance and growth in cucumber. This gene silencing resulted in decreased melatonin synthesis and increased sensitivity to salt stress (Qi et al. 2020). Moreover, virus-induced gene silencing (VIGS) strategies have been utilized to identify the roles of genes involved in melatonin production and melatonin-mediated plant resistance (Cheng et al. 2019; Thanh Huyen et al. 2024; Ling-Ling et al. 2023; Kumar et al. 2024b).

2.3.2 Transcriptomics

Sequencing of RNA transcripts is instrumental in identifying gene expression under various environmental conditions and developmental stages, thereby illuminating expression profiles and functional activity. Recent studies have reported insights into genomic expression profiles associated with melatonin application under different stress conditions. Transcriptome analyses have revealed the molecular mechanisms and regulatory roles of melatonin in various plant species, including rice (Xie et al. 2020), germinated rice seeds (Huangfu et al. 2021), okra (Zhan et al. 2021), banana seedlings (Wei et al. 2022), cotton seedlings (Duan et al. 2022), alfalfa (Li et al. 2022a), and alfalfa seedlings (Liu et al. 2024a) under salt stress; soybean (Cao et al. 2020), maize (Zhao et al. 2021), maize roots (Wang et al. 2023a), chrysanthemum (Luo et al. 2023), and dove tree (Liu et al. 2021) under drought stress; tomato under low night temperature (Yand et al. 2022); rhododendron under heat stress (Xu et al. 2023); plum fruit under post-harvest chilling injury (Xu et al. 2022); melon (specifically root development) under copper stress (Hu et al. 2020); and cotton under cadmium stress (Li et al. 2022b). mRNA sequencing (mRNA-seq) technology has been employed to explore the genomic functions of melatonin in Arabidopsis plants treated for 16 h with low (100 ppm) and high (1 mM) concentrations of melatonin. The 1 mM concentration significantly altered the expression of more genes compared to the 100 pM concentration. Exogenous application of melatonin can upregulate and downregulate nearly 360 genes associated with plant stress defense, with at least twofold changes in expression. This includes various transcription factors, stress receptors, kinases, and other downstream genes (Weeda et al. 2014). Additionally, significant alterations in the expression of genes related to hormone signal transduction pathways—including salicylic acid (SA), ethylene, jasmonic acid (JA), abscisic acid (ABA), and auxin—were observed. The expression of genes related to cell wall and redox pathways, particularly peroxidases, also changed significantly (Weeda et al. 2014).

2.3.3 Proteomics

Melatonin application in plants modulates proteomic profiles, enhancing stress tolerance through various mechanisms. Proteomic analyses have revealed that melatonin influences the abundance of proteins associated with stress responses, including those involved in antioxidant defense, protein folding, and signaling pathways. For instance, melatonin treatment upregulates heat shock proteins (HSPs), which act as molecular chaperones to prevent protein aggregation and assist in refolding denatured proteins under stress conditions (Aghdam and Arnao 2024). This upregulation contributes to improved plant resilience against environmental stressors. Additionally, melatonin enhances the activity of antioxidant enzymes such as superoxide dismutase (SOD) and catalase (CAT), which play crucial roles in scavenging reactive oxygen species (ROS), thereby mitigating oxidative damage induced by abiotic stresses (Bhardwaj et al. 2024). Furthermore, melatonin modulates the expression of proteins involved in the mitogen-activated protein kinase (MAPK) signaling pathway, which is integral to plant responses to various stresses (Pan et al. 2023). By influencing these proteins, melatonin helps activate downstream defense mechanisms, enhancing the plant's ability to cope with adverse conditions. These proteomic adjustments underscore the multifaceted role of melatonin in bolstering plant stress tolerance through the regulation of key proteins involved in defense and survival pathways.

Proteomic profiling involves a series of advanced techniques aimed at identifying, quantifying, and characterizing proteins to understand their roles in biological systems, particularly in plants under melatonin treatment. The process begins with efficient protein extraction from plant tissues, often using methods such as phenol extraction and TCA-acetone precipitation to isolate proteins while minimizing contamination (Zhang et al. 2022). Following extraction, protein quantification is performed using assays like the Bradford or BCA methods to provide accurate measurements of total protein concentrations for downstream analyses.

Protein separation techniques, such as one-dimensional (1D) or two-dimensional (2D) gel electrophoresis and liquid chromatography (LC), are then employed to isolate proteins based on their physical and chemical properties. Mass spectrometry (MS) is a critical component of proteomic profiling, with approaches like matrix-assisted laser desorption/ionization time-of-flight (MALDI-TOF) (Hu et al. 2024), electrospray ionization (ESI) (Zou et al. 2021), and tandem MS/MS used for high-throughput protein identification, de novo sequencing, and post-translational modification (PTM) analysis.

Quantification of proteins can be achieved using label-free methods or isotopic labeling techniques, such as isobaric tagging (TMT, iTRAQ) (Wang et al. 2022a, b) or stable isotope labeling by amino acids in cell culture (SILAC), which enable multiplex and precise quantification (Yin et al. 2021; Wang et al. 2022a). Bioinformatics tools are integral to interpreting large datasets, with platforms like UniProt and KEGG linking proteomic findings to biological pathways and processes, while tools like Cytoscape help visualize protein interaction networks. Advanced techniques, including shotgun and targeted proteomics, provide a more detailed analysis of protein landscapes, while phosphoproteomics offers insights into phosphorylation-dependent signaling pathways. Validation of findings often relies on complementary methods such as Western blotting, enzyme-linked immunosorbent assay (ELISA), or quantitative reverse transcription PCR (qRT-PCR), which confirm changes in protein abundance or expression (Zhang et al. 2023).

Additionally, studies on protein–protein interactions, through methods like yeast two-hybrid systems (Guo et al. 2022), co-immunoprecipitation (Cao et al. 2024), or protein microarrays, help unravel regulatory networks influenced by melatonin (Murch and Erland 2021). Collectively, these proteomic methods provide a comprehensive framework for exploring how melatonin modulates protein expression and activity, offering valuable insights into its role in enhancing plant stress tolerance, regulating growth, and improving defense mechanisms (Gupta 2023). This holistic approach highlights the critical importance of proteomic profiling in understanding the molecular pathways through which melatonin impacts plant biology.

Post-translational modifications (PTMs), including phosphorylation, acetylation, ubiquitination, and glycosylation, dynamically regulate the activity, stability, and localization of stress-related proteins, facilitating rapid adaptation to environmental changes (Zou et al. 2021; Bhardwaj et al. 2024).

2.3.4 Metabolomics

Metabolomics provides a comprehensive understanding of the metabolic adjustments triggered by melatonin in plants, highlighting its critical role in enhancing stress tolerance, regulating energy metabolism, promoting secondary metabolite production, and improving crop resilience. Through metabolomic studies, it has been shown that melatonin regulates metabolites involved in oxidative stress responses, including key antioxidants such as ascorbate, glutathione, and phenolic compounds, which play vital roles in scavenging reactive oxygen species (ROS) and mitigating oxidative damage under stress (Khan et al. 2020). Melatonin also influences the accumulation of primary metabolites such as sugars, organic acids, and amino acids, thereby maintaining energy homeostasis during environmental stress conditions (Zeng et al. 2022a). Furthermore, metabolomic profiling has demonstrated that melatonin enhances the biosynthesis of secondary metabolites like flavonoids, alkaloids, and phenolics, which significantly contribute to pathogen resistance and stress adaptation (Huang and Jin 2024). This technology has elucidated the role of melatonin in mitigating abiotic stresses such as drought, salinity, and extreme temperatures, as evidenced by the accumulation of stress-responsive metabolites like proline, betaine, and polyamines, as well as its influence on alleviating biotic stresses through increased production of antimicrobial compounds (Ayyaz et al. 2022; Kumar et al. 2023a, b).

In metabolomic profiling, a range of advanced analytical techniques is employed to comprehensively analyze the metabolites influenced by melatonin in plants. Gas Chromatography-Mass Spectrometry (GC-MS) (Nehela and Killiny 2020) is commonly used to detect and quantify volatile and primary metabolites such as organic acids, sugars, and amino acids, providing insights into fundamental metabolic adjustments. Liquid Chromatography-Mass Spectrometry (LC-MS) (Singh et al. 2023) is particularly effective for studying secondary metabolites, including flavonoids and phenolic compounds, which are often associated with stress responses and disease resistance. Nuclear Magnetic Resonance (NMR) spectroscopy (Mascellani Bergo et al. 2024), a nondestructive and highly accurate method, is utilized to quantify and identify a wide range of metabolites, offering both structural and functional information. Fourier Transform Infrared Spectroscopy (FTIR) (Santos et al. 2024) also enables rapid analysis of functional groups, allowing researchers to detect changes in metabolite composition induced by melatonin.

Metabolomics has identified key precursors and intermediates of melatonin biosynthesis, such as serotonin, N-acetylserotonin, and 5-hydroxytryptophan, providing deeper insights into its synthesis and regulation (Arnao and Hernández-Ruiz 2018). Studies on melatonin-induced metabolic shifts have also demonstrated contributions to crop productivity; for example, in rice, metabolomic profiling revealed increased levels of soluble sugars and amino acids under melatonin treatment, thereby promoting growth and resilience during drought stress (Zhang et al. 2015). Together, these findings underscore the value of metabolomics in uncovering the multifaceted roles of melatonin in plant stress biology and its potential for enhancing agricultural sustainability.

2.3.5 Epigenomics

Epigenomics examines how gene expression can change without altering the DNA sequence itself. These changes can impact various biological processes, including plant growth and development, stress responses to biotic and abiotic factors, and environmental adaptation. Plants increasingly face complex combinations of abiotic and biotic stresses due to escalating environmental challenges, significantly threatening their growth, development, and survival. Abiotic stresses, such as drought, extreme temperatures, salinity, heavy metal toxicity, and heat, disrupt essential physiological processes, reducing productivity and yield. Similarly, biotic stresses caused by pathogens—such as fungi, bacteria, and viruses—as well as damage from pests, further compromise plant health, exacerbating their vulnerability and reducing overall agricultural output (Ali et al. 2024).

Epigenomic analyses generally focus on mechanisms such as DNA methylation, histone modification, and noncoding RNAs (Wang and Yamaguchi 2024). These mechanisms can enhance the expression of disease-resistant genes, interact with other phytohormones, or regulate antioxidant enzymes. DNA methylation involves identifying differentially methylated regions (DMRs) using methods such as Whole-Genome Bisulfite Sequencing and Methylation-Sensitive Amplified Polymorphism (MSAP). Histone modification mapping employs techniques like Chromatin Immunoprecipitation Sequencing (ChIP-Seq) (Sharma et al. 2023) and mass spectrometry. Additionally, noncoding RNAs include small noncoding RNAs (snRNAs) and long noncoding RNAs (lncRNAs).

A study analyzing DNA methylation in watermelon infected by Cucumber Green Mottle Virus found that reduced RNA-directed DNA methylation (RdDM) of CHH sites and increased levels of 21- and 22-nucleotide small interfering RNAs (siRNAs) synergistically enhanced antiviral defense in watermelon (Sun et al. 2019). Several studies have demonstrated how the application of melatonin ameliorates plant diseases. For instance, a study conducted by Gao et al. (2020) on grape berries showed that exogenous application of melatonin influenced DNA methylation patterns, leading to changes in gene expression associated with resistance to *Botrytis cinerea* and a decreased fruit decay rate. The treatment resulted in reduced methylation levels affecting genes related to biotic stress responses and an increase in flavonoid biosynthesis, which correlated with enhanced expression of defense-related genes, thus increasing resistance to pathogens. In melatonin-treated cassava plants, a large number of lncRNAs were identified, and their potential functions were predicted via trans-acting, cis-acting, and miRNA targets. These findings provided a comprehensive overview of cassava lncRNAs in response to abiotic stresses (Ding et al. 2019).

Melatonin has been shown to interact with other phytohormones, such as auxin, abscisic acid, salicylic acid, and jasmonic acid, forming a complex regulatory network that increases plant resistance to diseases. The application of melatonin not only increases the concentrations of these phytohormones but also enhances sugar content, callose deposits, cellulose, xylose, mevalonate, gossypol, lignin, phenylpropanoids, antioxidant enzymes, and reactive oxygen and nitrogen species (ROS/RNS), all of which are involved in maintaining plant immunity (Zeng et al. 2022b). This regulation of other hormones also involves modifications in DNA methylation and histone acetylation, as well as interactions with noncoding RNAs (Tiwari et al.

2021a, b). The multifunctional nature of melatonin makes it a key component in the intricate network of plant defense.

Genes responsible for the plant defense mechanism during pathogen infection are identified through transcriptomic approaches. Direct treatment of melatonin on fungi and bacteria has been shown to alter cell ultrastructure and spore structure, resulting in higher mortality rates and impaired biofilm formation. Indirect effects include the stimulation of systemic acquired resistance (SAR), pathogen-associated molecular pattern-triggered immunity (PAMP-PTI), and effector-triggered immunity (ETI) (Tiwari et al. 2021b), as well as the regulation of phytohormone expression (Liu et al. 2024a). The significance of melatonin in the plant defense mechanism of Valencia sweet orange plants infected with *Candidatus* Liberibacter was explored by investigating the expression patterns of genes related to salicylic acid (SA), abscisic acid (ABA), and auxins (Nehela and Killiny 2020). Transcriptome analysis revealed factors responsible for the inhibitory effect of melatonin on *Phytophthora infestans*, suggesting major involvement of genes associated with amino acid metabolic processes following melatonin treatment. Additionally, differentially expressed genes related to virulence, stress tolerance, and fungicide resistance were identified (Zhang et al. 2017).

2.4 Applications of Omics-Based Strategies of Melatonin in Plant Disease Amelioration

Melatonin promotes stress tolerance in cereal crops by modulating critical physiological pathways, including growth modulation, water management, and the activation of antioxidant defenses. It enhances water intake and transport efficiency in challenging environments by improving hydraulic conductivity and stimulating root development (Qiao et al. 2020). Metabolomic alterations indicate that increased accumulation of indole and tryptophan, along with elevated levels of downstream metabolites involved in melatonin and indole acetate production, play crucial roles in plant–pathogen interactions and disease resistance (Zeng et al. 2022a, b). The effects of melatonin span a wide array of molecular and physiological processes, ultimately boosting cereal crop resilience under stress.

Melatonin increases resistance to rice stripe virus infection via a nitric oxide (NO)-dependent mechanism. In Arabidopsis, melatonin and NO work together to upregulate the expression of numerous salicylic acid (SA)-related genes (Shi et al. 2015). Further research utilizing genomics and proteomics approaches will enhance our understanding of melatonin's interactions with other signaling components, as well as the roles of SA, jasmonic acid (JA), ethylene (ET), NO, and melatonin receptors in immune signaling cascades (Lee et al. 2014). Increased melatonin accumulation in watermelon and other cucurbits has been shown to improve resistance to foliar diseases, including powdery mildew and soil-borne oomycetes. This resistance is achieved by altering the transcript levels of genes associated with pattern-triggered immunity (PTI) and effector-triggered immunity (ETI) (Mandal et al. 2018).

In cassava, the transcription factors MeRAV1 and MeRAV2 are important regulators of melatonin production, as they positively influence melatonin biosynthesis genes and increase endogenous melatonin levels. These transcription factors directly activate key cassava melatonin biosynthesis genes such as MeT5H, MeTDC2, and MeASMT1, promoting melatonin production. Additionally, MeRAV1 and MeRAV2 enhance plant defense by increasing resistance to cassava leaf blight. Further investigation into these regulators could improve cassava tolerance to biotic and abiotic stresses (Wei et al. 2018).

A study found that melatonin significantly decreased DNA methylation in grape berries, resulting in altered gene expression. This epigenetic modification increased flavonoid production and disease resistance by lowering the methylation levels of key gene promoters. These findings suggest that melatonin is essential for regulating metabolic pathways and enhancing plant defense mechanisms in grapevines (Gao et al. 2020). Recent transcriptomic findings demonstrate that melatonin treatment activates genes involved in ETI and PTI in both watermelon and Arabidopsis, highlighting melatonin's important role in improving plant immune responses by modulating critical defense-related genes (Mandal et al. 2018). Additionally, melatonin enhances rice resistance to *Xanthomonas oryzae* pv. oryzae, the pathogen responsible for leaf blight, by bolstering the plant's defense mechanisms and its ability to combat infection (Chen et al. 2020).

Melatonin improves disease resistance and flavonoid production in grape berries by influencing DNA methylation and gene expression, underscoring its involvement in plant defense and metabolism (Gao et al. 2020). Furthermore, exogenous melatonin effectively reduced potato late blight by suppressing *Phytophthora infestans* mycelial development, modifying cell ultrastructure, and decreasing stress tolerance. Notably, melatonin exhibited synergistic antifungal effects with fungicides, suggesting its potential to enhance fungicide efficacy while reducing the required doses. Transcriptome analysis revealed 66 differentially expressed genes involved in amino acid metabolism that were significantly affected by melatonin, providing insight into its role in plant defense systems (Zhang et al. 2017).

2.4.1 Integrated Omics Findings in Different Crops

Integrated omics-assisted techniques have significantly enhanced our understanding of biotic stress responses and the adaptive strategies that plants employ to survive and mitigate adverse conditions. Analyses of transcriptomic, proteomic, and metabolomic datasets provide insights into the essential genes, proteins, and metabolites involved in the biological and molecular processes that improve stress tolerance (Table 2.1). Combining these omics methods is crucial because biological information flows from DNA to the metabolome. With advancements in genomics, transcriptomics, and proteomics, we have gained valuable insights into numerous plant diseases and their underlying mechanisms. Recent developments in omics technologies have provided important molecular insights into the mechanisms enabling diverse stress tolerance in wheat (Singhal et al. 2021).

Table 2.1 Insights of omics technologies in melatonin research

Sl. no.	Omics type	Objective	Key findings	Tools/platform used	References
1	Transcriptomics	To identify the gene expression of cowpea wilt resistance on melatonin application	Exogenous application of melatonin promoted CAT, GDH-PX (glutathione peroxidase), and SA synthesis. It upregulated metabolic pathways of phenolic and lignin synthesis. Activated transcription factors like the WRKY and MYB families	RNA-Seq—Illumina sequencing	Gan et al. (2024)
2	Genomics	To identify gene family members of T5H, TDC, SNAT, and ASMT enzymes related to melatonin production in Arabidopsis, tomato, rice, and sorghum	Observed notable changes in the number of genes associated with each enzyme and their phylogenetic pattern	Genome databases like NCBI and Phytozome to retrieve genomic sequences and HMMeR, WoLF PSORT, and ChloroP for protein sequence analysis	Bhowal et al. (2021)
3	Transcriptomics	To examine the DEGs expressed in melatonin-treated cucumber plants damaged by cotton aphid	Enhanced MAPK and hormone signaling along with plant–pathogen interaction	RNA-Seq—Illumina sequencing	Liu et al. (2023)
4	Transcriptomics	To identify gene expression associated with melatonin production in Arabidopsis, tomato, rice, and sorghum	The expression profiles of genes involved differ based on the developmental stage and plant tissue. Presence of isoforms of the genes indicates their specificity	Genevestigator and expression atlas repositories to study the differential expression genes (DEGs)	Bhowal et al. (2021)

Sl. no.	Omics type	Objective	Key findings	Tools/platform used	References
5	Transcriptomics	To explore the significance of melatonin in *Botryosphaeria dothidea* infected pear plant	The expression analysis showed activation of phenylpropanoid metabolism and improvement in antioxidant enzymatic (polyphenol oxidase (PPO), SOD, and POD) activity through interaction of melatonin with phlorizin and JA	RNA-Seq	Xu et al. (2024a)
6	Transcriptomics	To identify gene expression profiles in the plant defense mechanism in melatonin-treated watermelon infected by *Podosphaera xanthii* and *Phytophthora capsici*	Melatonin application enhanced the plant immunity by upregulating defense-related genes and directly suppressing the pathogen	RNA-Seq—Illumina sequencing	Mandal et al. (2018)
7	Transcriptomics	To investigate gene expressions related to melatonin-induced plant resistance in cotton infected by *Verticillium dahlia*	Application of melatonin enhanced gene expression of gossypol, mevalonate (MVA) and phenylpropanoid pathways	RNA-Seq—Illumina sequencing	Cheng et al. (2019)

(continued)

Table 2.1 (continued)

Sl. no.	Omics type	Objective	Key findings	Tools/platform used	References
8	Transcriptomics	To understand the importance of melatonin in the antiviral defense system in tobacco infected by *Cucumber green mottle mosaic virus* (CGMMV)	Application of melatonin upregulated the expression of defense-related genes MYB1R1, CRISP1, and ERF RAP2-1	RNA-Seq—Illumina sequencing	Ling-Ling et al. (2023)
9	Transcriptomics	To investigate the significance of melatonin treatment in apple leaves under nutrient stress	Melatonin application reduced nutrient stress by improving nutrient absorption and utilization by upregulating metal-ion-transport-related genes and controlling glutathione metabolic activity	RNA-Seq—Illumina sequencing	Cao et al. (2022)
10	Transcriptomics	To examine the expression profiles of genes responsible for alleviating salt stress in cotton subjected to melatonin treatment	Melatonin upregulated photosynthesis and chlorophyll biosynthesis genes, altered phytohormone signaling pathways, triggered linoleic and α-linoleic acid pathway genes, enhanced ROS scavenging potential, and caused changes in sodium and potassium transporters to improve plant growth and biomass	RNA-Seq—Illumina sequencing—paired-end (PE150) strategy	Ren et al. (2022b)

Sl. no.	Omics type	Objective	Key findings	Tools/platform used	References
11	Transcriptomics	To investigate gene expression in tomato treated with melatonin under drought stress	Melatonin application downregulated genes of linoleic acid catabolic enzymes to enhance drought resistance and altered various hormone signal transduction pathways and transcription factors	RNA-Seq—DNA nanoball sequencing	Lu et al. (2022)
12	Transcriptomics	To explore the downstream transcription factors in melatonin-treated *Phaseolus vulgaris* L. under salt stress	Application of melatonin-regulated ABC transporters; sodium ion transmembrane transport; steroid and phenylpropane biosynthesis; and terpenoid, flavonoid, and tryptophan metabolism	RNA-Seq—Illumina sequencing	Yang et al. (2024)

(continued)

Table 2.1 (continued)

Sl. no.	Omics type	Objective	Key findings	Tools/platform used	References
13	Transcriptomics	To investigate the role of different genes expressed in melatonin-treated soybeans throughout various developmental stages under alkali stress	DEGs increased initially at the vegetative stage and later decreased during the pod filling stage. Observed regulation in several biosynthetic pathways like flavonoid and phenylpropanoid, and hormonal signaling gene families like AUX/IAA, GH3, AHP, GID1, and DELLA during the pod-filling stage. Also, upregulation of transcription factors like ERF, bHLH, NAC, C2H2, MYB, and MYB related during vegetative growth was observed	RNA-Seq—Illumina sequencing	Duan et al. (2024)

Sl. no.	Omics type	Objective	Key findings	Tools/platform used	References
14	Transcriptomics	To elucidate the expression of different genes in germinated maize seeds exposed to melatonin during salinity stress	Melatonin treatment upregulated expression of genes associated with hormone signal transduction (by regulation of transcription factors like WRKY, bHLH, MYB, and GRAS), photosynthesis, phenylpropanoid, cutin, suberine, and wax biosynthesis; starch, sucrose, and cyanoamino acid metabolism; and fatty acid elongation	Illumina sequencing—paired-end (PE150) strategy	Xu et al. (2024b)
15	Genomics	To perform genome-scale analysis of the B3 gene family in soybean	Melatonin treatment induced B3 genes to improve cold stress tolerance in soybean	TAIR—Arabidopsis genome database	Ren et al. (2022a)
16	Genomics	To investigate the genes involved in the melatonin synthesis pathway	Genes in association with melatonin biosynthesis, including TDC, T5H, SNAT, and ASMT were affected by abiotic stresses and are involved in circadian regulation	GreenPhyl V4—plant comparative genomics database	Ahn et al. (2021)

(continued)

Table 2.1 (continued)

Sl. no.	Omics type	Objective	Key findings	Tools/platform used	References
17	Genomics	To explore the role of the GhM2H gene related to melatonin metabolism in cotton	The melatonin level in cotton under salt stress was regulated by some genes from the GhM2H family and a few were induced by exogenous melatonin application	CottonFGD—cotton functional genomic database	Zhang et al. (2021)
18	Genomics	To evaluate the importance of the CsCOMT19 gene in melatonin synthesis in tea	The *cis*-acting elements were abundantly found in the promoter region of the CsCOMT19 gene, promising its role in plant growth as well as abiotic stress response	TPIA—tea plant genome database	Thanh Huyen et al. (2024)

Sl. no.	Omics type	Objective	Key findings	Tools/platform used	References
19	Proteomics	To evaluate how the exogenous melatonin enhances $ZnSO_4$ stress tolerance and regulates the isothiocyanate content of broccoli sprouts	Exogenous melatonin enhances glucoraphanin biosynthesis and its hydrolysis to isothiocyanates in broccoli sprouts under $ZnSO_4$ stress by upregulating glucoraphanin biosynthesis genes, increasing glucosinolate content, enhancing myrosinase activity, and elevating the abundance of related proteins. These findings underscore melatonin's role in promoting ITC accumulation and its potential in alleviating $ZnSO_4$ stress	iTRAQ (Isobaric Tags for Relative and Absolute Quantitation), LC-MS/MS	Yin et al. (2021)

(continued)

Table 2.1 (continued)

Sl. no.	Omics type	Objective	Key findings	Tools/platform used	References
20	Proteomics	To elucidate how melatonin confers enhanced polyamine metabolism and cell tolerance in *Vitis vinifera* against oxidative damage	Melatonin-treated grapes exhibited increased levels of nitrogen-containing compounds, including arginine (Arg), proline (Pro), gamma-aminobutyric acid (GABA), putrescine (Put), spermidine (Spd), and spermine (Spm). These compounds play crucial roles as antioxidants or molecular chaperones, helping to protect cellular membrane integrity and mitigate damage caused by oxidative stress	TMT—Tandem Mass Tags, LC-MS/MS	Wang et al. (2022a, b)

Sl. no.	Omics type	Objective	Key findings	Tools/platform used	References
21	Proteomics and metabolomics	To analyze the melatonin elevated resistance of *Brassica rapa* ssp. *Pekinensis* against the stem rot pathogen *Sclerotinia sclerotiorum* via modulation of ATP and glucosinolate biosynthesis	*Brassica rapa* leaves promoted thiamine formation as a defense mechanism against *Sclerotinia sclerotiorum* infection. Furthermore, melatonin administration enhanced this defense by inducing antioxidant activation in an ATP-dependent manner and stimulating glucosinolate biosynthesis, thereby inhibiting the development of *S. sclerotiorum*	Protein-nano liquid chromatography-mass spectrometry (LC-MS)/MS Metabolite-Gas Chromatography-Mass Spectrometry (GC-MS)	Teng et al. (2021)
22	Metabolomics	To study how the application of melatonin imparts resistance in pear against ring rot disease by increasing the amount of metabolites	Melatonin application enhanced resistance to ring rot caused by *Botryosphaeria dothidea* in pear fruit by increasing the activities of antioxidant enzymes, including peroxidase (POD), superoxide dismutase (SOD), and polyphenol oxidase (PPO). Additionally, it significantly elevated the contents of jasmonic acid and phlorizin	Liquid chromatography-mass spectrometry (LCMS)	Xu et al. (2024a)

(continued)

Table 2.1 (continued)

Sl. no.	Omics type	Objective	Key findings	Tools/platform used	References
23	Metabolomics	To analyze the melatonin involved in citrus response against Huanglongbing (HLB) disease caused by *Candidatus* Liberibacter *asiaticus* via modulation of stress-associated phytohormones and their biosynthetic genes	The infection caused the increase of phytomelatonin, jasmonic acid, and salicylic acid, and also increased the levels of phytohormones like auxin and abscisic acid, thereby conferring disease resistance in citrus	GC- MS	Nehela and Killiny (2020)
24	Metabolomics	To investigate the changes induced by exogenous melatonin application at different concentrations (10, 50, 150 μM) and its role in mitigating salinity stress in *Lactuca sativa* L. (lettuce)	High concentrations of melatonin enhanced salinity stress mitigation in a concentration-dependent manner. This effect was associated with the modulation of secondary metabolism, phytohormone levels, fatty acid composition, and amino acid biosynthesis across different melatonin concentrations	Ultra-high-pressure liquid chromatography (UHPLC) coupled to electrospray quadrupole-time-of-flight hybrid mass spectrometry (UHPLC-ESI/QTOF-MS)	Secomandi et al. (2025)
25	Metabolomics	To elucidate the mechanisms through which melatonin enhances alkali stress tolerance in soybean	High levels of glucosine and also phytohormones like cytokinin and gibberellin resulted in alkali stress tolerance	Ultra-high performance liquid chromatography-tandem mass spectrometry (UPLC-MS/MS)	Duan et al. (2024)

Sl. no.	Omics type	Objective	Key findings	Tools/platform used	References
26	Metabolomics	To analyze the exogenous application of melatonin promoting salt tolerance in smooth bromegrass seedlings	Regulated the profiles of the genes involved in proline metabolism, flavonoid biosynthesis, and melatonin biosynthesis pathway, and thereby induced salt tolerance	LC-MS	Song et al. (2024)
27	Metabolomics	To study how the metabolite profiling causes distinctive effects due to the application of exogenous melatonin on flavonoid biosynthesis in *Rosa rugosa*	Flavonoid biosynthesis was enhanced, with melatonin significantly upregulating key compounds such as quercitrin, 2″-O-galloylhyperin, and baimaside in flavonols, as well as peonidin-3-O-5-O-(6-O-coumaroyl)-diglucoside and peonidin-3-O-glucoside in anthocyanins	LC-MS/MS	Xu et al. (2024c)
28	Metabolomics	To uncover the potential mechanisms induced by melatonin application that contribute to water deficit tolerance in loquat (*Eriobotrya japonica* Lindl.)	Under water stress conditions, differential metabolites such as flavonoids, prenol lipids, organooxygen compounds, mini peptides, and fatty acyls made significant contributions	UHPLC-quadrupole time of flight (TOF)-MS/MS	Wang et al. (2023)

(continued)

Table 2.1 (continued)

Sl. no.	Omics type	Objective	Key findings	Tools/platform used	References
29	Metabolomics	To elucidate the metabolic regulation mechanism of melatonin for reducing cadmium accumulation and improving quality in rice	Melatonin caused the altered tryptophan metabolism and enhanced cross-talk between phytohormones	Fourier transform infrared (FTIR) spectroscopy	Liu et al. (2024b)
30	Epigenomics	To analyze disease resistance of grapevines against *Botrytis cinerea* in grapevine due to melatonin	Melatonin increases disease resistance and gene expression in grapevines	DNA methylation	Gao et al. (2020)
30	Epigenomics	To alleviate the lead stress in radish plants by treating with melatonin	Melatonin-induced DNA demethylation of metal transporter and antioxidant genes alleviated lead stress in radish plants, leading to enhanced antioxidant activity	DNA methylation-Whole-Genome Bisulfite Sequencing	Tang et al. (2021)

Sl. no.	Omics type	Objective	Key findings	Tools/platform used	References
31	Epigenomics	To investigate the changes in melatonin-deficient rice mutant plants associated with DNA methylation	The rice mutants exhibited premature leaf senescence during grain filling, accompanied by a reduction in grain yield. This deficiency triggered transcriptional reprogramming, particularly affecting genes related to chlorophyll and carbon metabolism, redox regulation, and transcriptional control	DNA methylation	et al. (2024)
32	Epigenomics	To analyze the role of melatonin-mediated DNA methylation in ethylene signal transduction and the ripening of tomato fruits	DNA methylation changes in the CpG islands of genes involved in ethylene biosynthesis and signaling pathways play a role in regulating the delayed ripening of melatonin-treated tomato fruits	DNA methylation	Shan et al. (2022)
33	Epigenomics	To analyze the development of resistance to cassava bacterial blight in cassava in the presence of melatonin	Histone deacetylase 9 modulates disease resistance by precisely regulating the histone deacetylation of melatonin biosynthetic genes and influencing melatonin accumulation	Histone deacetylation	Zhao et al. (2023)

(continued)

Table 2.1 (continued)

Sl. no.	Omics type	Objective	Key findings	Tools/platform used	References
34	Epigenomics	To identify the impact of melatonin in cadmium tolerance	Melatonin improved cadmium tolerance in rice by regulating cell wall and photosynthesis through long noncoding RNAs	Long noncoding RNA (lncRNA)	Qiu et al. (2024)

An integrated transcriptome and metabolome analysis revealed the response mechanism of wild rice (*Zizania latifolia*) to sheath rot infection. Genes associated with cell wall formation, glucose metabolism, and plant–pathogen interactions exhibited altered expression. Furthermore, 136 differentially accumulated metabolites were identified, including increased levels of tryptophan, melatonin, and indole acetate in infected plants (Chen et al. 2023). Nehela and Killiny (2020) employed integrated metabolomic and transcriptomic techniques to investigate melatonin's role in citrus defense against Huanglongbing (HLB) disease. In their study, melatonin was detected in *Citrus sinensis* leaves using gas chromatography-mass spectrometry. Infection by *Candidatus* Liberibacter *asiaticus* and infestation by *Diaphorina citri* increased melatonin levels and upregulated several biosynthetic genes (CsASMT, CsT5H, CsSNAT, CsTDC, and CsCOMT).

According to integrated transcriptome and metabolomic investigations, melatonin enhances the transcription of genes involved in amino acid and carbohydrate metabolism, leading to the accumulation of amino acids and sugar metabolites. This mechanism helps prolong postharvest senescence in cowpeas, emphasizing melatonin's significance in improving shelf life (Liu et al. 2024a). Additionally, melatonin effectively mitigates Sclerotinia stem rot in *Brassica rapa*, caused by the necrotic pathogen *Sclerotinia sclerotiorum*. Comprehensive proteomics and metabolomics investigations were conducted to elucidate the molecular pathways underlying melatonin-induced resistance to *S. sclerotiorum* in *B. rapa*. The integrative analysis of omics data revealed that melatonin administration regulated ATP and glucosinolate production, contributing to the defense against *S. sclerotiorum* infection (Teng et al. 2021).

Integrating metabolic profiling with transcriptome data helps bridge the gap between genetic inheritance and plant phenotypes, facilitating a better understanding of metabolic pathways and gene activities. A study that combined comparative metabolome, transcriptome, and proteome analysis found that low-temperature treatment significantly enhanced the production of secondary metabolites, such as flavonoids and phenolic acids, in tomato fruit (Liu et al. 2023).

2.5 Challenges and Perspectives

Melatonin plays a crucial role in enhancing plant immunity; however, several challenges limit its widespread application in agriculture. One significant hurdle is the complexity of melatonin signaling and its interactions with plant hormones such as salicylic acid, jasmonic acid, ethylene, and abscisic acid, which complicate the definition of its precise function in disease resistance. Additionally, the regulation of melatonin biosynthesis during pathogen attack remains unclear, and its effects can vary across different plant species and pathogens, necessitating species-specific investigations.

Practical challenges, such as high production costs, stability issues, and the need for optimized application methods, further hinder the large-scale use of melatonin in agriculture. While advancements in genomics, transcriptomics, and proteomics have improved our understanding of melatonin's role in plant defense, research in epigenetics and metabolomics remains limited. This gap restricts knowledge of melatonin's role in immune priming and secondary metabolite production.

To address these challenges, integrating multi-omics approaches can provide deeper insights into the mechanisms of melatonin. Additionally, computational tools such as systems biology and machine learning can help decipher the regulatory networks associated with melatonin. Genetic engineering and CRISPR-Cas technology offer potential solutions by enhancing melatonin biosynthesis in crops, thereby reducing the need for external applications.

Developing cost-effective melatonin-based biostimulants with optimized application strategies could also promote sustainable plant protection. Further exploration of melatonin's role in epigenetic modifications may unlock new strategies for priming plants against future infections. Moreover, large-scale field trials are necessary to validate melatonin's effectiveness under real-world conditions.

Overcoming these challenges and leveraging recent advancements could position melatonin as a powerful tool in sustainable agriculture, enhancing crop resilience, reducing reliance on pesticides, and contributing to global food security.

2.6 Conclusion

Melatonin has emerged as a promising bioactive molecule in plant disease management, offering multifaceted benefits such as enhancing immunity, reducing oxidative stress, and modulating defense-related pathways. Its interactions with plant hormones and involvement in immune signaling underscore its potential as a natural alternative to chemical pesticides. However, several challenges must be addressed before its widespread use in agriculture, including the complexity of melatonin's signaling networks, species-specific variability, and constraints on large-scale application.

Advancements in omics technologies—such as genomics, transcriptomics, proteomics, metabolomics, and epigenomics—have significantly improved our understanding of melatonin's role in plant–pathogen interactions. Integrating these approaches with computational tools like systems biology and machine learning can provide a comprehensive view of melatonin's regulatory mechanisms. Furthermore, genetic engineering techniques, such as CRISPR-Cas, can be employed to enhance endogenous melatonin production in crops, offering a long-term solution for disease resistance.

The development of cost-effective melatonin-based biostimulants and optimized application methods will also be crucial for practical field implementation. Future research should focus on bridging gaps in epigenetic and metabolomic studies to uncover melatonin's role in plant immune memory and secondary metabolite production. Additionally, large-scale field trials will be essential to validate laboratory findings and assess melatonin's effectiveness across diverse environmental conditions.

By overcoming these challenges and leveraging scientific advancements, melatonin has the potential to become a key player in sustainable agriculture, contributing to improved plant health, reduced pesticide dependency, and enhanced global food security.

References

Aghdam MS, Arnao MB (2024) Phytomelatonin: from intracellular signaling to global horticulture market. J Pineal Res 76(5):e12990

Ahn HR, Kim YJ, Lim YJ, Duan S, Eom SH, Jung KH (2021) Key genes in the melatonin biosynthesis pathway with circadian rhythm are associated with various abiotic stresses. Plan Theory 10(1):129

Ali M, Lamin-Samu AT, Izhar M, Farghal M, Abdul Mateen K, Ibadullah J, Haq SU, Khan A, Gong Z-H, Lu G (2021) Melatonin mitigates the infection of *Colletotrichum gloeosporioides* via modulation of the Chitinase gene and antioxidant activity in Capsicum annuum L. Antioxidants 10(1):7

Ali MA, Nasser MA, Abdelhamid AN, Ali IA, Saudy HS, Hassan KM (2024) Melatonin as a key factor for regulating and relieving abiotic stresses in harmony with phytohormones in horticultural plants—a review. J Soil Sci Plant Nutr 24(1):54–73

Altaf MA, Shahid R, Ren MX, Naz S, Altaf MM, Khan LU, Ahmad P (2022) Melatonin improves drought stress tolerance of tomato by modulating plant growth, root architecture, photosynthesis, and antioxidant defense system. Antioxidants 11(2):309

Arnao MB, Hernández-Ruiz J (2018) Melatonin and its relationship to plant hormones. Ann Bot 121(2):195–207

Arnao MB, Cano A, Hernandez-Ruiz J (2022) Phytomelatonin: an unexpected molecule with amazing performances in plants. J Exp Bot 73:5779–5800

Ayyaz A, Shahzadi AK, Fatima S, Yasin G, Zafar ZU, Athar HUR, Farooq MA (2022) Uncovering the role of melatonin in plant stress tolerance. Theor Exp Plant Physiol 34(3):335–346

Back K, Tan DX, Reiter RJ (2016) Melatonin biosynthesis in plants: multiple pathways catalyze tryptophan to melatonin in the cytoplasm or chloroplasts. J Pineal Res 61:426–437. https://doi.org/10.1111/jpi.12364

Becker S, Fanzo J (2023) Population and food systems: what does the future hold? Popul Environ 45(3):20

Bhardwaj R, Sati H, Oberoi HS, Pareek S (2024) The mechanisms of melatonin in low-temperature stress tolerance in postharvest fruits and vegetables. Future Postharvest Food 1:180

Bhatla SC, Lal MA, Lal MA, Kathpalia R, Sisodia R, Shakya R (2018) Biotic stress. In: Plant physiology, development and metabolism. Springer Nature, pp 1029–1095

Bhowal B, Bhattacharjee A, Goswami K, Sanan-Mishra N, Singla-Pareek SL, Kaur C, Sopory S (2021) Serotonin and melatonin biosynthesis in plants: genome-wide identification of the genes and their expression reveal a conserved role in stress and development. Int J Mol Sci 22(20):11034

Buttar HS, Singh A, Sirari A, Anupam KK, Kumar A, Lal MK, Tiwari RK, Kumar R (2023) Investigating the impact of fungicides and mungbean genotypes on the management of pod rot disease caused by *Fusarium equiseti* and *Fusarium chlamydosporum*. Front Plant Sci 14:1164245

Byeon Y, Back K (2015) Molecular cloning of melatonin 2-hydroxylase responsible for 2-hydroxymelatonin production in rice (*Oryza sativa*). J Pineal Res 58(3):343–351

Byeon Y, Lee HY, Back K (2015) Chloroplastic and cytoplasmic overexpression of sheep serotonin N-acetyltransferase in transgenic rice plants is associated with low melatonin production despite high enzyme activity. J Pineal Res 58(4):461–469

Cao L, Jin X, Zhang Y, Zhang M, Wang Y (2020) Transcriptomic and metabolomic profiling of melatonin-treated soybean (*Glycine max* L.) under drought stress during grain filling period through regulation of secondary metabolite biosynthesis pathways. PLoS One 15:e0239701

Cao Y, Du P, Ji J, He X, Zhang J, Shang Y, Liu H, Xu J, Bowen L (2022) Ionomic combined with transcriptomic and metabolomic analyses to explore the mechanism underlying the effect of melatonin in relieving nutrient stress in apple. Int J Mol Sci 23:9855

Cao H, Yang Q, Wang T, Du T, Song Z, Dong B, Chen T, Wei Y, Xue J, Meng D, Fu Y (2024) Melatonin-mediated CcARP1 alters F-actin dynamics by phosphorylation of CcADF9 to balance root growth and salt tolerance in pigeon pea. Plant Biotechnol J 22(1):98–11

Chaudhry S, Sidhu GPS (2022) Climate change regulated abiotic stress mechanisms in plants: a comprehensive review. Plant Cell Rep 41(1):1–31

Chen X, Laborda P, Liu F (2020) Exogenous melatonin enhances rice plant resistance against *Xanthomonas oryzae* pv. oryzae. Plant Dis 104:1701–1708

Chen L, Ma Y, He T, Chen T, Pan Y, Zhou D, Wang L (2023) Integrated transcriptome and metabolome analysis unveil the response mechanism in wild rice (*Zizania latifolia* griseb.) against sheath rot infection. Front Genet 14:1163464

Cheng L, He Q, Zhang F, Yu J, Li C, Zhao T, Zhang Y, Xie Q, Su B, Lei M, Zhu S, Chen J (2019) Melatonin enhances cotton immunity to Verticillium wilt via manipulating lignin and gossypol biosynthesis. Plant J 100(4):784–800

Debnath B, Islam W, Li M, Sun Y, Lu X, Mitra S, Hussain M, Liu S, Qiu D (2019) Melatonin mediates enhancement of stress tolerance in plants. Int J Mol Sci 20:1040

Devi R, Behera B, Raza MB, Mangal V, Altaf MA, Kumar R, Kumar A, Tiwari RK, Lal MK, Singh B (2021) An insight into microbes mediated heavy metal detoxification in plants: a review. J Soil Sci Plant Nutr 22:1–23

Ding Z, Wu C, Tie W, Yan Y, He G, Hu W (2019) Strand-specific RNA-seq based identification and functional prediction of lncRNAs in response to melatonin and simulated drought stresses in cassava. Plant Physiol Biochem 140:96–104

Duan W, Lu B, Liu L, Meng Y, Ma X, Li J, Zhang K, Sun H, Zhang Y, Dong H, Bai Z, Li C (2022) Effects of exogenous melatonin on root physiology, transcriptome and metabolome of cotton seedlings under salt stress. Int J Mol Sci 23:9456

Duan Y, Wang X, Jiao Y, Liu Y, Li Y, Song Y, Wang L, Tong X, Jiang Y, Wang S, Wang S (2024) Elucidating the role of exogenous melatonin in mitigating alkaline stress in soybeans across different growth stages: a transcriptomic and metabolomic approach. BMC Plant Biol 24(1):380

Fita A, Rodríguez-Burruezo A, Boscaiu M, Prohens J, Vicente O (2015) Breeding and domesticating crops adapted to drought and salinity: a new paradigm for increasing food production. Front Plant Sci 6:978

Fones HN, Bebber DP, Chaloner TM, Kay WT, Steinberg G, Gurr SJ (2020) Threats to global food security from emerging fungal and oomycete crop pathogens. Nat Food 1(6):332–342

Fujiwara T, Maisonneuve S, Isshiki M, Mizutani M, Chen L, Wong HL, Shimamoto K (2010) Sekiguchi lesion gene encodes a cytochrome P450 monooxygenase that catalyzes conversion of tryptamine to serotonin in rice. J Biol Chem 285(15):11308–11313

Gan Y, Tu Z, Yang Y, Cheng L, Wang N, Fan S, Wu C (2024) Enhancing cowpea wilt resistance: insights from gene coexpression network analysis with exogenous melatonin treatment. BMC Plant Biol 24:1–14

Gao S, Ma W, Lyu X, Cao X, Yao Y (2020) Melatonin may increase disease resistance and flavonoid biosynthesis through effects on DNA methylation and gene expression in grape berries. BMC Plant Biol 20:231

Gao YY, Chen HM, Chen DY, Hao GF (2023) Genetic and evolutionary dissection of melatonin response signaling facilitates the regulation of plant growth and stress responses [review]. J Pineal Res 74(2):35

Guo J, Bai Y, Wei Y, Dong Y, Zeng H, Reiter RJ, Shi H (2022) Fine-tuning of pathogenesis-related protein 1 (PR1) activity by the melatonin biosynthetic enzyme ASMT2 in defense response to cassava bacterial blight. J Pineal Res 72(2):e12784

Gupta R (2023) Melatonin: a promising candidate for maintaining food security under the threat of phytopathogens. Plant Physiol Biochem 198:107691

Hattori A, Migitaka H, Iigo M, Itoh M, Yamamoto K, Ohtani-Kaneko R, Hara M, Suzuki T, Reiter RJ (1995) Identification of melatonin in plants and its effects on plasma melatonin levels and binding to melatonin receptors in vertebrates. Biochem Mol Biol Int 35(3):627–634

Hossain ML, Li J (2024) Salinity challenges and adaptive strategies in salinization-affected coastal Bangladesh: implications for agricultural sustainability and water resource management. Environ Res Lett 19(11):114048

Hu Z, Fu Q, Zheng J, Zhang A, Wang H (2020) Transcriptomic and metabolomic analyses reveal that melatonin promotes melon root development under copper stress by inhibiting jasmonic acid biosynthesis. Horti Res 7:79

Hu H, Qiu K, Hao Q, He X, Qin L, Chen L, Yang C, Dai X, Liu H, Xu H, Guo H (2024) Electromagnetic field-assisted frozen tissue planarization enhances MALDI-MSI in plant spatial omics. Anal Chem 96(29):11809–11822

Huang S, Jin S (2024) Melatonin interaction with other phytohormones in the regulation of abiotic stresses in horticultural plants. Antioxidants 13(6):663

Huangfu L, Zhang Z, Zhou Y, Zhang E, Chen R, Fang H, Li P, Xu Y, Yao Y, Zhu M, Yin S (2021) Integrated physiological, metabolomic and transcriptomic analyses provide insights into the roles of exogenous melatonin in promoting rice seed germination under salt stress. Plant Growth Regul 95(1):19–31

Ianas O, Olinescu R, Bădescu I (1991) Melatonin involvement in oxidative processes. Endocrinologie 29:147–153

Jafari M, Shahsavar AR, Talebi M, Hesami M (2022) Exogenous melatonin protects lime plants from drought stress-induced damage by maintaining cell membrane structure, detoxifying ROS and regulating antioxidant systems. Horticulturae 8(3):257

Jagodzik P, Tajdel-Zielinska M, Ciesla A, Marczak M, Ludwikow A (2018) Mitogen-activated protein kinase cascades in plant hormone signaling. Front Plant Sci 9:1387

Junaid MD, Gokce AF (2024) Global agricultural losses and their causes. Bull Biol All Sci Res 2024(1):66–66

Kalia A, Gosal SK (2011) Effect of pesticide application on soil microorganisms. Arch Agron Soil Sci 57(6):569–596

Kalyabina VP, Esimbekova EN, Kopylova KV, Kratasyuk VA (2021) Pesticides: formulants, distribution pathways and effects on human health–a review. Toxicol Rep 8:1179–1192

Kamanga RM, Mndala L (2019) Crop abiotic stresses and nutrition of harvested food crops: a review of impacts, interventions and their effectiveness. Afr J Agric Res 14(3):118–135

Kang K, Kong K, Park S, Natsagdorj U, Kim YS, Back K (2011) Molecular cloning of a plant N-acetylserotoninmethyltransferase and its expression characteristics in rice. J Pineal Res 50(3):304–309

Khan A, Numan M, Khan AL, Lee IJ, Imran M, Asaf S, Al-Harrasi A (2020) Melatonin: awakening the defense mechanisms during plant oxidative stress. Plan Theory 9(4):407

Khondoker M, Mandal S, Gurav R, Hwang S (2023) Freshwater shortage, salinity increase, and global food production: a need for sustainable irrigation water desalination—a scoping review. Earth 4(2):223–240

Kumar S, Chandra R, Behera L, Keswani C, Sansinenea E (2022) Dual Trichoderma consortium mediated elevation of systemic defense response against early blight in potato. Eur J Plant Pathol 162:1–16

Kumar R, Altaf MA, Lal MK, Tiwari RK (2023a) Melatonin in plants: a regulator for plant growth and development. Springer Nature Singapore

Kumar R, Kaundal P, Tiwari RK, Lal MK, Kumari H, Kumar R, Naga KC, Kumar A, Singh B, Sagar V, Sharma S (2023b) Development of reverse transcription recombinase polymerase amplification (RT-RPA): a methodology for quick diagnosis of potato leafroll viral disease in potato. Int J Mol Sci 24(3):2511

Kumar S, Behera L, Kumari R, Bag D, Sowmya V, Keswani C, Minkina T, Bouket AC, Dutta P, Nehela Y, Rohini. (2024a) Unraveling the role of antimicrobial peptides in plant resistance against phytopathogens. Discov Sustain 5(1):243

Kumar S, Arutselvan R, Greeshma K, Bodhankar S, Akash AU, Prasad VSSK, Nehela Y, Singh UB, Bhupenchandra I, Sen A, Rajput LS (2024b) Unraveling the seed bio-priming contours for managing plant health. J Plant Growth Regul:1–50. https://doi.org/10.1007/s00344-024-11277-z

Lal MK, Sharma N, Adavi SB, Sharma E, Altaf MA, Tiwari RK, Kumar R, Kumar A, Dey A, Paul V, Singh B (2022) From source to sink: mechanistic insight of photoassimilates synthesis and partitioning under high temperature and elevated [CO2]. Plant Mol Biol 110(4):305–324

Lee HY, Back K (2017) Melatonin is required for H 2 O 2-and NO-mediated defence signalling through MAPKKK 3 and OXI 1 in *Arabidopsis thaliana*. J Pineal Res 62(2):e12379

Lee HY, Back K (2018) Melatonin plays a pivotal role in conferring tolerance against endoplasmic reticulum stress via mitogen-activated protein kinases and bZIP60 in *Arabidopsis thaliana*. Melatonin Res 1(1):94–108

Lee HY, Byeon Y, Back K (2014) Melatonin as a signal molecule triggering defense responses against pathogen attack in Arabidopsis and tobacco. J Pineal Res 57:262–268

Lerner AB, Case JD, Takahashi Y, Lee TH, Mori W (1958) Isolation of melatonin, the pineal gland factor that lightens melanocyteS1. J Am Chem Soc 80(10):2587–2587

Li S, Wang Y, Gao X, Lan J, Fu B (2022a) Comparative physiological and transcriptome analysis reveal the molecular mechanism of melatonin in regulating salt tolerance in alfalfa (*Medicago sativa* L.). Front Plant Sci 13:919177

Li L, Yan X, Li J, Wu X, Wang X (2022b) Metabolome and transcriptome association analysis revealed key factors involved in melatonin-mediated cadmium-stress tolerance in cotton. Front Plant Sci 13:995205

Ling-Ling Y, Qing-Lun L, Xiao-Yu H, Xing-Lin J, Wang H, Ya-Juan S, Lin-Lin C, Hong-Lian L, Yi-Qing L, Yang X, Shi Y (2023) A cysteine-rich secretory protein involves in phytohormone melatonin mediated plant resistance to CGMMV. BMC Plant Biol 23:1–15

Liu C, Chen L, Zhao R, Li R, Zhang S, Yu W, Sheng J, Shen L (2019) Melatonin induces disease resistance to *Botrytis cinerea* in tomato fruit by activating Jasmonic acid signaling pathway. J Agric Food Chem 67(22):6116–6124

Liu Q, Feng Z, Xu W, Vetukuri RR, Xu X (2021) Exogenous melatonin-stimulated transcriptomic alterations of *Davidia involucrata* seedlings under drought stress. Trees 35:1025–1038

Liu GF, Hu Q, Zhang X, Jiang JF, Zhang Y, Zhang ZX (2022) Melatonin biosynthesis and signal transduction in plants in response to environmental conditions [review]. J Exp Bot 73(17):5818–5827

Liu Y, Tang N, Lin D, Deng W, Li Z (2023) Integration of multi-omics analyses highlights the secondary metabolism response of tomato fruit to low temperature storage. Food Res Int 173:113316

Liu B, Chen W, Geng H, You IX, Qu XJ, Zhou YM (2024a) Melatonin and fungal metabolites synergistically enhance tomato tolerance to Cd and disease resistance by regulating phytohormones. Sci Hortic 327:112845

Liu Z, Ren X, Zhu W, Li Y, Li G, Liu C, Li D, Shi Y, Wang C, Zhu X, Sun H (2024b) Transcriptomic analysis of melatonin-mediated salt stress response in germinating alfalfa. Agriculture 14:661

Lu R, Liu Z, Shao Y, Sun F, Zhang Y, Cui J, Zhou T (2019) Melatonin is responsible for rice resistance to rice stripe virus infection through a nitric oxide-dependent pathway. Virol J 16:1–8

Lu Y, Bu S, Zhao S, Wang N, Xiao J, He F, Gao X (2022) Transcriptome and physiological analysis of increase in drought stress tolerance by melatonin in tomato. PLoS One 17:e0267594

Luo Y, Hu T, Huo Y, Wang L, Zhang L, Yan R (2023) Transcriptomic and physiological analyses reveal the molecular mechanism through which exogenous melatonin increases drought stress tolerance in chrysanthemum. Plan Theory 12:1489

Mandal MK, Suren H, Ward B, Boroujerdi A, Kousik C (2018) Differential roles of melatonin in plant-host resistance and pathogen suppression in cucurbits. J Pineal Res 65(3):1

Mascellani Bergo A, Leiss K, Havlik J (2024) Twenty years of 1H NMR plant metabolomics: a way forward toward assessment of plant metabolites for constitutive and inducible defenses to biotic stress. J Agric Food Chem 72(15):8332–8346

Murch SJ, Erland LA (2021) A systematic review of melatonin in plants: an example of evolution of literature. Front Plant Sci 12:683047

Nadathur SR, Wanasundara JPD, Scanlin L (2017) Proteins in the diet: challenges in feeding the global population. In: Sustainable protein sources. Academic Press, pp 1–19

Nayak L, Lal MK, Tiwari RK, Kumar R, Lal P, Das R, Behera B, Thakur P, Altaf MA, Kumar A (2023) A balancing act: exploring the interplay between HSPs and osmoprotectants in temperature stress responses. S Afr J Bot 162:64–71

Nehela Y, Killiny N (2020) Melatonin is involved in citrus response to the pathogen huanglongbing via modulation of phytohormonal biosynthesis. Plant Physiol 184(4):2216–2239

Ok Jin H, Back K (2022a) Exogenous gibberellin treatment enhances melatonin synthesis for melatonin-enriched rice production. Biomol Ther 12(2):198. https://doi.org/10.3390/biom12020198

Ok Jin H, Back K (2022b) Molecular regulation of antioxidant melatonin biosynthesis by Brassinosteroid acting as an endogenous elicitor of melatonin induction in rice seedlings. Antioxidants 11(5):918

Okant M, Kaya C (2019) The role of endogenous nitric oxide in melatonin-improved tolerance to lead toxicity in maize plants. Environ Sci Pollut Res Int 26:11864–11874

Pan Y, Xu X, Li L, Sun Q, Wang Q, Huang H, Tong Z, Zhang J (2023) Melatonin-mediated development and abiotic stress tolerance in plants. Front Plant Sci 14:1100827

Qi C, Zhang H, Liu Y, Wang X, Dong D, Yuan X, Li X, Zhang X, Li X, Zhang N, Guo Y-D (2020) CsSNAT positively regulates salt tolerance and growth of cucumber by promoting melatonin biosynthesis. Environ Exp Bot 175:104036

Qiao Y, Ren J, Yin L, Liu Y, Deng X, Liu P, Wang S (2020) Exogenous melatonin alleviates PEG-induced short-term water deficiency in maize by increasing hydraulic conductance. BMC Plant Biol 20:218

Qiu CW, Richmond M, Ma Y, Zhang S, Liu W, Feng X, Ahmed IM, Wu F (2024) Melatonin enhances cadmium tolerance in rice via long non-coding RNA-mediated modulation of cell wall and photosynthesis. J Hazard Mater 465:133251

Rastogi M, Verma S, Kumar S, Bharti S, Kumar G, Azam K, Singh V (2023) Soil health and sustainability in the age of organic amendments: a review. Int J Environ Clim Chang 13(10):2088–2102

Ren C, Wang H, Zhou Z, Jia J, Zhang Q, Liang C, Li W, Zhang Y, Yu G (2022a) Genome-wide identification of the B3 gene family in soybean and the response to melatonin under cold stress. Front Plant Sci 13:1091907

Ren W, Chen L, Xie ZM, Peng X (2022b) Combined transcriptome and metabolome analysis revealed pathways involved in improved salt tolerance of *Gossypium hirsutum* L. seedlings in response to exogenous melatonin application. BMC Plant Biol 22:552

Santos CS, Sousa C, Bagheri M, Pinho S, Vasconcelos MW (2024) Predicting iron deficiency and oxidative stress in *Glycine max* through Fourier transform infrared spectroscopy in a time-course experiment. Plant Soil 496(1):161–177

Savary S, Willocquet L (2020) Modeling the impact of crop diseases on global food security. Annu Rev Phytopathol 58(1):313–341

Savary S, Ficke A, Aubertot JN, Hollier C (2012) Crop losses due to diseases and their implications for global food production losses and food security. Food Secur 4(4):519–537

Secomandi E, Senizza B, De Gregorio MA, Miras-Moreno B, Rivero RM, Garcia-Perez P, Lucini L (2025) Phenotyping and metabolomics insights into the effect of melatonin in lettuce under non-stress and salinity conditions. Physiol Plant 177(1):e70055

Shan S, Wang Z, Pu H, Duan W, Song H, Li J, Zhang Z, Xu X (2022) DNA methylation mediated by melatonin was involved in ethylene signal transmission and ripening of tomato fruit. Sci Hortic 291:110566

Sharma M, Sidhu AK, Samota MK, Gupta M, Koli P, Choudhary M (2023) Post-translational modifications in histones and their role in abiotic stress tolerance in plants. Proteomes 11(4):38

Sharma N, Chauhan P, Tapwal A, Kumar S (2023a) Melatonin-mediated regulation of growth, production, and protection in forest plant species. In: Melatonin in plants: a regulator for plant growth and development. Springer Nature Singapore, Singapore, pp 267–290

Shi H, Chen Y, Tan DX, Reiter RJ, Chan Z, He C (2015) Melatonin induces nitric oxide and the potential mechanisms relate to innate immunity against bacterial pathogen infection in Arabidopsis. J Pineal Res 59:102–108

Shi HT, Wei YX, He CZ (2016) Melatonin-induced CBF DREB1s are essential for diurnal change of disease resistance and CCA1 expression in Arabidopsis. Plant Physiol Biochem 100:150–155

Singh DP, Maurya S, Yerasu SR, Bisen MS, Farag MA, Prabha R, Shukla R, Chaturvedi KK, Farooqi MS, Srivastava S, Rai A (2023) Metabolomics of early blight (*Alternaria solani*) susceptible tomato (*Solanum lycopersicum*) unfolds key biomarker metabolites and involved metabolic pathways. Sci Rep 13(1):21023

Singhal RK, Saha D, Skalicky M, Mishra UN, Chauhan J, Behera LP, El Sabagh A (2021) Crucial cell signaling compounds crosstalk and integrative multi-omics techniques for salinity stress tolerance in plants. Front Plant Sci 12:670369

Song W, Wang J, Wang X, Xi J, Cai W, Ma X, Zhang J, Fu B, Gao X (2024) Exogenous melatonin promotes salt tolerance in smooth bromegrass seedlings: physiological, transcriptomic, and metabolomic evidence. Physiol Plant 176(6):e14592

Sun Y, Fan M, He Y (2019) DNA methylation analysis of the *Citrullus lanatus* response to cucumber green mottle mosaic virus infection by whole-genome bisulfite sequencing. Genes 10(5):344

Tan DX, Manchester LC, Esteban-Zubero E, Zhou Z, Reiter RJ (2015) Melatonin as a potent and inducible endogenous antioxidant: synthesis and metabolism [review]. Molecules 20(10):18886–18906

Tang M, Xu L, Wang Y, Dong J, Zhang X, Wang K, Ying J, Li C, Liu L (2021) Melatonin-induced DNA demethylation of metal transporters and antioxidant genes alleviates lead stress in radish plants. Hortic Res 8:124

Teng Z, Yu Y, Zhu Z, Hong SB, Yang B, Zang Y (2021) Melatonin elevated *Sclerotinia sclerotiorum* resistance via modulation of ATP and glucosinolate biosynthesis in *Brassica rapa* ssp. pekinensis. J Proteome 243:104264

Thanh Huyen P, Tian X, Zhao H, Li T, Lu L (2024) Genome-wide characterization of COMT family and regulatory role of CsCOMT19 in melatonin synthesis in *Camellia sinensis*. BMC Plant Biol 24:1–16

Tiwari RK, Lal MK, Naga KC, Kumar R, Chourasia KN, Kumar D, Sharma S (2020) Emerging roles of melatonin in mitigating abiotic and biotic stresses of horticultural crops. Sci Hortic 272:109592

Tiwari RK, Lal MK, Kumar R, Chourasia KN, Naga KC, Kumar D, Das SK, Zinta G (2021a) Mechanistic insights on melatonin-mediated drought stress mitigation in plants. Physiol Plant 172(2):1212–1226

Tiwari RK, Lal MK, Kumar R, Mangal V, Altaf MA, Sharma S, Singh B, Kumar M (2021b) Insight into melatonin-mediated response and signaling in the regulation of plant defense under biotic stress. Plant Mol Biol 109:1–15

Umar OB, Ranti LA, Abdulbaki AS, Bola AL, Abdulhamid AK, Biola MR, Victor KO (2021) Stresses in plants: biotic and abiotic. In: Current trends in wheat research. IntechOpen, pp 1–8

United Nations (2015). https://sdgs.un.org/goals

Vollset SE, Goren E, Yuan CW, Cao J, Smith AE, Hsiao T, Bisignano C, Azhar GS, Castro E, Chalek J, Dolgert AJ (2020) Fertility, mortality, migration, and population scenarios for 195 countries and territories from 2017 to 2100: a forecasting analysis for the Global Burden of Disease Study. Lancet 396(10258):1285–1306

Wang X, Yamaguchi N (2024) Cause or effect: probing the roles of epigenetics in plant development and environmental responses. Curr Opin Plant Biol 81:102569

Wang R, Yang X, Xu H, Li T (2016) Research progress of melatonin biosynthesis and metabolism in higher plants. Plant Physiol J 52(5):615–627

Wang L, Zhou F, Liu X, Zhang H, Yan T, Sun Y, Shi K, Zheng X, Zhu Y, Shan D, Bai Y, Guo Y, Kong J (2022a) ELONGATED HYPOCOTYL 5-mediated suppression of melatonin biosynthesis is alleviated by darkness and promotes cotyledon opening. J Exp Bot 73(14):4941–4953

Wang X, Yang Z, Zhang Y, Cheng F, Xing X, Wen F, Hu Y, Chen C, Wei B, Bai P, Wang X (2022b) Tandem mass tag labeled quantitative proteomic analysis of differential protein expression on total alkaloid of *Aconitum flavum* Hand.-Mazz. against melophagus ovinus. Front Vet Science 9:951058

Wang D, Chen Q, Guo Q, Xia Y, Jing D, Liang G (2023) Metabolomic analyses reveal potential mechanisms induced by melatonin application for tolerance of water deficit in loquat (*Eriobotrya japonica* Lindl.). Sci Hortic 308:111569

Wang Y, Wang J, Guo H, Wu X, Hao M, Zhang R (2023a) Integrative transcriptome and metabolome analysis reveals the mechanism of exogenous melatonin alleviating drought stress in maize roots. Plant Physiol Biochem 199:107723

Watpade S, Naga KC, Pramanick KK, Tiwari RK, Kumar R, Shukla AK, Mhatre PH, Lal MK, Pal D, Manjunatha N (2023) First report of powdery mildew of pomegranate (*Punica granatum*) caused by *Erysiphe punicae* in India. J Plant Dis Prot 130:1–6. [online]

Weeda S, Zhang N, Zhao X, Ndip G, Guo Y, Buck GA, Fu C, Ren S (2014) Arabidopsis transcriptome analysis reveals key roles of melatonin in plant defense systems. PLoS One 9(3):e93462

Wei Y, Hu W, Wang Q, Zeng H, Li X, Yan Y, Reiter RJ, He C, Shi H (2017a) Identification, transcriptional and functional analysis of heat-shock protein 90s in banana (*Musa acuminata* L.) highlight their novel role in melatonin-mediated plant response to Fusarium wilt. J Pineal Res 62(1):e12367

Wei YX, Liu GY, Bai YJ, Xia FY, He CZ, Shi HT (2017b) Two transcriptional activators of acetylserotonin methyltransferase 2 and melatonin biosynthesis in cassava. J Exp Bot 68(17):4997–5006

Wei Y, Chang Y, Zeng H, Liu G, He C, Shi H (2018) RAV transcription factors are essential for disease resistance against cassava bacterial blight via activation of melatonin biosynthesis genes. J Pineal Res 64:1

Wei J, Li DX, Zhang JR, Shan C, Rengel Z, Song ZB, Chen Q (2018a) Phytomelatonin receptor PMTR 1-mediated signaling regulates stomatal closure in *Arabidopsis thaliana*. J Pineal Res 65(2):e12500

Wei J, Liang J, Liu D, Liu Y, Liu G, Wei S (2022) Melatonin-induced physiology and transcriptome changes in banana seedlings under salt stress conditions. Front Plant Sci 13:938262

Wei Y, Xu J, Miao S, Wei K, Peng L, Wang Y, Wei X (2023) Recent advances in the utilization of tea active ingredients to regulate sleep through neuroendocrine pathway, immune system and intestinal microbiota. Crit Rev Food Sci Nutr 63(25):7598–7626

Xie Z, Wang J, Wang W, Wang Y, Xu J, Li Z, Zhao X, Fu B (2020) Integrated analysis of the transcriptome and metabolome revealed the molecular mechanisms underlying the enhanced salt tolerance of rice due to the application of exogenous melatonin. Front Plant Sci 11:618680

Xu R, Wang L, Li K, Cao J, Zhao Z (2022) Integrative transcriptomic and metabolomic alterations unravel the effect of melatonin on mitigating postharvest chilling injury upon plum (cv.Friar) fruit. Postharvest Biol Technol 186:1

Xu YX, Zhang J, Wan ZY, Huang SX, Di HC, He Y, Jin SH (2023) Physiological and transcriptome analyses provide new insights into the mechanism mediating the enhanced tolerance of melatonin-treated rhododendron plants to heat stress. J Integr Agric 22:2397–2411

Xu H, Zhang S, Liang C, Li M, Wang R, Song J, Cui Z, Yang Y, Liu J, Li D (2024a) Melatonin enhances resistance to *Botryosphaeria dothidea* in pear by promoting jasmonic acid and phlorizin biosynthesis. BMC Plant Biol 24(1):470

Xu S, Wang S, Wang Z, Lu Y, Tao T, Huang Q, Lu Z, Wang H, Su Y, Gharib A, Zhou Y, Xu Y, Li P, Chen R, Yang Z (2024b) Integrative analyses of transcriptome, microRNA-seq and metabolome reveal insights into exogenous melatonin-mediated salt tolerance during seed germination of maize. Plant Growth Regul 103:689–704

Xu Y, Wang R, Ma Y, Li M, Bai M, Wei G, Wang J, Feng L (2024c) Metabolite and transcriptome profiling analysis provides new insights into the distinctive effects of exogenous melatonin on flavonoids biosynthesis in *Rosa rugosa*. Int J Mol Sci 25(17):9248

Yand X, Zhang Y, Liu T, Shi J, Qi M, Liu Y, Li T (2022) Integrated physiological, transcriptomic, and proteomic analyses reveal the regulatory role of melatonin in tomato plants' response to low night temperature. Antioxidants 11:2060

Yang LL, Li QL, Han XY, Jiang XL, Wang H, Shi YJ, Shi Y (2023) A cysteine-rich secretory protein involves in phytohormone melatonin mediated plant resistance to CGMMV. BMC Plant Biol 23(1):215

Yang X, Liu D, Liu C, Li M, Yan Z, Zhang Y, Feng G (2024) Possible melatonin-induced salt stress tolerance pathway in *Phaseolus vulgaris* L. using transcriptomic and metabolomic analyses. BMC Plant Biol 24:1–14

Yin Y, Liu Y, Cheng C, Yang Z, Luo Z, Fang W (2021) iTRAQ-based proteomic and physiological analyses of broccoli sprouts in response to exogenous melatonin with ZnSO 4 stress. RSC Adv 11(20):12336–12347

Yu P, Tang X, Chen B, Chen Z, Cui W, Xing Y, Li Y, Zhang F, Barroso JB, Rodriguez LG, Yao Y (2025) The melatonin synthase-encoding gene asmt mediates poplar resistance to drought stress and fungi *Dothiorella gregaria.* Gene 937:13

Zeng H, Bai Y, Wei Y, Reiter RJ, Shi H (2022a) Phytomelatonin as a central molecule in plant disease resistance. J Exp Bot 73(17):5874–5885

Zeng W, Mostafa S, Lu Z, Jin B (2022b) Melatonin-mediated abiotic stress tolerance in plants. Front Plant Sci 13:847175

Zhang HM, Zhang Y (2014) Melatonin: a well-documented antioxidant with conditional pro-oxidant actions. J Pineal Res 57(2):131–146

Zhan Y, Wu T, Zhao X, Wang Z, Chen Y (2021) Comparative physiological and full-length transcriptome analyses reveal the molecular mechanism of melatonin-mediated salt tolerance in okra (*Abelmoschus esculentus* L.). BMC Plant Biol 21:1–16

Zhang N, Sun Q, Zhang H, Cao Y, Weeda S, Ren S, Guo YD (2015) Roles of melatonin in abiotic stress resistance in plants. J Exp Bot 66(3):647–656

Zhang S, Zheng X, Reiter RJ, Feng S, Wang Y, Liu S, Jin L, Li Z, Datla R, Ren M (2017) Melatonin attenuates potato late blight by disrupting cell growth, stress tolerance, fungicide susceptibility and homeostasis of gene expression in *Phytophthora infestans*. Front Plant Sci 8:1993

Zhang Y, Wang J, Chen X, Lu X, Wang D, Wang J, Wang S, Chen C, Guo L, Malik WA, Fan Y (2021) Genome-wide identification and characteristic analysis of the downstream melatonin metabolism gene GhM2H in *Gossypium hirsutum* L. Biol Res 54:1–17

Zhang G, Yan Y, Zeng X, Wang Y, Zhang Y (2022) Quantitative proteomics analysis reveals proteins associated with high melatonin content in barley seeds under NaCl-induced salt stress. J Agric Food Chem 70(27):8492–8510

Zhang Z, Zhang X, Chen Y, Jiang W, Zhang J, Wang J, Wu Y, Wang S, Yang X, Liu M, Zhang Y (2023) Understanding the mechanism of red light-induced melatonin biosynthesis facilitates the engineering of melatonin-enriched tomatoes. Nat Commun 14(1):5525

Zhao C, Yang M, Wu X, Wang Y, Zhang R (2021) Physiological and transcriptomic analyses of the effects of exogenous melatonin on drought tolerance in maize (*Zea mays* L.). Plant Physiol Biochem 168:128–142

Zhao H, Ge Z, Zhou M, Zeng H, Wei Y, Liu G, Yan Y, Reiter RJ, He C, Shi H (2023) Histone deacetylase 9 regulates disease resistance through fine-tuning histone deacetylation of melatonin biosynthetic genes and melatonin accumulation in cassava. J Pineal Res 74(3):e12861

Zou J, Yu H, Yu Q, Jin X, Cao L, Wang M, Wang M, Ren C, Zhang Y (2021) Physiological and UPLC-MS/MS widely targeted metabolites mechanisms of alleviation of drought stress-induced soybean growth inhibition by melatonin. Ind Crop Prod 163:113323

Melatonin and Reactive Species in Alleviating Drought Stress

3

Md. Parvez Kabir, Md. Nasir Uddin, Arumoy Halder, Ariful Islam, and Akbar Hossain

Abstract

Drought stress is one of the major environmental stresses limiting global agricultural productivity. It negatively affects plant growth and development, reducing yields and causing substantial economic losses. Plants have evolved complex mechanisms involving signaling molecules, ROS, and regulatory pathways to cope with drought-induced stress. Notably, melatonin, a bioactive compound, has been identified as an important mediator of plant responses to stress. Melatonin regulates the signaling of reactive species, controls osmotic balance, and reinforces antioxidant defense systems, increasing drought tolerance. In the present study, we delve into the complex interaction between melatonin and ROS signaling in mitigating drought stress in plants. We highlight how melatonin modulates ROS signaling and enhances plant tolerance by activating stress-inducible genes, modulating antioxidative enzymes, and regulating water homeostasis. Moreover, this chapter covers current research on the synergistic effects of melatonin and ROS, leading plants to acclimate to drought stress. Finally, we also discuss the potential role of melatonin-based strategies as novel tools in crop improvement approaches to mitigate the effects of drought stress on future crop

M. P. Kabir · A. Hossain (✉)
Soil Science Division, Bangladesh Wheat and Maize Research Institute, Dinajpur, Bangladesh

M. N. Uddin
Sustainable Agricultural Systems, International Maize and Wheat Improvement Center, Dhaka, Bangladesh

A. Halder
Department of Soil Science, Gazipur Agricultural University, Gazipur, Bangladesh

A. Islam
Jute Agriculture Experimental Station, Bangladesh Jute Research Institute (BJRI), Manikgonj, Bangladesh

R. K. Tiwari et al. (eds.), *Melatonin: Signal Transduction Mechanisms and Defense Networks in Plants*, https://doi.org/10.1007/978-981-96-8869-2_3

improvement practices in drought-prone regions to promote agricultural sustainability.

Keywords
Drought stress · Melatonin · Reactive oxygen species (ROS) · Antioxidant defense · Crop improvement

3.1 Introduction

Drought stress is one of the major environmental stresses that adversely affects plant growth, development, and agricultural productivity worldwide (Begna 2020). As a result of climate change, droughts are occurring with relatively high frequency and severity, making it essential to understand how plants adapt to limited water availability to develop crops with drought tolerance traits and prevent food insecurity. Prime among such adaptive responses is the effective management of reactive oxygen species (ROS) and the adjustment of hormonal signaling pathways involved in the response to drought stress (Devireddy et al. 2021).

Reactive oxygen species (ROS) are highly oxidizing species formed as by-products of cellular metabolism and especially accumulate under stress such as drought (Sachdev et al. 2021). Although reactive oxygen species (ROS) are generally linked to cellular damage and oxidative stress, they can also act as signaling molecules that mediate certain physiological activities (Sies and Jones 2020). Under drought stress conditions, the accumulation of ROS can induce the activation of stress-responsive genes, the generation of antioxidants, and the balance of water levels, which help plants survive (Wu et al. 2022). Excessive ROS production may overwhelm a plant's protection systems, resulting in cellular damage and cell death (Berni et al. 2019).

Melatonin, a naturally occurring bioactive molecule, has emerged with significant prominence in recent years for its potential ability to improve plant stress tolerance, especially during drought (Ahmad et al. 2023). Initially, noted for its role in animals, melatonin has been discovered in diverse plant species, where it is a strong antioxidant and a major modulator of several relevant physiological functions (Zhao et al. 2019). This regulatory action to interact with ROS signaling pathways helps modulate their level of cellular homeostasis to mitigate oxidative damage. Melatonin improves plant drought stress tolerance by balancing the accumulation of reactive oxygen species (ROS) and activating antioxidant defense systems (Altaf et al. 2022).

In brief, the melatonin-ROS-drought signaling cascade in the plant response to drought stress is complicated. On the one hand, melatonin inhibits excess ROS and oxidative stress by acting as an antioxidant. Alternatively, melatonin may also function as a signaling molecule, downregulating stress-associated gene expression, increasing water uptake, and adjusting osmotic pressure (Faizan et al. 2024). Recent studies have indicated that melatonin may play a significant role in activating critical drought resistance-related proteins and pathways, ultimately leading to plant drought tolerance (Zhao et al. 2021). First, the collaboration between melatonin and ROS signaling endows plants with an advanced system to alleviate the detrimental effects of water deficit (Tiwari et al. 2021).

This chapter briefly outlines the current knowledge of the function of melatonin in reducing drought-induced stress, with special attention given to the interspecies involvement of ROS signaling. This study addresses how melatonin modulates ROS generation, the antioxidant defense machinery, and various stress pathways, along with the needed agronomic implications for enhancing crop drought tolerance. We conclude by focusing on strategies to utilize melatonin to improve resilience under the increasing occurrence of drought conditions.

3.2 Plant Growth Regulators in Response to Drought Stress

Drought stress is one of the major abiotic limitations for plant growth and crop yield worldwide. The response of plants to drought is controlled by a complex network of signaling pathways that involve multiple plant growth regulators (PGRs) (Sabagh et al. 2021). These hormones coordinate various physiological and molecular responses to meet plant challenges in coping with water deficit. Some of the most important PGRs are abscisic acid (ABA), gibberellins (GAs), auxins, cytokinins, ethylene, and jasmonic acid (JA). These hormones coordinate particular features of the drought response process, including stomatal control and the production of protective molecules; therefore, their activities often dictate the success of the plant's response to drought stress.

3.2.1 Abscisic Acid (ABA) in Response to Drought Stress

Abscisic acid (ABA) is a well-known master regulator of plant responses to drought stress. It is involved in the early perception of water stress and the induction of adaptive responses (Parwez et al. 2022). ABA modulates stomatal closure to decrease water loss, induces osmotic adjustment to sustain turgor pressure, and initiates the expression of stress-responsive genes, many of which protect cells from oxidative damage (Liu et al. 2022). ABA also promotes the production of protective proteins, such as dehydrins, that stabilize cellular structures when stressed (Sun et al. 2021).

Recent studies have demonstrated that ABA-mediated suppression of ROS is crucial in response to drought stress (Muhammad Aslam et al. 2022). The ROS production induced by ABA can activate antioxidant defense systems such as superoxide dismutase (SOD), catalase (CAT), and peroxidases. In this scenario, ABA also plays a dual role, acting as a stress signal and modulator of oxidative stress, allowing the plant to balance the accumulation of ROS and the expression of antioxidant defense to prevent oxidative damage (Li et al. 2022).

In addition, ABA works interactively with other plant hormones, including ethylene and gibberellins, to tailor the stress response (Parwez et al. 2022). The crosstalk between ABA and ROS has been widely investigated in the field, and recent results indicate that ROS act as secondary messenger molecules of the signaling pathway induced by hormones and allow the adjustment of gene expression and other mechanisms required for adaptation to drought conditions (Li et al. 2022).

3.2.2 Gibberellins (GAs) in Response to Drought Stress

Gibberellins are hormones that promote growth and regulate many processes, such as seed germination, stem elongation, and flowering (Castro-Camba et al. 2022). One of the strategies used by plants to avoid drought stress is reducing the synthesis of gibberellins, which generally occurs during drought stress as part of the plant's strategy to conserve energy and resources (Ilyas et al. 2021). Thus, the increase in GA is prevented, which inhibits overgrowth, and energy is diverted to stress tolerance mechanisms such as antioxidant pathway activation and the synthesis of osmolytes such as proline and glycine betaine.

The antagonistic interaction between gibberellins and ABA has been reported recently. Specifically, GA can repress ABA responses to drought, such as stomatal closure, implying that plants modulate these hormones to maximize stress tolerance (Muhammad Aslam et al. 2022). Furthermore, recent studies have shown that the levels of GAs can be modulated by ROS during drought, acting as intermediary signaling molecules in the crosstalk of hormones (Verma et al. 2019).

3.2.3 Cytokinins in Response to Drought Stress

The response of plants to drought stress is also regulated by cytokinins, which induce cell division and differentiation (Hai et al. 2020). These cytokinin levels are generally downregulated under drought conditions, resulting in reduced growth. However, cytokinins have become more effective in regulating other physiological processes to promote drought tolerance, including stomatal closure, osmotic adjustment, and antioxidant enzyme activity (Hai et al. 2020). This phenomenon has also been linked to the accumulation of osmoprotectants, such as proline, during water deficit to stabilize structures while limiting water stress (Semida et al. 2020).

Recent studies have shown that the interaction between melatonin and cytokinins can improve drought tolerance (Sun et al. 2023). Melatonin also regulates the biosynthesis and signaling pathways of cytokinins, increasing tolerance to drought by reducing transpiration and oxidative damage (Tiwari et al. 2021). Furthermore, cytokinins regulate the expression of genes involved in ROS scavenging, thus playing central roles in the antioxidant defense systems in plants in response to drought stress (Hai et al. 2020).

3.2.4 Auxins in Response to Drought Stress

Auxins are known to regulate growth and development in plants, particularly in elongating roots. Under drought stress conditions, auxins play a role in stimulating root development to reach deeper water sources (Wang et al. 2018). Auxin signaling also modulates plant adaptation to water stress by promoting the expression of stress-responsive proteins and regulating stomatal conductance (Verma et al. 2022).

Recent studies have shown that auxins are known to crosstalk with ROS signaling to mediate plant responses under drought stress (Parwez et al. 2022). Auxins are also regulated by the expression of genes involved in the synthesis of protective molecules, such as heat shock proteins and antioxidant enzymes, which play crucial roles in alleviating oxidative damage under drought conditions (Priya et al. 2019). In addition, interactions among nitric oxide (NO), auxins, and melatonin have been studied, with findings indicating that melatonin may facilitate auxin-related stress responses by stimulating root growth and modulating overall stress tolerance in plants (Zhang et al. 2022) (Fig. 3.1).

3.2.5 Ethylene in Response to Drought Stress

Ethylene is a gaseous plant hormone involved in the control of responses to various abiotic stresses, including drought. Drought conditions often result in the upregulation of ethylene, which can cause senescence, increase leaf abscission, and diminish growth (Nazir et al. 2024). Importantly, ethylene also plays a role in stress acclimatization by mediating the upregulation of stress-response genes, osmotic adjustment, and increased antioxidant defenses (Poór et al. 2022). Ethylene promotes stress tolerance under drought but is also involved in the induction of growth inhibition under prolonged drought conditions (Nazir et al. 2024).

As such, recent findings revealed that melatonin can modulate ethylene biosynthesis and its signaling pathways under drought stress conditions (Sun et al. 2023). The regulation of ethylene production by melatonin restores the balance between growth and stress tolerance in plants, resulting in an extended duration of water-deficit survival (Moustafa-Farag et al. 2020).

3.2.6 Jasmonic Acid (JA) in Response to Drought Stress

JA is a key regulator of plant defense responses and has been implicated in inducing drought tolerance (Wang et al. 2021). JA induces the expression of stress-responsive genes, including antioxidant enzyme-related genes, and enhances the ability of plants to tolerate oxidative injury. JA loopback is also crucial for stomatal closure, especially for reducing transpiration rates and regulating water use efficiency under drought (de Ollas and Dodd 2016).

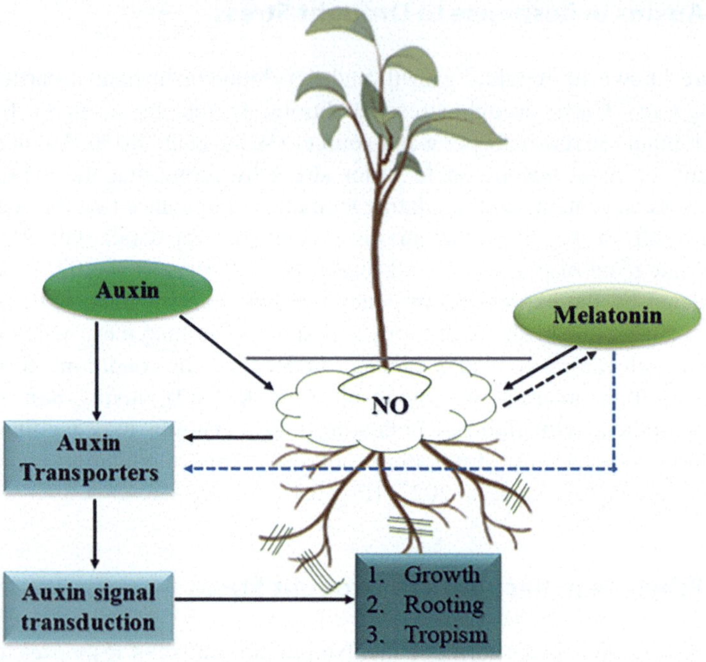

Fig. 3.1 Mutual regulation and balance between melatonin, auxin, and nitric oxide. Nitric oxide (NO) determines root system architecture and activity. The actions of melatonin and auxin are transmitted by different signaling pathways, which ultimately converge at checkpoints such as NO. (Source: Authors' own prepared, unpublished material)

Crosstalk between jasmonic acid (JA) and melatonin (MT) plays an important role in the mediation of drought tolerance (Luo et al. 2023). In addition, JA biosynthesis is regulated by melatonin, resulting in stronger antioxidant defense systems, which contribute to the enhancement of drought stress tolerance (Imran et al. 2021). Moreover, the increase in ROS production and antioxidant enzyme activity through the synergistic regulation of JA and melatonin contributes to elucidating the role of distinct signaling pathways in alleviating drought stress (Tiwari et al. 2021).

In summary, the plant response to drought stress is a complex process involving a network of plant growth regulators (PGRs). These hormones work together to control water retention, induce stress-responsive genes, and reduce oxidative damage. Plant growth regulators such as ABA, GAs, auxins, cytokinins, ethylene, and JA integrate with reactive oxygen species (ROS) as hubs for plant adaptation to drought. New research holds great promise, providing insight into how these hormones work with melatonin to improve drought tolerance. By modulating ROS signaling and crosstalk with various hormonal pathways, melatonin confers a mechanism that plants utilize to counter drought stress, thus representing a target for improving crop resilience.

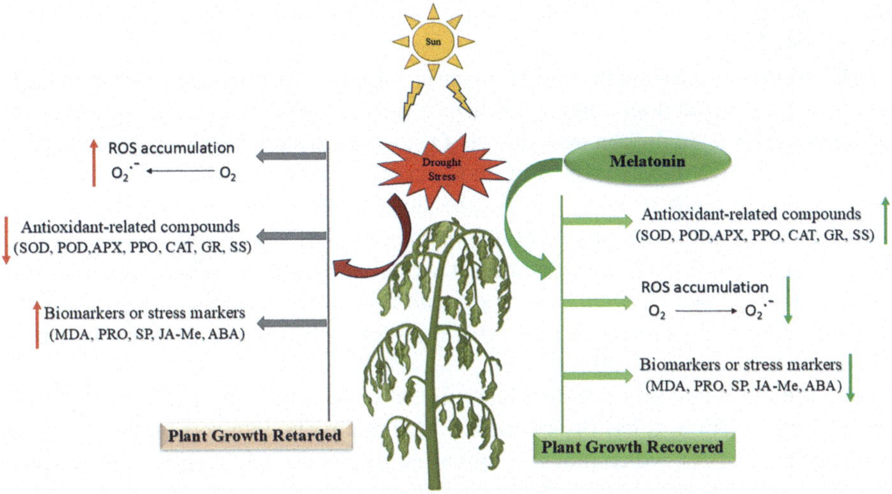

Fig. 3.2 Schematic diagram of the mechanism of melatonin action. Potential mechanism by which melatonin protects plants from drought stress. *SOD* superoxide dismutase, *POD* peroxidase, *APX* ascorbate peroxidase, *PPO* polyphenol oxidase, *CAT* catalase, *GR* glutathione reductase, *MDA* malondialdehyde, *PRO* proline, *SS* soluble sugar, *SP* soluble protein, *JA-Me* jasmonic acid methyl ester, *ABA* abscisic acid, ↑, upregulated; ↓, downregulated. Source: Authors' own prepared, unpublished material

3.3 Melatonins' Mechanisms for Drought Stress Reduction

Drought stress is a major environmental factor that negatively affects the yield of crops. It triggers a series of physiological and biochemical responses characterized by the inhibition of plant growth and development, including a reduction in photosynthesis, oxidative stress, and disruption of hormonal balance. Melatonin (N-acetyl-5-methoxytryptamine) is a small molecule that plays a significant role in promoting plant tolerance to drought conditions (Song et al. 2022). Melatonin is a bioactive compound with antioxidant activities that regulates multiple physiological processes to resist the adverse effects of drought (Imran et al. 2021). The mechanistic pathways by which melatonin relieves drought stress are quite complex and involve the regulation of ROS, antioxidant systems, water equilibrium, and hormonal signaling, as illustrated in Fig. 3.2.

3.3.1 Antioxidant Defense and ROS Scavenging

Melatonin ameliorates drought stress primarily by regulating reactive oxygen species (ROS) levels (Khan et al. 2019). ROS are highly reactive molecules that can accumulate in plant cells when exposed to various stress conditions, potentially causing oxidative damage to cellular components, including lipids, proteins, and nucleic acids. Oxidative stress can damage cellular integrity, affect metabolic processes, and decrease plant growth and production (Kerchev and Van Breusegem 2022).

Melatonin has strong antioxidant potential and directly quenches excess ROS. It serves as an electron donor and neutralizes highly reactive species such as superoxide anions (O_2•⁻), hydrogen peroxide (H_2O_2), and hydroxyl radicals (OH•), thereby blocking oxidative damage. In addition to acting as a direct scavenger of free radicals, melatonin tends to induce the expression of important antioxidant enzymes, such as superoxide dismutase (SOD), catalase (CAT), and peroxidase (POD), which can detoxify ROS and counteract oxidative stress in cells (Hacışevki and Baba 2018)). Under drought, melatonin increases the expression of antioxidant enzymes through its protective role, which improves the tolerance of plants to oxidative stress (Imran et al. 2021).

In addition, melatonin also mediates stress-related signaling pathways to regulate ROS production. It has been demonstrated that melatonin acts on the hormone abscisic acid (ABA), which plays an important role in drought response, to regulate oxidative stress-induced ROS production during drought stress and balance the plant antioxidant system (Hu et al. 2021). Melatonin is an endogenous regulator of ROS signaling under drought stress and has been implicated in establishing drought tolerance.

3.3.2 Regulation of Water Homeostasis and Osmotic Adjustment

Water loss through transpiration is one of the main consequences of drought stress, and maintaining water balance is critical for plant survival under water-deficient conditions. Melatonin helps maintain water homeostasis by inducing osmotic adjustment and increasing water use efficiency (Jahan et al. 2023).

Among the main properties of melatonin in water homeostasis is its effect on the behavior of stomata. When plants are under drought stress, their stomata are closed to save water, but this prevents carbon dioxide from entering the plant for photosynthesis. Melatonin is significantly involved in stomatal closure by interacting with ABA, a known inducer of stomatal closure, and is synthesized by plants under drought stress (Jensen et al. 2023). Melatonin regulates ABA signaling pathways, inducing stomatal closure, thus preventing the loss of water from the plant with no marked impairment in the rates of photosynthesis. This facilitates the regulation of stomatal conductance, contributing to plant drought stress adaptation by increasing water-use efficiency.

Moreover, melatonin enhances osmotic adjustment by upregulating the biosynthesis of compatible solutes (e.g., proline and soluble sugars) (Nie et al. 2023). These osmoprotectants assist in restoring cell structures to their normal state under dehydrating conditions while also preventing cellular turgor pressure and protecting enzymes from heat denaturation. Melatonin treatment has been demonstrated to increase the levels of many osmolytes, including proline, resulting in increased drought tolerance (Gul et al. 2022). Melatonin, therefore, enables plants to adapt to water-deficient conditions by regulating the synthesis of osmoprotectants and increasing water retention.

3.3.3 Adverse Impacts on the Photosynthetic System

Photosynthetic activity is negatively impacted by drought stress through damage to photosystem II (PSII) function and chlorophyll degradation (Liu et al. 2018). This decrease in photosynthetic capacity reduces the overall energy availability of the plant, making it less able to withstand stress. Melatonin has subsequently been implicated in protecting photosynthesis under drought stress by stabilizing the photosynthetic apparatus.

Melatonin has been reported to function as an antioxidant, which helps protect chlorophyll and thylakoid membranes from oxidative damage induced by ROS (Khan et al. 2019). Melatonin treatment preserves the chlorophyll content, maintains excellent nonphotochemical quenching efficiency of photosystem II, and improves overall photosynthetic activity under drought conditions (Huang et al. 2019). Melatonin reduces oxidative damage to the photosynthetic machinery, allowing energy production to be maintained in water-deficient conditions and allowing normal growth and survival.

3.3.4 Gene Expression and Stress Response-Related Protein Regulation

The molecular response of plants to drought stress is crucial for their survival. Melatonin can regulate the expression of drought-associated genes, including several stress-related genes that promote plant activation of protective mechanisms against drought (Ma et al. 2024). Melatonin is also known to regulate the expression of genes that play roles in osmotic adjustment, oxidative defense, and stress response.

Melatonin enhances the expression of dehydrins (proteins that protect plant cells during dehydration) (Chen et al. 2021). Dehydrins protect cellular membranes and proteins against damage during repeated cycles of dehydration and rehydration. Melatonin also modulates the expression of genes encoding osmoprotectants, such as proline and trehalose, which are vital for proper cellular function under water-deficient conditions (Qiao et al. 2020).

Moreover, melatonin modulates the activation of mitogen-activated protein kinase (MAPK) signaling pathways, which play pivotal roles in the expression of stress-responsive genes (Maity et al. 2022). Melatonin enhances the expression of genes critical for plant survival under drought stress, such as those involved in antioxidant defense, stress tolerance, and water conservation.

3.3.5 Root System Architecture

A well-developed root system is crucial for plants to utilize water from deeper soil layers during drought conditions. Melatonin has also been shown to promote root growth and development, enhancing plant water and nutrient uptake. Under drought stress, melatonin promotes both root elongation and root biomass (Liang et al. 2019). These changes in root system architecture help plants extract more water from the soil and increase drought tolerance.

Melatonin-mediated interactions with auxin signaling are believed to underline the effects of melatonin on root development. Melatonin is a plant hormone that enhances root growth and can promote auxin signals during drought stress to stimulate root elongation and branching (Zhang et al. 2022). Melatonin improves root system architecture and further assists plants in responding to limited water availability.

3.3.6 Synergistic Interactions with Other Hormones

The mechanism by which melatonin alleviates drought stress is closely associated with its crosstalk with other plant hormones, such as abscisic acid (ABA), ethylene, and jasmonic acid (JA). These hormones govern multiple aspects of the plant's stress response, from stomatal closure to antioxidant defense and gene expression (Sun et al. 2023).

Melatonin promotes ABA signaling, which is important for controlling stomata and stress-inducible genes. Under drought, melatonin and ABA interact synergistically to enable plants to conserve water while maintaining the process of photosynthesis (Ali et al. 2023). Moreover, melatonin also regulates the ethylene and JA signaling pathways induced during stress conditions, thus improving plant resilience against drought stress (Dzinyela et al. 2024).

In summary, melatonin plays a variety of roles in alleviating drought stress mechanisms in plants. Melatonin modulates ROS signaling, increases antioxidant activity, facilitates water conservation, mitigates damage to the photosynthetic apparatus, and optimizes root system architecture. Moreover, it also exerts its regulatory effects through its cross-regulatory interactions with other phytohormones, including ABA, ethylene, and JA, all of which are involved in a coordinated response to drought stress. With these various mechanisms in mind, melatonin holds great promise as a biostimulant that may increase crop drought tolerance and ultimately prove to be a useful agricultural tool in areas burdened with drought conditions.

3.4 Wider Agricultural Applications and Implications

Crops under drought conditions have reduced growth and yield. Thus, melatonin might be a low-cost supplement to mitigate drought stress and improve plant drought tolerance (DT) to increase agricultural production (Sadak et al. 2020). It is an endogenous potent antioxidant that is crucial for alleviating oxidative damage, enhancing water-use efficiency, and improving plant growth. The efficacy of melatonin in enhancing drought tolerance across various crops has been demonstrated through multiple studies, which highlights its role in improving plant growth, antioxidant activity, osmoprotection, and water-use efficiency under drought stress conditions, as summarized in Table 3.1.

3.4.1 Melatonin, a Biostimulant in Crop Management

One of the most attractive uses of melatonin in agriculture is as a biostimulant for enhancing crop drought tolerance. Biostimulants are natural or synthetic compounds that stimulate plant development and resilience against stress but do not need to act as fertilizers or pesticides. Owing to its ability to act as a biostimulant, melatonin has been applied in the agricultural environment under drought stress, as it increases the activity of antioxidant defense systems, modulates stomatal closure, promotes root growth, and regulates the accumulation of osmoprotectants such as proline (Sharma et al. 2024). Treatment with melatonin improves the yield of crops grown under water-deficient conditions, thereby becoming a promising biostimulant for crops growing in drought-prone areas, as supported by several studies (Ullah et al. 2024).

Molecular and field applications of melatonin have shown that it significantly enhances drought tolerance in many crops, including wheat, maize, rice, and soybean. For example, foliar applications of melatonin promote seed germination, increase the photosynthetic capacity, and increase root growth under drought-stress conditions (Yasmeen et al. 2022). More importantly, melatonin can be used by soil amendments or irrigation systems to better integrate into farming practices. Thus, melatonin-based biostimulants provide an efficient, low-cost, eco-friendly way to alleviate drought stress and increase plant growth in water-limited areas.

3.4.2 Combining Other Stress Tolerance Technologies

Melatonin provides distinct advantages in terms of drought tolerance, yet there are some potentially synergetic effects of melatonin in conjunction with other stress-tolerant technologies. The combination of melatonin with other biostimulants, namely, humic substances, seaweed extracts, and PGRs, can also impart synergistic effects that improve plant tolerance to a range of abiotic stresses, including drought, salinity, and extreme temperatures (Dias et al. 2022). The combination of melatonin with plant growth regulators, including abscisic acid (ABA) or cytokinins, may be a holistic approach to improve water retention, enhance osmotic balance, and ameliorate oxidative damage.

In addition, combining melatonin with novel agricultural strategies, including precision agriculture and controlled-environment agriculture, may facilitate the use of biostimulants in drought management. Precision agriculture advances in sensor

Table 3.1 Roles of melatonin in drought stress tolerance

Common name	Scientific name	Drought treatment	Melatonin treatment		Effects	References
			Concentration[a]	Application form		
Arabidopsis	*Arabidopsis thaliana*	Water withholding (21 d)	50 µM	Supplemented with nutrient solution	Stress-responsive genes and soluble sugars were enhanced	Shi et al. (2015)
Rice	*Oryza sativa*	Water drainage from vessels(5d)	100 µM	Pretreatment in growing distilled water	Plant growth, osmoprotectants proline, stress-responsive genes, and mitochondrial structure were enhanced; ROS and electroleakage were decreased	Lee and Back (2016)
Maize	*Zea mays*	40–45% field capacity (50d)	50 µM (foliar spray) and 100 µM (soil drench)	Foliar application and soil treatment	Photosynthesis and antioxidant enzymes were increased; ROS was decreased	Ahmad et al. (2019)
Wheat	*Triticum aestivum*	20% PEG 6000 (7d)	10 and100 µM (variety-dependent)	Seeds treatment	Germination percentage, germination index, germination potential, radicle length and number, plumule length, and lysine (germination-related amino acid) were enhanced	Li et al. (2019a, b, c)

Common name	Scientific name	Drought treatment	Melatonin treatment		Effects	References
			Concentration[a]	Application form		
Tartary Buckwheat	*Fagopyrum tataricum*	20% field capacity (15d)	100 μM	Foliar application	Water status, osmoprotection, secondary metabolites, antioxidant enzymes, and photosynthesis were increased; ROS was decreased	Hossain et al. (2020)
Barley	*Hordeum vulgare*	(Combined drought and cold)	1 mM	Foliar or soil application	Endogenous melatonin, ABA, water status, antioxidants, photosynthesis, and PSII efficiency were increased	Li et al. (2016)
Soybean	*Glycine max*	15% PEG6000 (3d)	100 μM	Foliar and root application	Plant growth and flowering, seed yield, gas exchange, PSII efficiency, and antioxidant enzymes were increased; MDA was decreased	Zhang et al. (2019)

(continued)

Table 3.1 (continued)

Common name	Scientific name	Drought treatment	Melatonin treatment		Effects	References
			Concentration[a]	Application form		
Cotton	*Gossypium hirsutum*	10% PEG6000 (7d)	100 μM	Seeds presoaking	Number and opening of stomata in cotton testa, germination parameters, antioxidant enzymes, osmoprotection, and GA3 were enhanced; ABA, ROS, and MDA were decreased	Bai et al. (2020)
Apple	*Malus domestica*	50% field capacity (3 months with sampling every month)	100 μM	Soil application	Chlorophyll, photosynthesis, photoprotection, antioxidant enzymes, GSH, and AsA contents were increased; oxidative damage, leaf senescence, senescence-associated gene12, pheophorbidea oxygenase-related gene, and ROS were decreased	Wang et al. (2013)
Grape	*Vitis vinifera*	Water withholding (18 d)	100 μM	Supplemented with irrigation	MDA and relative conductivity were decreased	Niu et al. (2019)

			Melatonin treatment			
Common name	Scientific name	Drought treatment	Concentration[a]	Application form	Effects	References
Kiwifruit	*Actinidia chinensis*	Water withholding (9 days)	100 μM	Irrigation pretreatment	Water-holding capacity, antioxidant enzyme-related genes, and GSH-AsA cycle-related genes were increased; ROS and MDA were decreased	Xia et al. (2020)
Tomato	*Solanum lycopersicum*	10% PEG (7d)	200 μM	Foliar application	Chlorophyll, p-coumaric acid content, and antioxidant enzymes were increased; MDA was decreased	Karaca and Cekic (2019)
Watermelon	*Citrullus lanatus*	Water withholding (4d)	150 μM	Root pretreatment	Wax accumulation and melatonin-ABA crosstalk were enhanced	Li et al. (2019a, b, c)
Coffee	*Coffea arabica*	40% of max moisture retention capacity(21d)	300 μM	Soil application	Root vigor, photoprotection, gas exchange, carboxylation efficiency, chlorophyll, and antioxidants were enhanced; MDA was decreased	Campos et al. (2019)

(continued)

Table 3.1 (continued)

Common name	Scientific name	Drought treatment	Melatonin treatment		Effects	References
			Concentration[a]	Application form		
Tea	*Camellia sinensis*	20% PEG6000 (2d)	100 μM	Foliar application pretreatment	Photosynthesis, GSH and AsA contents, antioxidant enzymes, and antioxidant enzyme-related genes were enhanced; ROS and MDA were decreased	Li et al. (2019a, b, c)
Tobacco	*Nicotiana benthamiana*	Water withholding (6d)	10 μM	Foliar application	MDA decreased; drought tolerance was enhanced	Lee and Back (2019)

[a] Only those doses of exogenous melatonin that had a greater positive impact on plant tolerance against drought stress were selected; these doses were increased or decreased compared with those of the control. *ROS* reactive oxygen species, *PSII* photosystem II, *GSH* glutathione, *AsA* ascorbate, *MDA* malondialdehyde, *ABA* abscisic acid, *GA3* gibberellic acid, *SOD* superoxide dismutase, *POD* peroxidase, *TSS* total soluble sugar, *PEG* polyethylene glycol

technologies and data analytics can assist farmers in monitoring factors such as soil moisture levels, plant health, and environmental stress factors that impact melatonin dynamics and help them apply melatonin precisely to plants at critical times. This may improve the cost effectiveness and environmental benefits of melatonin for enhancing drought tolerance.

3.4.3 Genetic Engineering for Increased Drought Resistance in Crops

In addition to its use as a biostimulant, melatonin can also be genetically manipulated to improve crop drought tolerance. Genetic modifications improve melatonin biosynthesis to cultivate crops that are tolerant to water stress for a long period of time (Moustafa-Farag et al. 2020). Several key enzymes responsible for melatonin biosynthesis, including tryptophan decarboxylase and serotonin N-acetyltransferase, have been identified in plants, and these pathways provide targets to increase endogenous melatonin levels in plants.

Selecting transgenic crops that can produce relatively high levels of melatonin can effectively help with drought tolerance through improved regulation of ROS homeostasis, thus maintaining cellular structural integrity, improving root growth, and optimizing the absorption of vital solutes (Zuo et al. 2014). For regions plagued with chronic water scarcity, this genetic tactic can be a great way for farmers to produce drought-tolerant crops; ensuring food security remains a concern as climates change. However, the commercial development of genetically modified (GM) crops with modified melatonin levels must address regulatory and public acceptance issues.

3.4.4 Policy Recommendations, Comparisons with Previous Studies, and Future Directions

Such policies for the research, development, and commercialization of melatonin must be established for melatonin to gain ground in agricultural practice. Governments and agricultural agencies should further research the molecular mechanisms of melatonin in terms of drought tolerance and other plant hormones. Additionally, regulations need to be developed so that melatonin-based products are safe, effective, and accessible for farmers in developed and developing countries.

Melatonin is not limited to its protective effect under different environment-stressing conditions; researchers have also attempted to investigate its interaction. These findings provide an advanced understanding of how melatonin may interact with other plant stresses, offering a more integrated understanding for developing plants that are more tolerant to varying environmental conditions. Furthermore, the role of melatonin in conferring greater tolerance to biotic constraints (pest or disease) may enable advanced crop management.

It is also important to form education and training programs addressing farmers' use of melatonin-based biostimulants, especially in areas where drought is a repeated challenge. Since farmers may not recognize these advantages, extension services and workshops are essential for raising awareness and advising farmers on the adoption of melatonin applications in their farming systems.

The multifaceted application of melatonin as a biostimulant in agriculture is a potential strategy for alleviating drought stress and increasing crop yield in water-deficient areas. This is now feasible and renewable and can be accomplished through integrative methods such as supplementing melatonin with other technologies that protect the plant from stresses, developing new genetic engineering approaches, and creating supportive policies, making melatonin a unique commodity in sustainable agriculture. With the constant increase in the world population and climate change contributing to limited access to water, the ability of melatonin to combat drought stress in crop plants will be crucial for successfully sustaining food security and global agricultural methodologies in the future.

3.5 Conclusion

Melatonin is a key regulator that helps increase stress tolerance in plants, especially under drought conditions. The alleviation of drought stress is also attributed to its potent antioxidant activity and its ability to modulate ROS signals and physiological processes such as water homeostasis, osmotic adjustment, and photosynthesis. Under drought stress, melatonin scavenges ROS to protect plant cells from oxidative damage and activates the antioxidant defense system. It also controls stomatal closure, increasing water use efficiency and the accumulation of compatible solutes to minimize the effects of water loss. Moreover, melatonin maintains the chlorophyll content and stabilizes the photosynthetic apparatus, thus sustaining the operation of photosystem II and energy production under drought stress. It further regulates the expression of stress-related genes and interacts with other plant hormones in the ecological process by increasing the ability of adaptive organisms to water scarcity. Melatonin also enhances root growth and system architecture, allowing plants to tap into deeper water reserves during drought. The evidence suggests that melatonin can alleviate plant drought stress. The regulation of stress-responsive pathways by melatonin results in more drought-resistant crops capable of overcoming extended periods of dehydration. Melatonin biostimulants provide promising opportunities to increase crop productivity and food security in water-limited regions. Further studies are needed to examine the molecular mechanisms underlying the impact of melatonin on the stress response of plants and assess the possibility of applying melatonin on a wider agricultural scale.

References

Ahmad S, Kamran M, Ding R, Meng X, Wang H, Ahmad I et al (2019) Exogenous melatonin confers drought stress by promoting plant growth, photosynthetic capacity and antioxidant defense system of maize seedlings. Peer J 7:e7793

Ahmad N, Naeem M, Ali H, Alabbosh KF, Hussain H, Khan I et al (2023) From challenges to solutions: the impact of melatonin on abiotic stress synergies in horticultural plants via redox regulation and epigenetic signalling. Sci Hortic 321:112369

Ali M, Pan Y, Liu H, Cheng Z (2023) Melatonin interaction with abscisic acid in the regulation of abiotic stress in Solanaceae family plants. Front Plant Sci 14:1271137

Altaf MA, Shahid R, Ren MX, Naz S, Altaf MM, Khan LU et al (2022) Melatonin improves drought stress tolerance of tomato by modulating plant growth, root architecture, photosynthesis, and antioxidant defense system. Antioxidants 11(2):309

Bai Y, Xiao S, Zhang Z, Zhang Y, Sun H, Zhang K et al (2020) Melatonin improves the germination rate of cotton seeds under drought stress by opening pores in the seed coat. Peer J 8:e9450

Begna T (2020) Effects of drought stress on crop production and productivity. Int J Res Stud Agric Sci 6(9):34–43

Berni R, Luyckx M, Xu X, Legay S, Sergeant K, Hausman JF et al (2019) Reactive oxygen species and heavy metal stress in plants: impact on the cell wall and secondary metabolism. Environ Exp Bot 161:98–106

Campos CN, Ávila RG, de Souza KRD, Azevedo LM, Alves JD (2019) Melatonin reduces oxidative stress and promotes drought tolerance in young *Coffea arabica* L. plants. Agric Water Manag 211:37–47

Castro-Camba R, Sánchez C, Vidal N, Vielba JM (2022) Plant development and crop yield: the role of gibberellins. Plan Theory 11(19):2650

Chen Z, Jia W, Li S, Xu J, Xu Z (2021) Enhancement of *Nicotiana tabacum* resistance against dehydration-induced leaf senescence via metabolite/phytohormone-gene regulatory networks modulated by melatonin. Front Plant Sci 12:686062

de Ollas C, Dodd IC (2016) Physiological impacts of ABA–JA interactions under water-limitation. Plant Mol Biol 91:641–650

Devireddy AR, Zandalinas SI, Fichman Y, Mittler R (2021) Integration of reactive oxygen species and hormone signalling during abiotic stress. Plant J 105(2):459–476

Dias MC, Araújo M, Silva S, Santos C (2022) Sustainable olive culture under climate change: the potential of biostimulants. Horticulturae 8(11):1048

Dzinyela R, Hwarari D, Opoku KN, Yang L, Movahedi A (2024) Enhancing drought stress tolerance in horticultural plants through melatonin-mediated phytohormonal crosstalk. Plant Cell Rep 43(11):272

Faizan M, Sultan H, Alam P, Karabulut F, Cheng SH, Rajput VD et al (2024) Melatonin and its cross-talk with other signalling molecules under abiotic stress. Plant Stress 11:100410

Gul N, Haq ZU, Ali H, Munsif F, Hassan SSU, Bungau S (2022) Melatonin pretreatment alleviated inhibitory effects of drought stress by enhancing antioxidant activities and accumulation of higher proline and plant pigments and improving maize productivity. Agronomy 12(10):2398

Hacışevki A, Baba B (2018) An overview of melatonin as an antioxidant molecule: a biochemical approach. IntechOpen

Hai NN, Chuong NN, Tu NHC, Kisiala A, Hoang XLT, Thao NP (2020) Role and regulation of cytokinins in plant response to drought stress. Plan Theory 9(4):422

Hossain MS, Li J, Sikdar A, Hasanuzzaman M, Uzizerimana F, Muhammad I et al (2020) Exogenous melatonin modulates the physiological and biochemical mechanisms of drought tolerance in tartary buckwheat (*Fagopyrum tataricum* (L.) Gaertn). Molecules 25(12):2828

Hu W, Zhang J, Yan K, Zhou Z, Zhao W, Zhang X et al (2021) Beneficial effects of abscisic acid and melatonin in overcoming drought stress in cotton (*Gossypium hirsutum* L.). Physiol Plant 173(4):2041–2054

Huang B, Chen YE, Zhao YQ, Ding CB, Liao JQ, Hu C et al (2019) Exogenous melatonin alleviates oxidative damages and protects photosystem II in maize seedlings under drought stress. Front Plant Sci 10:677

Ilyas M, Nisar M, Khan N, Hazrat A, Khan AH, Hayat K et al (2021) Drought tolerance strategies in plants: a mechanistic approach. J Plant Growth Regul 40:926–944

Imran M, Latif Khan A, Shahzad R, Aaqil Khan M, Bilal S, Khan A et al (2021) Exogenous melatonin induces drought stress tolerance by promoting plant growth and antioxidant defence system of soybean plants. AoB Plants 13(4):plab026

Jahan MS, Zhao CJ, Shi LB, Liang XR, Jabborova D, Nasar J, Zhou XB (2023) Physiological mechanism of melatonin attenuating to osmotic stress tolerance in soybean seedlings. Front Plant Sci 14:1193666

Jensen NB, Ottosen CO, Zhou R (2023) Exogenous melatonin alters stomatal regulation in tomato seedlings subjected to combined heat and drought stress through mechanisms distinct from ABA signalling. Plan Theory 12(5):1156

Karaca P, Cekic FÖ (2019) Exogenous melatonin-stimulated defense responses in tomato plants treated with polyethylene glycol. Int J Veg Sci 25(6):601–609

Kerchev PI, Van Breusegem F (2022) Improving oxidative stress resilience in plants. Plant J 109(2):359–372

Khan MN, Zhang J, Luo T, Liu J, Rizwan M, Fahad S et al (2019) Seed priming with melatonin coping drought stress in rapeseed by regulating reactive oxygen species detoxification: antioxidant defense system, osmotic adjustment, stomatal traits and chloroplast ultrastructure perseveration. Ind Crop Prod 140:111597

Lee HJ, Back K (2016) 2-Hydroxymelatonin promotes the resistance of rice plant to multiple simultaneous abiotic stresses (combined cold and drought). J Pineal Res 61(3):303–316

Lee HJ, Back K (2019) 2-Hydroxymelatonin confers tolerance against combined cold and drought stress in tobacco, tomato, and cucumber as a potent anti-stress compound in the evolution of land plants. Melatonin Res 2(2):35–46

Li X, Tan DX, Jiang D, Liu F (2016) Melatonin enhances cold tolerance in drought-primed wild-type and abscisic acid-deficient mutant barley. J Pineal Res 61(3):328–339

Li D, Zhang D, Wang H, Li H, Song S, Li H, Li R (2019a) Effects of melatonin on germination and amino acid content in different wheat varieties seeds under polyethylene glycol stress. BioRxiv 710954

Li H, Mo Y, Cui Q, Yang X, Guo Y, Wei C et al (2019b) Transcriptomic and physiological analyses reveal drought adaptation strategies in drought-tolerant and-susceptible watermelon genotypes. Plant Sci 278:32–43

Li J, Yang Y, Sun K, Chen Y, Chen X, Li X (2019c) Exogenous melatonin enhances cold, salt and drought stress tolerance by improving antioxidant defense in tea plant (*Camellia sinensis* (L.) O. Kuntze). Molecules 24(9):1826

Li S, Liu S, Zhang Q, Cui M, Zhao M, Li N et al (2022) The interaction of ABA and ROS in plant growth and stress resistances. Front Plant Sci 13:1050132

Liang D, Ni Z, Xia H, Xie Y, Lv X, Wang J et al (2019) Exogenous melatonin promotes biomass accumulation and photosynthesis of kiwifruit seedlings under drought stress. Sci Hortic 246:34–43

Liu J, Guo YY, Bai YW, Camberato JJ, Xue JQ, Zhang RH (2018) Effects of drought stress on the photosynthesis in maize. Russ J Plant Physiol 65:849–856

Liu H, Song S, Zhang H, Li Y, Niu L, Zhang J, Wang W (2022) Signalling transduction of ABA, ROS, and Ca2+ in plant stomatal closure in response to drought. Int J Mol Sci 23(23):14824

Luo M, Wang D, Delaplace P, Pan Y, Zhou Y, Tang W et al (2023) Melatonin enhances drought tolerance by affecting jasmonic acid and lignin biosynthesis in wheat (*Triticum aestivum* L.). Plant Physiol Biochem 202:107974

Ma Y, Tang M, Wang M, Yu Y, Ruan B (2024) Advances in understanding drought stress responses in Rice: molecular mechanisms of ABA Signalling and breeding prospects. Genes 15(12):1529

Maity S, Guchhait R, Pramanick K (2022) Melatonin mediated activation of MAP kinase pathway may reduce DNA damage stress in plants: a review. Biofactors 48(5):965–971

Moustafa-Farag M, Mahmoud A, Arnao MB, Sheteiwy MS, Dafea M, Soltan M et al (2020) Melatonin-induced water stress tolerance in plants: recent advances. Antioxidants 9(9):809

Muhammad Aslam M, Waseem M, Jakada BH, Okal EJ, Lei Z, Saqib HSA et al (2022) Mechanisms of abscisic acid-mediated drought stress responses in plants. Int J Mol Sci 23(3):1084

Nazir F, Peter P, Gupta R, Kumari S, Nawaz K, Khan MIR (2024) Plant hormone ethylene: a leading edge in conferring drought stress tolerance. Physiol Plant 176(1):e14151

Nie M, Ning N, Chen J, Zhang Y, Li S, Zheng L, Zhang H (2023) Melatonin enhances salt tolerance in sorghum by modulating photosynthetic performance, osmoregulation, antioxidant defense, and ion homeostasis. Open Life Sci 18(1):20220734

Niu X, Deqing C, Liang D (2019, July) Effects of exogenous melatonin and abscisic acid on osmotic adjustment substances of 'summer Black'grape under drought stress. In Iop conference series: earth and environmental science (vol. 295, no. 2, p. 012012). IOP Publishing

Parwez R, Aftab T, Gill SS, Naeem M (2022) Abscisic acid signalling and crosstalk with phytohormones in regulation of environmental stress responses. Environ Exp Bot 199:104885

Poór P, Nawaz K, Gupta R, Ashfaque F, Khan MIR (2022) Ethylene involvement in the regulation of heat stress tolerance in plants. Plant Cell Rep 41(3):675–698

Priya M, Dhanker OP, Siddique KH, HanumanthaRao B, Nair RM, Pandey S et al (2019) Drought and heat stress-related proteins: an update about their functional relevance in imparting stress tolerance in agricultural crops. Theor Appl Genet 132:1607–1638

Qiao Y, Ren J, Yin L, Liu Y, Deng X, Liu P, Wang S (2020) Exogenous melatonin alleviates PEG-induced short-term water deficiency in maize by increasing hydraulic conductance. BMC Plant Biol *20*:1–14

Sabagh AE, Mbarki S, Hossain A, Iqbal MA, Islam MS, Raza A et al (2021) Potential role of plant growth regulators in administering crucial processes against abiotic stresses. Frontiers Agron 3:648694

Sachdev S, Ansari SA, Ansari MI, Fujita M, Hasanuzzaman M (2021) Abiotic stress and reactive oxygen species: generation, signalling, and defense mechanisms. Antioxidants 10(2):277

Sadak MS, Abdalla AM, Abd Elhamid EM, Ezzo MI (2020) Role of melatonin in improving growth, yield quantity and quality of *Moringa oleifera* L. plant under drought stress. Bull Natl Res Cent 44:1–13

Semida WM, Abdelkhalik A, Rady MO, Marey RA, Abd El-Mageed TA (2020) Exogenously applied proline enhances growth and productivity of drought stressed onion by improving photosynthetic efficiency, water use efficiency and upregulating osmoprotectants. Sci Hortic 272:109580

Sharma P, Thakur N, Mann NA, Umar A (2024) Melatonin as plant growth regulator in sustainable agriculture. Sci Hortic 323:112421

Shi H, Qian Y, Tan DX, Reiter RJ, He C (2015) Melatonin induces the transcripts of CBF/DREB1s and their involvement in both abiotic and biotic stresses in Arabidopsis. J Pineal Res 59(3):334–342

Sies H, Jones DP (2020) Reactive oxygen species (ROS) as pleiotropic physiological signalling agents. Nat Rev Mol Cell Biol 21(7):363–383

Song R, Ritonga FN, Yu H, Ding C, Zhao X (2022) Plant melatonin: regulatory and protective role. Horticulturae 8(9):810

Sun Z, Li S, Chen W, Zhang J, Zhang L, Sun W, Wang Z (2021) Plant dehydrins: expression, regulatory networks, and protective roles in plants challenged by abiotic stress. Int J Mol Sci 22(23):12619

Sun H, Jia M, Wang Y, Lu H, Wang X (2023) The complexity of melatonin and other phytohormones crosstalk with other signalling molecules for drought tolerance in horticultural crops. Sci Hortic 321:112348

Tiwari RK, Lal MK, Kumar R, Chourasia KN, Naga KC, Kumar D et al (2021) Mechanistic insights on melatonin-mediated drought stress mitigation in plants. Physiol Plant 172(2):1212–1226

Ullah F, Saqib S, Khan W, Zhao L, Khan W, Li MY, Xiong YC (2024) Sodium nitroprusside and melatonin improve physiological vitality and drought acclimation via synergistically enhancing antioxidant response in dryland maize. J Plant Growth Regul 44:891–908

Verma G, Srivastava D, Tiwari P, Chakrabarty D (2019) ROS modulation in crop plants under drought stress. In: Reactive oxygen, nitrogen and sulfur species in plants: production, metabolism, signalling and defense mechanisms, pp 311–336

Verma S, Negi NP, Pareek S, Mudgal G, Kumar D (2022) Auxin response factors in plant adaptation to drought and salinity stress. Physiol Plant 174(3):e13714

Wang P, Sun X, Li C, Wei Z, Liang D, Ma F (2013) Long-term exogenous application of melatonin delays drought-induced leaf senescence in apple. J Pineal Res 54(3):292–302

Wang C, Zhao Y, Gu P, Zou F, Meng L, Song W et al (2018) Auxin is involved in lateral root formation induced by drought stress in tobacco seedlings. J Plant Growth Regul 37:539–549

Wang X, Li Q, Xie J, Huang M, Cai J, Zhou Q et al (2021) Abscisic acid and jasmonic acid are involved in drought priming-induced tolerance to drought in wheat. Crop J 9(1):120–132

Wu J, Wang J, Hui W, Zhao F, Wang P, Su C, Gong W (2022) Physiology of plant responses to water stress and related genes: a review. Forests 13(2):324

Xia H, Ni Z, Hu R, Lin L, Deng H, Wang J et al (2020) Melatonin alleviates drought stress by a nonenzymatic and enzymatic antioxidative system in kiwifruit seedlings. Int J Mol Sci 21(3):852

Yasmeen S, Wahab A, Saleem MH, Ali B, Qureshi KA, Jaremko M (2022) Melatonin as a foliar application and adaptation in lentil (*Lens culinaris* Medik.) crops under drought stress. Sustain For 14(24):16345

Zhang M, He S, Zhan Y, Qin B, Jin X, Wang M et al (2019) Exogenous melatonin reduces the inhibitory effect of osmotic stress on photosynthesis in soybean. PLoS One 14(12):e0226542

Zhang M, Gao C, Xu L, Niu H, Liu Q, Huang Y et al (2022) Melatonin and indole-3-acetic acid synergistically regulate plant growth and stress resistance. Cells 11(20):3250

Zhao D, Yu Y, Zhao Z, Sharma R, Reiter RJ (2019) Melatonin synthesis and function: evolutionary history in animals and plants. Front Endocrinol 10:441357

Zhao C, Yang M, Wu X, Wang Y, Zhang R (2021) Physiological and transcriptomic analyses of the effects of exogenous melatonin on drought tolerance in maize (*Zea mays* L.). Plant Physiol Biochem 168:128–142

Zuo B, Zheng X, He P, Wang L, Lei Q, Feng C et al (2014) Overexpression of MzASMT improves melatonin production and enhances drought tolerance in transgenic *Arabidopsis thaliana* plants. J Pineal Res 57(4):408–417

Melatonin: A Defense Molecule in Heat Stress Amelioration of Crop Plants

4

Naveed Mushtaq and Muhammad Ahsan Altaf

Abstract

The intensity of the recent rise in air temperature brought on by global warming is a significant issue that requires attention. Temperature is the most significant abiotic stressor that plants are subjected and numerous plant growth and development processes, including cellular membrane organization, food and water uptake, germination, antioxidant function, viability, photosynthesis, leaf senescence, starch accumulation, biomass output, and root growth, are impacted by temperature stress. The productivity could be significantly increased by comprehending the molecular and physiological mechanisms underlying abiotic stress tolerance. The use of biostimulants can significantly boost plant resistance to abiotic stresses without endangering the ecosystem. A multipurpose benefit like melatonin (MT) protects the plants against the damaging effects of heat stress (HS). By preserving the water content, protecting the membranes, and enhancing homeostasis (nutrients) and water efficiency (WUE), MT enhances plant growth and development during the HS. Moreover, by activating the antioxidant defense system, MT enhances gene expression, hormone and osmolyte interaction, and decreases the buildup of reactive oxygen species (ROS), all of which contribute to improved resistance to HS. To provide HS tolerance, high endogenous MT boosts antioxidant activity and gene expression. Therefore, understanding the specific roles of both exogenous and endogenous melatonin (MT) is essential for developing HS tolerance in plants. This chapter highlights the potential of MT in

N. Mushtaq (✉) · M. A. Altaf
Key Laboratory for Quality Regulation of Tropical Horticultural Crops of Hainan Province, School of Breeding and Multiplication (Sanya Institute of Breeding and Multiplication), Center of Nanfan and High-Efficiency Tropical Agriculture, Hainan University, Sanya, China

Key Laboratory for Quality Regulation of Tropical Horticultural Crops of Hainan Province, School of Tropical Agriculture and Forestry, Hainan University, Danzhou, China
e-mail: naveedmushtaq@hainanu.edu.cn

R. K. Tiwari et al. (eds.), *Melatonin: Signal Transduction Mechanisms and Defense Networks in Plants*, https://doi.org/10.1007/978-981-96-8869-2_4

enhancing plant responses to HS. It also explores MT's antioxidant properties, its interactions with other plant hormones, and successful genetic engineering approaches that have improved plant resilience to heat stress.

Keywords
Heat stress · Melatonin · Climate change · Water use efficiency

4.1 Introduction

The livelihoods of 2.8 billion rural people, along with plant vulnerability, crop phenotype, and agricultural production, are all under serious danger from the impacts of climate change (Sahoo et al. 2023). The increase in worldwide temperatures due to climate change and global warming negatively affects crop growth and productivity (Verma et al. 2024). Global warming, along with its influence on climate patterns, is a key component of the ongoing climate shift. According to recent climate projections, rising temperatures, unpredictable rainfall, and increasing CO_2 emissions could significantly reduce the yields of many economically important crops (Bouabdelli et al. 2022). To better understand biodiversity, ecological processes, and the innate functioning of plant systems under changing environmental conditions, it is essential to explore the link between crop phenology and temperature.

Heat stress (HS) is a major challenge that significantly reduces crop yields and adversely affects plant growth and development (Cohen et al. 2020; Hassan et al. 2020; Lal et al. 2022). One of the key consequences of HS is the excessive production of reactive oxygen species (ROS) (Su et al. 2024; Mushtaq et al. 2024), which damage vital cellular components such as proteins, lipids, and DNA (Lal et al. 2022). This oxidative stress severely impairs photosynthetic efficiency (Nayak et al. 2023). ROS also disrupt enzymatic activity and compromise membrane integrity, leading to altered growth patterns and reduced productivity (Yasmeen et al. 2024). Furthermore, HS affects carbohydrate metabolism (Iqbal et al. 2024), ultimately lowering yield by altering the expression of genes involved in these metabolic pathways (Balla et al. 2019; Rehmani et al. 2013).

A major factor contributing to reduced plant development under HS is the denaturation of proteins and enzymes, along with decreased uptake of water and nutrients (Cohen et al. 2020). To counteract the damaging effects of heat-induced ROS, plants activate an efficient antioxidant defense system (Balla et al. 2019; Fahad et al. 2017). They also accumulate protective compounds such as osmolytes (e.g., proline, glycine betaine), soluble sugars (e.g., fructans, mannitol, raffinose, trehalose), and various hormones (Cohen et al. 2020). Higher accumulation of these compounds enhances plant resilience to abiotic stress.

Melatonin (MT), a multifunctional hormone, plays a vital role in supporting plant growth and development under various abiotic stresses. MT helps maintain physiological balance and protects plants from stress-induced damage (Altaf et al. 2022; Barman et al. 2024; Mushtaq et al. 2022). It serves as a powerful antioxidant, scavenging ROS, safeguarding the photosynthetic machinery from oxidative stress, and improving photosynthetic efficiency (Huang and Jin 2024; Waadt et al. 2022). Under HS, MT enhances photosynthetic performance by regulating stomatal conductance, increasing chlorophyll content, boosting Rubisco activity, and improving PSII efficiency (Tiwari et al. 2020, 2021a, b). Moreover, MT has been shown to enhance heat tolerance in crops like tomato (Khan et al. 2023) and wheat (Buttar et al. 2020) by preserving membrane stability, strengthening the antioxidant system, and improving water relations. It also upregulates stress-responsive genes, thereby supporting the photosynthetic system and increasing resistance to HS (Buttar et al. 2020).

Moreover, MT plays a role in a variety of activities, including development, fruit ripening, leaf senescence, and the reduction of oxidative damage brought on by stress (Sun et al. 2020). In addition, it interacts with many hormones and signaling molecules to mitigate the impacts of abiotic stressors (He and He 2020; Kaya et al. 2019). Besides, MT increases plant growth and biomass production under HS stress by lowering ROS, cell damage, and electrolyte leakage (Jahan et al. 2021a). The function of MT in plants cultivated under HS has recently been thoroughly investigated. Thus, we methodically described the many pathways of MT-induced HS tolerance in plants in this review. Additionally, we talked about the MT-mediated antioxidant system and MT biosynthesis engineering to increase plants' resistance to HS. To further illustrate the significance of MT crosstalk for HS tolerance in plants, we also included information on its interactions with various osmolytes and hormones, as well as potential future research avenues. Consequently, this review will give readers fresh perspectives on the function of MT under HS. Overall, the development of plants with enhanced MT synthesis for enhanced heat tolerance would greatly benefit from this knowledge.

4.2 Plant Responses During the Heat Stress

Plants are subject to many abiotic stresses that have an adverse effect on their development and growth (Devi et al. 2021). HS is a severe abiotic stressor that negatively affects plants at every stage of their life cycle (Table 4.1). The initial phase of each plant is seed germination. By limiting the availability of water and nutrients and interfering with enzyme activity, HS dramatically lowers seed germination (Nahar et al. 2010; Chebrolu et al. 2016). Significant yield losses result from HS's reduction of root length, poor seedling growth, and poor stand establishment (Kumar et al. 2011; Choudhary et al. 2024). Additionally, HS causes a variety of physiology-related changes in plants, such as stem and leaf burning, leaf searing, fruit discoloration, leaf rolling, senescence, abscission, chlorosis, and necrosis (Akter and Islam 2017; Zahra et al. 2022). It also decreases cell counts, cell size, growth, and biomass output (Jian et al. 2024).

HS also influences plant osmotic adjustments by raising evapotranspiration, which influences plant solute production that is crucial for osmotic adjustments (Fig. 4.1) (Mumithrakamatchi et al. 2024). Along with increasing respiration and photorespiration, HS also slows down photosynthesis, which lowers assimilation

Table 4.1 The impact of heat stress on the physiology, development, and yield of several crops

Heat stress	Crops	HS and plant stages	Effects on plants	References
43–45 °C (Temp)	Rice	Rice reproductive stage	Reduced seed rate, grain quality and chlorophyll level	Cheabu et al. (2018)
45 °C (Temp)	Maize	Maize seedling stage	Reduced photosynthetic activity and electron transport	Doğru (2021)
28 °C (Temp)	Wheat	Wheat grain filling stage	Lessened grain and protein contents	Laddomada et al. (2021)
32 °C (Temp)	Brassica	First open flower stage	Decreased pod number, and chlorophyll and pollen quality	Chen et al. (2020)
36 °C (Temp)	Soybean	For 3 weeks	Reduced oil concentration in seeds, protein contents, and protein content	Ortiz et al. (2021)
35 °C (Temp)	Barley	Anthesis stage	Reduced grains size, and pod number, and grain weight	Dawood et al. (2020)
38 °C (Temp)	Tomato	Seedling stage	Lessened stomatal conductance, chlorophyll, transpiration, and biomass	Haque et al. (2021)
32 °C (Temp)	Groundnut	Pod filling stage	HS reduced photosynthesis, damage membrane, and seed quality	Aravind et al. (2024)
32–33 °C (Temp)	Lentil	Flowering stage	Effect cell membranes and lessened chlorophyll and grain protein level	Sita et al. (2018)

output and consequently stunts growth and biomass production (Kan et al. 2023). Additionally, HS disrupts cell metabolism by impeding the water balance brought about by a decrease in root water uptake and a rise in plant leaf water loss (Jagadish et al. 2021). Significant yield losses result from HS's reduction of cell water potential and induction of ROS generation, which damages the cellular membrane and increases electrolyte leakage (Ullah et al. 2021).

One of the main causes of heat-induced declines in growth and yield is heat stress, which also reduces the uptake of water and nutrients (Li et al. 2023a). Although this decrease in nutrient uptake is dependent on the plant species, soil nutrient status, and stress circumstances, heat stress also modifies the plant source–sink relationship and nutrient accumulation in plants (Abdelrahman et al. 2019). Additionally, HS disrupts the enzymes that are involved in the metabolism of nutrients, which lowers their uptake and accumulation in plants (Hungria and Kaschuk 2013; Puglia 2023). Furthermore, because HS inhibits root growth, biomass, and

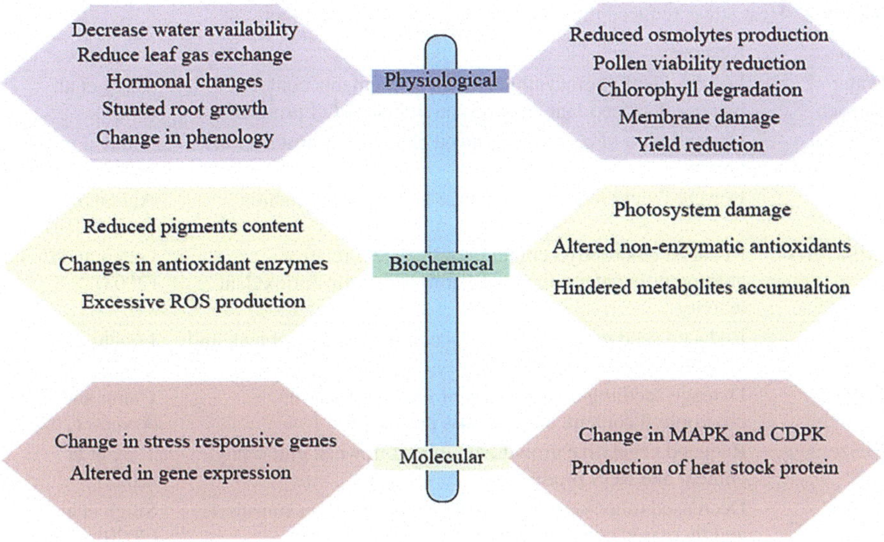

Fig. 4.1 The effect of heat stress on the plants

nutrient uptake by roots, it also lowers nutrient acquisition (Li et al. 2023a, b). Furthermore, HS limits the movement of carbohydrates from shoot to root, depletes labile carbon, impairs protein function, and significantly reduces nutrient uptake.

One of the most crucial plant functions that is thought to be very susceptible to HS is photosynthesis (Killi et al. 2020). Both C3 and C4 plants' photosynthetic efficiency is significantly impacted by HS (Yang et al. 2011). A substantial decrease in photosynthesis occurs under HS as a result of heat injury's effects on the carbon metabolism, thylakoid responses, chloroplast modifications, and thylakoid disarray (Cheabu et al. 2018). A decrease in photosynthesis is also a result of HS's lowering of PS-II activities, the Fv/Fm, photosynthetic pigment synthesis, conductance, leaf water contents, and intercellular CO_2 concentration (Greer and Weedon 2011; Marchand et al. 2005). In addition to decreasing source and sink activity, high temperatures also significantly stunt growth and biomass production. With decreasing carbon absorption and the division and dispersion of carbon and nitrogen in plants, HS has an impact on the source and sink relationships in plants (Calderini et al. 2006). The metabolism of leaf proteins and starches is adversely affected by these changes, which is thought to be the reason for the decline in production and quality (Yang et al. 2011). All phases of plant growing stages are susceptible to HS, but the reproductive stage is considered to be the most vulnerable, and even a small temperature increase during this phase can result in considerable yield losses (Lobell et al. 2011). Depending on the plant variety and the degree of stress, HS during the phase of reproduction causes floral buds to shrink and flowers to drop (Sato et al. 2006). Additionally, HS decreases fruit development and grains retained with stigma, overall during the reproduction induces seed sterility and subsequently lowers final yield (Maheswari et al. 2011).

Table 4.2 Heat stress reduced the growth and development of plants

Crop	Response	References
Water spinach	Lowered leaf photosynthetic activity, pigments content; increased antioxidant enzyme activity and ROS production	Wang et al. (2023)
Tomato	Upregulation of heat stock related genes and transcription factors	Aldubai et al. (2022)
Rice	Lowered yield reduction, damage flowering production	Aghamolki et al. (2014)
Strawberry	Reduced chlorophyll content, lowered leaf relative chlorophyll content, increased non-enzymatic antioxidant activity	Gulen and Eris (2003)
Canola	Reduced seed germination, reduced pollen development, and damage reproductive fitness	Lohani et al. (2022)
Pepper	Damage seedling health, reduced seed germination, chlorophyll content, and biomass production	Utami and Aryanti (2021)
Maize	Reduced vegetative growth, number of flowers, chlorophyll content and ROS production	Teng et al. (2022)
Potato	Decreased tuber yield, leaf photosynthesis, sugar transport, and increased enzyme activity	Singh et al. (2020)
Chinese Cabbage	Enhanced secondary metabolites production, proline uptake, antioxidant enzyme activity and ROS accumulation	
Melon	Reduced leaf gas exchange elements, chlorophyll content, damage photosystem damage, increased antioxidant enzyme activity and ROS production	Weng et al. (2022)
Wheat	Declined yield related traits, small green size, reduced reductive growth and development	

Until 50 °C, temperature increases prompt an immense rise in respiration; nevertheless, following that heat-induced damage to respiratory processes leads to a significant decrease in respiration (Munjonji et al. 2021). HS also results in a significant increase in ROS, which damages cellular membranes, proteins, lipids, and DNA, as well as destroys enzymatic activities in plants (Iqbal et al. 2021a) (Table 4.2).

4.3 Induction of Antioxidants with Melatonin Under Heat Stress

For this purpose, to counteract the generation of ROS brought on by heat, plants, nevertheless, activate an antioxidant defense system made up of several antioxidant enzymes like CAT, APX, GR, DHAR, POD, and SOD (Calderini et al. 2006). In order to counteract the effects of HS, plants additionally collect a variety of osmolytes, including sugars, proline, glyciene-betaine, and other hormones (Ejaz et al. 2020). Through cellular membrane protection, antioxidant activity, membrane stability, and plant water relations, the buildup of these osmolytes enhances plant survival (Cheabu et al. 2018). Plant hormones are essential for how plants react to various stressors. Hormone synthesis, breakdown, and distribution to various plant organs are all evidently impacted by HS, suggesting that they also have a major impact on how plants react to HS (Hassan et al. 2020).

4.4 Heat Stress Upregulated the Melatonin Biosynthesis Process in Plants

MT is a crucial signaling molecule that affects plants in a variety of stress conditions. The plant mitochondria and chloroplasts are the location for the synthesis of MT (Tan et al. 2012), while fruit plants and several crops have also been found to synthesize it (Arnao and Hernández-Ruiz 2012). Though the endogenous MT concentration varies greatly among plant species, organs, stress levels, and developmental stages present in plants (Arnao and Hernández-Ruiz 2009), the level of MT in plants is controlled by seasonal and circadian cycles (Beilby et al. 2015). For instance, during the mature stages of tomato and morning glory plants, the concentration of MT increases dramatically (Van Der Merwe et al. 2015). Rice and tomato plants grown in the field acquired more exogenous MT than plants grown in growing chambers, indicating that light conditions also have a substantial impact on endogenous MT (Byeon et al. 2012; Boccalandro et al. 2011). Under stress conditions, MT, a significant antioxidant, interacts with ROS and lowers their concentration (Arnao and Hernández-Ruiz 2012). This suggests that an increase in ROS generation is associated with an increase in endogenous MT during stress (Arnao and Hernández-Ruiz 2012). Following the application of salt and osmotic stress, there was a considerable rise in the concentration of MT in grapevine, barley, and lupin (Arnao and Hernández-Ruiz 2009). Similarly, after being exposed to HS, endogenous MT significantly increased in rice plants (Byeon et al. 2012). Through boosting the expression of heat shock protein, N-acetylserotonin methyltransferase (ASMT) enzymes efficiently raised endogenous MT and protected the plants against the destructive effects of HS (Hardeland 2016).

Another study displayed that the HS enhanced the production of endogenous MT in plants, which was connected to a rise in the activity of the enzymes that are involved in the biosynthesis of MT (Xu et al. 2016). Furthermore, when exposed to HS, *Phacelia tanacetifolia* also showed a significant rise in MT, which shielded the plants from the harmful effects of HS by promoting antioxidant activities (Arnao and Hernández-Ruiz 2015; Hardeland 2016). These results showed that endogenous MT production was stimulated by stress, indicating its function in plant responses to various stressful situations (Hardeland 2016). The activities of the enzymes and genes involved in MT production are related to MT accumulation in plants. In plants cultivated under stress, MT synthesis was markedly enhanced by the upregulation of genes and enzymatic activity involved in MT synthesis (Arnao and Hernández-Ruiz 2012; Beilby et al. 2015; Byeon et al. 2012).

4.5 Melatonin Enhanced the Plant Performance in the HS Environment

A major abiotic stressor, HS, has become much more intense recently as a result of climate change and global warming. HS is a severe issue since it causes metabolic, physiological, and biochemical disturbances as well as leaf senescence. And consequently results in a significant decline in growth and ultimate output (Jahan et al. 2021a, b). Under stress conditions, MT, a crucial signaling molecule, enhances plant functioning and causes notable improvements in growth and development (Jahan et al. 2021a, b; Liang et al. 2015). Furthermore, MT is a well-known antioxidant that scavenges reactive oxygen species (ROS), shields plants from oxidative damage brought on by stress, and enhances plant performance (Liang et al. 2015; Ahammed et al. 2018a).

4.5.1 Melatonin Affects the Plant's Growth and Development

Plant growth and development are severely inhibited by HS, which also results in a large decrease in ultimate production (Fig. 4.2). A naturally occurring chemical called MT significantly improves plant performance in stressful situations (Arnao and Hernández-Ruiz 2017). Under HS, germination (seeds) is drastically decreased,

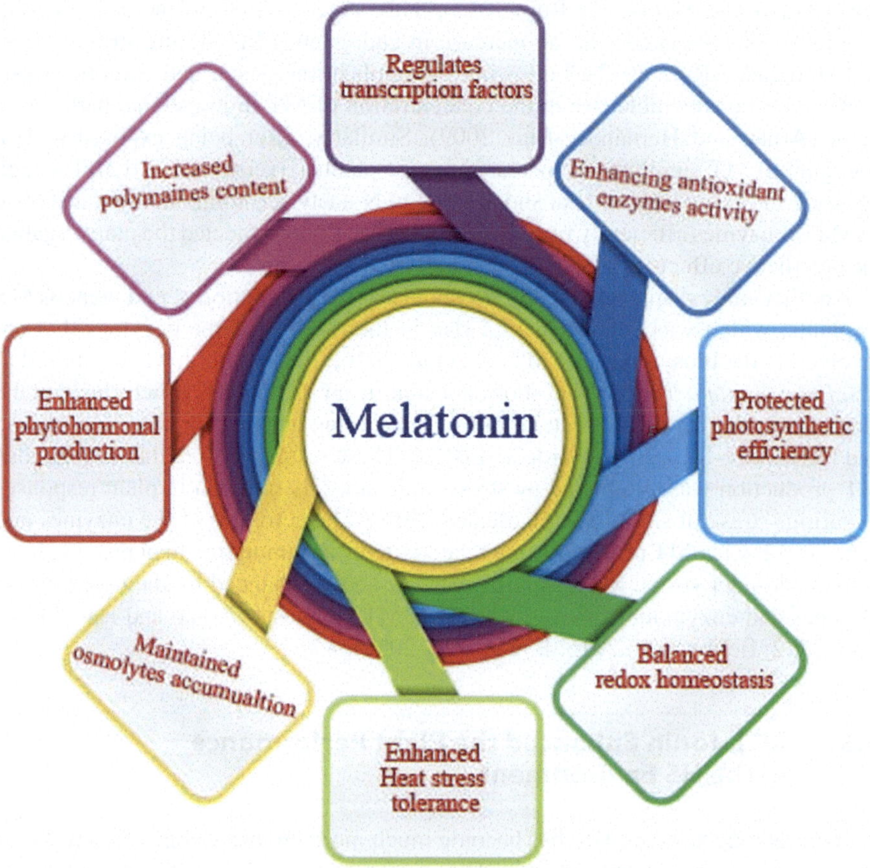

Fig. 4.2 The melatonin-induced heat-stress tolerance mechanisms in plants

which lowers the established stand and the target population. Because of its higher antioxidant activities under HS than the control, MT treatment (1000 μM) increased germination in *Arabidopsis thaliana* by 60% (Hernández et al. 2015). Furthermore, MT decreased leaf senescence, enhanced root architecture, and significantly increased plant growth under HS (Jahan et al. 2019; Sha et al. 2023).

The MT application significantly detoxifies ROS and regulates antioxidant activities, which lessens oxidative stress brought on by heat and enhances plant growth and development (Yu et al. 2023). Moreover, an exogenous supply of MT boosts the production of the P5CS gene and HSPs, which are involved in ROS detoxification and enhance plant development under HS (Yu et al. 2023). Additionally, MT enhanced the levels of amino acids and polyamines (PAs: putrescine, spermidine, and spermine) as well as elevated the expression of genes responsive to stress, all of which enhanced the plants' ability to withstand HS (Jahan et al. 2019; Li et al. 2020).

Growth and development were significantly improved by exogenous MT supplementation, which also raised endogenous nitric oxide (NO) levels, nitrate reductase activities, and elevated the transcription of MYB and WRKy in plants (Jahan et al. 2019; Yu et al. 2023). Another study found that by boosting the expression of MT biosynthesis genes and the accumulation of amino acids, an exogenous spray of MT (100 μM) considerably improved photosynthesis and biomass production (Li et al. 2020). ROS generation brought on by HS stress damages the photosynthetic machinery and lowers the overall photosynthetic efficiency of plants. Growth under HS was significantly improved by the exogenous application of MT (50 mM), which decreased ROS production, MDA, H_2O_2, and electrolyte leakage (EL) and increased protein and chlorophyll level (Table 4.2) by inducing gene expression and the activities of antioxidant enzymes (CAT, POD, and SOD) (Yu et al. 2023). According to a different study, using MT (20 μM) greatly increased biomass output and root and shoot growth by raising total chlorophyll synthesis and soluble proteins (TSP), and enhanced the antioxidant activities (CAT, POD, and SOD) in tall fescue (*Festuca arundinaceous*), hence reducing the formation of MDA and H_2O_2 (Alam et al. 2021). Additionally, an exogenous supply of MT sustains the uptake of nutrients and water, enhancing photosynthetic efficiency and leading to notable advancements in growth and development (Chaturvedi et al. 2017).

4.5.2 Melatonin Sustains Membrane Strength and Water Level

The integrity and functionality of the membrane system are impacted by the substantial disruption of membrane protein structure caused by HS. The EL from the cell membrane is measured to assess membrane malfunction in response to HS (Chaturvedi et al. 2017). A rise in EL proposes that the detrimental effects of HS have caused a severe loss of membrane integrity (Chaturvedi et al. 2017; Wahid and Close 2007). Nonetheless, when exposed to HS, MT significantly lowers the EL and preserves membrane integrity. By preserving membrane integrity, MT treatment greatly enhanced antioxidant activities and drastically decreased EL and MDA accumulation (Li et al. 2023a). Another investigation found that MT treatment greatly improved membrane stability under HS and decreased MDA and H_2O_2 levels (Ahammed et al. 2018a, b). Lipid peroxidation was decreased by the overexpression of transcripts and antioxidant activity (APX and SOD); though, under HS, MT deficiency can result in a marked increase in lipid peroxidation (Qi et al. 2018).

MT reduced MDA and EL (Table 4.3) by healing the damaged membrane and lowering heat-induced oxidative damage by balancing ROS in high-temperature

Table 4.3 Melatonin enhanced the heat stress tolerance by regulating the plant's physiological and molecular functions

Crop	Findings	References
Wheat	Enhanced proline content, antioxidant enzyme activity, photosynthetic activity; reduced H_2O_2 and MDA accumulation in leaves	Buttar et al. (2020)
Tall Fescue	Improved chlorophyll content, antioxidant enzyme activity, soluble sugar and carotenoid content; decreased MDA and H_2O_2 level	Alam et al. (2021)
Maize	Increased proline concentration, SOD, POD, CAT activity, soluble sugar content, and upregulated related gene expression	Cao et al. (2024)
Celery	Reduced electrical conductivity, MDA production; increased chlorophyll content, upregulated pyruvate metabolism in leaves	Li et al. (2022)
Tomato	Protected leaf photosynthetic system, maintain photosystem II reaction center, increased photosynthetic pigments concentration	Jahan et al. (2021a, b)
Chrysanthemum	Decreased H2O2, MDA, EL level; increased carotenoids and flavonoids content, proline and soluble protein content	Xing et al. (2021)
Rice	Balanced pollen viability, increased plant growth, enzyme activity and enhanced heat stress tolerance	Barman et al. (2024)
Sweet potato	Maintained photosystem II efficiency, increased enzyme activity, chlorophyll content, reduced H_2O_2 and MDA production in leaves	Kumar et al. (2024)
Kiwifruit	Regulates carotenoids and melatonin biosynthesis genes and enhanced heat stress tolerance	Xia et al. (2021)
Potato	Protected leaf photosynthetic system, increased pigments molecules concentration, maintained photosystem II	Qiulan et al. (2023)
Carnation	Upregulated heat-stressed genes, reduced ROS production, enhanced antioxidant enzyme activity	Elmongy and Abd El-Baset (2024)
Kiwifruit	Improved proline content, regulates AsA-GSH cycle, enhanced heat stress tolerance	Liang et al. (2018)
Strawberry	Maintained relative water content, increased chlorophyll content, pigment molecules concentration, and enzyme activity	Manafi et al. (2022)
Canola	Regulating antioxidant enzymes system, maintained redox homeostasis, protected leaf photosynthetic system	Imran et al. (2021)

circumstances. Plants subjected to HS had a dramatic increase in MDA, which damaged the membrane and increased EL (Jahan et al. 2019; Liang et al. 2018). By boosting the expression of the ASMT gene, which in turn decreased the EL and efficiency of PS-II by enhancing antioxidant activities, the foliar addition of MT (10 μM) decreased heat-induced photo-inhibition (Xu et al. 2016). Heat-induced water deficit impairs plant growth and photosynthetic efficiency while upsetting plant water relations. MT treatment enhances thermotolerance under heat stress by preserving higher RCW, minimal oxidative damage, through increased water uptake and reduced water loss (Haydari et al. 2019).

4.5.3 Melatonin Improves Water Use Efficiency and Nutrient Uptake

The initial line of defense against heat stress is thought to be the closure of stomata. Because of a decrease in stomata conductance, plant water usage efficiency (WUE) is greatly reduced under HS. Plant photosynthetic efficiency is impacted by any changes in stomata mobility because it denatures the proteins associated with the photosystem (Hassan et al. 2020). In order to maintain higher WUE and photosynthetic efficiency under HS circumstances, MT application significantly raised stomata conductance and decreased canopy temperature, which increased water loss (Barman et al. 2024). Due to a decrease in stomata motions, HS also decreased transpiration; however, MT application accelerated transpiration in response to heat-induced stomata closure, preserving a greater WUE in plants exposed to HS (Chaturvedi et al. 2017). Because nutrients are essential to the physiological, biochemical, and metabolic processes of plants, they must be absorbed. Due to the reduction in membrane fluidity, which lowers the turgor pressure of cells, HS circumstances lessen the amount of water that enters through the roots (Beheshti et al. 2023). When under stress, MT treatment increases the vapor pressure differential between the atmosphere and the leaf surface, allowing the roots of the plant to absorb more water and nutrients (Iqbal et al. 2021a). When MT is applied, root activity is maintained and water intake is enhanced, which raises the uptake of nutrients like N, Mg, Ca, P, and K and enhances plant performance and resistance to HS (Elsayed et al. 2021). Some studies show that using MT significantly raised Ca under HS. In plants developing under HS conditions, the increase in Ca absorption shielded the cellular membranes and decreased the buildup of heat-induced MDA and H_2O_2 (Dabaghkar et al. 2024). Additionally, these authors observed that the use of MT significantly enhanced the absorption of additional nutrients, such as K, Mg, P, and N, leading to a notable enhancement in growth performance and resistance to HS (Sehar et al. 2023).

4.5.4 Melatonin Protects PSII Apparatus and Increases Photosynthesis Activity

One of the most significant plant functions that is adversely impacted by HS is photosynthesis (Hassan et al. 2020). While MT treatment considerably detoxifies ROS, maintains optimal chlorophyll synthesis, and shields the chlorophyll from degradation under HS, HS induces chlorophyll degradation and significantly disrupts chlorophyll synthesis (Meena et al. 2017). The concentration of carotenoids and chlorophyll was decreased by HS, but the concentration of both substances was much increased by MT foliar spray, as shown by the plants' greener leaves (Alam et al. 2021). Moreover, via activating the Calvin cycle enzyme gene (sedoheptulose-1,7-bisphosphatase), MT supplementation promotes photosynthetic carbon uptake in plants, leading to notable increases in photosynthetic efficiency (Wu et al. 2022). Chlorophyll and carotenoid concentrations were reduced by high temperatures; however, exogenous MT supplementation significantly restored the aforementioned photosynthetic pigments under HS in comparison to the control (Jahan et al. 2019) (Fig. 4.3). Additionally, the efficiency of PS-II, electron transport rate, photochemical quenching coefficient, and chlorophyll fluorescence were all markedly enhanced by MT supplementation, which led to a notable rise in photosynthetic efficiency and plant development (Jahan et al. 2019).

The boosting of MT production and sugar metabolism through an exogenous supply of MT can mitigate heat-induced photo-inhibition (Jahan et al. 2019). The endogenous MT concentration and MT biosynthesis gene expression were both increased by MT treatment, which may have contributed to the higher chlorophyll levels under HS (Byeon et al. 2012). The RuBisCo is the main hub of any stress circumstances and is thought to be a rate-limiting enzyme for RuBP carboxylation.

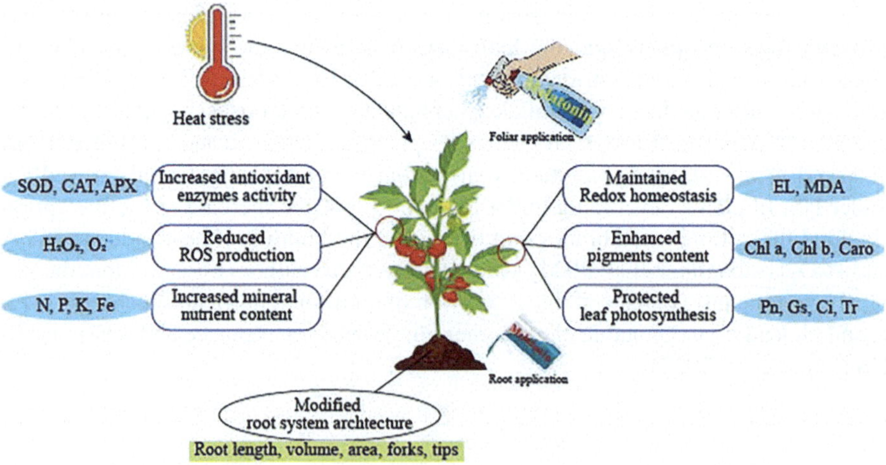

Fig. 4.3 The effect of melatonin during the heat stress in the plants and prompting of protection mechanisms to remove the ROS and protect the pigments and functioning in plants

RuBisCo activities and the expression of genes like rbcS and rbsL associated with RuBisCo activity were both markedly decreased by HS. These genes' expression was markedly enhanced by MT treatment, which also preserved increased photosynthetic efficiency under HS (Jahan et al. 2021a, b). A useful technique for assessing how well plants photosynthesize under different stress situations is chlorophyll fluorescence. Melatonin therapy plays a functional role in mending heat-induced damage to the photosynthetic machinery by boosting photochemical quenching (qP) and decreasing nonphotochemical quenching (NPQ), which helps offset the effects of heat stress (HS) (Jahan et al. 2019). Plant growth under both normal and stressed situations depends on optimal energy flow, which is produced by effective photosynthetic electron transport (Ye et al. 2012). By preserving electron transport and photosynthetic activity, exogenous melatonin application greatly aids plant development under HS (Ahammed et al. 2018a).

4.5.5 Melatonin Preserves the Osmolytes and Other Hormone Crosstalk

Osmolytes play a vital role in protecting plants from the damaging effects of heat stress (HS) by supporting physiological functions and improving stress adaptation (Kumar et al. 2024). Melatonin (MT) application has been shown to boost endogenous MT levels, leading to enhanced PSII activity and upregulation of photosynthesis-related genes (Talaat 2023). In *Lolium perenne*, exogenous MT increased both MT and cytokinin levels, reduced ABA accumulation, and protected proteins from oxidative damage under HS (Xu et al. 2016). In cucumber seedlings, MT treatment lowered ammonium buildup and boosted enzymes involved in nitrogen metabolism, thereby improving nitrate content under HS (Jiang et al. 2024). Similarly, MT improved tall fescue growth and increased total soluble protein content during HS (Alam et al. 2021).

Under heat stress, MT application significantly reduced ABA levels while increasing gibberellic acid (GA) and MT content. It downregulated ABA biosynthesis genes and signaling transcripts, while promoting GA and MT biosynthesis, thus mitigating ABA accumulation and regulating stress responses (Jahan et al. 2019). The rise in GA and MT also helped delay heat-induced senescence. Among the key stress-related hormones, ABA and salicylic acid (SA) play central roles. ABA is linked to ROS accumulation—especially H_2O_2—in stressed cells (Liu et al. 2010). MT reduces ABA by upregulating catabolic genes (*CYP707A1*, *CYP707A2*) and downregulating the biosynthesis gene *NCED* (Ye et al. 2012). Several studies have also reported that MT increases cytokinin levels while reducing ABA under HS (Ma et al. 2020).

Although SA levels initially rise under HS, they tend to decline with prolonged stress. MT counters this by upregulating the SA biosynthesis gene *PAL2*, increasing SA content and improving heat tolerance (Imran et al. 2021). Additionally, exogenous MT boosts IAA and GA levels while decreasing ABA accumulation (Pu et al. 2021). It also significantly enhances methyl jasmonate (MeJA) levels, reducing

H_2O_2 buildup (Li et al. 2021). MT has been shown to increase nitric oxide (NO) levels, which supports cellular homeostasis and boosts antioxidant defenses, reducing oxidative stress (Kaya et al. 2020; Irshad et al. 2021). There is also a positive correlation between MT and IAA, with IAA application increasing endogenous MT levels (Wang et al. 2015).

Furthermore, MT enhances the expression of key stress-responsive genes such as *TaWRKY26*, *TaWRKY39*, and *TaMYB80*, contributing to improved thermotolerance in plants (Buttar et al. 2020). Likewise, several authors observed that by raising endogenous MT concentrations, exogenous MT produced heat tolerance (Xu et al. 2016; Shi et al. 2015). The endogenous MT under HS rises significantly as a result of the application of MT, which also enhances the expression of MT synthesis enzymes (Xu et al. 2016). In order to give plants heat resistance, MT treatment also encourages the buildup of osmolytes. For instance, MT treatment has been shown to increase the accumulation of Pro, Tre, and TSS in maize roots and sprouts as well as to encourage HS tolerance in maize plants (Li et al. 2018). Another study in strawberry plants showed that the Pro contents increased in response to heat stress, and this increase was exacerbated by MT (50 μM), induced antioxidant activities, Pro accumulation, and ultimately increased thermos-tolerance (Manafi et al. 2021).

4.5.6 Melatonin Regulates the Accumulation of Key Secondary Metabolites

Secondary metabolites are essential for plants' signaling transduction, which is thought to help offset the impacts of various stressors (Jan et al. 2021). According to earlier research, MT interacts with the polyamines (Pas) signaling pathways to positively regulate plant development and abiotic stressors (Ke et al. 2018). MT works with the PA and NO production pathways to reduce heat-induced oxidative damage. Better PA synthesis under HS is favored by the upregulation of PA biosynthesis genes (ADC1/2, SAMDC1/2, SPMS, and SPDS1/2/3/5/6) by exogenous MT input (Jahan et al. 2019). According to a different study, MT treatment significantly increased the heat tolerance of long-term HS plants via modulating PA metabolism. Through the metabolism of L-arginine and PAs, MT and NO may be able to fight against various stressors (Alam et al. 2018). However, PAs also control the nitrate reductase (NR) and nitric oxide synthases (NOS) pathways (Rosales et al. 2011). MT enhances NO activity under HS, as evidenced by the upregulation of NO levels, NOS and NR activities, and gene expression upon administration (Jahan et al. 2019). Under stressed circumstances, the phenolic and flavonoid compounds have superior antioxidant and ROS scavenging properties. When compared to controls, the application of MT dramatically increased the antioxidant (DPPH), total phenolic (TFC), and total flavonoid (TFC) activities under HS by 90.50%, 73.10%, and 54.30%, respectively (Ke et al. 2018), which raised the HS stress in plants. By raising antioxidant activities, an exogenous supply of MT considerably raised the amount of free PAs (Put, putrescine; Spd, spermidine; Spm, spermine) in plants and improved their resistance to HS (Jahan et al. 2019).

4.5.7 Melatonin Strengthens the ROS and Antioxidant Defense System and Detoxifies ROS

One of the main causes of plant damage is heat-induced oxidative stress, which harms plant lipids, proteins, and DNA (Distéfano et al. 2024). However, in order to combat the harm caused by heat-induced oxidative stress, plants activate an exceptional antioxidant defense mechanism (Alam et al. 2021). Meanwhile, with the boosting antioxidant activity under HS, MT is regarded as a dynamic antioxidant that broadly promotes cellular redox homeostasis (Bose and Howlader 2020). CAT and GPX enzymes typically scavenge H_2O_2 at various cellular locations and are thought to have a key role in controlling plant cells' H_2O_2 homeostasis (Table 4.3).

The activities of CAT, GR, and GPX were markedly enhanced by an exogenous supply of MT (70 μM), which also lessened the heat-induced damage effects in maize seedlings (Li et al. 2018). Similarly, by triggering the antioxidant defense system, MT directly scavenges ROS (Fan et al. 2018). Furthermore, by acting as a signaling molecule and enhancing the antioxidants' actions in plants, MT raises HS tolerance (Bose and Howlader 2020). Additionally, MT supplementation decreased ROS generation, enhanced thermotolerance, and maintained greater activity of GSH, AsA, CAT, GR, GPX, and SOD as well as ratios of AsA + DHA and GSH + GSSG (Li et al. 2014).

In a different study, it was observed that exogenous MT significantly increased thermo-tolerance and enhanced the activities of APX, CAT, GPX, and SOD in comparison to control. By boosting the activities of APX, CAT, GPX, and SOD under HS stress, exogenous MT administration raised endogenous MT and the expression of MT biosynthesis genes and reduced ROS buildup (Wang et al. 2020). AsA and GSH activities in plants were enhanced by MT foliar spraying, indicating that MT application enhanced antioxidant activities under HS (Tang et al. 2020). One significant strategy to stop plants from producing free radicals is to use the AsA-GSH cycle. Under heat and drought stress, MT treatment reduced H_2O_2 generation by increasing AsA and GSH production (Li et al. 2014). Other studies have discovered that MT treatment enhanced GSH production and APX and GR activities, detoxifying ROS and lowering MDA buildup in wheat plants cultivated under HS (Buttar et al. 2020). Additionally, MT treatment reduces H_2O_2 and TBARS levels, which have been connected to increased antioxidant activity. The increase in antioxidant activity, mediated by H2S, can enhance photosynthesis and carbohydrate metabolism, providing the required energy and carbon skeleton for plant development (Iqbal et al. 2021b). Furthermore, via boosting the expression of their genes, MT treatments activated APX, CAT, GSH, Gly-I, Gly-II, GR, and SOD, which in turn enhanced plants' thermotolerance (Li et al. 2018). The development of HS tolerance in plants was made possible by the increase in antioxidant enzyme activity. One of the most significant physiological processes of MT-induced HS tolerance in plants is the activation of the antioxidant defense system. Consequently, it is proposed that the MT and antioxidant signaling pathways have a dynamic interaction that aids in the development of HS tolerance in plants.

4.5.8 Melatonin Induced the Defensive-Associated Genes

Melatonin (MT) treatment significantly boosts the expression of stress-responsive genes, such as *FaTHsfA2a*, *HSP90*, and *FaHsfB1a*, providing plants with enhanced heat tolerance (Shi et al. 2015). In *Arabidopsis*, MT treatment notably increased the expression of A1 heat-shock factors (HSFA1s) (Shi et al. 2015). It also activated transcripts of heat-responsive genes like *HSFA2*, *HSP90*, and *HSP101*, which are critical for heat stress (HS) tolerance. The strong link between HS and heat shock proteins (HSPs) is evident in studies showing that overexpressing *HSP70* significantly improves HS tolerance (Wang et al. 2020). In chrysanthemum leaves, exogenous MT enhanced thermotolerance and upregulated several HSPs, including *HSFB3*, *HSFA1a*, *HSFA2b*, *HSP23*, *HSP70*, *HSP80*, and *HSP90* (Xing et al. 2021). *HsfA2* and *HSP90* are key players in ROS detoxification through the H_2O_2 signaling pathway. In tomato seedlings, MT application increased *HsfA2* and *HSP90* expression, providing them with greater heat tolerance (Jahan et al. 2019).

Another study showed that the HSP90 coordinates the DNA-binding enhancement process, and the entire process is connected to MT-induced HS tolerance (Katano et al. 2018). HsfA2 also plays a significant role in H_2O_2 signaling and enhances HS memory subsistence. After a significant improvement in thermotolerance, exogenous MT supplementation in tomato seedlings enhanced MT production and HSP expression (Xu et al. 2016). In a similar vein, improvements in HS were largely due to the increased expression of FaTHsfA2a, HSP90, and FaHsfB1a in strawberry plants treated with MT (Manafi et al. 2021). The transcription factors MYB and WRKY are crucial in controlling intricate and dynamic traits under HS. Exogenous MT significantly increased wheat plants' heat resistance and enhanced the expression of stress-responsive genes (TaWRKY39 and TaWRKY26) (Buttar et al. 2020). In a similar manner, MT significantly improved HS tolerance and raised TaMYB80 gene expression following six hours of HS. Furthermore, MT enhanced endogenous MT and heat shock factor (HSP) expression, which enhanced HSP accumulation after a significant rise in HS tolerance (Shi et al. 2015; Xu et al. 2016).

4.6 Melatonin Transgenic Enhanced the Heat Resistance

Under various stress conditions, the induction of different hormones and osmolytes enhances the growth and development of the plant. Worldwide research is being made to create transgenic plants with higher MT levels in order to confer HS resistance because of the potential properties of MT. The MT application lowers ROS and preserves improved plant growth and development under stress environments. For instance, a tomato line with over-expression of the SlSNAT gene enhanced PS-II and Fv/Fm growth and efficiency, decreased heat-induced damage, and protected the ribulose bisphosphate carboxylase and oxygenase proteins (Wang et al. 2020). By enhancing the antioxidant activities, endogenous MT, and decreasing ABA accumulation, GhM2H gene overexpression significantly increased cotton's

HS tolerance (Ghareeb et al. 2022). Furthermore, after a notable boost in growth and resistance to heat stresses, the transgenic watermelon melatonin plants with ClCOMT1 genes resulted in a notable rise in endogenous MT levels (Chang et al. 2021. Additionally, HS tolerance was increased in apple plants containing the ASMT gene. The expression of ASMT genes significantly increased in apple plants exposed to several stresses, like cold, drought, and heat stressors. This enhanced endogenous MT production increased the plants' resistance to these conditions by raising antioxidant activities (Wang et al. 2021).

Researchers produced MT-deficient tomato plants by blocking the genes that are involved in the manufacture of melatonin (MT), such as COMT1 (Caffeic Acid O-Methyltransferase-1) (Ahammed et al. 2018a). By interfering with light contacts, carbon fixation, PS-II reaction efficiency, and electron transport, they discovered that inhibiting COMT1 decreased heat stress (HS). But the heat-induced photosynthetic limitations were mitigated by MT treatment, indicating that MT is essential for preserving photosynthetic efficiency under stress (Ahammed et al. 2018a). Furthermore, in mustard sprouts, endogenous MT levels were markedly increased, and HS tolerance was improved by upregulating genes involved in MT synthesis and antioxidant activity, as well as serotonin N-acetyltransferase (BjSNAT-1) (Cheng et al. 2020; Alam et al. 2021).

References

Abdelrahman M, Burritt DJ, Gupta A, Tsujimoto H, Tran L-SP (2019) Heat stress effects on source–sink relationships and metabolome dynamics in wheat. J Exp Bot 71:543–554. https://doi.org/10.1093/jxb/erz296

Aghamolki MTK, Yusop MK, Oad FC, Zakikhani H, Jaafar HZ, Kharidah S, Musa MH (2014) Heat stress effects on yield parameters of selected rice cultivars at reproductive growth stages. J Food Agric Environ 12(2):741–746

Ahammed GJ, Xu W, Liu A, Chen S (2018a) COMT1 silencing aggravates heat stress-induced reduction in photosynthesis by decreasing chlorophyll content, photosystem II activity, and electron transport efficiency in tomato. Front Plant Sci 9. https://doi.org/10.3389/fpls.2018.00998

Ahammed GJ, Xu W, Liu A, Chen S (2018b) Endogenous melatonin deficiency aggravates high temperature-induced oxidative stress in *Solanum lycopersicum* L. Environ Exp Bot 161:303–311. https://doi.org/10.1016/j.envexpbot.2018.06.006

Akter N, Islam MR (2017) Heat stress effects and management in wheat. A review. Agron Sustain Dev 37. https://doi.org/10.1007/s13593-017-0443-9

Alam MN, Zhang L, Yang L, Islam MR, Liu Y, Luo H, Yang P, Wang Q, Chan Z (2018) Transcriptomic profiling of tall fescue in response to heat stress and improved thermotolerance by melatonin and 24-epibrassinolide. BMC Genomics 19:224. https://doi.org/10.1186/s12864-018-4588-y

Alam MN, Yang L, Yi X, Wang Q, Robin AHK (2021) Role of melatonin in inducing the physiological and biochemical processes associated with heat stress tolerance in Tall Fescue (*Festuca arundinaceous*). J Plant Growth Regul 41:2759–2768. https://doi.org/10.1007/s00344-021-10472-6

Aldubai AA, Alsadon AA, Migdadi HH, Alghamdi SS, Al-Faifi SA, Afzal M (2022) Response of tomato (*Solanum lycopersicum* L.) genotypes to heat stress using morphological and expression study. Plan Theory 11(5):615

Altaf MA, Shahid R, Ren M-X, Naz S, Altaf MM, Khan LU, Tiwari RK, Lal MK, Shahid MA, Kumar R, Nawaz MA, Jahan MS, Jan BL, Ahmad P (2022) Melatonin improves drought stress tolerance of tomato by modulating plant growth, root architecture, photosynthesis, and antioxidant defense system. Antioxidants 11:309. https://doi.org/10.3390/antiox11020309

Aravind B, Shreeraksha RJ, Poornima R, Ravichandran D, Krishnaraj PU, Chimmad VP, Mirajkar KK, Bagewadi B, Janila P, Pandey MK, Varshney RK, Nayak SN (2024) Impact of heat stress on physiological characteristics and expression of heat shock proteins (HSPs) in groundnut (*Arachis hypogaea* L.). Physiol Mol Biol Plants 30:1691–1706. https://doi.org/10.1007/s12298-024-01520-y

Arnao MB, Hernández-Ruiz J (2009) Chemical stress by different agents affects the melatonin content of barley roots. J Pineal Res 46:295–299. https://doi.org/10.1111/j.1600-079x.2008.00660.x

Arnao MB, Hernández-Ruiz J (2012) Growth conditions influence the melatonin content of tomato plants. Food Chem 138:1212–1214. https://doi.org/10.1016/j.foodchem.2012.10.077

Arnao MB, Hernández-Ruiz J (2015) Functions of melatonin in plants: a review. J Pineal Res 59:133–150. https://doi.org/10.1111/jpi.12253

Arnao MB, Hernández-Ruiz J (2017) Melatonin and its relationship to plant hormones. Ann Bot 121:195–207. https://doi.org/10.1093/aob/mcx114

Balla K, Karsai I, Bónis P, Kiss T, Berki Z, Horváth Á, Mayer M, Bencze S, Veisz O (2019) Heat stress responses in a large set of winter wheat cultivars (*Triticum aestivum* L.) depend on the timing and duration of stress. PLoS One 14:e0222639. https://doi.org/10.1371/journal.pone.0222639

Barman D, Kumar R, Ghimire OP, Ramesh R, Gupta S, Nagar S, Pal M, Dalal M, Chinnusamy V, Arora A (2024) Melatonin induces acclimation to heat stress and pollen viability by enhancing antioxidative defense in rice (*Oryza sativa* L.). Environ Exp Bot 220:105693

Beheshti F, Abdipour M, Hosseinifarahi M, Kelidari A (2023) Characterization of lavender essential oil and its chemical composition intercropped with Marigold as affected by polyamines and Arbuscular Mycorrhizal fungi inoculation. Commun Soil Sci Plant Anal 55:343–364. https://doi.org/10.1080/00103624.2023.2268657

Beilby MJ, Turi CE, Baker TC, Tymm FJ, Murch SJ (2015) Circadian changes in endogenous concentrations of indole-3-acetic acid, melatonin, serotonin, abscisic acid and jasmonic acid in Characeae (*Chara australis* Brown). Plant Signal Behav 10:e1082697. https://doi.org/10.1080/15592324.2015.1082697

Boccalandro HE, González CV, Wunderlin DA, Silva MF (2011) Melatonin levels, determined by LC-ESI-MS/MS, fluctuate during the day/night cycle in Vitis vinifera cv Malbec: evidence of its antioxidant role in fruits. J Pineal Res 51:226–232. https://doi.org/10.1111/j.1600-079x.2011.00884.x

Bose SK, Howlader P (2020) Melatonin plays multifunctional role in horticultural crops against environmental stresses: a review. Environ Exp Bot 176:104063. https://doi.org/10.1016/j.envexpbot.2020.104063

Bouabdelli S, Zeroual A, Meddi M, Assani A (2022) Impact of temperature on agricultural drought occurrence under the effects of climate change. Theor Appl Climatol 148:191–209. https://doi.org/10.1007/s00704-022-03935-7

Buttar ZA, Wu SN, Arnao MB, Wang C, Ullah I, Wang C (2020) Melatonin suppressed the heat stress-induced damage in wheat seedlings by modulating the antioxidant machinery. Plan Theory 9(7):809

Byeon Y, Park S, Kim Y, Park D, Lee S, Back K (2012) Light-regulated melatonin biosynthesis in rice during the senescence process in detached leaves. J Pineal Res 53:107–111. https://doi.org/10.1111/j.1600-079x.2012.00976.x

Calderini DF, Reynolds MP, Slafer GA (2006) Source - sink effects on grain weight of bread wheat, durum wheat, and triticale at different locations. Aust J Agric Res 57:227. https://doi.org/10.1071/ar05107

Cao L, Fahim AM, Liang X, Fan S, Song Y, Liu H, Ye F, Ma C, Zhang D, Lu X (2024) Melatonin enhances heat tolerance via increasing antioxidant enzyme activities and osmotic regulatory substances by upregulating zmenol expression in maize (*Zea mays* L.). Antioxidants 13(9):1144

Chang J, Guo Y, Yan J, Zhang Z, Yuan L, Wei C, Zhang Y, Ma J, Yang J, Zhang X, Li H (2021) The role of watermelon caffeic acid O-methyltransferase (ClCOMT1) in melatonin biosynthesis and abiotic stress tolerance. Hortic Res 8:210. https://doi.org/10.1038/s41438-021-00645-5

Chaturvedi AK, Bahuguna RN, Shah D, Pal M, Jagadish SVK (2017) High temperature stress during flowering and grain filling offsets beneficial impact of elevated CO2 on assimilate partitioning and sink-strength in rice. Sci Rep 7:8227. https://doi.org/10.1038/s41598-017-07464-6

Cheabu S, Moung-Ngam P, Arikit S, Vanavichit A, Malumpong C (2018) Effects of heat stress at vegetative and reproductive stages on spikelet fertility. Rice Sci 25:218–226. https://doi.org/10.1016/j.rsci.2018.06.005

Chebrolu KK, Fritschi FB, Ye S, Krishnan HB, Smith JR, Gillman JD (2016) Impact of heat stress during seed development on soybean seed metabolome. Metabolomics 12. https://doi.org/10.1007/s11306-015-0941-1

Chen S, Stefanova K, Siddique KHM, Cowling WA (2020) Transient daily heat stress during the early reproductive phase disrupts pod and seed development in *Brassica napus* L. Food Energy Secur 10. https://doi.org/10.1002/fes3.262

Cheng C, Liu Y, Fang W, Tao J, Yang Z, Yin Y (2020) iTRAQ-based proteomic and physiological analyses of mustard sprouts in response to heat stress. RSC Adv 10:6052–6062. https://doi.org/10.1039/c9ra10089j

Choudhary S, Bhat TM, Alwutayd KM, El-Moneim DA, Naaz N (2024) Salicylic acid enhances thermotolerance and antioxidant defense in *Trigonella foenum* graecum L. under heat stress. Heliyon 10:e27227. https://doi.org/10.1016/j.heliyon.2024.e27227

Cohen I, Zandalinas SI, Huck C, Fritschi FB, Mittler R (2020) Meta-analysis of drought and heat stress combination impact on crop yield and yield components. Physiol Plant 171:66–76. https://doi.org/10.1111/ppl.13203

Dabaghkar Y, Eghlima G, Babashpour-Asl M, Mohammadi M, Ghorbanpour M (2024) Exploring the impact of exogenous melatonin on agro-morphological characteristics, carvacrol, and rosmarinic acid production in *Satureja rechingeri* Jamzad under drought stress. Chem Biol Technol Agric 11. https://doi.org/10.1186/s40538-024-00643-4

Dawood MFA, Moursi YS, Amro A, Baenziger PS, Sallam A (2020) Investigation of heat-induced changes in the grain yield and grains metabolites, with molecular insights on the candidate genes in barley. Agronomy 10:1730. https://doi.org/10.3390/agronomy10111730

Devi R, Behera B, Raza MB, Mangal V, Altaf MA, Kumar R, Kumar A, Tiwari RK, Lal MK, Singh B (2021) An insight into microbes mediated heavy metal detoxification in plants: a review. J Soil Sci Plant Nutr 22:1–23

Distéfano AM, Bauer V, Cascallares M, López GA, Fiol DF, Zabaleta E, Pagnussat GC (2024) Heat stress in plants: sensing, signalling and ferroptosis. J Exp Bot 76:1357. https://doi.org/10.1093/jxb/erae296

Doğru A (2021) Effects of heat stress on photosystem II activity and antioxidant enzymes in two maize cultivars. Planta 253:85. https://doi.org/10.1007/s00425-021-03611-6

Ejaz S, Fahad S, Anjum MA, Nawaz A, Naz S, Hussain S, Ahmad S (2020) Role of Osmolytes in the mechanisms of antioxidant defense of plants. Sustain Agric Rev:95–117. https://doi.org/10.1007/978-3-030-38881-2_4

Elmongy MS, Abd El-Baset MM (2024) Melatonin application induced physiological and molecular changes in carnation (*Dianthus caryophyllus* L.) under heat stress. Horticulturae 10(2):122

Elsayed S, El-Naby SA, EEl-Gohary A, Hendawy S (2021) Mitigation of heat stress effects on Chamomile and its essential oil using melatonin or gibberellic acid and some agricultural treatments. Egypt J Chem 0. https://doi.org/10.21608/ejchem.2021.80586.3993

Fahad S, Bajwa AA, Nazir U, Anjum SA, Farooq A, Zohaib A, Sadia S, Nasim W, Adkins S, Saud S, Ihsan MZ, Alharby H, Wu C, Wang D, Huang J (2017) Crop production under drought and heat stress: plant responses and management options. Front Plant Sci 8. https://doi.org/10.3389/fpls.2017.01147

Fan J, Xie Y, Zhang Z, Chen L (2018) Melatonin: a multifunctional factor in plants. Int J Mol Sci 19:1528. https://doi.org/10.3390/ijms19051528

Ghareeb RY, El-Din NGE-DS, Maghraby DME, Ibrahim DSS, Abdel-Megeed A, Abdelsalam NR (2022) Nematicidal activity of seaweed-synthesized silver nanoparticles and extracts against *Meloidogyne incognita* on tomato plants. Sci Rep 12:3841. https://doi.org/10.1038/s41598-022-06600-1

Greer DH, Weedon MM (2011) Modelling photosynthetic responses to temperature of grapevine (*Vitis vinifera* cv. Semillon) leaves on vines grown in a hot climate. Plant Cell Environ 35:1050–1064. https://doi.org/10.1111/j.1365-3040.2011.02471.x

Gulen H, Eris A (2003) Some physiological changes in strawberry (Fragaria× ananassa 'Camarosa') plants under heat stress. J Hortic Sci Biotechnol 78(6):894–898

Haque MS, Husna MT, Uddin MN, Hossain MA, Sarwar AKMG, Ali OM, Latef AAHA, Hossain A (2021) Heat stress at early reproductive stage differentially alters several physiological and biochemical traits of three tomato cultivars. Horticulturae 7:330. https://doi.org/10.3390/horticulturae7100330

Hardeland R (2016) Melatonin in plants – diversity of levels and multiplicity of functions. Front Plant Sci 7. https://doi.org/10.3389/fpls.2016.00198

Hassan MU, Chattha MU, Khan I, Chattha MB, Barbanti L, Aamer M, Iqbal MM, Nawaz M, Mahmood A, Ali A, Aslam MT (2020) Heat stress in cultivated plants: nature, impact, mechanisms, and mitigation strategies—a review. Plant Biosyst an International Journal Dealing with All Aspects of Plant Biology 155:211–234. https://doi.org/10.1080/11263504.2020.1727987

Haydari M, Maresca V, Rigano D, Taleei A, Shahnejat-Bushehri AA, Hadian J, Sorbo S, Guida M, Manna C, Piscopo M, Notariale R, De Ruberto F, Fusaro L, Basile A (2019) Salicylic acid and melatonin alleviate the effects of heat stress on essential oil composition and antioxidant enzyme activity in Mentha × piperita and Mentha arvensis L. Antioxidants 8:547. https://doi.org/10.3390/antiox8110547

He H, He L (2020) Crosstalk between melatonin and nitric oxide in plant development and stress responses. Physiol Plant 170:218–226. https://doi.org/10.1111/ppl.13143

Hernández IG, Gomez FJV, Cerutti S, Arana MV, Silva MF (2015) Melatonin in Arabidopsis thaliana acts as plant growth regulator at low concentrations and preserves seed viability at high concentrations. Plant Physiol Biochem 94:191–196. https://doi.org/10.1016/j.plaphy.2015.06.011

Huang S, Jin S (2024) Melatonin interaction with other Phytohormones in the regulation of abiotic stresses in horticultural plants. Antioxidants 13:663. https://doi.org/10.3390/antiox13060663

Hungria M, Kaschuk G (2013) Regulation of N2 fixation and NO3−/NH4+ assimilation in nodulated and N-fertilized *Phaseolus vulgaris* L. exposed to high temperature stress. Environ Exp Bot 98:32–39. https://doi.org/10.1016/j.envexpbot.2013.10.010

Imran M, Khan MA, Shahzad R, Bilal S, Khan M, Yun B-W, Khan AL, Lee I-J (2021) Melatonin ameliorates Thermotolerance in soybean seedling through balancing redox homeostasis and modulating antioxidant defense, Phytohormones and polyamines biosynthesis. Molecules 26:5116. https://doi.org/10.3390/molecules26175116

Iqbal N, Fatma M, Gautam H, Umar S, Sofo A, D'ippolito I, Khan NA (2021a) The crosstalk of melatonin and hydrogen sulfide determines photosynthetic performance by regulation of carbohydrate metabolism in wheat under heat stress. Plan Theory 10:1778. https://doi.org/10.3390/plants10091778

Iqbal N, Umar S, Khan NA, Corpas FJ (2021b) Nitric oxide and hydrogen sulfide coordinately reduce glucose sensitivity and decrease oxidative stress via Ascorbate-Glutathione cycle in heat-stressed wheat (*Triticum aestivum* L.) plants. Antioxidants 10:108. https://doi.org/10.3390/antiox10010108

Iqbal M, Raja NI, Mashwani Z-U-R, Yasmeen F, Hussain M, Ejaz M, Abasi F, Ehsan M, Ikram M, Proćków J (2024) Insight into carbohydrate metabolism, protein quantification and mineral regulation in wheat (*Triticum aestivum* L.) by the action of green synthesized silver nanoparticles (AgNPs) against heat stress. J Biomol Struct Dyn:1–15. https://doi.org/10.1080/07391102.2024.2311333

Irshad A, Rehman RNU, Kareem HA, Yang P, Hu T (2021) Addressing the challenge of cold stress resilience with the synergistic effect of Rhizobium inoculation and exogenous melatonin application in *Medicago truncatula*. Ecotoxicol Environ Saf 226:112816. https://doi.org/10.1016/j.ecoenv.2021.112816

Jagadish SVK, Way DA, Sharkey TD (2021) Plant heat stress: concepts directing future research. Plant Cell Environ 44:1992–2005. https://doi.org/10.1111/pce.14050

Jahan MS, Shu S, Wang Y, Chen Z, He M, Tao M, Sun J, Guo S (2019) Melatonin alleviates heat-induced damage of tomato seedlings by balancing redox homeostasis and modulating polyamine and nitric oxide biosynthesis. BMC Plant Biol 19:414. https://doi.org/10.1186/s12870-019-1992-7

Jahan MS, Guo S, Sun J, Shu S, Wang Y, El-Yazied AA, Alabdallah NM, Hikal M, Mohamed MHM, Ibrahim MFM, Hasan MM (2021a) Melatonin-mediated photosynthetic performance of tomato seedlings under high-temperature stress. Plant Physiol Biochem 167:309–320. https://doi.org/10.1016/j.plaphy.2021.08.002

Jahan MS, Shu S, Wang Y, Hasan MM, El-Yazied AA, Alabdallah NM, Hajjar D, Altaf MA, Sun J, Guo S (2021b) Melatonin pretreatment confers heat tolerance and repression of heat-induced senescence in tomato through the modulation of ABA- and GA-mediated pathways. Front Plant Sci 12. https://doi.org/10.3389/fpls.2021.650955

Jan R, Asaf S, Numan M, Lubna N, Kim K-M (2021) Plant secondary metabolite biosynthesis and transcriptional regulation in response to biotic and abiotic stress conditions. Agronomy 11:968. https://doi.org/10.3390/agronomy11050968

Jian H, Gao Z, Guo Y, Xu X, Li X, Yu M, Liu G, Bian D, Cui Y, Du X (2024) Supplemental irrigation mitigates yield loss of maize through reducing canopy temperature under heat stress. Agric Water Manag 299:108888. https://doi.org/10.1016/j.agwat.2024.108888

Jiang T, Feng Y, Zhao M, Meng L, Li J, Zhang X, Ai X, Bi H (2024) Regulation mechanism of melatonin on photosynthesis of cucumber under high temperature stress. Veg Res 4:0. https://doi.org/10.48130/vegres-0024-0037

Kan Y, Mu X-R, Gao J, Lin H-X, Lin Y (2023) The molecular basis of heat stress responses in plants. Mol Plant 16:1612–1634. https://doi.org/10.1016/j.molp.2023.09.013

Katano K, Honda K, Suzuki N (2018) Integration between ROS regulatory systems and other signals in the regulation of various types of heat responses in plants. Int J Mol Sci 19:3370. https://doi.org/10.3390/ijms19113370

Kaya C, Okant M, Ugurlar F, Alyemeni MN, Ashraf M, Ahmad P (2019) Melatonin-mediated nitric oxide improves tolerance to cadmium toxicity by reducing oxidative stress in wheat plants. Chemosphere 225:627–638. https://doi.org/10.1016/j.chemosphere.2019.03.026

Kaya C, Ashraf M, Alyemeni MN, Ahmad P (2020) The role of nitrate reductase in brassinosteroid-induced endogenous nitric oxide generation to improve cadmium stress tolerance of pepper plants by upregulating the ascorbate-glutathione cycle. Ecotoxicol Environ Saf 196:110483. https://doi.org/10.1016/j.ecoenv.2020.110483

Ke Q, Ye J, Wang B, Ren J, Yin L, Deng X, Wang S (2018) Melatonin mitigates salt stress in wheat seedlings by modulating polyamine metabolism. Front Plant Sci 9. https://doi.org/10.3389/fpls.2018.00914

Khan HMT, Balal RM, Hussain Z, Javed SA, Jaffar MT, Alsahli AA (2023) Exogenous application of melatonin mitigate the heat stress in different tomato (*Solanum lycopersicum* L.) cultivars. J King Saud Univ Sci 36:103086. https://doi.org/10.1016/j.jksus.2023.103086

Killi D, Raschi A, Bussotti F (2020) Lipid peroxidation and chlorophyll fluorescence of photosystem II performance during drought and heat stress is associated with the antioxidant capacities of C3 sunflower and C4 maize varieties. Int J Mol Sci 21:4846. https://doi.org/10.3390/ijms21144846

Kumar S, Kaur R, Kaur N, Bhandhari K, Kaushal N, Gupta K, Bains TS, Nayyar H (2011) Heat-stress induced inhibition in growth and chlorosis in mungbean (*Phaseolus aureus* Roxb.) is partly mitigated by ascorbic acid application and is related to reduction in oxidative stress. Acta Physiol Plant 33:2091–2101. https://doi.org/10.1007/s11738-011-0748-2

Kumar S, Yu R, Liu Y, Liu Y, Khan MN, Liu Y, Wang M, Zhu G (2024) Exogenous melatonin enhances heat stress tolerance in sweetpotato by modulating antioxidant defense system, osmotic homeostasis and stomatal traits. Hortic Plant J 11:431. https://doi.org/10.1016/j.hpj.2023.12.006

Laddomada B, Blanco A, Mita G, D'Amico L, Singh RP, Ammar K, Crossa J, Guzmán C (2021) Drought and heat stress impacts on phenolic acids accumulation in durum wheat cultivars. Food Secur 10:2142. https://doi.org/10.3390/foods10092142

Lal MK, Sharma N, Adavi SB, Sharma E, Altaf MA, Tiwari RK, Kumar R, Kumar A, Dey A, Paul V, Singh B (2022) From source to sink: mechanistic insight of photoassimilates synthesis and partitioning under high temperature and elevated [CO2]. Plant Mol Biol 110(4):305–324

Li Z-G, Yi X-Y, Li Y-T (2014) Effect of pretreatment with hydrogen sulfide donor sodium hydrosulfide on heat tolerance in relation to antioxidant system in maize (*Zea mays*) seedlings. Biologia 69:1001–1009. https://doi.org/10.2478/s11756-014-0396-2

Li Z-G, Xu Y, Bai L-K, Zhang S-Y, Wang Y (2018) Melatonin enhances thermotolerance of maize seedlings (*Zea mays* L.) by modulating antioxidant defense, methylglyoxal detoxification, and osmoregulation systems. Protoplasma 256:471–490. https://doi.org/10.1007/s00709-018-1311-4

Li X, Li M-H, Deng W-W, Ahammed GJ, Wei J-P, Yan P, Zhang L-P, Fu J-Y, Han W-Y (2020) Exogenous melatonin improves tea quality under moderate high temperatures by increasing epigallocatechin-3-gallate and theanine biosynthesis in *Camellia sinensis* L. J Plant Physiol 253:153273. https://doi.org/10.1016/j.jplph.2020.153273

Li H, Guo Y, Lan Z, Xu K, Chang J, Ahammed GJ, Ma J, Wei C, Zhang X (2021) Methyl jasmonate mediates melatonin-induced cold tolerance of grafted watermelon plants. Hortic Res 8:57. https://doi.org/10.1038/s41438-021-00496-0

Li M, Zhou J, Du J, Li X, Sun Y, Wang Z, Lin Y, Zhang Y, Wang Y, He W, Wang X (2022) Comparative physiological and transcriptomic analyses of improved heat stress tolerance in celery (*Apium graveolens* L.) caused by exogenous melatonin. Int J Mol Sci 23(19):11382

Li Q, Bian Y, Li R, Yang Z, Liu N, Yu J (2023a) Chitosan-enhanced heat tolerance associated with alterations in antioxidant defense system and gene expression in creeping bentgrass. Grass Res 3:0. https://doi.org/10.48130/gr-2023-0007

Li Z, Zhou M, Zeng W, Zhang Y, Liu L, Liu W, Peng Y (2023b) Root metabolites remodeling regulated by γ-aminobutyric acid (GABA) improves adaptability to high temperature in creeping bentgrass. Plant Soil 500:181–195. https://doi.org/10.1007/s11104-023-05905-y

Liang C, Zheng G, Li W, Wang Y, Hu B, Wang H, Wu H, Qian Y, Zhu X, Tan D, Chen S, Chu C (2015) Melatonin delays leaf senescence and enhances salt stress tolerance in rice. J Pineal Res 59:91–101. https://doi.org/10.1111/jpi.12243

Liang D, Gao F, Ni Z, Lin L, Deng Q, Tang Y, Wang X, Luo X, Xia H (2018) Melatonin improves heat tolerance in kiwifruit seedlings through promoting antioxidant enzymatic activity and glutathione S-transferase transcription. Molecules 23(3):584

Liu Y, Ye N, Liu R, Chen M, Zhang J (2010) H2O2 mediates the regulation of ABA catabolism and GA biosynthesis in Arabidopsis seed dormancy and germination. J Exp Bot 61:2979–2990. https://doi.org/10.1093/jxb/erq125

Lobell DB, Bänziger M, Magorokosho C, Vivek B (2011) Nonlinear heat effects on African maize as evidenced by historical yield trials. Nat Clim Chang 1:42–45. https://doi.org/10.1038/nclimate1043

Lohani N, Singh MB, Bhalla PL (2022) Short-term heat stress during flowering results in a decline in Canola seed productivity. J Agron Crop Sci 208(4):486–496

Ma G, Zhang M, Xu J, Zhou W, Cao L (2020) Transcriptomic analysis of short-term heat stress response in *Pinellia ternata* provided novel insights into the improved thermotolerance by spermidine and melatonin. Ecotoxicol Environ Saf 202:110877. https://doi.org/10.1016/j.ecoenv.2020.110877

Maheswari M, Yadav SK, Shanker AK, Kumar MA, Venkateswarlu B (2011) Overview of plant stresses: mechanisms, adaptations and research pursuit. In: Springer eBooks, pp 1–18. https://doi.org/10.1007/978-94-007-2220-0_1

Manafi H, Baninasab B, Gholami M, Talebi M, Khanizadeh S (2021) Exogenous melatonin alleviates heat-induced oxidative damage in strawberry (Fragaria × ananassa Duch. cv. Ventana) plant. J Plant Growth Regul 41:52–64. https://doi.org/10.1007/s00344-020-10279-x

Manafi H, Baninasab B, Gholami M, Talebi M, Khanizadeh S (2022) Exogenous melatonin alleviates heat-induced oxidative damage in strawberry (Fragaria× ananassa Duch. cv. Ventana) plant. J Plant Growth Regul 41:52–64

Marchand FL, Mertens S, Kockelbergh F, Beyens L, Nijs I (2005) Performance of High Arctic tundra plants improved during but deteriorated after exposure to a simulated extreme temperature event. Glob Chang Biol 11:2078–2089. https://doi.org/10.1111/j.1365-2486.2005.01046.x

Meena KK, Sorty AM, Bitla UM, Choudhary K, Gupta P, Pareek A, Singh DP, Prabha R, Sahu PK, Gupta VK, Singh HB, Krishanani KK, Minhas PS (2017) Abiotic stress responses and microbe-mediated mitigation in plants: the omics strategies. Front Plant Sci 8. https://doi.org/10.3389/fpls.2017.00172

Mumithrakamatchi AK, Alagarswamy S, Anitha K, Djanaguiraman M, Kalarani MK, Swarnapriya R, Marimuthu S, Vellaikumar S, Kanagarajan S (2024) Melatonin imparts tolerance to combined drought and high-temperature stresses in tomato through osmotic adjustment and ABA accumulation. Front Plant Sci 15. https://doi.org/10.3389/fpls.2024.1382914

Munjonji L, Ayisi KK, Mafeo TP, Maphanga T, Mabitsela KE (2021) Seasonal variation in soil CO2 emission and leaf gas exchange of well-managed commercial *Citrus sinensis* (L.) orchards. Plant Soil 465:65–81. https://doi.org/10.1007/s11104-021-04986-x

Mushtaq N, Iqbal S, Hayat F, Raziq A, Ayaz A, Zaman W (2022) Melatonin in micro-tom tomato: improved drought tolerance via the regulation of the photosynthetic apparatus, membrane stability, Osmoprotectants, and root system. Life 12:1922. https://doi.org/10.3390/life12111922

Mushtaq N, Altaf MA, Shu H, Lu X, Cheng S, El-Sheikh MA, Ahmad P, Fu H, Wang Z (2024) The induction of polyamines metabolism pathway and membrane stability with silicon alleviate the vanadium toxicity in pepper plants. J Hazard Mater 484:136665. https://doi.org/10.1016/j.jhazmat.2024.136665

Nahar K, Ahamed KU, Fujita M (2010) Phenological variation and its relation with yield in several wheat (*Triticum aestivum* L.) cultivars under normal and late sowing mediated heat stress condition. Notulae Scientia Biologicae 2:51–56. https://doi.org/10.15835/nsb234723

Nayak L, Lal MK, Tiwari RK, Kumar R, Lal P, Das R, Behera B, Thakur P, Altaf MA, Kumar A (2023) A balancing act: exploring the interplay between HSPs and osmoprotectants in temperature stress responses. S Afr J Bot 162:64–71

Ortiz AC, De Smet I, Sozzani R, Locke AM (2021) Field-grown soybean shows genotypic variation in physiological and seed composition responses to heat stress during seed development. Environ Exp Bot 195:104768. https://doi.org/10.1016/j.envexpbot.2021.104768

Pu Y-J, Cisse EHM, Zhang L-J, Miao L-F, Nawaz M, Yang F (2021) Coupling exogenous melatonin with Ca2+ alleviated chilling stress in *Dalbergia odorifera* T. Chen. Trees 35:1541–1554. https://doi.org/10.1007/s00468-021-02134-7

Puglia GD (2023) Reactive oxygen and nitrogen species (RONS) signalling in seed dormancy release, perception of environmental cues, and heat stress response. Plant Growth Regul 103:9–32. https://doi.org/10.1007/s10725-023-01094-x

Qi Z-Y, Wang K-X, Yan M-Y, Kanwar M, Li D-Y, Wijaya L, Alyemeni M, Ahmad P, Zhou J (2018) Melatonin alleviates high temperature-induced pollen abortion in *Solanum lycopersicum*. Molecules 23:386. https://doi.org/10.3390/molecules23020386

Qiulan H, Xi W, Ying L, Wei Z, Jing L, Munir S, Meirong H (2023) Elucidating the molecular mechanisms of exogenous melatonin for improving heat tolerance in *Solanum tuberosum* L. seedlings. Sci Hortic 322:112423

Rehmani MIA, Wei G, Hussain N, Ding C, Li G, Liu Z, Wang S, Ding Y (2013) Yield and quality responses of two indica rice hybrids to post-anthesis asymmetric day and night open-field warming in lower reaches of Yangtze River delta. Field Crop Res 156:231–241. https://doi.org/10.1016/j.fcr.2013.09.019

Rosales EP, Iannone MF, Groppa MD, Benavides MP (2011) Polyamines modulate nitrate reductase activity in wheat leaves: involvement of nitric oxide. Amino Acids 42:857–865. https://doi.org/10.1007/s00726-011-1001-4

Sahoo G, Mishra P, Wani AM, Sharma A, Mishra D, Patra D, Mishra I, Behera M (2023) Impact of climate change on livelihood security and biodiversity – issues and mitigation strategies. In: Springer climate, pp 1–27. https://doi.org/10.1007/978-3-031-19059-9_1

Sato S, Kamiyama M, Iwata T, Makita N, Furukawa H, Ikeda H (2006) Moderate increase of mean daily temperature adversely affects fruit set of *Lycopersicon esculentum* by disrupting specific physiological processes in male reproductive development. Ann Bot 97:731–738. https://doi.org/10.1093/aob/mcl037

Sehar Z, Fatma M, Khan S, Mir IR, Abdi G, Khan NA (2023) Melatonin influences methyl jasmonate-induced protection of photosynthetic activity in wheat plants against heat stress by regulating ethylene-synthesis genes and antioxidant metabolism. Sci Rep 13:7468. https://doi.org/10.1038/s41598-023-34682-y

Sha Y, Liu Z, Hao Z, Huang Y, Shao H, Feng G, Chen F, Mi G (2023) Root growth, root senescence and root system architecture in maize under conservative strip tillage system. Plant Soil 495:253–269. https://doi.org/10.1007/s11104-023-06322-x

Shi H, Tan D, Reiter RJ, Ye T, Yang F, Chan Z (2015) Melatonin induces class A1 heat-shock factors (HSFA1s) and their possible involvement of thermotolerance in Arabidopsis. J Pineal Res 58:335–342. https://doi.org/10.1111/jpi.12219

Singh B, Kukreja S, Goutam U (2020) Impact of heat stress on potato (*Solanum tuberosum* L.): present scenario and future opportunities. J Hortic Sci Biotechnol 95(4):407–424

Sita K, Sehgal A, Bhandari K, Kumar J, Kumar S, Singh S, Siddique KH, Nayyar H (2018) Impact of heat stress during seed filling on seed quality and seed yield in lentil (*Lens culinaris*Medikus) genotypes. J Sci Food Agric 98:5134–5141. https://doi.org/10.1002/jsfa.9054

Su R, Luo J, Wang Y, Xiao Y, Liu X, Deng H, Lu X, Chen Q, Chen G, Tang W, Zhang G (2024) GDSL lipase gene HTA1 negatively regulates heat tolerance in rice seedlings by regulating reactive oxygen species accumulation. Antioxidants 13:592. https://doi.org/10.3390/antiox13050592

Sun C, Liu L, Wang L, Li B, Jin C, Lin X (2020) Melatonin: a master regulator of plant development and stress responses. J Integr Plant Biol 63:126–145. https://doi.org/10.1111/jipb.12993

Talaat NB (2023) Drought stress alleviator melatonin reconfigures water-stressed barley (*Hordeum vulgare* L.) plants' photosynthetic efficiency, antioxidant capacity, and endogenous phytohormone profile. Int J Mol Sci 24:16228. https://doi.org/10.3390/ijms242216228

Tan D, Manchester LC, Liu X, Rosales-Corral SA, Acuna-Castroviejo D, Reiter RJ (2012) Mitochondria and chloroplasts as the original sites of melatonin synthesis: a hypothesis related to melatonin's primary function and evolution in eukaryotes. J Pineal Res 54:127–138. https://doi.org/10.1111/jpi.12026

Tang Q, Li C, Ge Y, Li X, Cheng Y, Hou J, Li J (2020) Exogenous application of melatonin maintains storage quality of jujubes by enhancing anti-oxidative ability and suppressing the activity of cell wall-degrading enzymes. LWT 127:109431. https://doi.org/10.1016/j.lwt.2020.109431

Teng LI, Zhang XP, Qing LIU, Jin LIU, Chen YQ, Peng SUI (2022) Yield penalty of maize (*Zea mays* L.) under heat stress in different growth stages: a review. J Integr Agric 21(9):2465–2476

Tiwari RK, Lal MK, Naga KC, Kumar R, Chourasia KN, Kumar D, Sharma S (2020) Emerging roles of melatonin in mitigating abiotic and biotic stresses of horticultural crops. Sci Hortic 272:109592

Tiwari RK, Lal MK, Kumar R, Chourasia KN, Naga KC, Kumar D, Das SK, Zinta G (2021a) Mechanistic insights on melatonin-mediated drought stress mitigation in plants. Physiol Plant 172(2):1212–1226

Tiwari RK, Lal MK, Kumar R, Mangal V, Altaf MA, Sharma S, Singh B, Kumar M (2021b) Insight into melatonin-mediated response and signaling in the regulation of plant defense under biotic stress. Plant Mol Biol 109:1–15

Ullah A, Nadeem F, Nawaz A, Siddique KHM, Farooq M (2021) Heat stress effects on the reproductive physiology and yield of wheat. J Agron Crop Sci 208:1–17. https://doi.org/10.1111/jac.12572

Utami D, Aryanti E (2021) Impact of heat stress on germination and seedling growth of chili pepper (*Capsicum annuum* L.). In: IOP conference series: earth and environmental science, vol 637, No. 1. IOP Publishing, p 012032

Van Der Merwe R, Labuschagne MT, Herselman L, Hugo A (2015) Effect of heat stress on seed yield components and oil composition in high- and mid-oleic sunflower hybrids. S Afr J Plant Soil 32:121–128. https://doi.org/10.1080/02571862.2015.1018354

Verma KK, Song X, Kumari A, Jagadesh M, Singh SK, Bhatt R, Singh M, Seth CS, Li Y (2024) Climate change adaptation: challenges for agricultural sustainability. Plant Cell Environ 48:2522. https://doi.org/10.1111/pce.15078

Waadt R, Seller CA, Hsu P-K, Takahashi Y, Munemasa S, Schroeder JI (2022) Plant hormone regulation of abiotic stress responses. Nat Rev Mol Cell Biol 23:680–694. https://doi.org/10.1038/s41580-022-00479-6

Wahid A, Close TJ (2007) Expression of dehydrins under heat stress and their relationship with water relations of sugarcane leaves. Biol Plant 51:104–109. https://doi.org/10.1007/s10535-007-0021-0

Wang LY, Liu JL, Wang WX, Sun Y (2015) Exogenous melatonin improves growth and photosynthetic capacity of cucumber under salinity-induced stress. Photosynthetica 54:19–27. https://doi.org/10.1007/s11099-015-0140-3

Wang X, Zhang H, Xie Q, Liu Y, Lv H, Bai R, Ma R, Li X, Zhang X, Guo Y-D, Zhang N (2020) SlSNAT interacts with HSP40, a molecular Chaperone, to regulate melatonin biosynthesis and promote Thermotolerance in tomato. Plant Cell Physiol 61:909–921. https://doi.org/10.1093/pcp/pcaa018

Wang H, Song C, Fang S, Wang Z, Song S, Jiao J, Wang M, Zheng X, Bai T (2021) Genome-wide identification and expression analysis of the ASMT gene family reveals their role in abiotic stress tolerance in apple. Sci Hortic 293:110683. https://doi.org/10.1016/j.scienta.2021.110683

Wang X, Altaf MA, Hao Y, Wang Z, Zhu G (2023) Effect of heat stress on root architecture, photosynthesis, and antioxidant profile of water spinach (*Ipomoea aquatica* Forsk) seedlings. Horticulturae 9(8):923

Weng J, Rehman A, Li P, Chang L, Zhang Y, Niu Q (2022) Physiological and transcriptomic analysis reveals the responses and difference to high temperature and humidity stress in two melon genotypes. Int J Mol Sci 23(2):734

Wu P, Ma Y, Ahammed GJ, Hao B, Chen J, Wan W, Zhao Y, Cui H, Xu W, Cui J, Liu H (2022) Insights into melatonin-induced photosynthetic electron transport under low-temperature stress in cucumber. Front Plant Sci 13. https://doi.org/10.3389/fpls.2022.1029854

Xia H, Zhou Y, Deng H, Lin L, Deng Q, Wang J, Lv X, Zhang XA, Liang D (2021) Melatonin improves heat tolerance in *Actinidia deliciosa* via carotenoid biosynthesis and heat shock proteins expression. Physiol Plant 172(3):1582–1593

Xing X, Ding Y, Jin J, Song A, Chen S, Chen F, Fang W, Jiang J (2021) Physiological and transcripts analyses reveal the mechanism by which melatonin alleviates heat stress in chrysanthemum seedlings. Front Plant Sci 12:673236

Xu W, Cai S, Zhang Y, Wang Y, Ahammed GJ, Xia X, Shi K, Zhou Y, Yu J, Reiter RJ, Zhou J (2016) Melatonin enhances thermotolerance by promoting cellular protein protection in tomato plants. J Pineal Res 61:457–469. https://doi.org/10.1111/jpi.12359

Yang F, Jørgensen AD, Li H, Søndergaard I, Finnie C, Svensson B, Jiang D, Wollenweber B, Jacobsen S (2011) Implications of high-temperature events and water deficits on protein profiles in wheat (*Triticum aestivum*L. cv. Vinjett) grain. Proteomics 11:1684–1695. https://doi.org/10.1002/pmic.201000654

Yasmeen H, Rasheed R, Ashraf MA, Zafar S, Ali S (2024) Allantoin regulated oxidative defense, secondary metabolism and ions homeostasis in maize (*Zea mays* L.) under heat stress. Physiol Mol Biol Plants 30:1719–1739. https://doi.org/10.1007/s12298-024-01519-5

Ye Z-P, Robakowski P, Suggett DJ (2012) A mechanistic model for the light response of photosynthetic electron transport rate based on light harvesting properties of photosynthetic pigment molecules. Planta 237:837–847. https://doi.org/10.1007/s00425-012-1790-z

Yu Y, Qiu H, Wang H, Wang C, He C, Xu M, Zhang W, Zhang Z, Su L (2023) Melatonin improves saponin biosynthesis and primary root growth in Psammosilene tunicoides hairy roots through multiple hormonal signaling and transcriptional pathways. Ind Crop Prod 200:116819. https://doi.org/10.1016/j.indcrop.2023.116819

Zahra N, Hafeez MB, Ghaffar A, Kausar A, Zeidi MA, Siddique KHM, Farooq M (2022) Plant photosynthesis under heat stress: effects and management. Environ Exp Bot 206:105178. https://doi.org/10.1016/j.envexpbot.2022.105178

5 Insights into the Role of Melatonin in Regulating Ionic Homeostasis and Redox Networks Under Salinity Stress

Abazar Ghorbani, Abolghassem Emamverdian, Nasim Roodbari, Azade Masoumi, Xin-Xin Zhang, and Moxian Chen

Abstract

Salinity is a significant non-living environmental factor that negatively impacts plant growth and biomass. This chapter explores the multifaceted role of melatonin in enhancing salinity resilience through its regulatory effects on ionic homeostasis, redox networks, and stress-responsive pathways. By modulating ion transporters, enhancing antioxidant defenses, and regulating key genetic and molecular networks, melatonin alleviates ionic and oxidative stress in plants. Advances in omics technologies have provided deeper insights into melatonin's influence on stress adaptation, while biotechnological applications such as priming, integration with biostimulants, and nanotechnology have demonstrated its potential in improving crop resilience. Future research directions emphasize bridging the gap between laboratory findings and field applications to harness melatonin's potential for sustainable agriculture.

A. Ghorbani · X.-X. Zhang · M. Chen (✉)
National Key Laboratory of Green Pesticide, Key Laboratory of Green Pesticide and Agricultural Bioengineering, Ministry of Education, Center for R&D of Fine Chemicals of Guizhou University, Guiyang, China

A. Emamverdian
Co-Innovation Center for Sustainable Forestry in Southern China, Nanjing Forestry University, Nanjing, China

N. Roodbari
Department of Biology, Kahnooj Branch, Islamic Azad University, Kahnooj, Iran

A. Masoumi
Department of Biology, Faculty of Sciences, Mashhad Branch, Islamic Azad University, Mashhad, Iran

R. K. Tiwari et al. (eds.), *Melatonin: Signal Transduction Mechanisms and Defense Networks in Plants*, https://doi.org/10.1007/978-981-96-8869-2_5

Keywords
Melatonin · Salinity stress · Ionic homeostasis · Redox networks · Omics technologies · Sustainable agriculture

5.1 Introduction

Salt stress is a major environmental factor that significantly limits agricultural output on a global scale. With an estimated 20% of irrigated farmland and 2% of dryland agriculture affected by salinity, understanding plant adaptation mechanisms to high salt concentrations has become more urgent (Ghorbani et al. 2018a; Zhou et al. 2023). Salinity harms plants through a combination of ionic and osmotic stress, secondary oxidative damage, disrupting cellular processes, and reducing growth and productivity. These stresses arise when excessive salts, primarily NaCl, accumulate in the soil, interfering with water uptake and ionic balance in cells (Ghorbani et al. 2019; Atta et al. 2023; Chen et al. 2023).

Plants in saline situations suffer from a twofold challenge: to keep the water potential at an adequate level to avoid dehydration and to keep the ionic balance to avoid the toxicity from excessive Na and Cl ions (Hao et al. 2021; Sha et al. 2024). Moreover, salinity induces excessive reactive oxygen species (ROS), which interact with cellular bio-components (Ghorbani et al. 2025; Pehlivan et al. 2025). Despite their damaging potential, ROS act as signaling molecules that modulate numerous physiological and molecular pathways in plants (Atta et al. 2023; Delai et al. 2024; Cao et al. 2024). Thus, a balance between the generation and detoxification of ROS is very critical for plant survival during conditions of salinity stress.

Nowadays, melatonin, which was first discovered in animals, has also become a subject discussed in plant studies as a major plant mediator in the stress response. Melatonin has contributed to the maintenance of ion homeostasis and redox balance as an antioxidant and signaling molecule (Altaf et al. 2022). Its mode of action was suggested to involve up-regulation of the expression of ion transporters for Na and K, decrease of ROS via its scavenging action, and activation of antioxidant defense. In addition, melatonin influences stress-responsive hormonal pathways and interacts with redox signaling networks, which drive precise responses in plants during salinity challenges (Khan et al. 2022).

This chapter summarizes the regulatory mechanisms underlying ionic homeostasis and redox networks, focusing mainly on the salinity-stress-ameliorating action of melatonin. It explains how melatonin integrates ion transport and antioxidant systems and how recent genetic and biotechnological advances may raise salt tolerance to another level. In this way, it further elaborates on how melatonin enhances the resilience of crops in saline environments from a mechanistic viewpoint.

5.2 Salinity Stress and Its Physiological Impacts

Plants can absorb soil water when their water potential is lower than that of the soil. The presence of excessive soluble salts in the soil, like Na, reduces the root zone water potential and, accordingly, reduces the water availability, thus reducing its absorption by plants (Zhou et al. 2023; Sepehry Javan et al. 2024). Consequently, salinity imposes osmotic stress or water deficit, which is a key obstacle to those growing in saline environments. Water deficit adversely affects water use efficiency, disrupts water relations, and reduces cell turgor in plants (Huang et al. 2017). Additionally, it causes disruption of photosynthetic pigments and stomatal closure (Zahedi et al. 2022). Salinity-induced stomatal closure decreases CO_2 availability for fixation, and the CO_2-fixing enzyme activity is impaired under salinity. In chloroplasts, the proton motive force needed for ATP synthesis relies on the exact timing of changes in pH and electrochemical potential across the thylakoid membranes. However, sodium can upset the balance as it carries a positive charge and changes the pH (Ghorbani et al. 2019; Atta et al. 2023). Consequently, photosynthesis is significantly inhibited under salt environments due to reduced gas exchange coupled with an impaired photosynthetic system.

Several mechanisms mediate the absorption of water and solutes from the soil into the xylem vessels of the roots. However, the transport of Na into plant roots requires specific channels and transport proteins, even though plants lack sodium transport systems. Studies have shown that high-affinity potassium transporters (HKTs) and non-selective cation channels (NSCCs) are involved in the transport of sodium across plasma membranes (Su et al. 2015). Wei et al. (2021) noted that HKT2;1 elevated sodium uptake during salinity, which causes sodium accumulation in rice plants. Furthermore, voltage-independent channels, non-selective outward-rectifying channels (NORCs), and/or low-affinity potassium channels (AKT1) contribute to the sodium influx across plant plasma membranes (Pottosin and Dobrovinskaya 2014). The above phenomenon triggers rapid sodium accumulation in the plant cytosol under salinity. For example, a brief exposure of only 2 min to elevated levels of NaCl results in a substantial content of sodium in the roots of Arabidopsis (Essah et al. 2003). As K and Na utilize the same transport systems, increased Na levels in the soil can inhibit K uptake into plant cells, causing K deficiency. Due to the high molecular similarity between them, cytosolic Na accumulation often replaces K in various K-dependent enzymes. However, Na cannot replicate K's functions, leading to severe metabolic disruptions and ionic stress in plants (Fig. 5.1).

Chloride (Cl) ions may cause toxicity in plants as well, as excessive levels of Cl are often detected in sensitive tissues during salinity. Cl uptake may antagonize nitrate nutrition, which is a vital N source for higher plants (Ashraf et al. 2018). However, some evidence suggests that Cl may act as a beneficial macronutrient with tightly regulated transport mechanisms. In maize, Wen et al. (2017) identified NPF6.4 and NPF6.6 as Cl influx transporters within plant cells. In saline soils, Cl and Na taken up by the roots are transported over long distances to shoots through the xylem, ultimately accumulating in photosynthetic leaves (Keisham et al. 2018).

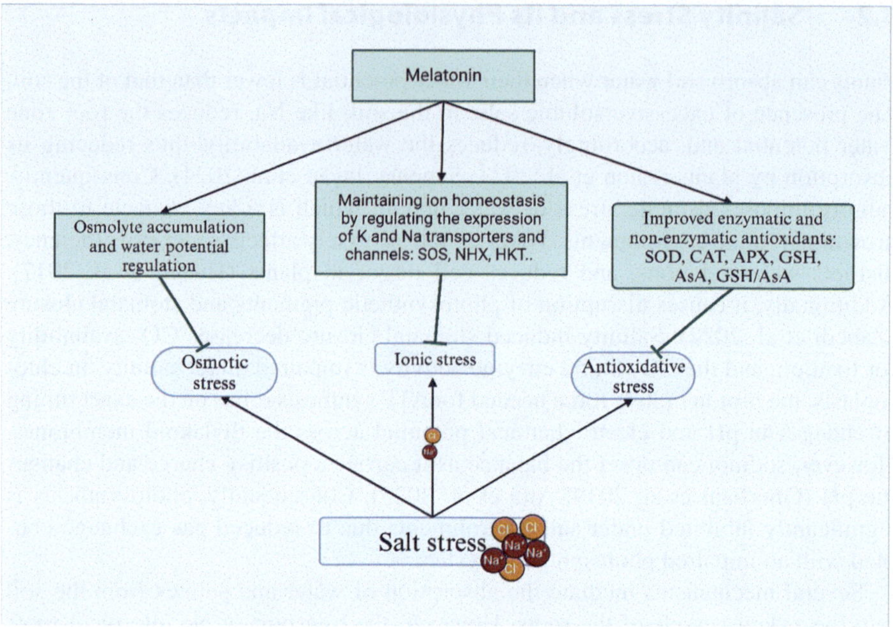

Fig. 5.1 Mechanisms of melatonin in mitigating salt stress in plants: regulation of osmotic, ionic, and oxidative stress through osmolyte accumulation, ion homeostasis, and antioxidative defense systems

This accumulation ultimately exerts harmful impacts on plant biomass and performance. Recent research has shown that cytokinin signaling mediates Cl removal from shoots, thereby enhancing maize's ability to withstand salinity. This provides compelling evidence that Cl induces toxicity during high salinity, though this is not linked to nitrate deficiency caused by Cl accumulation (Yin et al. 2023).

Salinity also triggers the rapid production of ROS, like hydrogen peroxide (H_2O_2), singlet oxygen, superoxide, and hydroxyl radical, due to plant metabolic disruptions (Nanehkaran et al. 2024). The extreme reactivity of ROS causes oxidative damage to plant bio-macromolecules, leading to severe oxidative stress and redox imbalance (Gill and Tuteja 2010). ROS are generated in multiple plant organelles under salinity stress. Salinity-caused osmotic stress reduces the availability of CO_2 through stomatal closure, thus increasing singlet oxygen and superoxide in chloroplasts (Hameed et al. 2021). Peroxisomes also become an additional source of ROS due to increased photorespiration and, hence, greater production of H_2O_2 under salinity conditions (Corpas et al. 2019). Electron leakage from the ETC initiates mitochondrial ROS production, causing in the generation of superoxide (Sharma et al. 2012), further converted to H_2O_2 by Mn-SOD. Alternatively, ROS may also play roles as signaling molecules in the case of responses and defense mechanisms to salt stress.

5.3 Regulation of Ionic Homeostasis During Salinity

Salinity significantly impacts plants by disrupting their ionic balance, mainly by accumulating toxic Cl and Na ions in the cytoplasm. Ionic homeostasis is required for the salinity survival of plants. Here, the function of ion channels and transporters, salt exclusion and compartmentalization mechanisms, and melatonin-mediated ionic transport and homeostasis in diminishing the harmful effects of salinity are discussed (Fig. 5.1).

5.3.1 Role of Ion Transporters and Channels

Plants maintain ionic balance as a response to salinity stress by transporting ions via ion transporters and channels. These transporters are mainly implicated in the control, uptake, further transport, and compartmentalization of K and Na to maintain cellular function and dissipate the toxicity associated with these ions.

The Na/H antiporters in the tonoplast and plasma membrane are important to hinder the creation of cytosolic Na toxicity by its efflux outside the cell or sequestration into the vacuole. The plasma membrane Na/H antiporter SOS1 is an example that, under salinity stress, is induced to drive the efflux of excessive cytosolic Na into either the apoplast or the extracellular environment (Ali et al. 2023a, b). The SOS1 activity is controlled by the SOS signaling pathway, which consists of SOS2, a protein kinase, and SOS3, a Ca-binding protein that detects increased cytosolic Ca levels caused by salinity. Sequestration of sodium ions into vacuoles is mediated by antiporters such as NHX1 (for Na/H exchanger 1) in the tonoplast. Compartmentalization avoids sodium toxicity in the cytoplasm and may allow maintaining osmotic balance to favor the retention of water inside plant cells (Solis et al. 2022).

Maintenance of a low cytosolic Na/K ratio is indispensable for the proper function of enzymatic activities and cellular homeostasis under saline conditions. High-affinity K transporters (HKTs) are key regulators to limit Na levels in sensitive tissues (Dave et al. 2022). For example, HKT1;1 mediates the transport of Na from shoots back to roots and thus decreases Na toxicity in photosynthetic tissues. Under conditions when K is low, voltage-gated potassium channels, like AKT1 (Arabidopsis K Transporter 1), are crucial for K uptake (Luo et al. 2024). These channels ensure that sufficient K is available despite Na competing for the same transport pathways. In some plants, SKOR (Stelar K Outward Rectifier) channels regulate K loading into the xylem to facilitate its movement to shoots, where it would counterbalance the deleterious effects of Na (Dave et al. 2022).

Non-selective cation channels (NSCCs) and low-affinity cation transporters may also inadvertently allow the uptake of sodium. However, their contribution under salinity stress is reduced by the preferential activation of sodium-specific transporters coupled with the inhibition of NSCC activity (Ali et al. 2023a, b).

5.3.2 Mechanisms of Salt Exclusion and Compartmentalization

Plants have evolved highly specialized processes to prevent ionic toxicity through the exclusion of excessive salts or compartmentalization within defined tissues or organelles.

Salt exclusion refers to the prevention of excessive accumulation of Na and Cl in both shoots and roots. The major antiporter that facilitates Na exclusion from the cytosol, located in the plasma membrane, is SOS1 (Meng et al. 2018). In addition, there is reduced transport of Na to shoots by the HKTs such as HKT1;5, retrieving Na from the xylem sap back into the root cells to be sequestered. Stomata regulation also makes an indirect contribution toward salt exclusion by means of controlling the transpiration stream to lower the transportation of salt from the roots to the shoots (Shao et al. 2021). Sequestration into the vacuole is one important way of mitigating the toxicity of Na and Cl in the cytoplasm. Among these, NHX antiporters, especially NHX1, mediate the active transportation of Na into the vacuoles using the proton gradient established by V-ATPases and V-PPases; this sequestration detoxifies the cytosol, contributing to osmotic adjustment and sustaining turgor pressure during salinity (Meng et al. 2018).

That is, the differential build-up of Na and Cl in older leaves or less metabolically active tissues protects young, photosynthetically active tissues. Most halophytes have specialized salt-excreting glands or salt bladders that excrete the extra salts onto the surface of the leaves. A good K/Na ratio is crucial for enzymatic activities and metabolic stability. Plants achieve this by increasing K absorption through selective transporters like AKT1 and inhibiting the non-selective uptake of Na. Compatible solutes like glycine betaine and proline are produced, which stabilize enzymes and membranes against ionic stress (Hao et al. 2021).

5.3.3 Melatonin-Mediated Adjustment of Ionic Movement and Homeostasis

The multireactional molecule melatonin turned out to be one of the important regulators of ionic homeostasis in conditions of salinity stress (Ghorbani et al. 2023a). Besides its well-documented antioxidant activity, its role has been extended to maintaining ion transporter activities, which influence ionic balance. Melatonin-induced gene expression of Na/H antiporters (NHX1 and SOS1). Increased activity of these transporters enhances the cytoplasmic efflux of Na and its sequestration into vacuoles, decreasing Na toxicity (Efimova et al. 2022). Indeed, exogenous melatonin elevated the expression of SOS1 in *Malus hupehensis* and *Brassica napus*, thereby enhancing their salt resilience due to the elevated exclusion of Na^+ (Li et al. 2012; Zhao et al. 2018) (Table 5.1).

Melatonin participates in sustaining K homeostasis by regulating K transporters and channels. Melatonin treatment upregulates the expression of AKT1, leading to K uptake under salinity stress. This maintains a high K/Na ratio, which is indispensable for both cellular function and ionic stability (Li et al. 2019, 2012). It also

Table 5.1 Summary of melatonin's role in ionic homeostasis under salinity stress

Plant Species	Salinity level	Melatonin Concentration	Observed changes	References
Oryza sativa L.	150 mM	50–200 μM	Promoted Na efflux and K influx, improved K/Na	Yan et al. (2020)
	100 mM	100 μM	Upregulated *SOS* and *NHX* expression	Khan et al. (2024a, b)
Solanum lycopersicum	150 mM	150 μM	Upregulated expression of *NHX1-4*, enhanced K/Na ratio	Ghorbani et al. (2023a, b, c, d)
	200 mM	30 μM	Improved K/Na	Siddiqui et al. (2021)
	50 mM	500 μM	Increased expression of PIP, decreased Na, enhanced K content	Masoumi et al. (2024)
Gossypium hirsutum	150 mM	50–200 μM	Reduced Na and Cl levels, improved K/Na ratio	Jiang et al. (2021)
Sorghum bicolor	75 and 150 μM	50–200 μM	Improved K and Ca, decreased Na	Helal et al. (2024)
Withania coagulans	200 mM	100 and 200 μM	Reduced Na, increased K	Dehvari-Nagan et al. (2023)
Malus hupehensis	200 mol/L	0.1 mmol/L	Adjusted expression of *CBL1, CIPK23, HKT1, SOS1, NHX2, NHX4, and NHX6,* increased K content, declined Na level	Li et al. (2016)
Phaseolus vulgaris	2.6 and 5.1 g/L	20, 100, and 500 μM	Enhanced K/Na ratio	Alinia et al. (2022)
	50 mM	20 μM	Enhanced K Ca, decreased Na, enhanced K/Na ratio	El-Beltagi et al. (2023)
Abelmoschus esculentus	150 mM	50, 100 and 200 μM	Improved K/Na ratio	Wang et al. (2024a, b)
Trigonella foenum-graecum	150 and 300 mM	30, 60, and 90 ppm	Increased K, decreased Cl, and Na contents	Mohamadi Esboei et al. (2022)
Lactuca sativa	50 and 100 mM	50, 100, and 150 μM	Decreased Cl and Na	EL-Bauome et al. (2024)
Triticum aestivum	100 mM	100 μM	Upregulated *SOS1, SOS2, SOS3*, and *NHX1* expression	Khan et al. (2024a, b)

engages calcium signaling by activating SOS components, which include SOS1 and SOS2, to enhance Na efflux. The participation of melatonin in calcium signaling puts it as a modulator in ionic transport networks (Guo et al. 2025).

Melatonin indirectly regulates oxidative stress to maintain ionic homeostasis (Ghorbani et al. 2023b). The antioxidant activity of melatonin diminishes the impacts of ROS on ion transporters and membranes; hence, it maintains the function of transport systems under saline conditions (Li et al. 2020). Melatonin modulates transcription by interacting specifically with stress-responsive transcription factors that drive the expressions of ion transporter genes. In addition, melatonin has altered the hormonal pathway associated with salinity responses, including ABA and ethylene. For instance, ABA signaling mediated by melatonin induced the closing of stomata, hence reducing salt transportation to shoots (Xu et al. 2019).

5.4 Redox Networks and Melatonin in Salinity Stress Responses

Salinity stress seriously upsets plant growth through impairment of cellular functions due mainly to oxidative damage mediated by overproduction of ROS (Ghorbani et al. 2023c). Melatonin has lately been identified as one of the main regulators of the redox state of plants during different abiotic stressors, including salinity. In this respect, salinity-mediated generation of ROS markedly reduced photosynthetic efficiency in tomato plants in the work of Ghorbani et al. (2018b), as reflected in the stunted growth of plants. Likewise, in a recent study on *Medicago sativa* by Guo et al. (2025), salt stress mediated an enhancement in the level of ROS, more so H_2O_2, which in turn directly caused oxidative damage to chloroplast and mitochondrial membranes, disrupting cellular respiration and photosynthesis.

With the evolution of powerful antioxidant defense mechanisms in plants, enzymatic and non-enzymatic, counteract oxidative damage (Ghorbani et al. 2024a). To neutralize ROS by their conversion into less harmful molecules, the key enzymes involved are catalase (CAT), superoxide dismutase (SOD), and peroxidase (POD). SOD catalyzes the conversion of superoxide anion to H_2O_2, which is subsequently detoxified by CAT to water and oxygen. Non-enzymatic antioxidants like ascorbate and glutathione (GSH) act as electron donors to reduce ROS and guard cellular integrity (Noctor et al. 2021).

Recent studies have highlighted the significant role of antioxidant enzymes in enhancing plant salt tolerance (Table 5.2). In a study by Kostopoulou et al. (2014), salt stress led to a considerable increase in the activity of SOD, CAT, and POD in *Citrus aurantium*, which helped alleviate oxidative damage and enhance salt tolerance. Additionally, GSH and ascorbate were found to synergistically enhance antioxidant defense systems. In eggplant, elevated levels of GSH were shown to protect against oxidative damage caused by salinity, while ascorbate also contributed to the reduction of ROS levels, boosting the plant's overall stress tolerance (Zhang et al. 2024).

Melatonin has emerged as an excellent agent for scavenging ROS and stimulating antioxidant machinery in plants under salinity stress. As a very effective free radical scavenger, melatonin quenches ROS directly, which finally reduces oxidative damage and sustains cell viability. Many reports show that melatonin enhances antioxidant enzyme activity and non-enzymatic antioxidants, thus reducing salt-induced oxidative stress. Melatonin treatment in salt-stressed fenugreek plants drastically reduced the generation of ROS while increasing the activities of SOD and POD in the work of Mohamadi Esboei et al. (2022), therefore conferring a high

Table 5.2 Summary of melatonin's role in redox homeostasis under salinity stress

Plant species	Salinity level	Melatonin concentration	Observed changes	References
Gladiolus grandiflorus	5 dS m^{-1}	0.2–0.8 mM	Reduced MDA, and H_2O_2, increased AsA content and activities of POD, SOD, and CAT	Zulfiqar et al. (2024)
Oryza sativa L.	100 mM	100 μM	Declined H_2O_2 and MDA contents, increased GSH content and activity of CAT	Khan et al. (2024a, b)
	150 mM	1 μM	Decreased ROS and MDA content, upregulated expression of *APX1, APX, GPOD, Mn-SOD, and Cu/Zn-SOD*	Ubaidillah et al. (2024)
Lactuca sativa	50 and 100 mM	50, 100, and 150 μM	Decreased H_2O_2, MDA, $O_2^{\bullet-}$ and EL, downregulated activity of GR, POD, and SOD	EL-Bauome et al. (2024)
Sorghum bicolor	75 and 150 μM	50–200 μM	Decreased H_2O_2, MDA, and MG, improved APX, POX, SOD, and CAT activities	Helal et al. (2024)
Lonicera japonica	150 mM	60 μM	Declined H_2O_2, MDA, $O_2^{\bullet-}$ and EL, IMPROVED CAT, SOD, POD, and APX	Song et al. (2024)
Cyperus esculentus	200 mM	150 μM	Declined H_2O_2, MDA, $O_2^{\bullet-}$ and EL, improved CAT, SOD, and POD	Wang et al. (2024a, b)
Triticum aestivum	100 mM	100 μM	Lowered H_2O_2 and TBARS, increased GSH, induced CAT, SOD, and GR activities	Khan et al. (2024a, b)
	6 and 12 dS m^{-1}	70 μM	Decreased ROS and MG, improved DPPH, GSH, and AsA, and GSH/GSSG and AsA/DHA ratios, induced APX, GR, DHAR, MDHAR, GPX, Gly I, Gly II, and GST activity	Talaat and Todorova (2022)
Glycine max	100 mM	0.01, 0.05, and 0.1 mM	Decreased H_2O_2, MDA, MG, an $O_2^{\bullet-}$, increased GSH and AsA content, enhanced the activity of SOD, GR, APX, CAT, and Gly I	Alharbi et al. (2021)
Medicago sativa	200 mM	0.1–0.4 mM	Declined MDA content, improved CAT, SOD, and POD	Guo et al. (2023)
Gossypium hirsutum	150 mM	50, 100, 200, and 500 μM	Decreased ROS and MDA, improved GSH and AsA contents, induced SOD, POS, APX, and CAT activities	Jiang et al. (2020)

(continued)

Table 5.2 (continued)

Plant species	Salinity level	Melatonin concentration	Observed changes	References
Zea mays	200 mM	30, 60, and 90 μM	Lowered ROS, MDA, and EL, improved activity of SOD, POD, CAT, and APX	Ahmad et al. (2021)
	150 mM	20 and 100 μM	Decreased H_2O_2, MDA, MG, an $O_2^{\bullet-}$, enhanced activity of SOD, GPX, GR, APX, CAT, and POD	Chen et al. (2018)
	8.9 dS m^{-1}	0.5 and 1.0 mM	Reduced H_2O_2 and MDA, enhanced POD, CAT, and SOD activities	Ali et al. (2024)

degree of protection against oxidative stress. The increase in GSH and ascorbate due to melatonin enhanced antioxidant defense networking in the plant. In another work, Muhammad et al. (2024) reported that under melatonin treatment, *Viola tricolor* seedlings exposed to salt stress suppressed lipid peroxidation and enhanced antioxidant enzyme activities, hence suggesting that melatonin may participate in maintaining the cellular redox balance, leading to the improvement of salt tolerance in plants.

Melatonin can influence different redox signaling pathways, thus modulating the plant response under salinity. It has been reported to influence signaling molecules such as NO and H_2S, which have been implicated in the maintenance of redox homeostasis and the activation of stress-responsive pathways (Ghorbani et al. 2024a, b). Modulation of antioxidant gene expressions by melatonin interactions with such signaling molecules enhances the overall stress tolerance in plants. As reported by Zhao et al. (2018), melatonin enhanced the production of NO under salt-induced stress in *Brassica napus*, followed by the induction of antioxidant defense systems and reduced oxidative damage. Further, it was observed that the H_2S signaling pathway was influenced by melatonin through the activation of key genes related to stress and thus enhanced tolerance to oxidative damage. Very recently, melatonin was reported to interact with MAPK and plant hormones such as ABA and JA in coordinating responses against salinity stress. This would, therefore, suggest that melatonin acts both as a direct antioxidant and modulates complex networks of signaling pathways to improve salinity tolerance in plants (Khan et al. 2024a, b; Jiang et al. 2024).

5.5 Crosstalk Between Ionic Homeostasis, Redox Regulation, and Melatonin

The intricate interplay between ionic homeostasis, redox regulation, and melatonin has emerged as a pivotal mechanism in mitigating the deleterious effects of salinity stress in plants. Salinity disrupts ionic equilibrium and induces oxidative stress, necessitating an integrated response to sustain cellular homeostasis. Melatonin, a multifunctional molecule, has been recognized as a critical mediator in orchestrating the crosstalk between these networks, with significant implications for plant resilience.

5.5.1 Melatonin's Influence on Ion Transport and ROS Signaling

Melatonin alters ion transport mechanisms and thus contributes significantly to the ionic homeostasis imposed by salinity stress (Ghasemi-Omran et al. 2021). Salinity generally triggers abnormal increases in Na ion influx and can be associated with decreases in K ions, which impair cellular functions. Melatonin, therefore, under such conditions, can decrease the accumulation of ion imbalance by modulating the activity of ion channels and transporters. Melatonin enhances the activity of the Na/H antiporter SOS1, which in turn extrudes Na from cells and hence decreases cytosolic sodium toxicity, as cited by Efimova et al. (2022). Melatonin promotes K+ acquisition and retention, for example, by the regulation of HTK1 and inward-rectifying K channels described in Li et al. 2020.

It also controls the vacuolar compartmentalization of Na. In relation to this, Li et al. (2020) showed that under melatonin supplementation, vacuolar NHX1 genes are upregulated, helping to sequester excessive Na ions into the vacuole and reducing cytosolic toxicity. Melatonin enhances aquaporin expression, as Fu et al. (2024) showed, which helps improve water transport and cellular turgor in saline environments.

Meanwhile, melatonin has a great effect on the ROS signaling pathway. Overproduction of ROS by salinity may cause damage to the cell. However, melatonin mitigates this process by acting as a direct scavenger of ROS and modulating antioxidant systems. Surprisingly, Li et al. (2022) have presented evidence that melatonin triggers antioxidant enzyme expressions like SOD, CAT, and POD, contributing to a reduction in oxidative damage. These double roles of melatonin, such as modulating ionic transport and ROS signaling, assign an integrated part for melatonin in adapting to stresses. In the same vein, recently, Guo et al. (2025) have noted that melatonin interacts with ROS-responsive transcription factors such as WRKY and NAC, resulting in enhanced transcription of the stress-responsive genes.

5.5.2 Feedback Mechanisms Integrating Ionic and Redox Homeostasis

There is an interplay of ionic and redox homeostasis where melatonin may play the role of a central coordinator. Melatonin controls the level of Na^+ and K^+; thus, it indirectly controls the level of ROS because disruption of ionic homeostasis enhances oxidative stress. In turn, the redox status might modify the performance of ion transporters and ion channels, setting up a bidirectional relationship.

Melatonin's capability to conduct such processes is reflected in studies focused on the signaling cascades it participates in. Such are studies by Guo et al. (2025), where they observe that melatonin modulates calcium signaling, which in turn plays this dual role of a second messenger in both ionic and redox regulations. Melatonin maintains intracellular Na/K homeostasis by increasing Ca-mediated activation of SOS1 and the plasma membrane H-ATPase and, at the same time, enhances NADPH oxidase activities, an important enzyme in the ROS signal cascade. The coordination maintains that the ROS produced as part of the stress signaling would not cause cellular damage. Correspondingly, Vafadar et al. (2020) showed that Ca signaling regulated by melatonin impinges on ABA pathways, thus integrating ionic and redox responses.

Also, several links to antioxidant capacity in association with melatonin have been reported, including ionic transport efficiency. For instance, Li et al. (2020) showed that the activation of ROS-scavenging enzymes via melatonin mediation decreases the oxidative modification of ion channels/transporters, thus maintaining their activity under salinity. Protection of ion homeostasis further forms a positive feed-forward cycle in enhancing stress tolerance. On the other hand, Liu et al. (2024) found that melatonin enhances the transcription of the stress-responsive SnRK2 kinase, which is important for both ionic and redox homeostasis under salt conditions. A deeper integration into transcriptional networks reflects the melatonin-mediated crosstalk. Melatonin modulates the expression of genes responsible for ionic transport, such as SOS1 and NHX1, and antioxidant pathways, including SOD and CAT; transcriptomic evidence has recently been provided by Yang et al. (2024). Indications are that melatonin orchestrates the expression of those stress-responsive genes that are implicated in balancing ionic and redox states to optimize cellular responses to salinity stress.

5.6 Genetic and Molecular Insights into Melatonin's Role in Salinity Tolerance

The multifunctional role of melatonin in increasing the tolerance of salinity stretches from physiological and biochemical regulation to genetic and molecular dimensions. Most recently, the advances in genetic and omics research have disclosed critical genes and regulatory networks of melatonin and possible biotechnological applications for stress mitigation.

The mechanisms of melatonin-induced salinity tolerance are closely related to its role in regulating key stress-responsive genes and their associated networks. Melatonin treatment causes an alteration in the significant expression of genes responsible for ionic transport, antioxidant defense, and stress signaling. For example, Zhan et al. (2019) demonstrated that *SOS1*, *NHX1*, and *HKT1*, key components maintaining ionic balance in salt-stressed plants, were upregulated under the influence of melatonin. Similarly, in the work of Liu et al. (2020), it was determined that melatonin enhances the expression of RBOH, which is responsible for controlled ROS production with the purpose of signaling.

The transcriptional regulation by melatonin often involves the activation of transcription factors such as WRKY, NAC, and DREB. Zhao et al. (2021) highlighted that melatonin induces WRKY46 and NAC42, which subsequently activate downstream stress-responsive genes like P5CS (Pyrroline-5-Carboxylate Synthase) involved in proline biosynthesis, contributing to osmotic adjustment. Additionally, studies by Ali et al. (2023a, b) showed that melatonin modulates the expression of genes linked to ABA pathways, integrating hormonal crosstalk into the stress response.

5.6.1 Advances in Omics Studies Revealing Melatonin's Functions

Furthermore, omics technologies, such as transcriptomics, proteomics, and metabolomics, have shed light on the molecular mechanism underlying melatonin-induced salinity tolerance. Several transcriptomic studies have established that melatonin triggers a wide array of genes relating to the adaptation of stresses. For example, Yang et al. (2024) conducted RNA sequencing in melatonin-treated *Phaseolus vulgaris* under salinity stress and identified approximately 500 differentially expressed genes related to ionic homeostasis, ROS detoxification, and secondary metabolism.

Further, proteomics studies have revealed post-translational modifications regulated by melatonin. Zhang et al. (2017) have shown that melatonin increased the phosphorylation of proteins involved in SOS signaling and antioxidant enzymes, thus modulating their activity under stress responses. Furthermore, approaches in metabolomics have unraveled some specific metabolic pathways altered by melatonin. Yang et al. 2024, indicated that melatonin increases the biosynthesis of polyamines and flavonoids, both contributing to the mitigation of oxidative damage and stabilization of cellular structure.

5.6.2 Genetic Engineering Approaches to Enhance Melatonin Biosynthesis

The integration of multi-omics datasets enabled the construction of regulatory networks that delineate the multifaceted role that melatonin plays in salinity tolerance. For example, through systems biology approaches, Guo et al. (2025) identified key hubs and network interactions that centered on melatonin signaling, which coordinated with Ca and ABA pathways, thus pushing it even further as the master regulator in adaptation to stresses.

The possibility of improving plant stress tolerance by genetic engineering of melatonin biosynthesis pathways has consequently attracted considerable attention (Ghorbani et al. 2024b). Overexpression of important genes of melatonin biosynthesis, such as TDC, T5H, SNAT, and ASMT, with the aim of increasing endogenous melatonin content in plants, has been tried to raise the level of tolerance of salinity in plants (Ahn et al. 2021). Overexpression of these genes showed promising signs of enhancement in salinity tolerance. For instance, Liu et al. (2019) mentioned that the overexpression of SNAT1 in tomatoes increased melatonin content and increased the expression of stress-responsive genes, maintaining a better ionic and redox balance under saline conditions. Similarly, in another study, Zhu et al. (2022) mentioned that the overexpression of T5H in mulberry enhanced biosynthesis of melatonin, further improved antioxidant capacity, and osmotic adjustment. Recently, negative regulators involved in melatonin biosynthesis were also targeted using the CRISPR/Cas9 technique (Sagharyan et al. 2025). Integration of the transgenic approach or genome editing with omics-based analysis would be a strong platform for developing improved crops with enhanced melatonin-mediated stress resilience. Therefore, there is immense hope for mitigating the impacts of salinity stress in agriculture.

These genetic and molecular details of the functioning of melatonin in reference to salinity tolerance establish it as a suitable target for biotechnological interventions. Important gene identifications and their regulatory networks, besides applications of omics technologies, have remarkably improved the understanding with respect to melatonin functions. Furthermore, genetic engineering approaches for the improvement of melatonin biosynthesis emerge as a promising way to develop stress-tolerant crops that could perform well under saline conditions.

5.7 Biotechnological Approaches to Enhance Salinity Tolerance

Biotechnological innovations in the methods of applying melatonin to improve salinity tolerance in crops from direct methods of application to advanced nanotechnology-based delivery.

Salinity stress has considerably been mitigated with the aid of melatonin treatments, such as seed priming and foliage application. According to Bakyani et al. (2021), the exogenous application of melatonin caused improved ionic and redox homeostasis, hence showing improved growth and declining Na/K imbalance in the melatonin-treated *Camelina sativa* grown under salinity stress. Besides, Hussain et al. (2024) showed the potency of melatonin seed priming for improving seed germination and vigor under high salinity.

A combination of melatonin with PGPR and biostimulants exerts synergistic action through increasing resistance in plants against various abiotic stresses. For example, Imran et al. (2023) discovered that the combination of melatonin with PGPR could modulate antioxidant activity and nutrient assimilation in plants growing under abiotic stresses, taking tomato plants as a model. In this respect, Wei et al. (2023) pointed out that melatonin applied together with plant biostimulants effectively improved the physiological and biochemical responses of plants.

Nanotechnology opens new perspectives for effective delivery mechanisms for melatonin. Nanocarriers, including liposomes and nanoparticles, improve melatonin stability and bioavailability (Li et al. 2024a, b). Evidence by Gohari et al. (2022) showed that nanoparticles loaded with melatonin improved salinity tolerance in spearmint by specific delivery into root tissues, thus enhancing salt-stress-responsive pathways.

5.8 Future Perspectives and Research Directions

Despite these great advances, important knowledge gaps remain in the role of melatonin in the biology of stress. Exactly how melatonin interacts with other hormones and signaling pathways through detailed molecular mechanisms needs study. This will require further field-scale testing of melatonin-based approaches to confirm practical efficacy.

We foresee a huge potential for agricultural sustainability with melatonin. Low-cost melatonin formulation, its synergistic use with emerging technologies, and genome editing may be drastically changing the concept of crop management in saline conditions.

5.9 Conclusions

Melatonin may now be considered the central molecule that interlinks physiological, biochemical, genetic, and technological aspects regarding salinity stress tolerance. Advanced biotechnological approaches coupled with new delivery methods will be required for maximum exploitation of melatonin in enhancing crop resilience. Due to its distinctive capacity to modulate ionic homeostasis, ROS detoxification, and gene expression, it has become one of the prime molecules of interest in the area of stress biology.

Now, translation from laboratory findings into field applications is needed. The collaboration of researchers, field agriculturists, and decision-makers will help attract the potential value of melatonin to overcome some of the global challenges like salinity and food insecurity. Second, the development of inexpensive, eco-friendly melatonin-based formulations would advance sustainable agriculture with new hope for successful crop cultivation in difficult environments.

References

Ahmad S, Cui W, Kamran M et al (2021) Exogenous application of melatonin induces tolerance to salt stress by improving the photosynthetic efficiency and antioxidant defense system of maize seedling. J Plant Growth Regul 40:1270–1283. https://doi.org/10.1007/s00344-020-10187-0

Ahn R, Kim J, Lim J, Duan S, Eom H, Jung H (2021) Key genes in the melatonin biosynthesis pathway with circadian rhythm are associated with various abiotic stresses. Plan Theory 10(1):129. https://doi.org/10.3390/plants10010129

Alharbi BM, Elhakem AH, Alnusairi GSH, Soliman MH, Hakeem KR, Hasan MM, Abdelhamid MT (2021) Exogenous application of melatonin alleviates salt stress-induced decline in growth and photosynthesis in *Glycine max* (L.) seedlings by improving mineral uptake, antioxidant and glyoxalase system. Plant Soil Environ 67(4):208–220. https://doi.org/10.17221/659/2020-PSE

Ali M, Pan Y, Liu H, Cheng Z (2023a) Melatonin interaction with abscisic acid in the regulation of abiotic stress in Solanaceae family plants. Front Plant Sci 14:1271137. https://doi.org/10.3389/fpls.2023.1271137

Ali A, Petrov V, Yun D, Gechev T (2023b) Revisiting plant salt tolerance: novel components of the SOS pathway. Trends Plant Sci 28(9):1060–1069. https://doi.org/10.1016/j.tplants.2023.04.003

Ali M, Malik Z, Abbasi GH, Irfan M, Ahmad S, Ameen M, Ali A, Sohaib M, Rizwan M, Ali S (2024) Potential of melatonin in enhancing antioxidant defense system and yield of maize (*Zea mays* L.) hybrids under saline condition. Sci Hortic 325:112665. https://doi.org/10.1016/j.scienta.2023.112665

Alinia M, Kazemeini SA, Dadkhodaie A, Sepehri M, Mahjenabadi VA, Amjad SF, Poczai P, Bassouny MA, Abdelhafez AA (2022) Co-application of ACC deaminase-producing rhizobial bacteria and melatonin improves salt tolerance in common bean (*Phaseolus vulgaris* L.) through ion homeostasis. Sci Rep 12(1):1–13. https://doi.org/10.1038/s41598-022-26084-3

Altaf MA, Shahid R, Ren MX, Naz S, Altaf MM, Khan LU, Tiwari RK, Lal MK, Shahid MA, Kumar R, Nawaz MA (2022) Melatonin improves drought stress tolerance of tomato by modulating plant growth, root architecture, photosynthesis, and antioxidant defense system. Antioxidants 11(2):309

Ashraf M, Shahzad SM, Imtiaz M, Rizwan MS, Arif MS, Kausar R (2018) Nitrogen nutrition and adaptation of glycophytes to saline environment: a review. Arch Agron Soil Sci 64:1181–1206

Atta K, Mondal S, Gorai S, Singh AP, Kumari A, Ghosh T, Roy A, Hembram S, Gaikwad DJ, Mondal S, Bhattacharya S, Jha UC, Jespersen D (2023) Impacts of salinity stress on crop plants: improving salt tolerance through genetic and molecular dissection. Front Plant Sci 14:1241736. https://doi.org/10.3389/fpls.2023.1241736

Bakyani MR, Alinia M, Kazemeini SA, Abadía J, Dadkhodaie A (2021) Foliar application of melatonin improves the salt tolerance, ion and redox homeostasis and seed oil fatty acid profile in *Camelina sativa*. Plan Theory 11(22):3113. https://doi.org/10.3390/plants11223113

Cao H, Ghorbanpour A, Ghorbani A et al (2024) The impact of silicon on the morphological and physio-chemical characteristics of *Calendula officinalis* exposed to arsenic-induced toxicity. Russ J Plant Physiol 71:79. https://doi.org/10.1134/S1021443724603811

Chen E, Mao J, Sun Q, Huang B, Ding B, Gu Y, Liao Q, Hu C, Zhang W, Yuan S, Yuan M (2018) Exogenous melatonin enhances salt stress tolerance in maize seedlings by improving antioxidant and photosynthetic capacity. Physiol Plant 164(3):349–363. https://doi.org/10.1111/ppl.12737

Chen Y, Leng YN, Zhu FY et al (2023) Water-saving techniques: physiological responses and regulatory mechanisms of crops. Adv Biotechnol 1:3

Corpas FJ, Del Rı'o LA, Palma JM (2019) Plant peroxisomes at the crossroad of NO and H_2O_2 metabolism. J Integr Plant Biol 61:803–816

Dave A, Agarwal P, Agarwal PK (2022) Mechanism of high affinity potassium transporter (HKT) towards improved crop productivity in saline agricultural lands. *3*. Biotech 12(2):51. https://doi.org/10.1007/s13205-021-03092-0

Dehvari-Nagan P, Abbaspour H, Asare MH et al (2023) Melatonin confers NaCl tolerance in *Withania coagulans* L. by maintaining Na^+/K^+ homeostasis, strengthening the antioxidant defense system and modulating withanolides synthesis-related genes. Russ. J Plant Physiol 70:52. https://doi.org/10.1134/S1021443723600125

Delai C, Abid R, Tian T, Liu R, Xiong Y, Ma S, Ghorbani A (2024) A comprehensive review of integrated management strategies for damping-off disease in chili. Front Microbiol 15:1479957. https://doi.org/10.3389/fmicb.2024.1479957

Efimova MV, Danilova ED, Zlobin IE, Kolomeichuk LV, Murgan OK, Boyko EV, Kuznetsov VV (2022) Priming potato plants with melatonin protects stolon formation under delayed salt stress by maintaining the photochemical function of photosystem II, ionic homeostasis and activating the antioxidant system. Int J Mol Sci 24(7):6134. https://doi.org/10.3390/ijms24076134

EL-Bauome HA, Doklega SM, Saleh SA, Mohamed AS, Suliman AA, Abd El-Hady MAM (2024) Effects of melatonin on lettuce plant growth, antioxidant enzymes and photosynthetic pigments under salinity stress conditions. Folia Hortic 36(1):1–17

El-Beltagi HS, El-Yazied AA, El-Gawad HGA, Kandeel M, Shalaby TA, Mansour AT, Al-Harbi NA, Al-Qahtani SM, Alkhateeb AA, Ibrahim MFM (2023) Synergistic impact of melatonin and putrescine interaction in mitigating salinity stress in snap bean seedlings: reduction of oxidative damage and inhibition of polyamine catabolism. Horticulturae 9(2):285. https://doi.org/10.3390/horticulturae9020285

Essah PA, Davenport R, Tester M (2003) Sodium influx and accumulation in Arabidopsis. Plant Physiol 133:307–318

Fu Y, Li P, Si Z, Ma S, Gao Y (2024) Seeds priming with melatonin improves root hydraulic conductivity of wheat varieties under drought, salinity, and combined stress. Int J Mol Sci 25(9):5055. https://doi.org/10.3390/ijms25095055

Ghasemi-Omran VO, Ghorbani A, Sajjadi-Otaghsara SA (2021) Melatonin alleviates NaCl-induced damage by regulating ionic homeostasis, antioxidant system, redox homeostasis, and expression of steviol glycosides-related biosynthetic genes in *in vitro* cultured *Stevia rebaudiana* Bertoni. In Vitro Cell Dev Biol-Plant 57:319–331. https://doi.org/10.1007/s11627-021-10161-9

Ghorbani A, Razavi SM, Ghasemi Omran VO, Pirdashti H (2018a) *Piriformospora indica* inoculation alleviates the adverse effect of NaCl stress on growth, gas exchange and chlorophyll fluorescence in tomato (*Solanum lycopersicum* L.). Plant Biol 20(4):729–736. https://doi.org/10.1111/plb.12717

Ghorbani A, Razavi SM, Omran VOG et al (2018b) *Piriformospora indica* alleviates salinity by boosting redox poise and antioxidative potential of tomato. Russ J Plant Physiol 65:898–907. https://doi.org/10.1134/S1021443718060079

Ghorbani A, Omran VOG, Razavi SM et al (2019) *Piriformospora indica* confers salinity tolerance on tomato (*Lycopersicon esculentum* Mill.) through amelioration of nutrient accumulation, K^+/Na^+ homeostasis and water status. Plant Cell Rep 38:1151–1163. https://doi.org/10.1007/s00299-019-02434-w

Ghorbani A, Emamverdian A, Chen MX, Naz S et al (2023a) Illustrating recent development in melatonin-heavy metal research in plant. In: Kumar R, Altaf MA, Lal MK, Tiwari RK (eds) Melatonin in plants: a pleiotropic molecule for abiotic stresses and pathogen infection. Springer, Singapore. https://doi.org/10.1007/978-981-99-6741-4_6

Ghorbani A, Emamverdian A, Pishkar L, Chashmi KA, Salavati J, Zargar M, Chen M (2023b) Melatonin-mediated nitric oxide signaling enhances adaptation of tomato plants to aluminum stress. S Afr J Bot 162:443–450. https://doi.org/10.1016/j.sajb.2023.09.031

Ghorbani A, Ollah V, Chen M (2023c) The effect of glycine betaine on nitrogen and polyamine metabolisms, expression of glycoside-related biosynthetic enzymes, and K/Na balance of stevia under salt stress. Plan Theory 12(8):1628. https://doi.org/10.3390/plants12081628

Ghorbani A, Pishkar L, Saravi KV, Chen M (2023d) Melatonin-mediated endogenous nitric oxide coordinately boosts stability through proline and nitrogen metabolism, antioxidant capacity, and Na^+/K^+ transporters in tomato under NaCl stress. Front Plant Sci 14:1135943. https://doi.org/10.3389/fpls.2023.1135943

Ghorbani A, Emamverdian A, Pehlivan N et al (2024a) Nano-enabled agrochemicals: mitigating heavy metal toxicity and enhancing crop adaptability for sustainable crop production. J Nanobiotechnol 22:91. https://doi.org/10.1186/s12951-024-02371-1

Ghorbani A, Pehlivan N, Zargar M, Chen M (2024b) Synergistic role of melatonin and hydrogen sulfide in modulating secondary metabolites and metal uptake/sequestration in arsenic-stressed tomato plants. Sci Hortic 331:113159. https://doi.org/10.1016/j.scienta.2024.113159

Ghorbani A, Khalofah A, Razavi SM, Chen M, & Maggi F (2025) Innovative and sustainable catechin-based nanocomposites for enhancing salinity tolerance and secondary metabolite production in *Stevia rebaudiana* (Bertoni) Bertoni. Ind Crops Prod 231:121134. https://doi.org/10.1016/j.indcrop.2025.121134

Gill SS, Tuteja N (2010) Reactive oxygen species and antioxidant machinery in abiotic stress tolerance in crop plants. Plant Physiol Biochem 48:909–930

Gohari G, Farhadi H, Panahirad S, Zareei E, Labib P, Jafari H, Mahdavinia G, Hassanpouraghdam MB, Ioannou A, Kulak M, Fotopoulos V (2022) Mitigation of salinity impact in spearmint plants through the application of engineered chitosan-melatonin nanoparticles. Int J Biol Macromol 224:893–907. https://doi.org/10.1016/j.ijbiomac.2022.10.175

Guo X, Shi Y, Zhu G, Zhou G (2023) Melatonin mitigated salinity stress on alfalfa by improving antioxidant defense and osmoregulation. Agronomy 13(7):1727. https://doi.org/10.3390/agronomy13071727

Guo S, Wang X, Li X, Ma Y, Yang J, Fu B, Li S (2025) Melatonin and calcium synergistically improve salt tolerance in alfalfa (*Medicago sativa.* L). Ind Crop Prod 224:120322. https://doi.org/10.1016/j.indcrop.2024.120322

Hameed A, Ahmed MZ, Hussain T, Aziz I, Ahmad N, Gul B, Nielsen BL (2021) Effects of salinity stress on chloroplast structure and function. Cells 10:2023

Hao S, Wang Y, Yan Y, Liu Y, Wang J, Chen S (2021) A review on plant responses to salt stress and their mechanisms of salt resistance. Horticulturae 7(6):132. https://doi.org/10.3390/horticulturae7060132

Helal NM, Saudy HS, Hamada MMA, El-Yazied AA, El-Gawad HGA, Mukherjee S, Al-Qahtani SM, Al-Harbi NA, El-Sayed SM, Ibrahim MFM (2024) Potentiality of melatonin for reinforcing salinity tolerance in sorghum seedlings via boosting photosynthetic pigments, ionic and osmotic homeostasis and reducing the carbonyl/oxidative stress markers. J Soil Sci Plant Nutr 24:4243–4260

Huang L, He B, Han L, Liu J, Wang H, Chen Z (2017) A global examination of the response of ecosystem water-use efficiency to drought based on MODIS data. Sci Total Environ 601-602:1097–1107

Hussain S, Nisar F, Gul B, Hameed A (2024) Seed priming with melatonin improved salinity tolerance of halophytes during early life-cycle stages. Plant Growth Regul 103:351–368. https://doi.org/10.1007/s10725-023-01110-0

Imran M, Mpovo CL, Khan MA, Shaffique S, Ninson D, Bilal S, Khan M, Kwon H, Kang M, Yun W, Lee J (2023) Synergistic effect of melatonin and *Lysinibacillus fusiformis* L. (PLT16) to mitigate drought stress via regulation of hormonal, antioxidants system, and physio-molecular responses in soybean plants. Int J Mol Sci 24(10):8489. https://doi.org/10.3390/ijms24108489

Jiang D, Lu B, Liu L, Duan W, Chen L, Li J, Zhang K, Sun H, Zhang Y, Dong H, Li C, Bai Z (2020) Exogenous melatonin improves salt stress adaptation of cotton seedlings by regulating active oxygen metabolism. Peer J 8:e10486. https://doi.org/10.7717/peerj.10486

Jiang D, Lu B, Liu L et al (2021) Exogenous melatonin improves the salt tolerance of cotton by removing active oxygen and protecting photosynthetic organs. BMC Plant Biol 21:331. https://doi.org/10.1186/s12870-021-03082-7

Jiang S, Lan Z, Zhang Y, Kang X, Zhao L, Wu X, Gao H (2024) Mechanisms by which exogenous substances enhance plant salt tolerance through the modulation of ion membrane transport and reactive oxygen species metabolism. Antioxidants 13(9):1050. https://doi.org/10.3390/antiox13091050

Keisham M, Mukherjee S, Bhatla SC (2018) Mechanisms of sodium transport in plants-progresses and challenges. Int J Mol Sci 19:647

Khan TA, Saleem M, Fariduddin Q (2022) Recent advances and mechanistic insights on melatonin-mediated salt stress signaling in plants. Plant Physiol Biochem 188:97–107. https://doi.org/10.1016/j.plaphy.2022.08.007

Khan S, Alvi AF, Fatma M, Al-Hashimi A, Sofo A, Khan NA (2024a) Relative effects of melatonin and hydrogen sulfide treatments in mitigating salt damage in wheat. Front Plant Sci 15:1406092. https://doi.org/10.3389/fpls.2024.1406092

Khan Z, Jan R, Asif S, Farooq M, Jang H, Kim G, Kim N, Kim M (2024b) Exogenous melatonin induces salt and drought stress tolerance in rice by promoting plant growth and defense system. Sci Rep 14:1214. https://doi.org/10.1038/s41598-024-51369-0

Kostopoulou Z, Therios I, Roumeliotis E, Kanellis AK, Molassiotis A (2014) Melatonin combined with ascorbic acid provides salt adaptation in *Citrus aurantium* L. Seedlings. Plant Physiol Biochem 86:155–165. https://doi.org/10.1016/j.plaphy.2014.11.021

Li C, Wang P, Wei Z, Liang D, Liu C, Yin L, Jia D, Fu M, Ma F (2012) The mitigation effects of exogenous melatonin on salinity-induced stress in *Malus hupehensis.* J Pineal Res 53:298–306. https://doi.org/10.1111/j.1600-079X.2012.00999.x

Li C, Liang B, Chang C, Wei Z, Zhou S, Ma F (2016) Exogenous melatonin improved potassium content in Malus under different stress conditions. J Pineal Res 61(2):218–229. https://doi.org/10.1111/jpi.12342

Li J, Liu J, Zhu T, Zhao C, Li L, Chen M (2019) The role of melatonin in salt stress responses. Int J Mol Sci 20(7):1735. https://doi.org/10.3390/ijms20071735

Li J, Yuan F, Liu Y et al (2020) Exogenous melatonin enhances salt secretion from salt glands by upregulating the expression of ion transporter and vesicle transport genes in Limonium bicolor. BMC Plant Biol 20:493. https://doi.org/10.1186/s12870-020-02703-x

Li J, Liu Y, Zhang M et al (2022) Melatonin increases growth and salt tolerance of *Limonium bicolor* by improving photosynthetic and antioxidant capacity. BMC Plant Biol 22:16. https://doi.org/10.1186/s12870-021-03402-x

Li P, Liang D, Yang E, Yang E, Zeb M, Huang H, Sun H, Zhang W, Peng C, Zhao Y, Ma W (2024a) Bio-nanopore technology for biomolecules detection. Adv Biotechnol 2:45. https://doi.org/10.1007/s44307-024-00051-7

Li T, Cai H, Lai Y, Yao H, Li D (2024b) A simple and effective method to remove pigments from heterologous secretory proteins expressed in *Pichia pastoris*. Adv Biotechnol 2:5. https://doi.org/10.1007/s44307-024-00013-z

Liu D, Sun S, Liu L, Shi D, Chen Y, Zhao K (2019) Overexpression of the melatonin synthesis-related gene SlCOMT1 improves the resistance of tomato to salt stress. Molecules 24(8):1514. https://doi.org/10.3390/molecules24081514

Liu J, Shabala S, Zhang J, Ma G, Chen D, Shabala L, Zeng F, Chen H, Zhou M, Venkataraman G, Zhao Q (2020) Melatonin improves rice salinity stress tolerance by NADPH oxidase-dependent control of the plasma membrane K^+ transporters and K^+ homeostasis. Plant Cell Environ 43(11):2591–2605. https://doi.org/10.1111/pce.13759

Liu Z, Ren X, Zhu W, Li Y, Li G, Liu C, Li D, Shi Y, Wang C, Zhu X, Sun H (2024) Transcriptomic analysis of melatonin-mediated salt stress response in germinating alfalfa. Agriculture 14(5):661. https://doi.org/10.3390/agriculture14050661

Luo M, Chu J, Wang Y, Chang J, Zhou Y, Jiang X (2024) A high-affinity potassium transporter (MeHKT1) from cassava (*Manihot esculenta*) negatively regulates the response of transgenic *Arabidopsis* to salt stress. BMC Plant Biol 24:372. https://doi.org/10.1186/s12870-024-05084-7

Masoumi Z, Haghighi M, Mozafarian M (2024) Effects of foliar spraying with melatonin and chitosan Nano-encapsulated melatonin on tomato (*Lycopersicon esculentum* L. cv. Falcato) plants under salinity stress. BMC Plant Biol 24:961. https://doi.org/10.1186/s12870-024-05672-7

Meng X, Zhou J, Sui N (2018) Mechanisms of salt tolerance in halophytes: current understanding and recent advances. Open Life Sci 13:149. https://doi.org/10.1515/biol-2018-0020

Mohamadi Esboei M, Ebrahimi A, Amerian MR, Alipour H (2022) Melatonin confers fenugreek tolerance to salinity stress by stimulating the biosynthesis processes of enzymatic, non-enzymatic antioxidants, and diosgenin content. Front Plant Sci 1:890613. https://doi.org/10.3389/fpls.2022.890613

Muhammad HMD, Naz S, Ahmad R, Ali E, Anwar M, Altaf MA, Alansi S, Alsahli AA, Abou Fayssal S (2024) Melatonin triggers salinity tolerance in pansy (Viola tricolor) by regulation of defense system. J King Saud Univ Sci 36(8):103286. https://doi.org/10.1016/j.jksus.2024.103286

Nanehkaran FM, Razavi SM, Ghasemian A et al (2024) Foliar applied potassium nanoparticles (K-NPs) and potassium sulfate on growth, physiological, and phytochemical parameters in *Melissa officinalis L.* under salt stress. Environ Sci Pollut Res 31:31108–31122. https://doi.org/10.1007/s11356-024-33306-w

Noctor G, Mhamdi A, Chaouch S (2021) The role of glutathione in oxidative stress and its interplay with other antioxidants in plants. Antioxidants 10(7):1106

Pehlivan N, Altaf MT, Emamverdian A, Ghorbani A (2025) Beyond the lab: Future-proofing agriculture for climate resilience and stress management. Front Plant Sci 16:1565850. https://doi.org/10.3389/fpls.2025.1565850

Pottosin I, Dobrovinskaya O (2014) Non-selective cation channels in plasma and vacuolar membranes and their contribution to Kþtransport. J Plant Physiol 171:732–742

Sagharyan M, Mohammadbagherlou S, Samari E, Zargar M, Ghorbani A, Chen M (2025) Synthetic biology and metabolic engineering strategies in identifying and producing plant natural products; with emphasis on the CRISPR/Cas systems. Ind Crops Prod 230:121060. https://doi.org/10.1016/j.indcrop.2025.121060

Sepehry Javan Z, Razavi SM, Khalofah A et al (2024) The ameliorating effects of cinnamic acid-based nanocomposite against salt stress in peppermint. Environ Sci Pollut Res 31:45055–45073. https://doi.org/10.1007/s11356-024-34158-0

Sha S, Cai G, Liu S et al (2024) Roots to the rescue: how plants harness hydraulic redistribution to survive drought across contrasting soil textures. Adv Biotechnol 2:43. https://doi.org/10.1007/s44307-024-00050-8

Shao Y, Cheng Y, Pang H, Chang M, He F, Wang M, Davis DJ, Zhang S, Betz O, Fleck C, Dai T, Madahhosseini S, Wilkop T, Jernstedt J, Drakakaki G (2021) Investigation of salt tolerance mechanisms across a root developmental gradient in almond rootstocks. Front Plant Sci 11:595055. https://doi.org/10.3389/fpls.2020.595055

Sharma P, Jha AB, Dubey RS, Pessarakli M (2012) Reactive oxygen species, oxidative damage, and antioxidative defense mechanism in plants under stressful conditions. J Bot 2012:217037

Siddiqui MH, Khan MN, Mukherjee S, Basahi RA, Alamri S, Alsubaie QD, Ali HM, Al-Munqedhi BMA, Almohisen IA (2021) Exogenous melatonin-mediated regulation of K^+/Na^+ transport, H^+-ATPase activity and enzymatic antioxidative defence operate through endogenous hydrogen sulphide signalling in NaCl-stressed tomato seedling roots. Plant Biol 23(5):797–805. https://doi.org/10.1111/plb.13296

Solis CA, Yong T, Zhou M, Venkataraman G, Shabala L, Holford P, Shabala S, Chen H (2022) Evolutionary significance of NHX family and NHX1 in salinity stress adaptation in the genus Oryza. Int J Mol Sci 23(4):2092. https://doi.org/10.3390/ijms23042092

Song C, Manzoor MA, Ren Y et al (2024) Exogenous melatonin alleviates sodium chloride stress and increases vegetative growth in Lonicera japonica seedlings via gene regulation. BMC Plant Biol 24:790. https://doi.org/10.1186/s12870-024-05506-6

Su Y, Luo W, Lin W, Ma L, Kabir MH (2015) Model of cation transportation mediated by high-affinity potassium transporters (HKTs) in higher plants. Biol Proced Online 17:1–13

Talaat NB, Todorova D (2022) Antioxidant machinery and glyoxalase system regulation confers salt stress tolerance to wheat (*Triticum aestivum* L.) plants treated with melatonin and salicylic acid. J Soil Sci Plant Nutr 22:3527–3540. https://doi.org/10.1007/s42729-022-00907-8

Ubaidillah M, Farooq M, Kim K (2024) Enhancing salt tolerance in rice genotypes through exogenous melatonin application by modulating growth patterns and antistress agents. Sci Rep 14(1):1–14. https://doi.org/10.1038/s41598-024-77161-8

Vafadar F, Amooaghaie R, Ehsanzadeh P, Ghanati F, Sajedi RH (2020) Crosstalk between melatonin and Ca^{2+}/CaM evokes systemic salt tolerance in *Dracocephalum kotschyi.* J Plant Physiol 252:153237. https://doi.org/10.1016/j.jplph.2020.153237

Wang N, Luo X, Wang Z, Liu J (2024a) Mitigating effect of exogenous melatonin on salt and drought stress in *Cyperus esculentus* L. during the tillering stage. Agronomy 14(5):1009. https://doi.org/10.3390/agronomy14051009

Wang F, Wan C, Wu W, Yang S, Chen X (2024b) Exogenous melatonin (MT) enhances salt tolerance of okra (*Abelmoschus esculentus* L.) plants by regulating proline, photosynthesis, ion homeostasis and ROS pathways. Vegetos 37:224–238. https://doi.org/10.1007/s42535-023-00568-7

Wei H, Wang X, He Y, Xu H, Wang L (2021) Clock component OsPRR73 positively regulates rice salt tolerance by modulating OsHKT2;1-mediated sodium homeostasis. EMBO J 40:e105086

Wei H, He W, Kuang Y, Wang Z, Wang Y, Hu W, Tang M, Chen H (2023) Arbuscular mycorrhizal symbiosis and melatonin synergistically suppress heat-induced leaf senescence involves in abscisic acid, gibberellin, and cytokinin-mediated pathways in perennial ryegrass. Environ Exp Bot 213:105436. https://doi.org/10.1016/j.envexpbot.2023.105436

Wen Z, Tyerman SD, Dechorgnat J, Ovchinnikova E, Dhugga KS, Kaiser BN (2017) Maize NPF6 proteins are homologs of Arabidopsis CHL1 that are selective for both nitrate and chloride. Plant Cell 29:2581–2596

Xu L, Xiang G, Sun Q, Ni Y, Jin Z, Gao S, Yao Y (2019) Melatonin enhances salt tolerance by promoting MYB108A-mediated ethylene biosynthesis in grapevines. Hortic Res 6(1):1–14. https://doi.org/10.1038/s41438-019-0197-4

Yan F, Wei H, Li W, Liu Z, Tang S, Chen L, Ding C, Jiang Y, Ding Y, Li G (2020) Melatonin improves K^+ and Na^+ homeostasis in rice under salt stress by mediated nitric oxide. Ecotoxicol Environ Saf 206:111358. https://doi.org/10.1016/j.ecoenv.2020.111358

Yang X, Liu D, Liu C et al (2024) Possible melatonin-induced salt stress tolerance pathway in *Phaseolus vulgaris* L. using transcriptomic and metabolomic analyses. BMC Plant Biol 24:72. https://doi.org/10.1186/s12870-023-04705-x

Yin P, Liang X, Zhao H, Xu Z, Chen L, Yang X et al (2023) Cytokinin signaling promotes salt tolerance by modulating shoot chloride exclusion in maize. Mol Plant 16:1031–1047

Zahedi SM, Hosseini MS, Naghmeh D, Abadı'a J, Germ M, Gholami R, Abdelrahman M (2022) Evaluation of drought tolerance in three commercial pomegranate cultivars using photosynthetic pigments, yield parameters and biochemical traits as biomarkers. Agric Water Manag 261:107357

Zhan H, Nie X, Zhang T, Li S, Wang X, Du X, Tong W, Song W (2019) Melatonin: a small molecule but important for salt stress tolerance in plants. Int J Mol Sci 20(3):709. https://doi.org/10.3390/ijms20030709

Zhang N, Zhang H, Sun Q, Cao Y, Li X, Zhao B, Wu P, Guo Y (2017) Proteomic analysis reveals a role of melatonin in promoting cucumber seed germination under high salinity by regulating energy production. Sci Rep 7(1):1–15. https://doi.org/10.1038/s41598-017-00566-1

Zhang Y, Yang C, Liu S, Xie Z, Zhang J, Ji F (2024) Alleviation of salt toxicity by melatonin supplementation in eggplant involves the up-regulation of ascorbate–glutathione and glyoxalase system. Sci Hortic 326:112720. https://doi.org/10.1016/j.scienta.2023.112720

Zhao G, Zhao Y, Yu X, Kiprotich F, Han H, Guan R, Wang R, Shen W (2018) Nitric oxide is required for melatonin-enhanced tolerance against salinity stress in rapeseed (*Brassica napus* L.) seedlings. Int J Mol Sci 19:1912. https://doi.org/10.3390/ijms19071912

Zhao C, Yang M, Wu X, Wang Y, Zhang R (2021) Physiological and transcriptomic analyses of the effects of exogenous melatonin on drought tolerance in maize (*Zea mays* L.). Plant Physiol Biochem 168:128–142. https://doi.org/10.1016/j.plaphy.2021.09.044

Zhou H, Shi H, Yang Y, Feng X, Chen X, Xiao F, Lin H, Guo Y (2023) Insights into plant salt stress signaling and tolerance. J Genet Genomics 51(1):16–34. https://doi.org/10.1016/j.jgg.2023.08.007

Zhu B, Zheng S, Fan W, Zhang M, Xia Z, Chen X, Zhao A (2022) Ectopic overexpression of mulberry MnT5H2 enhances melatonin production and salt tolerance in tobacco. Front Plant Sci 13:1061141. https://doi.org/10.3389/fpls.2022.1061141

Zulfiqar F, Nafees M, Moosa A, Ferrante A, Darras A (2024) Melatonin induces proline, secondary metabolites, sugars and antioxidants activity to regulate oxidative stress and ROS scavenging in salt stressed sword lily. Heliyon 10(11):e32569. https://doi.org/10.1016/j.heliyon.2024.e32569

How Melatonin Regulates Ion Chelation, Immobilization, and Cellular Functions Under Heavy Metal Stress?

6

Malini Muthu Karpagam and James Godwin

Abstract

Heavy metal (HM) accumulation is a growing environmental concern that affects the growth and development of plants and poses a risk to human health. When taken up by plants, HM causes severe damage, including oxidative damage, hampered physiological processes, disrupted ion homeostasis, and metabolic imbalance. Plants tackle HM stress through innate defense mechanisms, such as immobilization of HM ions, their chelation by phytochelatins and metallothioneins, limited foliar translocation, metal sequestration and compartmentalization, metal exclusion, and free radicals scavenging by antioxidant enzymes. Melatonin (MT), a natural plant hormone, has been observed to exhibit multifaceted protection under HM stress. MT enhances plant tolerance to HM stress by improving redox and nutrient homeostasis, metabolic functions, and osmotic balance. It is imperative to elucidate the mechanisms by which both exogenous and endogenous MT mitigate HM toxicity in plants to recognize potential knowledge gaps that should be addressed in the future. This chapter focuses on providing better insights into the MT-mediated HM stress response and tolerance mechanisms in plants, including MT-mediated HM chelation, immobilization and sequestration, oxidative damage reduction, and transcriptional and epigenetic regulation of HM stress.

M. M. Karpagam (✉)
Smith College, Northampton, MA, USA
e-mail: mmk@smith.edu

J. Godwin
Donald Danforth Plant Science Center, St. Louis, MO, USA

R. K. Tiwari et al. (eds.), *Melatonin: Signal Transduction Mechanisms and Defense Networks in Plants*, https://doi.org/10.1007/978-981-96-8869-2_6

Keywords
Reactive oxygen species (ROS) · Antioxidants · Oxidative stress · Ion chelation · Vacuolar sequestration · Melatonin · Heavy metals · Epigenetics · Gene regulation

6.1 Introduction

Heavy metal (HM) stress is a critical abiotic stress that is significantly hampering plant development and survival (Wang et al. 2024b). HM contaminants generally originate from activities like mining, industrial discharge, and overuse of chemical fertilizers and pesticides, which end up accumulating in soil and water systems (Hasan et al. 2015; Kalaivanan and Ganeshamurthy 2016; Weissmannová and Pavlovský 2017; Yang et al. 2023a; Cheng et al. 2024).

The term "heavy metal" collectively refers to metal elements with atomic density higher than 4 g cm^{-3}, or five times or more, greater than water. HMs can be categorized as essential and non-essential groups (Hawkes 1997; Nagajyoti et al. 2010). Essential HMs such as copper (Cu), zinc (Zn), iron (Fe), cobalt (Co), manganese (Mn), nickel (Ni), and molybdenum (Mo) play biochemical and physiological functions. However, excess concentration can cause toxic effects in plants (Shah et al. 2010; DalCorso 2012; Ramezani et al. 2021). In contrast, non-essential HMs such as cadmium (Cd), chromium (Cr), arsenic (As), silver (Ag), mercury (Hg), lead (Pb), and antimony (Sb) are toxic to plants even at low concentration (DalCorso 2012; Gu et al. 2017; Ghorbani et al. 2023). The sessile nature of plants makes them more vulnerable to HM toxicity, as they cannot escape the contaminated environment (Devi et al. 2021). HMs are ubiquitously present in the water and soil systems (Tiwari et al. 2018; Moustafa-Farag et al. 2020a). HM ions in soil primarily contact the root surface and get absorbed via roots, while in aquatic systems, the entire plant gets exposed. Furthermore, HMs deposited on the foliar surfaces are directly absorbed through leaves (Shah et al. 2010). Once translocated into plant tissue, HMs mainly induce reactive oxygen species (ROS) production in cellular organelles like peroxisomes, mitochondria, and chloroplasts, which induces oxidative stress (Chen and Su 2018; Ghorbani et al. 2023; Wang et al. 2024b). ROS production disrupts cellular homeostasis and damages macromolecules. HMs also bind to sulfhydryl groups in proteins, affecting the enzymatic functionality and structural integrity and displacing essential elements like Cu and Fe, resulting in nutrient deficiencies (Foyer and Noctor 2008; DalCorso 2012). For instance, in Cu-stressed cucumber plants, the leaf K uptake is reduced, photosynthesis is inhibited, and cell expansion is retarded (Alaoui-Sossé et al. 2004). Additionally, HM exposure alters physiological and biochemical characteristics. For example, Cd exposure affects water balance, nutrient uptake (Drazic et al. 2006; Scebba et al. 2006), and photosynthesis (Vassilev et al. 1998; Shah et al. 2010). Altogether, these impairments affect plant growth and development.

It is noteworthy that HMs are non-degradable by soil microorganisms. Thus, they tend to over-accumulate in soil-crop systems (Lin et al. 2022a). This overaccumulation, besides causing physiological damage in crops, also poses a prominent risk to human health by entering the food chain and human body (Cocârţă et al. 2016; Edelstein and Ben-Hur 2018; Zhang et al. 2023). For instance, while Cd toxicity affects normal plant functioning, it also causes negative impacts on animal and human health via bioaccumulation in the food chain (Rizwan et al. 2016; Gu et al. 2021). These concerns make it imperative to find ways to mitigate HM stress in plants and reduce its accumulation in plants.

6.2 Melatonin-Mediated Heavy Metal Stress Tolerance

HM contamination is a growing environmental concern that not only harms plant growth and development but also affects human health. Plants have evolved inherent strategies to combat the risk of HM stress, including immobilization of HM ions, chelation of HM ions by phytochelatins and metallothioneins, detoxification, compartmentalization of metal ions, and the cell structure repair (Sanità di Toppi and Gabbrielli 1999; Hall 2002; Park et al. 2012; Hasan et al. 2015). In addition, plants also rely on an antioxidant defense system to counteract the harmful ROS generated under HM stress (Jozefczak et al. 2014). Studies have demonstrated that external application of agents like melatonin, glutathione, hydrogen sulfide, salicylic acid, or priming with hydrogen peroxide and nitroprusside to provide tolerance against HM stress (Savvides et al. 2016; Moustafa-Farag et al. 2020a; Tiwari et al. 2020; Hoque et al. 2021).

Melatonin (MT) (N-Acetyl-5-methoxytryptamine) is a well-known phytohormone synthesized in chloroplasts and mitochondria (Kováčik et al. 2017; Tiwari et al. 2021a, b). MT displays numerous tolerance mechanisms in the face of HM stress, including improved antioxidant activity, regulating osmotic balance, protecting cell membranes, enhanced root growth, and nutrient homeostasis (Jahan et al. 2020; Altaf et al. 2021, 2022a, b). As MT plays a multifaceted role in plant growth and stress mitigation, various studies have investigated its role over diverse HMs, which comprises Cd (Hasan et al. 2015; Gu et al. 2017), Cu (Posmyk et al. 2008; Cao et al. 2019), Fe (Ahammed et al. 2020), B (Sarafi et al. 2017), Ni (Jahan et al. 2020), and vanadium (V) (Altaf et al. 2022b).

The endogenous MT levels in plants were observed to surge on exposure to HM (Wang et al. 2024b). For example, MT levels in algae were observed to increase significantly under Zn, Cd, and Pb stress, indicating its role in counterbalancing toxicity (Tal et al. 2011). Likewise, Zn sulfate exposure increased endogenous MT production in barley and lupine (Arnao and Hernández-Ruiz 2009, 2013).

As a response to HM stress, key MT biosynthesis genes also get activated. For instance, a study conducted on tomato Cai et al. (2017) observed that the expression of the MT biosynthesis gene (*COMT1*) upregulated the heat shock proteins (HSPs) under Cd exposure. Another study in rice reported that Cd stress increased MT synthesis and stress resistance by enhancing the activity of MT biosynthesis enzymes such as tryptophan decarboxylase (*TDC*) and N-acetyl 5-hydroxytryptamine methyltransferase (*ASMT*) (Byeon et al. 2015). Additionally, Cd-induced chloroplast damage led to MT accumulation, thus protecting photosynthetic machinery (Lee et al. 2018).

The external MT application has also proved to enhance stress resilience in HM by improving the internal defense mechanisms (Moustafa-Farag et al. 2020b; Sun et al. 2021). In watermelon, external MT application minimized V-translocation from roots to shoots, improved antioxidant activity, and altered root architecture for improved nutrient absorption (Nawaz et al. 2018). In pea plants, MT application protected the photosynthetic apparatus under Cu stress while improving its overall survival (Tan et al. 2007b; Wang et al. 2024b). Meanwhile, tomato plants exhibited improved photosynthetic efficiency, root growth, decreased cell membrane damage, and nutrient homeostasis under Ni toxicity (Jahan et al. 2020; Altaf et al. 2021).

All these studies highlight that MT offers multidimensional protection under HM stress (Fig. 6.1). This chapter focuses on recent progress in understanding the mechanisms by which MT alleviates HM toxicity in plants. It focuses on its roles in HM chelation and immobilization, oxidative damage reduction, and transcription and epigenetic regulation under HM stress.

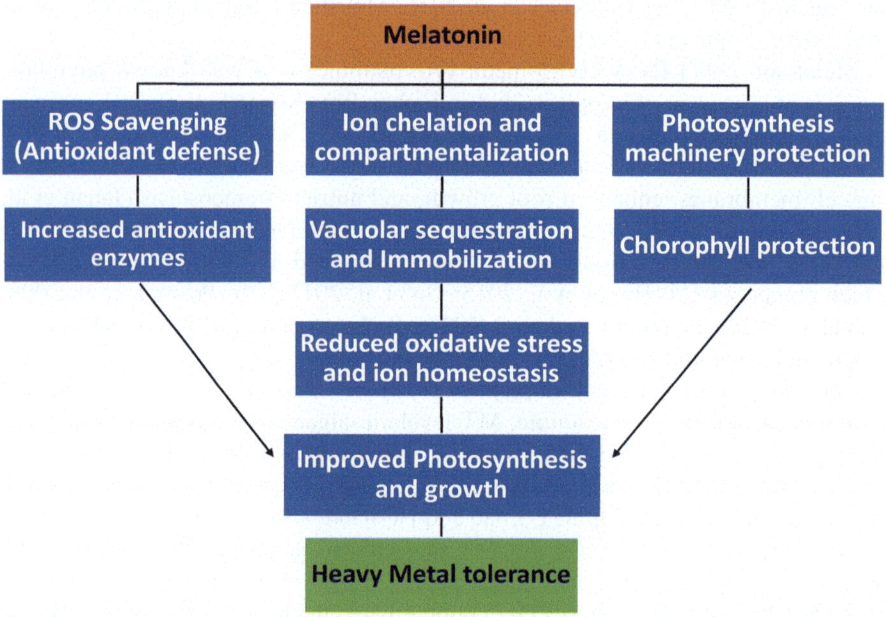

Fig. 6.1 Simplified illustration of the MT-mediated pathway alleviating HM toxicity in plants. Abbreviations: *ROS* Reactive Oxygen Species

6.3 Melatonin in Heavy Metal Stress Management

6.3.1 Melatonin-Regulated Heavy Metal Chelation and Immobilization

Plants respond to HM uptake through a well-synchronized mechanism, which starts with HM chelation and consequent compartmentalization into vacuoles (Limson et al. 1998; Clemens 2006; Wang et al. 2024a, b). A class of small HM-binding peptides called phytochelatin (PC) plays a key role in chelation. It is synthesized by PC synthase using glutathione (GSH) as a substrate (Limson et al. 1998; Clemens 2006). PCs chelate with HMs to form a phytochelatin-metal complex in the cytoplasm, which subsequently gets transported into the vacuole, thereby reducing disruptions in the cytoplasm (de Oliveira et al. 2014; de Oliveira et al. 2016).

MT promotes HM detoxification by increasing the biosynthesis of PCs and GSH. For example, MT enhanced the PC accumulation in tomato plants under Cd stress, increasing the proportion of Cd compartmentalized into vacuoles and bound to cell walls. This process alleviates Cd toxicity by constraining Cd translocation within the plant system (Hasan et al. 2015; Li et al. 2016; Cai et al. 2017).

The first layer of defense against HM uptake is the immobilization of HMs by binding them to the cell wall matrix (Krzesłowska 2011; Altaf et al. 2023). This process is expedited by MT-mediated upregulation of cell wall biosynthesis genes such as *CesA*, *CSL*, and *PME*, which are connected to cellulose, hemicellulose, and demethylesterification of pectin. In particular, the carboxyl groups in the cell wall components increase the HM-binding sites, enabling HM immobilization (Shi et al. 2008; Cao et al. 2019). For instance, foliar application of MT in tobacco plants improved Cd stress tolerance by suppressing Cd transporter gene expression while enhancing Cd fixation to cell walls and vacuoles (Wang et al. 2019).

In *Malus hupehensis*, exogenous MT improves Al^{3+} tolerance through citric acid-mediated chelation followed by vacuolar sequestration, which involves *MdSTOP1*-regulated transporters and maintaining ion balance through H^+ efflux (Wang et al. 2023). In cucumber seedlings, exogenous MT improves Cu resistance by chelating Cu ions via enhanced GSH and PC levels, thereby improving Cu sequestration while mitigating oxidative stress (Cao et al. 2019). Through coordinated chelation, immobilization, and vacuolar sequestration, MT plays a crucial role in enhancing plant tolerance to HM toxicity and sustaining cellular homeostasis (Altaf et al. 2023; Wang et al. 2024b).

6.3.2 Melatonin-Mediated Antioxidant Defense Mechanisms Under Heavy Metal Stress

HM stress disturbs the dynamic balance between the generation of ROS, reactive nitrogen species (NOS), and detoxification by the antioxidative protection mechanism in plants, thereby leading to oxidative stress (Hoque et al. 2021). An excessive level of HM ions in the cytosol produces excessive free radicals, including hydrogen peroxide (H_2O_2), nitric oxide radicals (NO^-), superoxide anion (O_2^-), and hydroxy radicals ($\cdot OH^-$) (Chan and Shi 2015; Savvides et al. 2016). Excessive endogenous levels of ROS and NOS are detrimental to cells, causing significant damage to cellular biomolecules such as proteins, nucleic acids, and membrane lipids. They can also disrupt chloroplast structures, ultimately leading to cell death (Birben et al. 2012; Ahmad et al. 2015).

As a potent antioxidant, MT can enhance physiological activity in plants, readily traversing cell membranes to reach the cytoplasm (Tiwari et al. 2020, 2021a, b). Although the plasma membrane typically regulates the absorption of small molecules, MT's amphipathic nature enables it to traverse the membrane easily with minimal resistance (Zhang et al. 2015; Kumar et al. 2023, 2024). Numerous works have explored the pathways involved in MT synthesis and the scavenging of ROS in plants during HM stress (Tan et al. 2007a; Cheng et al. 2024). It is reported that endogenous levels of MT are higher under HM conditions (Khan et al. 2020; Wang et al. 2024a, b). The ability of MT to scavenge ROS was first identified two decades ago (Ianas et al. 1991). MT's antioxidant capacity is twice as potent as vitamin E, four times greater than GSH, and 14 times stronger than mannitol (Pieri et al. 1995). In plants, MT functions effectively as an antioxidant molecule using direct and indirect mechanisms during HM stress (Khan et al. 2020; Wang et al. 2024a). Indirectly, MT is transmitted as a signal for ROS-mediated pathways by regulation of H_2O_2 levels while also acting as a direct antioxidant to maintain low ROS levels (Tan et al. 2007b). MT enhances the production of antioxidative enzymes as a defense against oxidative damage under HM stress (Tiwari et al. 2024). They rely on enzymatic antioxidants such as superoxide dismutase (SOD), glutathione peroxidase (GPX), catalase (CAT), and enzymes of the ascorbate-glutathione (AsA-GSH) cycle, including ascorbate peroxidase (APX), monodehydroascorbate reductase (MDHAR), dehydroascorbate reductase (DHAR), and glutathione reductase (GR). Nonenzymatic antioxidants include ascorbic acid (AsA), reduced GSH, carotenoids, tocopherols, and phenolic compounds, all of which play a crucial role in free radical scavenging (Khan et al. 2020). MT-mediated antioxidant activity is strongly influenced by the application rate of MT, the concentration of HM, and the plant species (Cheng et al. 2024). For example, soybean subjected to Al-stress displayed a notable increase in growth and antioxidant activity with a 1 mM MT treatment compared to treatments with 100- or 200 mM MT (Zhang et al. 2017). In wheat, MT decreases the H_2O_2 levels while enhancing the antioxidant enzymes APX, CAT, GSH, POD, and SOD exposed to Cd stress (Ni et al. 2018). The mitigation of HM-induced oxidative stress through ROS scavenging and enhanced antioxidative activity has been reported in different plants, including wheat, mustard, tomato, watermelon, rapeseed, and cucumber, under HM stress caused by Cd, Al, Cu, Pb, Se, Ni, Fe, As, Zn, and V (Ahmad et al. 2010; Faizan et al. 2024; Jahan et al. 2020; Khan et al. 2020; Ulhassan et al. 2019; Wang et al. 2024a; Zhang et al. 2015). We have compiled a table listing the antioxidant enzymes that are upregulated during HM stress in different plants with the application of MT (Table 6.1).

Table 6.1 Antioxidant enzymes upregulated during HM stress with the application of MT in plants

Plant	Heavy metal	Elevated antioxidant enzymes	References
Triticum aestivum	Aluminum	SOD, APX, CAT, and POD	Sun et al. (2020)
Triticum aestivum	Boron	SOD, CAT, GR, APX, and LOX	Al-Huqail et al. (2020)
Triticum aestivum	Chromium	SOD, CAT, APX, and GR	Seleiman et al. (2020)
Triticum aestivum	Cadmium	SOD, APX, GSH, CAT, and POD	Ni et al. (2018)
Nicotiana tabacum	Cadmium	SOD, APX, and CAT	Wang et al. (2019)
Melissa officinalis	Cadmium	POD	Hodi et al. (2021)
Cathranthus roseus	Cadmium	POD and CAT	Nabaei and Amooaghaie (2019)
Trigonella foenum graecum	Lead	SOD and CAT	Xalxo and Keshavkant (2019)
Solanum lycopersicum	Lanthanum	SOD, APX, and GR	Siddiqui et al. (2019)
Solanum lycopersicum	Nickel	SOD, APX, CAT, and GR	Jahan et al. (2020)
Solanum lycopersicum	Aluminum	SOD, APX, and GR	Ghorbani et al. (2023)
Solanum lycopersicum	Chromium	SOD, CAT, APX, and GR	Raja et al. (2023)
Solanum lycopersicum	Copper	CAT, APX, SOD, and GR	Zhang et al. (2022)
Solanum lycopersicum	Cadmium	CAT, SOD, APX, and GR	Song et al. (2024)
Cucumus sativus	Iron	SOD, POD, and CAT	Ahammed et al. (2020)
Cucumus sativus	Copper	POD, SOD, APX, and GR	Cao et al. (2019)
Cucumis sativus	Cadmium	SOD, CAT, and POD	Kang et al. (2024)
Cucumis melo	Copper	CAT, POD, and SOD	Hu et al. (2020)
Citrullus lanatus	Vanadium	SOD and CAT	Nawaz et al. (2018)
Brassica napus	Selenium	OD, APX, GR, and CAT	Ulhassan et al. (2019)
Brassica napus	Chromium	SOD, APX, CAT, and POD	Ayyaz et al. (2021)
Brassica napus	Cobalt	SOD, CAT, APX, and GR	Ali et al. (2024)
Rhus typhina	Nickel	SOD	Qu et al. (2024)
Capsicum annuum	Nickel	SOD, CAT, APX, GR, GST, and POD	Altaf et al. (2022a)
Capsicum annuum	Cadmium	SOD, CAT, and POD	Saqib et al. (2024)
Cicer arietinum	Cadmium	SOD, CAT, and APX	Sakouhi et al. (2023)
Hordeum vulgare	Cadmium	SOD and POD	Danilova et al. (2023)
Carthamus tinctorius	Lead	CAT, GPX, and GR	Namdjoyan et al. (2020)
Zea mays	Chromium	POD, SOD, APX, GR, and CAT	Yang et al. (2023b)
Zea mays	Copper	CAT and APX	Raza Khan et al. (2023)
Cyphomandra betacea	Cadmium	SOD and POD	Lin et al. (2018)

(continued)

Table 6.1 (continued)

Plant	Heavy metal	Elevated antioxidant enzymes	References
Oryza sativa	Copper	SOD, POD, APX, CAT, and GR	Li et al. (2022a)
Oryza sativa	Arsenic	SOD and CAT	Nazarian and Ghanati (2020)
Avena nuda	Cadmium	SOD, POD, and CAT	Wang et al. (2022)
Spinacia oleracea	Arsenic	SOD, POD, and CAT	Asif et al. (2020)
Spinacia oleracea	Boron	SOD, CAT, and APX	Moussa and Algamal (2017)
Raphanus sativus	Cadmium	GR, SOD, POD, CAT, and APX	Xu et al. (2020)
Raphanus sativus	Lead	APX and GR	Tang et al. (2021)
Vicia faba	Arsenic	SOD and AsA-GSH	Siddiqui et al. (2020)
Strawberry (Fragaria × ananassa)	Cadmium	SOD, POD, CAT, and APX	Wu et al. (2021)
Perilla frutescens	Cadmium	SOD and POD	Xiang et al. (2019)
Camellia sinensis	Arsenic	APX, SOD, and POD	Li et al. (2021)
Glycine max	Aluminum	SOD and POD	Zhang et al. (2017)
Malva parviflora	Cadmium	SOD, CAT, and GPX	Tousi et al. (2020)

6.3.3 Melatonin Protects Photosynthesis Machinery Under Heavy Metal Stress

Disruption of photosynthetic machinery due to HM stress has been well documented in different plant species (Shomali et al. 2024; Wang et al. 2024b). MT's capacity to sustain photosynthesis under adverse conditions is a key factor in promoting plant growth. Together with salicylic acid (SA), MT reduces Cd absorption, boosts photosynthetic pigment levels, enhances the activity of the AsA-GSH cycle, and modulates the glyoxalase system, thereby improving plant tolerance to Cd stress (Amjadi et al. 2021). Similarly, combined SA and MT treatment proves highly effective in detoxifying As toxicity, optimizing nitrogen metabolism in pepper plants, and significantly promoting overall growth and development (Kaya et al. 2022). In tomato seedlings, exogenous MT and sulfur application exhibit a synergistic role in mitigating the toxic effects of lanthanum (La) by enhancing Rubisco activity, reducing ROS formation, and hampering damage to photosynthetic pigments (Siddiqui et al. 2019). Foliar application of MT and methyl-jasmonate alleviates Cr-induced damage to growth and prevents photoinhibition and chlorophyll synthesis in tomatoes (Qin et al. 2024). MT reduces the toxic effects of Cd in rice plants by boosting total chlorophyll levels, including chlorophyll a and chlorophyll b, while improving photosynthetic efficiency (Fv/Fm), demonstrating its positive influence on photosynthetic performance (Munir et al. 2024). MT restored the plant height, SPAD value, stem diameter, leaf area, stomatal conductance, and net photosynthetic rate caused by Cd stress, thereby enhancing biomass accumulation in various cotton organs (Yu et al. 2024). MT treatment under Cd stress in pepper seedlings significantly restored growth parameters, pigment concentration, and gas exchange traits, leading to enhanced growth and biomass production (Saqib et al. 2024). Another report on *Catharanthus roseus* (*L.*) has shown that under Cd stress, foliar application of MT effectively mitigates stress-induced damage by significantly reducing electrolyte leakage, enhancing chlorophyll content, and restoring the dynamic balance of essential cations in the leaves (Nabaei and Amooaghaie 2019). Under Cd stress conditions, MT treatment led to an increase in root and above-ground dry biomass of Perilla (*Perilla frutescens* (L.) Britt.). Simultaneously, the contents of chlorophyll a, chlorophyll b, and total chlorophyll rose significantly (Xiang et al. 2019). In cucumber seedlings, exogenous MT increased the net photosynthetic rate with the enhancement of stomatal conductance and transpiration rate under Cd stress. Further analysis of the chlorophyll fluorescence parameters indicated that the reduced photochemical energy conversion in PSII and the photochemical fluorescence quenching coefficient were alleviated by MT treatment under Cd stress (Wu et al. 2019). Another study demonstrated that MT enhances photosynthetic efficiency, nutrient uptake, and root activity while reducing osmotic damage under Ni stress (Jahan et al. 2020). Foliar spray of MT improved photosynthesis, yield, and biomass under Cr stress (Sun et al. 2025). Pepper plants exposed to excess boron (B) for an extended period exhibited toxicity symptoms, reduced photosynthesis, lower dry weight, and significant B content in leaves and fruits. These effects were alleviated, and photosynthesis was restored with the application of MT and resveratrol (Sarafi et al. 2017). MT improved photosynthetic efficiency and protected stomatal aperture in maize seedlings under copper oxide nanoparticles (CuO NPs) (Raza Khan et al. 2023). In mallow plants, MT improved the photochemical efficiency of PSII by mitigating the toxic effects of Cd on the oxygen-evolving complex (OEC) at the electron donor site and by decreasing non-photochemical quenching (NPQ).

Further, MT helps preserve chlorophyll content in plants exposed to Cd, increasing chlorophyll biosynthesis precursors and slowing its degradation rate (Zoufan et al. 2023). Under B toxicity, foliar application of MT significantly improved the biosynthesis of photosynthetic pigments (Chl *a* and Chl *b*) and the activity of a key enzyme involved in chlorophyll synthesis, δ-ALAD. Additionally, it reduced chlorophyll degradation and the activity of the chlorophyll-degrading enzyme, chlorophyllase (Chlase) (Al-Huqail et al. 2020). Another report showed that the growth in strawberry seedlings was inhibited under Cd stress, and the seedling biomass and chlorophyll content were decreased. However, MT treatment slowed the inhibitory effect of seedling growth and significantly increased the seedling biomass and chlorophyll content (Wu et al. 2021). Numerous reports have shown that MT enhances photosynthetic capacity and offers adequate protection for photosynthesis through multiple mechanisms (Wang et al. 2024b). These include downregulation of genes involved in chlorophyll degradation and upregulating those related to chlorophyll synthesis (Lin et al. 2022b) and increasing the abundance of photosynthesis-related proteins (Wang et al. 2022). Besides that, MT plays a role in regulating the thylakoid membrane fluidity, facilitating the recovery of D1 protein synthesis and photosynthetic electron transport chain, and participating in the ABA signal pathway or calcium signal transduction under different stress (Khan et al. 2019; Lin et al. 2022b; Shomali et al. 2024; Wang et al. 2021a, 2022; Zhou et al. 2016). Further research is needed to understand the direct molecular mechanisms by which MT is involved in the photosynthetic machinery of plants under HM stress.

6.3.4 Melatonin-Induced Transcriptional Changes Under Heavy Metal Stress

When plants are exposed to HM stress, signal molecules such as ROS, transcription factors, hormones, protein kinases, and associated signaling cascades play an essential role in transmitting information to help mitigate the stress (Zhang et al. 2015; Feng et al. 2016; Ali et al. 2022; Altaf et al. 2023). MT regulates the expression of several genes during HM stress, further impacting the plant's adaptive mechanisms under these stress conditions. Numerous studies have shown that MT treatment under heavy metal (HM) stress leads to the upregulation of redox-related genes, such as *SOD*, *APX*, *POD*, and *GR* (Altaf et al. 2023; Wang et al. 2024b). Exogenous MT alleviates damage in radish plants caused by Cd stress by increasing the expression of transporter genes such as *yellow stripe 1-like* (*YSL*), *heavy metal ATPase* (*HMA*), and *ATP-binding cassette* (*ABC*), as well as the gene encoding the metal-binding protein *RsMT1* (Xu et al. 2020). MT alleviated As-induced disruptions by upregulating biosynthetic genes (*GA3ox*, *TDC*, *SNAT*, and *ASMT*) to enhance gibberellic acid and MT levels and increasing abscisic acid through *NCED3* induction and *ABA8ox1* suppression. It also improved tolerance by reducing As accumulation via the regulation of transporter genes (*OsNramp1*, *OsPT2*, *OsPT8*, and *OsLsi1*) in two rice indica cultivars (Samanta et al. 2021). Similarly, MT regulates the expression of genes, including *CLHl* (chlorophyllase-related gene), *PAO* (pheophorbide, an oxygenase-related gene), and *NHX1/AKT1* (genes associated with ion homeostasis) (Wang et al. 2024b). MT application upregulates the expression of anthocyanin biosynthesis genes (*ANS* and *CHS*) to promote anthocyanin production and mitigate As and Cr toxicity in tea and tomato (Li et al. 2021; Sun et al. 2023). Another report showed that MT reduced the Cd content of Chinese cabbage seedlings by decreasing the expression of Cd transport gene *IRT1,* thus improving the tolerance to Cd (Wang et al. 2021b). Under Fe stress, MT altered the Fe uptake in cucumber by increasing the expression *FRO2* and *IRT1* (Ahammed et al. 2020). In tomato seedlings, MT treatment increases the expression of respiratory burst oxidase (*RBOH*) under Ni stress (Jahan et al. 2020). Besides, MT accumulated in soybean roots on exposure to Al treatment coincides with the upregulation of MT biosynthesis genes *NSI-X1* and *NSI-X2*, thus suggesting its critical role in alleviating Al toxicity (Zhang et al. 2017). Further, Cd tolerance in tomato is enhanced by the transcription factor, HsfA1a, which directly upregulates the MT biosynthetic gene *COMT1* through binding to its promoter and increased MT biosynthesis (Cai et al. 2017). Although the roles played by transcription factors under various HM stresses have been comprehensively reviewed previously (Ali et al. 2022; Li et al. 2022b), the exact mechanisms through which MT influences transcription factors and thereby regulates gene expression remain largely unclear and require further investigation.

6.3.5 Melatonin-Induced Epigenetic Regulation Under Heavy Metal Stress

DNA methylation is an epigenetic modification linked to transcriptional regulation and gene expression (Bouyer et al. 2017). In plant genomes, cytosine methylation occurs in three patterns: CG, CHG, and CHH (where H represents A, T, or C). The latest study has revealed that HM exposure influences DNA methylation levels (Talarico et al. 2024). For example, CG and non-CG methylation patterns significantly varied between Cd-exposed and Cd-free rice (*Oryza sativa* ssp. *japonica* cv. Nipponbare) genomes (Feng et al. 2016). Particularly, numerous genes associated with stress response, metal transportation, and transcription factors were found to be hypermethylated, and most of the differentially methylated genes displayed transcriptional changes under Cd stress (Feng et al. 2016). Studies indicate that Zn, Pb, and Cd can either independently or in combination differentially regulate the DNA methyltransferase expression and alter DNA methylation levels in maize (Shafiq et al. 2020). Notably, Cd exposure in *Posidonia oceanica* resulted in the upregulation of the DNA methyltransferase gene, *CMT1,* causing DNA hypermethylation (Greco et al. 2012). In radish plants, Pb stress escalated genome-wide DNA methylation levels, while MT treatment led to a global reduction in DNA methylation (Tang et al. 2021). Several key antioxidant genes, such as *RsAPX2*, *RsPOD52*, and *RsGST,* showed reduced methylation levels under MT treatment, allowing radish plants to mitigate lead-induced ROS and alleviate oxidative stress. Additionally, MT treatment induced low methylation in HM transporter genes (e.g., *RsABCF5*, *RsYSL7*, and *RsHMT*) and transcription factors (e.g., *RsWRKY41* and *RsMYB2*), contributing to reduced Pb accumulation in radish roots (Tang et al. 2021). This raises the question of whether MT-induced expression of low-methylated metal transporter genes and genes encoding antioxidant genes is a general mechanism in plants under HM stress. The underlying epigenetic mechanism remains unclear, particularly whether MT directly interacts with DNA methyltransferases or demethylases or indirectly modulates their activity. Investigating these interactions could reveal MT's role in modulating epigenetic mechanisms, such as DNA methylation or histone modifications during HM stress.

6.4 Conclusion and Future Directions

The rapid increase in anthropogenic and technological activities has significantly contributed to widespread HM contamination. HMs have both acute and chronic effects on both the fauna and flora. Therefore, there is an essential requirement for sustainable, eco-friendly solutions in crop production under HM stress. Existing literature strongly supports MT's multifaceted role as a potent mitigator of HM stress in plants, demonstrating improvements in photosynthetic efficiency, nutrient homeostasis, ion chelation and compartmentalization, antioxidant capacity, and overall growth performance. While these findings spotlight MT's role as a more sustainable and eco-friendly phyto-protectant, critical gaps remain to be addressed. There is limited information on how to fine-tune MT dosages for various plant species and environmental conditions. Besides that, understanding how to ensure stable biosynthesis of MT and genetic regulation under field conditions remains a significant challenge. The complexity of endogenous MT signaling is primarily defined by the limited identification of receptors identified in plant species and its crosstalk with other molecular pathways, highlighting the need for integration of genomic, transcriptomic, and proteomic approaches. Furthermore, questions remain concerning the balance between enhanced HM tolerance and the maintenance of crop quality, especially in horticultural plants. Future research should focus on developing integrated approaches that combine molecular breeding, gene editing, and optimized agronomic practices to regulate endogenous MT levels precisely. A mechanistic understanding of MT interaction with other signaling molecules under diverse HM stress will refine its application strategies and ensure they do not compromise yield or quality. Additionally, the intricate interplay through which MT orchestrates epigenetic modifications, transcriptional regulation, and redox balance presents an exciting frontier for understanding and improving plant HM tolerance.

References

Ahammed GJ, Wu M, Wang Y, Yan Y, Mao Q, Ren J, Ma R, Liu A, Chen S (2020) Melatonin alleviates iron stress by improving iron homeostasis, antioxidant defense, and secondary metabolism in cucumber. Sci Hortic 265:109205

Ahmad P, Jaleel CA, Salem MA, Nabi G, Sharma S (2010) Roles of enzymatic and nonenzymatic antioxidants in plants during abiotic stress. Crit Rev Biotechnol 30:161–175

Ahmad P, Sarwat M, Bhat NA, Wani MR, Kazi AG, Tran LSP (2015) Alleviation of cadmium toxicity in *Brassica juncea* L. (Czern. & Coss.) by calcium application involves various physiological and biochemical strategies. PLoS One 10:e0114571

Alaoui-Sossé B, Genet P, Vinit-Dunand F, Toussaint M-L, Epron D, Badot P-M (2004) Effect of copper on growth in cucumber plants (*Cucumis sativus*) and its relationships with carbohydrate accumulation and changes in ion contents. Plant Sci 166:1213–1218

Al-Huqail AA, Khan MN, Ali HM, Siddiqui MH, Al-Huqail AA, AlZuaibr FM, Al-Muwayhi MA, Marraiki N, Al-Humaid LA (2020) Exogenous melatonin mitigates boron toxicity in wheat. Ecotoxicol Environ Saf 201:110822

Ali S, Gill RA, Shafique MS, Ahmar S, Kamran M, Zhang N, Riaz M, Nawaz M, Fang R, Ali B et al (2022) Role of phytomelatonin responsive to metal stresses: an omics perspective and future scenario. Front Plant Sci 13:936747

Ali S, Ali B, Sajid IA, Ahmad S, Yousaf MA, Ulhassan Z, Zhang K, Ali S, Zhou W, Mao B (2024) Synergistic effects of exogenous melatonin and zinc oxide nanoparticles in alleviating cobalt stress in Brassica napus: insights from stress-related markers and antioxidant machinery. Environ Sci Nano 12:368. https://doi.org/10.1039/D4EN00821A

Altaf MA, Shahid R, Ren M-X, Mora-Poblete F, Arnao MB, Naz S, Anwar M, Altaf MM, Shahid S, Shakoor A et al (2021) Phytomelatonin: an overview of the importance and mediating functions of melatonin against environmental stresses. Physiol Plant 172:820–846

Altaf MA, Hao YY, He CY, Mumtaz MA, Shu HY, Fu HZ, Wang ZW (2022a) Physiological and biochemical responses of pepper (L.) seedlings to nickel toxicity. Front Plant Sci 13:950392. https://doi.org/10.3389/fpls.2022.950392

Altaf MA, Shahid R, Ren M-X, Khan LU, Altaf MM, Jahan MS, Nawaz MA, Naz S, Shahid S, Lal MK et al (2022b) Protective mechanisms of melatonin against vanadium Phytotoxicity in tomato seedlings: insights into nutritional status, photosynthesis, root architecture system, and antioxidant machinery. J Plant Growth Regul 41:3300–3316

Altaf MA, Sharma N, Srivastava D, Mandal S, Adavi S, Jena R, Bairwa RK, Gopalakrishnan AV, Kumar A, Dey A et al (2023) Deciphering the melatonin-mediated response and signalling in the regulation of heavy metal stress in plants. Planta 257(115):1–17

Amjadi Z, Namdjoyan S, Abolhasani Soorki A (2021) Exogenous melatonin and salicylic acid alleviates cadmium toxicity in safflower (*Carthamus tinctorius* L.) seedlings. Ecotoxicology 30:387–401

Arnao MB, Hernández-Ruiz J (2009) Chemical stress by different agents affects the melatonin content of barley roots. J Pineal Res 46:295–299

Arnao MB, Hernández-Ruiz J (2013) Growth conditions determine different melatonin levels in *Upinus albus* L. J Pineal Res 55:149–155

Asif M, Pervez A, Irshad U, Mehmood Q, Ahmad R (2020) Melatonin and plant growth-promoting rhizobacteria alleviate the cadmium and arsenic stresses and increase the growth of L. Plant Soil Environ 66:234–241

Ayyaz A, Farooq MA, Dawood M, Majid A, Javed M, Athar H-R, Bano H, Zafar ZU (2021) Exogenous melatonin regulates chromium stress-induced feedback inhibition of photosynthesis and antioxidative protection in *Brassica napus* cultivars. Plant Cell Rep 40:2063–2080

Birben E, Sahiner UM, Sackesen C, Erzurum S, Kalayci O (2012) Oxidative stress and antioxidant defense. World Allergy Organ J 5:9

Bouyer D, Kramdi A, Kassam M, Heese M, Schnittger A, Roudier F, Colot V (2017) DNA methylation dynamics during early plant life. Genome Biol 18:1–12

Byeon Y, Yool Lee H, Choi D-W, Back K (2015) Chloroplast-encoded serotonin N-acetyltransferase in the red alga *Pyropia yezoensis*: gene transition to the nucleus from chloroplasts. J Exp Bot 66:709–717

Cai SY, Zhang Y, Xu YP, Qi ZY, Li MQ, Ahammed GJ, Xia XJ, Shi K, Zhou YH, Reiter RJ et al (2017) HsfA1a upregulates melatonin biosynthesis to confer cadmium tolerance in tomato plants. J Pineal Res 62:e12387

Cao YY, Qi CD, Li S, Wang Z, Wang X, Wang J, Ren S, Li X, Zhang N, Guo YD (2019) Melatonin alleviates copper toxicity via improving copper sequestration and ROS scavenging in cucumber. Plant Cell Physiol 60:562–574

Chan Z, Shi H (2015) Improved abiotic stress tolerance of bermudagrass by exogenous small molecules. Plant Signal Behav 10:e991577

Chen T, Su Y (2018) Uptake by rice seedlings and in-plant degradation of atrazine as influenced by the oxidative stress induced by added arsenic or phosphate deficiency. Hum Ecol Risk Assess Int J 24:1550–1564

Cheng X, Liu X, Liu F, Yang Y, Kou T (2024) Facing heavy metal stress, what are the positive responses of melatonin in plants: a review. Agronomy 14:2094. https://doi.org/10.3390/agronomy14092094

Clemens S (2006) Evolution and function of phytochelatin synthases. J Plant Physiol 163:319–332

Cocârţă DM, Neamţu S, Reşetar Deac AM (2016) Carcinogenic risk evaluation for human health risk assessment from soils contaminated with heavy metals. Int J Environ Sci Technol 13:2025–2036

DalCorso G (2012) In: Furini A (ed) Heavy metal toxicity in plants BT – plants and heavy metals. Springer, Dordrecht, pp 1–25

Danilova ED, Zlobin IE, Efimova MV (2023) Short-term pretreatment with melatonin increases the resistance of barley plants to the subsequent action of cadmium. Russ J Plant Physiol 70:95. https://doi.org/10.1134/S1021443723600800

de Oliveira LM, Ma LQ, Santos JAG, Guilherme LRG, Lessl JT (2014) Effects of arsenate, chromate, and sulfate on arsenic and chromium uptake and translocation by arsenic hyperaccumulator Pteris vittata L. Environ Pollut 184:187–192
de Oliveira LM, Gress J, De J, Rathinasabapathi B, Marchi G, Chen Y, Ma LQ (2016) Sulfate and chromate increased each other's uptake and translocation in as-hyperaccumulator *Pteris vittata*. Chemosphere 147:36–43
Devi R, Behera B, Raza MB, Mangal V, Altaf MA, Kumar R, Kumar A, Tiwari RK, Lal MK, Singh B (2021) An insight into microbes mediated heavy metal detoxification in plants: a review. J Soil Sci Plant Nutr 22:1–23
Drazic G, Mihailovic N, Lojic M (2006) Cadmium accumulation in *Medicago sativa* seedlings treated with salicylic acid. Biol Plant 50:239–244
Edelstein M, Ben-Hur M (2018) Heavy metals and metalloids: sources, risks and strategies to reduce their accumulation in horticultural crops. Sci Hortic 234:431–444
Faizan M, Sultan H, Alam P, Karabulut F, Cheng SH, Rajput VD, Minkina T, Hayat S, Khan MN, Nie L (2024) Melatonin and its cross-talk with other signaling molecules under abiotic stress. Plant Stress 11:100410
Feng SJ, Liu XS, Tao H, Tan SK, Chu SS, Oono Y, Zhang XD, Chen J, Yang ZM (2016) Variation of DNA methylation patterns associated with gene expression in rice (*Oryza sativa*) exposed to cadmium. Plant Cell Environ 39:2629–2649
Foyer CH, Noctor G (2008) Redox regulation in photosynthetic organisms: signaling, acclimation, and practical implications. Antioxid Redox Signal 11:861–905
Ghorbani A, Emamverdian A, Chen M-X, Naz S, Muhammad HMD, Altaf MA, Ahmad R (2023) Illustrating recent development in melatonin-heavy metal research in plant BT – melatonin in plants: a pleiotropic molecule for abiotic stresses and pathogen infection. Springer Nature, Singapore, pp 95–111
Greco M, Chiappetta A, Bruno L, Bitonti MB (2012) In *Posidonia oceanica* cadmium induces changes in DNA methylation and chromatin patterning. J Exp Bot 63:695–709
Gu Q, Chen Z, Yu X, Cui W, Pan J, Zhao G, Xu S, Wang R, Shen W (2017) Melatonin confers plant tolerance against cadmium stress via the decrease of cadmium accumulation and reestablishment of microRNA-mediated redox homeostasis. Plant Sci 261:28–37
Gu Q, Wang C, Xiao Q, Chen Z, Han Y (2021) Melatonin confers plant cadmium tolerance: an update. Int J Mol Sci 22:11704. https://doi.org/10.3390/ijms222111704
Hall JL (2002) Cellular mechanisms for heavy metal detoxification and tolerance. J Exp Bot 53:1–11
Hasan MK, Ahammed GJ, Yin L, Shi K, Xia X, Zhou Y, Yu J, Zhou J (2015) Melatonin mitigates cadmium phytotoxicity through modulation of phytochelatins biosynthesis, vacuolar sequestration, and antioxidant potential in *Solanum lycopersicum* L. Front Plant Sci 6:601
Hawkes SJ (1997) What is a "heavy metal"? J Chem Educ 74:1374
Hodi E, Galijaevi S, Balaban M, Rekanovic S, Makic H, Kukavica B, Mihajlovic D (2021) The protective role of melatonin under heavy metal-induced stress in Melissa Officinalis L. Turk J Chem 45:737–748
Hoque MN, Tahjib-Ul-arif M, Hannan A, Sultana N, Akhter S, Hasanuzzaman M, Akter F, Hossain MS, Sayed MA, Hasan MT et al (2021) Melatonin modulates plant tolerance to heavy metal stress: morphological responses to molecular mechanisms. Int J Mol Sci 22:11445
Hu Z, Fu Q, Zheng J, Zhang A, Wang H (2020) Transcriptomic and metabolomic analyses reveal that melatonin promotes melon root development under copper stress by inhibiting jasmonic acid biosynthesis. Hortic Res 7:79. https://doi.org/10.1038/S41438-020-0293-5
Ianas O, Olinescu R, Badescu I (1991) Melatonin involvement in oxidative processes. Endocrinologie 29:147–153
Jahan MS, Guo S, Baloch AR, Sun J, Shu S, Wang Y, Ahammed GJ, Kabir K, Roy R (2020) Melatonin alleviates nickel phytotoxicity by improving photosynthesis, secondary metabolism and oxidative stress tolerance in tomato seedlings. Ecotoxicol Environ Saf 197:110593
Jozefczak M, Keunen E, Schat H, Bliek M, Hernández LE, Carleer R, Remans T, Bohler S, Vangronsveld J, Cuypers A (2014) Differential response of Arabidopsis leaves and roots

to cadmium: glutathione-related chelating capacity vs antioxidant capacity. Plant Physiol Biochem 83:1–9

Kalaivanan D, Ganeshamurthy AN (2016) Mechanisms of heavy metal toxicity in plants BT – abiotic stress physiology of horticultural crops. In: Shivashankara KS, Laxman RH (eds) NKS Rao. Springer India, New Delhi, pp 85–102

Kang X, Pei ZQ, Xu TT, Dong CY, Bai X, Ma C, Zhu Q, Chai CH, Wang J, Zheng S et al (2024) Exogenous melatonin enhanced cadmium stress tolerance of cucumber seedlings (*Cucumis sativus* L.). Biologia (Bratisl) 79:2025–2042

Kaya C, Sarıoglu A, Ashraf M, Alyemeni MN, Ahmad P (2022) The combined supplementation of melatonin and salicylic acid effectively detoxifies arsenic toxicity by modulating phytochelatins and nitrogen metabolism in pepper plants. Environ Pollut 297:118727

Khan MN, Zhang J, Luo T, Liu J, Rizwan M, Fahad S, Xu Z, Hu L (2019) Seed priming with melatonin coping drought stress in rapeseed by regulating reactive oxygen species detoxification: antioxidant defense system, osmotic adjustment, stomatal traits and chloroplast ultrastructure perseveration. Ind Crop Prod 140:111597

Khan A, Numan M, Khan AL, Lee IJ, Imran M, Asaf S, Al-Harrasi A (2020) Melatonin: awakening the defense mechanisms during plant oxidative stress. Plan Theory 9:407

Kováčik J, Rotková G, Bujdoš M, Babula P, Peterková V, Matúš P (2017) Ascorbic acid protects *Coccomyxa subellipsoidea* against metal toxicity through modulation of ROS/NO balance and metal uptake. J Hazard Mater 339:200–207

Krzesłowska M (2011) The cell wall in plant cell response to trace metals: polysaccharide remodeling and its role in defense strategy. Acta Physiol Plant 33:35–51

Kumar R, Altaf MA, Lal MK, Tiwari RK (eds) (2023) Melatonin in plants: a regulator for plant growth and development. Springer Nature, Singapore

Kumar R, Altaf MA, Lal MK, Tiwari RK (eds) (2024) Melatonin in plants: a pleiotropic molecule for abiotic stresses and pathogen infection. Springer

Lee K, Lee HY, Back K (2018) Rice histone deacetylase 10 and Arabidopsis histone deacetylase 14 genes encode N-acetylserotonin deacetylase, which catalyzes conversion of N-acetylserotonin into serotonin, a reverse reaction for melatonin biosynthesis in plants. J Pineal Res 64:e12460

Li M-Q, Hasan MK, Li C-X, Ahammed GJ, Xia X-J, Shi K, Zhou Y-H, Reiter RJ, Yu J-Q, Xu M-X et al (2016) Melatonin mediates selenium-induced tolerance to cadmium stress in tomato plants. J Pineal Res 61:291–302

Li X, Ahammed GJ, Zhang XN, Zhang L, Yan P, Zhang LP, Fu JY, Han WY (2021) Melatonin-mediated regulation of anthocyanin biosynthesis and antioxidant defense confer tolerance to arsenic stress in *Camellia sinensis* L. J Hazard Mater 403:123922

Li RQ, Wu LQ, Shao YF, Hu QW, Zhang HL (2022a) Melatonin alleviates copper stress to promote rice seed germination and seedling growth via crosstalk among various defensive response pathways. Plant Physiol Biochem 179:65–77

Li S, Han X, Lu Z, Qiu W, Yu M, Li H, He Z, Zhuo R (2022b) MAPK cascades and transcriptional factors: regulation of heavy metal tolerance in plants. Int J Mol Sci 23:4463

Limson J, Nyokong T, Daya S (1998) The interaction of melatonin and its precursors with aluminium, cadmium, copper, iron, lead, and zinc: an adsorptive voltammetric study. J Pineal Res 24:15–21

Lin L, Li J, Chen F, Liao M, Tang Y, Liang D, Xia H, Lai Y, Wang X, Chen C et al (2018) Effects of melatonin on the growth and cadmium characteristics of *Cyphomandra betacea* seedlings. Environ Monit Assess 190:1–8

Lin H, Wang Z, Liu C, Dong Y (2022a) Technologies for removing heavy metal from contaminated soils on farmland: a review. Chemosphere 305:135457

Lin S, Song XF, Mao HT, Li SQ, Gan JY, Yuan M, Zhang ZW, Yuan S, Zhang HY, Su YQ et al (2022b) Exogenous melatonin improved photosynthetic efficiency of photosystem II by reversible phosphorylation of thylakoid proteins in wheat under osmotic stress. Front Plant Sci 13:966181

Moussa HR, Algamal SMA (2017) Does exogenous application of melatonin ameliorate boron toxicity in spinach plants? Int J Veg Sci 23:233–245

Moustafa-Farag M, Elkelish A, Dafea M, Khan M, Arnao MB, Abdelhamid MT, El-Ezz AA, Almoneafy A, Mahmoud A, Awad M et al (2020a) Role of melatonin in plant tolerance to soil stressors: salinity, pH and heavy metals. Molecules 25:5359. https://doi.org/10.3390/molecules25225359

Moustafa-Farag M, Mahmoud A, Arnao MB, Sheteiwy MS, Dafea M, Soltan M, Elkelish A, Hasanuzzaman M, Ai S (2020b) Melatonin-induced water stress tolerance in plants: recent advances. Antioxidants (Basel) 9:809. https://doi.org/10.3390/antiox9090809

Munir R, Yasin MU, Afzal M, Jan M, Muhammad S, Jan N, Nana C, Munir F, Iqbal H, Tawab F et al (2024) Melatonin alleviated cadmium accumulation and toxicity by modulating phytohormonal balance and antioxidant metabolism in rice. Chemosphere 346:140590

Nabaei M, Amooaghaie R (2019) Nitric oxide is involved in the regulation of melatonin-induced antioxidant responses in Catharanthus roseus roots under cadmium stress. Botany 97:681–690. https://doi.org/10.1139/cjb-2019-0107

Nagajyoti PC, Lee KD, Sreekanth TVM (2010) Heavy metals, occurrence and toxicity for plants: a review. Environ Chem Lett 8:199–216

Namdjoyan S, Soorki AA, Elyasi N, Kazemi N, Simaei M (2020) Melatonin alleviates lead-induced oxidative damage in safflower seedlings. Ecotoxicology 29:108–118

Nawaz MA, Jiao Y, Chen C, Shireen F, Zheng Z, Imtiaz M, Bie Z, Huang Y (2018) Melatonin pretreatment improves vanadium stress tolerance of watermelon seedlings by reducing vanadium concentration in the leaves and regulating melatonin biosynthesis and antioxidant-related gene expression. J Plant Physiol 220:115–127

Nazarian M, Ghanati F (2020) The role of melatonin in reinforcement of antioxidant system of rice plant (*Oryza sativa* L.) under arsenite toxicity? Plant Physiol Rep 25:395–404

Ni J, Wang Q, Shah FA, Liu W, Wang D, Huang S, Fu S, Wu L (2018) Exogenous melatonin confers cadmium tolerance by counterbalancing the hydrogen peroxide homeostasis in wheat seedlings. Molecules 23:799

Park J, Song W-Y, Ko D, Eom Y, Hansen TH, Schiller M, Lee TG, Martinoia E, Lee Y (2012) The phytochelatin transporters AtABCC1 and AtABCC2 mediate tolerance to cadmium and mercury. Plant J 69:278–288

Pieri C, Moroni F, Marra M, Marcheselli F, Recchioni R (1995) Melatonin is an efficient antioxidant. Arch Gerontol Geriatr 20:159–165

Posmyk MM, Kuran H, Marciniak K, Janas KM (2008) Presowing seed treatment with melatonin protects red cabbage seedlings against toxic copper ion concentrations. J Pineal Res 45:24–31

Qin C, Lian H, Alqahtani FM, Ahanger MA (2024) Chromium mediated damaging effects on growth, nitrogen metabolism and chlorophyll synthesis in tomato can be alleviated by foliar application of melatonin and jasmonic acid priming. Sci Hortic 323:112494

Qu T, Ma Y, Yun M, Zhao C (2024) Transcriptome analysis revealed the possible reasons for the change of Ni resistance in *Rhus typhina* after spraying melatonin. Plants (Basel) 13:1287. https://doi.org/10.3390/plants13101287

Raja V, Qadir SU, Kumar N, Alsahli AA, Rinklebe J, Ahmad P (2023) Melatonin and strigolactone mitigate chromium toxicity through modulation of ascorbate-glutathione pathway and gene expression in tomato. Plant Physiol Biochem 201:107872. https://doi.org/10.1016/J.PLAPHY.2023.107872

Ramezani M, Enayati M, Ramezani M, Ghorbani A (2021) A study of different strategical views into heavy metal(oid) removal in the environment. Arab J Geosci 14:2225

Raza Khan A, Fan X, Salam A, Azhar W, Ulhassan Z, Qi J, Liaquat F, Yang S, Gan Y (2023) Melatonin-mediated resistance to copper oxide nanoparticles-induced toxicity by regulating the photosynthetic apparatus, cellular damages and antioxidant defense system in maize seedlings. Environ Pollut 316:120639

Rizwan M, Ali S, Adrees M, Rizvi H, Zia-ur-Rehman M, Hannan F, Qayyum MF, Hafeez F, Ok YS (2016) Cadmium stress in rice: toxic effects, tolerance mechanisms, and management: a critical review. Environ Sci Pollut Res 23:17859–17879

Sakouhi L, Werghi S, Ben Massoud M, Murata Y, Chaoui A (2023) Comparative study of different melatonin application effects on antioxidant system in cadmium-exposed chickpea seedlings. J Soil Sci Plant Nutr 23:5513. https://doi.org/10.1007/s42729-023-01418-w

Samanta S, Banerjee A, Roychoudhury A (2021) Exogenous melatonin regulates endogenous phytohormone homeostasis and thiol-mediated detoxification in two indica rice cultivars under arsenic stress. Plant Cell Rep 40(8):1585–1602

Sanità di Toppi L, Gabbrielli R (1999) Response to cadmium in higher plants. Environ Exp Bot 41:105–130

Saqib M, Shahzad U, Abid Z, Zulfiqar F, Tiwari RK, Lal MK, Khan WA, El-Sheikh MA, Altaf MA (2024) Melatonin alleviates cadmium phytotoxicity through regulation of growth, photosynthesis, and antioxidant potential in two pepper genotypes. Acta Physiol Plant 46:1–11

Sarafi E, Tsouvaltzis P, Chatzissavvidis C, Siomos A, Therios I (2017) Melatonin and resveratrol reverse the toxic effect of high boron (B) and modulate biochemical parameters in pepper plants (*Capsicum annuum* L.). Plant Physiol Biochem 112:173–182

Savvides A, Ali S, Tester M, Fotopoulos V (2016) Chemical priming of plants against multiple abiotic stresses: mission possible? Trends Plant Sci 21:329–340

Scebba F, Arduini I, Ercoli L, Sebastiani L (2006) Cadmium effects on growth and antioxidant enzymes activities in *Miscanthus sinensis*. Biol Plant 50:688–692

Seleiman MF, Ali S, Refay Y, Rizwan M, Alhammad BA, El-Hendawy SE (2020) Chromium resistant microbes and melatonin reduced Cr uptake and toxicity, improved physio-biochemical traits and yield of wheat in contaminated soil. Chemosphere 250:126239

Shafiq S, Ali A, Sajjad Y, Zeb Q, Shahzad M, Khan AR, Nazir R, Widemann E (2020) The interplay between toxic and essential metals for their uptake and translocation is likely governed by DNA methylation and histone deacetylation in maize. Int J Mol Sci 21:6959

Shah FUR, Ahmad N, Masood KR, Peralta-Videa JR, Ahmad F u D (2010) In: Ashraf M, Ozturk M, Ahmad MSA (eds) Heavy metal toxicity in plants BT – plant adaptation and phytoremediation. Springer, Dordrecht, pp 71–97

Shi J, Wu B, Yuan X, Cao YY, Chen X, Chen Y, Hu T (2008) An X-ray absorption spectroscopy investigation of speciation and biotransformation of copper in *Elsholtzia splendens*. Plant Soil 302:163–174

Shomali A, Das S, Sarraf M, Johnson R, Janeeshma E, Kumar V, Aliniaeifard S, Puthur JT, Hasanuzzaman M (2024) Modulation of plant photosynthetic processes during metal and metalloid stress, and strategies for manipulating photosynthesis-related traits. Plant Physiol Biochem 206:108211

Siddiqui MH, Alamri S, Alsubaie QD, Ali HM, Ibrahim AA, Alsadon A (2019) Potential roles of melatonin and sulfur in alleviation of lanthanum toxicity in tomato seedlings. Ecotoxicol Environ Saf 180:656–667

Siddiqui MH, Alamri S, Khan MN, Corpas FJ, Al-Amri AA, Alsubaie QD, Ali HM, Kalaji HM, Ahmad P (2020) Melatonin and calcium function synergistically to promote the resilience through ROS metabolism under arsenic-induced stress. J Hazard Mater 398:122882. https://doi.org/10.1016/j.jhazmat.2020.122882

Song C, Manzoor MA, Mao D, Ren X, Zhang W, Zhang Y (2024) Photosynthetic machinery and antioxidant enzymes system regulation confers cadmium stress tolerance to tomato seedlings pretreated with melatonin. Sci Hortic 323:112550

Sun C, Lv T, Huang L, Liu X, Jin C, Lin X (2020) Melatonin ameliorates aluminum toxicity through enhancing aluminum exclusion and reestablishing redox homeostasis in roots of wheat. J Pineal Res 68:e12642. https://doi.org/10.1111/JPI.12642

Sun C, Liu L, Wang L, Li B, Jin C, Lin X (2021) Melatonin: a master regulator of plant development and stress responses. J Integr Plant Biol 63:126–145

Sun S, Liu A, Li Z, Guo T, Chen S, Ahammed GJ (2023) Anthocyanin synthesis is critical for melatonin-induced chromium stress tolerance in tomato. J Hazard Mater 453:131456

Sun C, Xu L, Gao Q, Sun S, Liu X, Zhang Z, Tian Z, Dai T, Sun J (2025) Foliar spraying melatonin reduces the threat of chromium-contaminated water to wheat production by improving photosynthesis, limiting Cr translocation and reducing oxidative stress. Ecotoxicol Environ Saf 290:117485

Tal O, Haim A, Harel O, Gerchman Y (2011) Melatonin as an antioxidant and its semi-lunar rhythm in green macroalga Ulva sp. J Exp Bot 62:1903–1910

Talarico E, Zambelli A, Araniti F, Greco E, Chiappetta A, Bruno L (2024) Unravelling the epigenetic code: DNA methylation in plants and its role in stress response. Epigenomes 8:30

Tan DX, Manchester LC, Helton P, Reiter RJ (2007a) Phytoremediative capacity of plants enriched with melatonin. Plant Signal Behav 2:514

Tan DX, Manchester LC, Terron MP, Flores LJ, Reiter RJ (2007b) One molecule, many derivatives: a never-ending interaction of melatonin with reactive oxygen and nitrogen species? J Pineal Res 42:28–42

Tang M, Xu L, Wang Y, Dong J, Zhang X, Wang K, Ying J, Li C, Liu L (2021) Melatonin-induced DNA demethylation of metal transporters and antioxidant genes alleviates lead stress in radish plants. Hortic Res 8:1–14

Tiwari G, Duraivadivel P, Sharma S, Hariprasad P (2018) 1-Aminocyclopropane-1-carboxylic acid deaminase producing beneficial rhizobacteria ameliorate the biomass characters of *Panicum maximum* Jacq. By mitigating drought and salt stress. Sci Rep 8:17513

Tiwari RK, Lal MK, Naga KC, Kumar R, Chourasia KN, Subhash S, Kumar D, Sharma S (2020) Emerging roles of melatonin in mitigating abiotic and biotic stresses of horticultural crops. Sci Hortic 272:109592

Tiwari RK, Lal MK, Kumar R, Chourasia KN, Naga KC, Kumar D, Das SK, Zinta G (2021a) Mechanistic insights on melatonin-mediated drought stress mitigation in plants. Physiol Plant 172:1212–1226

Tiwari RK, Lal MK, Kumar R, Mangal V, Altaf MA, Sharma S, Singh B, Kumar M (2021b) Insight into melatonin-mediated response and signaling in the regulation of plant defense under biotic stress. Plant Mol Biol 109:385–399

Tiwari RK, Altaf MA, Kumar R, Lal MK (2024) Evolution of melatonin as an oxidative stress mitigator in plant. In: Melatonin in plants: a pleiotropic molecule for abiotic stresses and pathogen infection. Springer Nature, Singapore, pp 1–20

Tousi S, Zoufan P, Ghahfarrokhie AR (2020) Alleviation of cadmium-induced phytotoxicity and growth improvement by exogenous melatonin pretreatment in mallow (*Malva parviflora*) plants. Ecotoxicol Environ Saf 206:111403

Ulhassan Z, Huang Q, Gill RA, Ali S, Mwamba TM, Ali B, Hina F, Zhou W (2019) Protective mechanisms of melatonin against selenium toxicity in *Brassica napus*: insights into physiological traits, thiol biosynthesis and antioxidant machinery. BMC Plant Biol 19:507

Vassilev A, Yordanov I, Tsonev T (1998) Effects of Cd2+ on the physiological state and photosynthetic activity of young barley plants. Photosynthetica 34:293–302

Wang M, Duan S, Zhou Z, Chen S, Wang D (2019) Foliar spraying of melatonin confers cadmium tolerance in *Nicotiana tabacum* L. Ecotoxicol Environ Saf 170:68–76

Wang D, Chen Q, Chen W, Guo Q, Xia Y, Wang S, Jing D, Liang G (2021a) Physiological and transcription analyses reveal the regulatory mechanism of melatonin in inducing drought resistance in loquat (*Eriobotrya japonica* Lindl.) seedlings. Environ Exp Bot 181:104291

Wang T, Song J, Liu Z, Liu Z, Cui J (2021b) Melatonin alleviates cadmium toxicity by reducing nitric oxide accumulation and IRT1 expression in Chinese cabbage seedlings. Environ Sci Pollut Res 28:15394–15405

Wang Y, Guo D, Wang J, Tian B, Li Y, Sun G, Zhang H (2022) Exogenous melatonin alleviates NO_2 damage in tobacco leaves by promoting antioxidant defense, modulating redox homeostasis, and signal transduction. J Hazard Mater 424:127265

Wang C, Bian C, Li J, Han L, Guo D, Wang T, Sun Z, Ma C, Liu X, Tian Y et al (2023) Melatonin promotes Al3+ compartmentalization via H+ transport and ion gradients in *Malus hupehensis*. Plant Physiol 193:821–839

Wang X, Cao M, Li H, Liu Y, Fan S, Zhang N, Guo Y (2024a) Strategies and prospects for melatonin to alleviate abiotic stress in horticultural plants. Hortic Plant J 10:601–614

Wang Y, Shen K, Ge S, Li X (2024b) Physiological and molecular mechanisms of melatonin-induced tolerance to heavy metal stress BT – melatonin in plants: role in plant growth, development, and stress response. Springer Nature, Singapore, pp 153–165

Weissmannová HD, Pavlovský J (2017) Indices of soil contamination by heavy metals–methodology of calculation for pollution assessment (minireview). Environ Monit Assess 189:616

Wu X, Li H, Luo H, Lin D, Zhou X (2019) Effect of exogenous melatonin on photosynthetic characteristics in cucumber seedlings under cadmium stress: a rapid detection method for the cadmium resistance. E3S Web Conf 136:07023

Wu S, Wang Y, Zhang J, Gong X, Zhang Z, Sun J, Chen X, Wang Y (2021) Exogenous melatonin improves physiological characteristics and promotes growth of strawberry seedlings under cadmium stress. Hortic Plant J 7:13–22

Xalxo R, Keshavkant S (2019) Melatonin, glutathione and thiourea attenuates lead and acid rain-induced deleterious responses by regulating gene expression of antioxidants in *Trigonella foenum* graecum L. Chemosphere 221:1–10

Xiang G, Lin L, Liao M, Tang Y, Liang D, Xia H, Wang J, Wang X, Sun G, Zhang H et al (2019) Effects of melatonin on cadmium accumulation in the accumulator plant *Perilla frutescens*. Chem Ecol 35:553–562

Xu L, Zhang F, Tang M, Wang Y, Dong J, Ying J, Chen Y, Hu B, Li C, Liu L (2020) Melatonin confers cadmium tolerance by modulating critical heavy metal chelators and transporters in radish plants. J Pineal Res 69:e12659

Yang H, Fang R, Luo L, Yang W, Huang Q, Yang C, Hui W, Gong W, Wang J (2023a) Potential roles of melatonin in mitigating the heavy metals toxicity in horticultural plants. Sci Hortic 321:112269

Yang X, Ren J, Lin X, Yang Z, Deng X, Ke Q (2023b) Melatonin alleviates chromium toxicity in maize by modulation of cell wall polysaccharides biosynthesis, glutathione metabolism, and antioxidant capacity. Int J Mol Sci 24:3816

Yu SY, Wang N, Zhu LX, Xu WJ, Zhang YJ, Sun HC, Zhang K, Li AC, Bai ZY, Liu LT et al (2024) Melatonin mitigates cadmium toxicity by promoting root development, delaying root senescence, and regulating cadmium transport in cotton. Ecotoxicol Environ Saf 283:116786

Zhang N, Sun Q, Zhang H, Cao Y, Weeda S, Ren S, Guo YD (2015) Roles of melatonin in abiotic stress resistance in plants. J Exp Bot 66:647–656

Zhang J, Zeng B, Mao Y, Kong X, Wang X, Yang Y, Zhang J, Xu J, Rengel Z, Chen Q (2017) Melatonin alleviates aluminium toxicity through modulating antioxidative enzymes and enhancing organic acid anion exudation in soybean. Funct Plant Biol 44:961–968

Zhang T, Wang Y, Ma X, Ouyang Z, Deng L, Shen S, Dong X, Du N, Dong H, Guo Z et al (2022) Melatonin alleviates copper toxicity via improving ROS metabolism and antioxidant defense response in tomato seedlings. Antioxidants (Basel). 11:758. https://doi.org/10.3390/antiox11040758

Zhang H, Qiu Y, Ji Y, Wu X, Xu X, Wu P (2023) Melatonin promotes seed germination via regulation of ABA signaling under low temperature stress in cucumber. J Plant Growth Regul 42:2232–2245

Zhou X, Zhao H, Cao K, Hu L, Du T, Baluška F, Zou Z (2016) Beneficial roles of melatonin on redox regulation of photosynthetic electron transport and synthesis of D1 protein in tomato seedlings under salt stress. Front Plant Sci 7:224452

Zoufan P, Zare Bavani MR, Tousi S, Rahnama A (2023) Effect of exogenous melatonin on improvement of chlorophyll content and photochemical efficiency of PSII in mallow plants (*Malva parviflora* L.) treated with cadmium. Physiol Mol Biol Plants 29:145–157

Melatonin Interplay with Phytohormones and Signaling Molecules Under Waterlogging Stress

7

Mohammad Shah Jahan, Maha M. Hamada, and Mohamed F. M. Ibrahim

Abstract

Waterlogging is a pivotal abiotic stressor that causes a severe threat to global crop production. Waterlogging stress negatively impacts plants, which induces oxidative damage while raising heavy metal contents and energy crisis, which leads to hindering metabolic processes and stunted plant growth. Melatonin is a low molecular weight indoleamine promising growth hormone that participates in multiple plant physiological and cellular activities. Melatonin acts as a strong free-radical scavenger and oxidizing agent as well to optimize redox homeostasis in plants by directly scavenging reactive oxygen species and reactive nitrogen species. The protective function of melatonin is to serve as a strong antioxidant within the plant cells. This chapter offers an extensive discussion about how melatonin boosts plant tolerance to waterlogging stress, enlightening the melatonin biosynthesis process, antioxidant functions, and its interaction with phytohormones and genes associated with hypoxic and low-oxygen conditions during waterlogging stress. This chapter also extends our present understanding of how melatonin-driven mechanisms increase plant tolerance to waterlogging stress.

Keywords

Nitromela · Neurotransmitter · Hypoxia · Anoxia · ROS · Phytohormones

M. S. Jahan (✉)
Department of Horticulture, Sher-e-Bangla Agricultural University, Dhaka, Bangladesh

M. M. Hamada
Department of Agronomy, Faculty of Agriculture, Ain Shams University, Cairo, Egypt

M. F. M. Ibrahim
Department of Agricultural Botany, Faculty of Agriculture, Ain Shams University, Cairo, Egypt

R. K. Tiwari et al. (eds.), *Melatonin: Signal Transduction Mechanisms and Defense Networks in Plants*, https://doi.org/10.1007/978-981-96-8869-2_7

7.1 Introduction

Waterlogging is a key abiotic stressor that has a substantial impact on agricultural productivity across the globe. Approximately 16% of the global arable lands are affected by waterlogging and flooding (Zhou et al. 2020), and waterlogged areas are increasing day by day, which poses a significant threat to global food security (Donat et al. 2016). Prolonged soil waterlogging can significantly reduce by 25–60% of the major crop yield (Tian et al. 2021). Waterlogging stress significantly diminishes oxygen (O_2) content in the root rhizosphere, consequently restricting O_2 diffusion as well as gas exchange properties in plant cells, therefore substantially affecting the growth and development of plants and abruptly disrupting various plant metabolic activities (Xu et al. 2022). Waterlogging stress significantly impairs the photosynthetic process, thus restricting chlorophyll biosynthesis, and finally, biomass production declines in plants. During the initial phase of waterlogging, Gas exchange attributes like stomatal conductance and transpiration rate at the early stage of waterlogging decrease, whereas elevates CO_2 diffusion, thereby inhibiting photosynthesis processes (Yan et al. 2018). Extended waterlogging stress impairs chlorophyll production and reduces light absorption capacity, ultimately leading to a marked decline in the photosynthesis process (Grzesiak et al. 2016). Simultaneously, the Rubisco (ribulose 1,5-bisphosphate carboxylase/oxygenase) enzyme activity declines, exerting an additional influence on photosynthesis (Wang et al. 2014). Waterlogging stress also drastically declines nutrient absorption efficiency in plants and thus affects photosynthesis efficiency. Consequently, the photosynthesis process is considered as the vital parameter for justifying waterlogging tolerance (Dickin and Wright 2008). Under optimal physiological state, pyruvate is synthesized via glycolysis, resulting in the generation of CO_2 and H_2O, and finally produces adenosine triphosphate (ATP). Under waterlogged conditions, the oxygen concentration rapidly decreases in soil, and the dissolved oxygen is quickly utilized by plant roots and rhizosphere microorganisms as well, leading to create hypoxic or potentially anoxic conditions (Panozzo et al. 2019). As a result, plants switch from aerobic respiration to anaerobic processes, particularly the ethanolic fermentation process (Yamane et al. 2023). Moreover, when the concentrations of alcohol and aldehyde reach a certain threshold level, their excessive accumulation can inflict substantial harm on plant cells, ultimately causing plant death (Xu et al. 2022). The energy-deficient transitional phase in waterlogged plants triggers MDA and ROS production, and excessive ROS directly or indirectly affects cell membrane structure, nucleic acids, and protein contents, finally reducing plant growth (Arbona et al. 2008). Most of the crops cannot withstand prolonged waterlogging (Voesenek et al. 2006). ROS components production and removal under standard conditions occur in dynamic equilibrium within plant systems (Miller et al. 2008). Plants can remove ROS components through enzymatic and non-enzymatic dynamics (Komatsu et al. 2013). Plants under waterlogging stress reduce ROS homeostasis, resulting in excess production of ROS components, cell membrane permeability increases, and eventually, cell death (Zhao et al. 2022). The *RBOH* gene expression plays a considerable role in the formation and signaling of RBOH-dependent ROS production in Arabidopsis thaliana (Morales et al. 2016). Furthermore, the production of ROS via RBOH-mediated pathways plays an important function in the proliferation of flood-induced aerenchyma development in multiple plants (Yamauchi et al. 2017).

Melatonin is an evolutionary conserved chemical substance that critically contributes to diverse aspects of plant surveillance. It has been discovered as a vital protector during stress conditions. Recently, melatonin has been recognized as a potential biostimulant for the mitigation of diverse stressors (Mukherjee et al. 2024). In addition, melatonin also plays a role as a front-line defender under oxidative stress (Majumder et al. 2024). Moreover, melatonin assists diverse plant biological functions including photosynthesis metabolism, slowing leaf senescence's mechanism, and regulating carbohydrate, lipid, protein, nitrogen, and carbon metabolic processes as well (Arnao and Hernández-Ruiz 2019; Jahan et al. 2023). In addition to its antioxidant functions, melatonin also shows critical roles in hormonal regulation, interaction with other signaling molecules, transcriptional regulation, and epigenetics regulators as well (Arnao and Hernández-Ruiz 2019). Hussain et al. (2024) narrated that melatonin positively regulates the chlorophyll mechanism by triggering hormonal and transcription factors. The essential secondary metabolite synthesis is enhanced with melatonin application, and these metabolites reduce oxidative damage in plants (Sandoval et al. 2024). Recent studies suggest that RBOH-induced apoplastic ROS may act as downstream effectors in melatonin-derived stress tolerance (Ahammed et al. 2024). Additionally, melatonin plays an important function in regulating programmed cell death in plants and concurrently bolsters their tolerance to stress conditions (Jalili et al. 2022; Jahan et al. 2024). In recent years, researchers around the globe have shown more interest in the exploration of melatonin functions under waterlogging conditions. Melatonin increases waterlogging tolerance in diverse plant species by protecting the photosynthesis process, regulating aerobic respiration, and balancing ROS homeostasis in *Malusbaccata* (Linn) seedlings (Zheng et al. 2017) and reprogramming polyamine and ethylene metabolism in *Medicago sativa* (Zhang et al. 2019). Previous research findings suggested that the regulatory functions of melatonin with phytohormones in plants, operating both under optimal and stress-induced conditions and enhancing growth-promoting hormones, specifically auxins, cytokinins, and gibberellins, thus fostering the vegetative and reproductive growth of plants (Mukherjee et al. 2024; Jahan et al. 2023). Jahan et al. (2021) reported that melatonin treatment elevated GA content and reduced ABA content in tomato seedlings under high temperatures. Melatonin application augments salt stress tolerance in grapevine by upregulating ethylene production via stimulation of 1-aminocyclopropanr-1-carboxylic acid (AAC) synthesis (Xu et al. 2019). The powerful functions of melatonin in alleviating waterlogging stress in plants are summarized in Fig. 7.1.

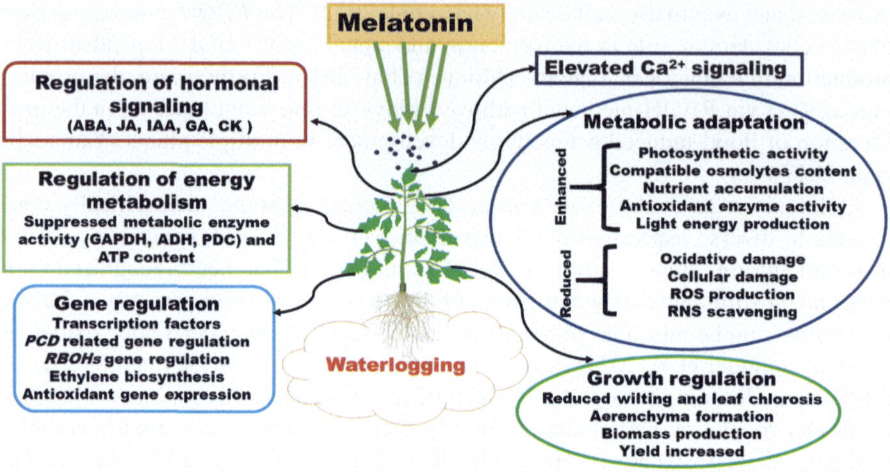

Fig. 7.1 The potent functions of melatonin in plants during waterlogging stress. (www.biorender.com)

7.2 Melatonin-Induced Waterlogging Strategies

7.2.1 Melatonin Boosts Photosynthesis Capacity and Regulates Energy Metabolism During Waterlogging Stress

Photosynthesis is a fundamental biological mechanism that regulates the growth and productivity of plants. Waterlogging stress-induced stomatal limitation poses a substantial impediment to the photosynthetic processes in plants (Hussain et al. 2022). The root-to-leaf signaling cascade triggered by waterlogging events progressively impairs CO_2 assimilation, reduces stomatal conductance, weakens light harvesting and utilization processes, and accelerates photosynthetic pigment degradation, resulting in reduced both the concentration and activity of carbon reductase within the photosynthetic apparatus (Reddy et al. 2004). Waterlogging stress substantially impairs photosynthetic electron transport efficacy, leading to excessive energy generation and the inhibition of photosynthetic processes. Photosynthetic organisms employ two primary mechanisms to manage excess energy absorption by chloroplasts: photochemical quenching (dissipation of surplus energy through photosynthetic processes) and non-photochemical quenching (facilitating the release of excess energy as thermal radiation), which work in concert to maintain energy homeostasis within the photosynthetic system. This process induces heightened excitation pressure on both photosystems (PSI and PSII) in the thylakoid membrane, consequently triggering ROS generation (Razi and Muneer 2021). A large number of melatonin intermediates are produced in the chloroplasts, cytoplasm, and endoplasmic reticulum, which either facilitate or delay the following enzymatic processes. For instance, the SNAT enzyme is localized in chloroplasts, and ASMT/COMT is localized in the cytoplasm. Melatonin is synthesized both in mitochondria and chloroplasts, where it helps to mitigate cell membrane damage, protect chloroplast integrity and functional mitochondrial structure, and trigger cell proliferation. In addition, melatonin facilitates the chlorophyll biosynthesis process, and the maximum photochemical efficiency (Fv/Fm) of the PSII improves through the augmentation of chlorophyll content (Wang et al. 2021). Under stress conditions, melatonin under stress circumstances augments photosynthesis-related gene expression and accelerates glucose and lipid metabolism along with ascorbic acid synthesis as well (Jahan et al. 2023). Melatonin treatment downregulates the chlorophyll degradation gene *PAO* and upregulates the expression of *RBCS2* and *AREB* (Szafrańska et al. 2017; Cherono et al. 2021). Melatonin alleviates the physiological and biochemical damages induced by water stress in tomato seedlings and *Zanthoxylum armatum,* and it facilitates enhanced nutrient absorption, promotes the synthesis of photosynthetic pigments, balanced carbon flow through the Calvin cycle, Rubisco, and carbonic anhydrase activities, and finally, total soluble carbohydrates were stimulated in plant tissues (Murchie and Niyogi 2011; Su et al. 2023). Waterlogging stress inhibited the key physiological attributes like SPAD, gas exchange parameters, and photosystem II values, and melatonin treatments restored their values in wheat (Ma et al. 2022).

Waterlogging stress impeded gas exchange, hence increasing CO_2 partial pressure in the root zone. Excessive water stress imbalance redox potential in soil and H^+ causes less cation absorption, such as K^+ and Mg^{2+} in plants, whereas Mg^{2+}, Ca^{2+}, Fe^{2+}, and other cations are elevated. Flooding stress induces hypoxic conditions in plants, prompting changes in aerobic respiration to anaerobic respiration. The terminal electron acceptor in mitochondrial electron transport is inhibited by hypoxia, which impedes both ATP synthesis and NADH oxidation processes. An inadequate input of ATP leads to an elevated respiration process and ROS generation. Melatonin, analogous to other phytohormones, has physiological, biochemical, and cellular effects including alterations in membrane permeability caused by intracellular Ca^{2+} and ion transporters (Arnao and Hernández-Ruiz 2019). Melatonin treatment can induce the production of glycerol, and the subsequent increase in sugar levels ultimately elevates endogenous nitric oxide (NO) concentrations. Pyruvate decarboxylase and alcohol dehydrogenase activities are notably alleviated through melatonin and KNO_3 admission, which simultaneously elevate the levels of soluble protein, escalate organic substances consumption, and restrict energy production, leading to an imbalance in the supply-demand relationship. Consequently, plants require increased quantities of glucose, sucrose, ATP, and NAD^+ to enhance ATP production and fulfill their fundamental energy requirements. In recent years, melatonin has increased the concentration of free proline, sucrose, and fructose accumulation in plant tissues by modifying the expression of certain genes like as*OsP5CS*, *OsSUS7*, and *OsSPS1* (Lu et al. 2022). One of the most significant markers to alleviate waterlogging stress is ethanol fermentation, and ADH and PDC play vital roles in the ethanol fermentation process (Pan et al. 2021). Waterlogging-tolerant seedlings can enhance the ethanol production rate through the regulation of expression of *ADH* and *PDC* gene, instantly serving as energy for plant growth during hypoxia (Ismond et al. 2003). Melatonin treatment increased ADH and PDC activity along with elevated ATP and nutrient content in tomato seedlings under waterlogging stress (Jahan et al. 2024). Melatonin has been demonstrated to promote waterlogging tolerance in wheat varieties through the suppression of ADH, PDC, and LDH enzyme activities (Ma et al. 2022). In addition, waterlogging tolerance in *Prunuspersica* increased with the melatonin application by regulating antioxidant metabolism, promotion of aerenchyma development, and anaerobic respiratory processes (Gu et al. 2021). The potential functions of melatonin under waterlogging/hypoxia stress are summarized in Table 7.1.

7.2.2 Melatonin Triggers Defense Mechanism in Plants Under Waterlogging Stress

The formation and removal of ROS components in plants under normal conditions present in a dynamic equilibrium position; however, environmental stress disrupts this balance, leading to increased ROS production (Alizadeh-Vaskasi et al. 2018). During the initial stages of flooding, an excess of water fills the rhizosphere; the plant roots and soil microbes immediately utilize the leftover oxygen, resulting in hypoxic conditions (Pan et al. 2021). The primary products of oxidative stress are singlet oxygen ($^1O_2$) and superoxides. The aggregation of numerous chemical compounds, including anion radicals (O_2^-), and H_2O_2, along with hydroxyl radicals (OH) as well (Gill et al. 2019), induced the efflux of root cytosol, reduction of root pressure and deterioration of cell membrane permeability and these effects were further intensified by waterlogging, culminating in irreversible root cell death (Bramley et al. 2011). MDA is frequently utilized to assess the magnitude of cellular peroxidation in plants (Rai-Kalal et al. 2021). Melatonin treatment in waterlogged tomato seedlings significantly reduced oxidative damage (less production of H_2O_2 and $O_2^{\bullet -}$) along with lowered production of MDA and electrolyte leakage (Jahan et al. 2024). Another study showed that melatonin-sprayed wheat cultivars substantially inhibited oxidative components like MDA and $O_2^{\bullet -}$production in plants exposed to waterlogging stress (Ma et al. 2022). Melatonin exhibits significant efficacy in mitigating lower ROS accumulation in root structures, inhibits higher membrane lipid peroxidation formation, and minimizes cellular membrane damage, thereby slowing the root senescence process (Arnao and Hernández-Ruiz 2015). Melatonin and tryptophan trigger an increase in phenolic substance content in white lupin leaves, and these phenolic substances help to reduce oxidative damage under water stress (Sadak and Ramadan 2021). Antioxidant components comprise SOD, POD, CAT, APX, GR, and so on (Xia et al. 2020), and melatonin has strong antioxidant properties, directly scavenging excess ROS and indirectly regulating antioxidant defense systems through the participation of their catalytic processes (Fig. 7.2). Plants mitigate ROS-induced cellular damage via the induction of a potent antioxidant defense system (Cheng et al. 2016) that increases stress tolerance in plants (Mittler 2002). Melatonin treatment significantly elevated POD and SOD activities, consequently augmenting the resilience of plants to abiotic stressors (Zhang et al. 2022). Melatonin treatment optimizes aerobic respiration in roots and shields the photosystem, thus enhancing waterlogging tolerance in wheat (Zheng et al. 2017).

Table 7.1 Melatonin functions under waterlogging stress in different species

Plant species	Treatment duration	Melatonin concentration	Method of application	Melatonin functions	References
Apple	9 days	50, 100, 200, 600 μM	Foliar spray or root irrigation	Increased melatonin concentration and antioxidant enzyme activity, photosynthesis efficiency, and aerobic respiration. Decreased oxidative damage and lowered leaf wilting and chlorosis.	Zheng et al. (2017)
Alfalfa	10 days	100 μM	Foliar spray	Slowed leaf senescence process. Restricted ROS components production. Increased photosynthesis efficiency. Reprogramming polyamines and ethylene biosynthesis process. Increased melatonin production.	Zhang et al. (2019)
Maize	7 days	100 μM	Seed soaking, foliar spray	Increased plant biomass and photosynthetic activity. Decreased oxidative damage. Inhibited pyruvate and decarboxylase activity. Triggered enzymatic and protein functions.	Ahmad et al. (2022)

(continued)

Table 7.1 (continued)

Plant species	Treatment duration	Melatonin concentration	Method of application	Melatonin functions	References
Cotton	10 days	100 µM	Foliar spray	Increased biological and seed cotton yield. Increased phytohormones content like IAA, GA and decreased ABA. Melatonin biosynthesis gene expression upregulated, reduced oxidative damage, and pyruvate and decarboxylase activity.	Zhang et al. (2024)
Cabbage	10 days	50,100,150,200 µM	Irrigation	Increased biomass production, photosynthetic activity, and enzymatic activity as well.	Can (2023)
Kiwi	9 days	10,50,100,200, 300 µM	Irrigation	Increased antioxidant activity and reduced ROS production. Modulation of amino acid contents. Increased proline, ACC, and GABA content.	Huo et al. (2022)
Tomato	10 days	50, 100, 200 µM	Foliar spray	Increased MT, IAA, JA, GA_3 content and decreased ABA, ACC content. Suppressed *RBOH* gene expression, inhibited PCD.	Jahan et al. (2024)
Wheat	7 days	100 µM	Foliar spray	Increased dry matter accumulation and yield. Increased water status. Enhanced gas exchange parameter. Reduced oxidative stress.	Ma et al. (2022)

Clematis tientaiensis and Clematis lanuginosa	28 days	50,100, 200 μM	Foliar spray	Increased photosynthesis, chlorophyll content, antioxidant enzyme activity, proline and soluble sugars content. Decreased H_2O_2 and relative conductivity. Modulations of transcription factors.	Chen et al. (2024)
Peach	12 days	100, 200, 400 μM	Irrigation	Reduced wilting and shedding of leaf. Increased Chl content and stomatal aperture and antioxidant enzyme content. Enhanced aerenchyma and suppressed metabolic enzyme activity. Elevated expression of Ca^{2+} signaling and hypoxia-related ERF VII transcription factor.	Gu et al. (2021)
Zanthoxylumarmatum	3 days	0,400,800 μM	Irrigated to soil	Promoted Chl and carotenoid content, gas exchange attributes, increased stomatal functions, electron transport in PSII, and photosynthesis efficacy.	Su et al. (2023)
Durum wheat	35 days	100 μM	Foliar spray	Increased biomass production, increased CO_2 assimilation, protected photosynthetic elements. Reduced oxidative damage.	Quaratiello et al. (2024)

(continued)

Table 7.1 (continued)

Plant species	Treatment duration	Melatonin concentration	Method of application	Melatonin functions	References
Cauliflower	20 days	50,100 μM	Foliar spray	Improved growth attributes, Chl content, photosystem II, and Chl fluorescence values.	Seymen et al. (2023)
Apples	9 days	200,400 μM	Irrigation	Increased ROS scavenging capacity. Nitric oxide (NO) content and nitrate reductase (NR) activity increased.	Cao et al. (2024)
Pepper	60 days	500 μM	Foliar spray	Elevated NR activity, hormonal changes (ABA), amino acid changes, and *PIP* gene expression.	Khosravi et al. (2023)
Soybean, wheat, maize, and alfalfa	6 h	30 μM	Priming seeds	Upregulated GA biosynthesis genes, ABA catabolism genes, and downregulated ROS biosynthesis genes expression.	Luo et al. (2024)

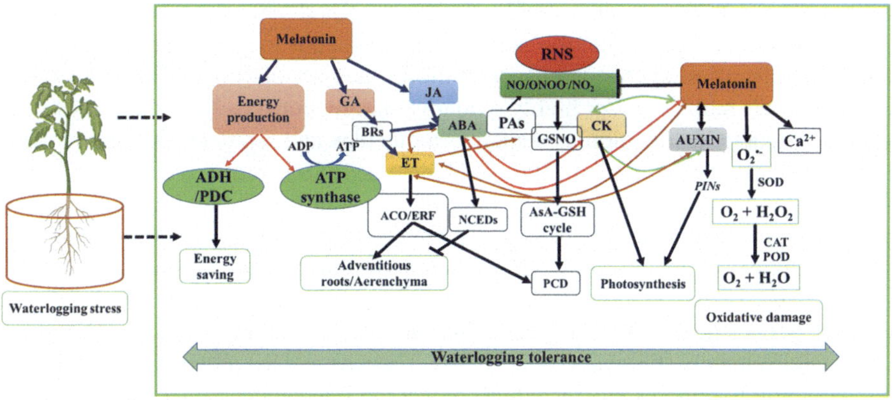

Fig. 7.2 Interactions of melatonin with other phytohormones and signaling molecules during waterlogging stress. (www.biorender.com)

Jahan et al. (2024) demonstrated that the melatonin foliar treatment effectively enhanced the antioxidant enzymes like SOD, POD, and CAT activities and their encoding transcription abundance in tomato leaves under waterlogging stress, consequently mitigating the deleterious effects through the detoxification of excess ROS. Melatonin can also upregulate various subtypes of enzymes, including ZnSOD and FeSOD (Campos et al. 2019). Chen et al. (2024) reported that melatonin application in *Clematis* elevated waterlogging tolerance through the boosting of the antioxidant defense system, lowering ROS component production, and decreasing REC content, and these all help to protect cell membrane integrity and reduce oxidative damage. As a heme component, POD and CAT largely found in peroxisomes, serve an essential role in catalyzing hydrogen peroxide into oxygen and water. Melatonin application stimulated CAT biosynthesis under submergence stress (Zheng et al. 2017). The fundamental AsA-GSH cycle enzymes are APX, GR, DHAR, and MDHAR and these enzymes synergistically work with other antioxidant enzymes to eliminate excess ROS and maintain ROS homeostasis. AsA is a strong ROS scavenging agent in plants (Li et al. 2013), on the other hand, GR increases the reduction of oxidized glutathione (GSSG) to GSH by using electrons provided by NADPH. Concurrently, NADPH acts as a reducing agent and is oxidized to form $NADP^+$(Wang et al. 2020). Under water stress, the content of GSH as well as the GSH/GSSG ratio is higher in melatonin-applied plants than in controlled ones (Bidabadi et al. 2020).

7.2.3 Melatonin Regulates RNS Signaling Under Waterlogging Stress

Generally, the optimal growth and development as well as yield performance of crops are significantly hampered by diverse stressors. Reactive nitrogen species (RNS) are an array of radical and non-radical compounds generated within the plant cells during extreme weather conditions in a method referred to as nitrosactive stress (Reiter et al. 2009; Saddhe et al. 2019). Analogous to ROS, RNS demonstrate dual characteristics: at elevated concentrations, they are detrimental to cell activity owing to their radical and/or highly oxidative nature; conversely, certain RNS possess the capacity to function as signaling molecules and regulate essential physiological processes. The fundamental component of RNS is nitric oxide (NO), which functions as an essential signaling molecule for transmitting and modulating the cellular redox state under hypoxic conditions (Samant et al. 2024). NO is predominantly synthesized by cytosolic NR, mitochondrial complexes III and IV, and AOX (alternative oxidase) (Gupta et al. 2011). Nitric oxide reacts with superoxide to produce peroxynitrite ($ONOO^-$), and it inhibits excess accumulation of H_2O_2 during stress environments (Fernando et al. 2019). The role of $ONOO^-$ in signaling under hypoxia largely remains elucidated. Moreover, an imbalanced redox state produces the HNO (nitroxyl) under waterlogging conditions (Arasimowicz-Jelonek et al. 2023). Like melatonin, as a signaling molecule, NO responds to both internal and external cues, acting as a direct free-radical scavenger and mitigating oxidative stress through processes that do not rely on receptor-mediated interactions (Fig. 7.2). Melatonin has the ability to regulate NO production or eliminate excess NO formation (Liu et al. 2019). Melatonin can be subjected to NO nitrosation under diverse pH states in the presence of oxygen to generate N-nitroso melatonin (NOMET) (Liu et al. 2019).

Melatonin and NO act as important signaling molecules that contribute to mitigating abiotic stresses including waterlogging stress. Melatonin and/or NO application significantly boost plant tolerance to diverse stressors via their elevated accumulation. The content of NO is reduced under waterlogging stress (Zhang et al. 2021). The reduced content of NO is often related to stunted growth of plants as well as increased plant tolerance to stress (Krishnan et al. 2018). Melatonin and NO play similar functions during stressful environments. Zhang et al. (2024) reported that melatonin and NO increases waterlogging tolerance in cotton seedlings through compensatory and quiescence adaptation techniques. As a result, further research is required to identify the particular mechanisms by which melatonin and NO modulate stress responses. Some studies demonstrated that melatonin and NO increased salt tolerance in sunflower (Arora and Bhatla 2017), rapeseed seedlings (Zhao et al. 2018), and protected pepper seedlings under iron deficiency and salinity stress (Kaya et al. 2020). To fully understand the actual mechanism between these two potent molecules, a further in-depth study is perquisites.

7.3 Melatonin Interactions with Phytohormones During Waterlogging Stress

Phytohormones serve as essential signaling molecules in plants, exhibiting rapid responsiveness to environmental fluctuations and playing a critical role in triggering plant growth and development. The advantageous function of melatonin in stress responses can be largely attributed to its modulation of stress-related gene expression. Considerable interplay exists between melatonin and various plant growth regulators. Such interactions manifest in an antagonistic, synergistic, and parallel manner. Waterlogging stress changes hormonal biosynthesis processes, translocation, and signal processes as well, leading to metabolic adjust of plants under stress conditions (Wurms et al. 2023). The variations of the concentration of endogenous hormones in plants function as a regulatory physiological indicator under hypoxic conditions. Plant tolerance to waterlogging is modulated by the complex interplay of multiple phytohormones (Fig. 7.2).

Ethylene known as the gaseous hormone plays a vital function in various plant physiological processes in particular accelerating senescence and abscission in the leaf, promoting cellulose activity, and proliferating adventitious and aerenchyma root formation (Shiono et al. 2019). Under waterlogging conditions, a significant amount of ethylene is produced that contributes to facilitating the substantial uptake of nutrients and water through roots (Vwioko et al. 2017). Simultaneously, ethylene induces ethylene response factors (ERFs) during stress conditions (Tamang et al. 2014). Under hypoxia conditions, ethylene either synergistically or antagonistically interplays with other phytohormones like ABA, GA, CK, JA, and so on, thus helping to adjust to a hypoxic environment and produce aerenchyma as well (Shimamura et al. 2014).

Abscisic acid (ABA) is well known for its critical role in closing stomata while decreasing transpiration rates. Phytohormone ABA notably inhibits later root formation during waterlogging stress (Kim et al. 2018). The ABA content in the waterlogging-tolerant cucumber cultivar decreased, whereas the waterlogging-sensitive cucumber cultivar slightly altered the ABA concentration in roots under hypoxic circumstances (Xu et al. 2022). Gibberellin is a classical phytohormone, basically contributing not only to cell elongation and division but also to stem elongation (Phukan et al. 2016). Ethylene plays a critical role in optimizing the antagonistic relationship between ABA and GA, and this optimal function might contribute to stem elongation, thereby facilitating oxygen acquisition during a flooding environment (Rzewuski and Sauter 2008).

A significant amount of studies reported that waterlogging stress increases ABA, JAs, and BRs content and decreases the content of CKs and GAs; conversely, melatonin application decreases ABA content while increasing CKs and GAs contents (Verma et al. 2016; Moustafa-Farag et al. 2020). The leaf senescence, as well as chlorophyll breakdown, significantly inhibits with CKs treatment under water stress situations (Liang et al. 2019). In addition, melatonin treatment significantly elevates cytokinin biosynthesis genes (*IPT3*, *IPT5*, and *AHK4*) and GAs biosynthesis genes *GA20ox* and *GA3ox* expression as well during stress conditions in cucumber (Wang et al. 2021). Moreover, melatonin also helps to activate BR biosynthetic genes, and this gene induction fosters stomatal dynamics, protecting cellular integrity and enhancing water uptake (Hwang and Back 2018). Furthermore, the application of melatonin notably elevates JAs concentration under hypoxia conditions and helps to increase plant stress tolerance (Cohen et al. 2021).

Nitric oxide (NO), as a signaling molecule, serves a crucial role in biological systems, influencing hormonal and stress-related signal transduction pathways, regulating ROS scavenging mechanisms, and controlling PCD while facilitating aerenchyma tissue formation (Pucciariello and Perata 2017). Under waterlogging environments, the activity of NR increases, leading to notable NO production. Under normoxic conditions, melatonin application increased the expression levels of *OsAUX1*, *OsARF19*, *OsPIN1a ~ 1d*, *OsPIN8*, *OsPIN9*, and *OsGH3-2* in rice seedling roots, whereas,under hypoxic conditions, it predominantly boosted *OsPIN2*, *OsPIN1b*, and *OsGH3-2*expression levels demonstrating that melatonin possibly activates auxin signaling pathways directly or through indirect mechanisms (Liu et al. 2023).

The transcription factors (TFs) function as essential gene expression modulators to drive plant responses to diverse abiotic stressors (He et al. 2023). Melatonin treatment under waterlogging stress caused differential expression of 23 and 14 transcription factors in *C. lanuginosa* and *C. tientaiensis*, respectively, and most of the genes belonged to *MYB*, *FAR1*, *bHLH*, Zinc finger protein, *LOB*, *ERF*, *WRKY*, and *B3* gene families and these TFs have a vital role in regulating waterlogging tolerance in *Clematis tiantianensis* (Chen et al. 2024). The findings of this research further demonstrate that *ABR1* significantly functions to mitigate waterlogging stress in *Clematis* while melatonin elevates the *ABR1* expression, thus fostering the plant's ability to withstand waterlogging conditions. The *WRKY* gene family acts as a critical element that helps plants respond to hypoxic stress conditions in various plant organisms. The WRKY4 transcription factor functions as a dual regulatory protein in *Clematis*by, potentially promoting photosynthesis and antioxidant activity, which could increase waterlogging tolerance in plants (Chen et al. 2024).

7.4 Conclusions and Future Research Direction

Waterlogging stress negatively regulates the growth and developmental processes of plants, influencing various aspects, including morphological characteristics, cellular structure, water status, and gene expression profiles. When dealing with hypoxic stress, phytohormones implicate a crucial function in modulating various essential physiological functions, including regulation of antioxidant metabolism, carbon metabolism (specifically carbohydrate production), stomatal behavior, and the progression of leaf senescence. Melatonin contributes a bifunctional role, as it works as an antioxidant agent, increasing plant ability to mitigate oxidative stress while simultaneously acting as a growth regulator, substantially contributing to increased plant immunity to survive under waterlogging stress. Future advanced research methodologies may elucidate the potential interactions between melatonin and various signaling molecules, including NO, H_2S, and other phytohormones. The precise signaling cascade mechanism through which melatonin-mediated waterlogging stress tolerance remains to be fully characterized. The miracle substance melatonin promotes plant stress-related transcription factors while enhancing antioxidant capacity and photosynthesis efficacy and altering metabolite levels in plants. The application of melatonin precursors and metabolite compounds enhances plant resistance to hypoxia stress and leads to increased agricultural production under flooding conditions. However, comprehensive studies utilizing diverse omics approaches and meticulous optimization of dosages and application methods on a plant-to-plant basis will yield valuable information for assessing large-scale applications under field conditions.

References

Ahammed GJ, Li Z, Chen J, Dong Y, Qu K, Guo T, Wang F, Liu A, Chen S, Li X (2024) Reactive oxygen species signaling in melatonin-mediated plant stress response. Plant Physiol Biochem 207:108398

Ahmad S, Wang GY, Muhammad I, Chi YX, Zeeshan M, Nasar J, Zhou XB (2022) Interactive effects of melatonin and nitrogen improve drought tolerance of maize seedlings by regulating growth and physiochemical attributes. Antioxidants 11(2):359

Alizadeh-Vaskasi F, Pirdashti H, Cherati Araei A, Saadatmand S (2018) Waterlogging effects on some antioxidant enzymes activities and yield of three wheat promising lines. Acta Slov 111:621–631

Arasimowicz-Jelonek M, Floryszak-Wieczorek J, Suarez S, Doctorovich F, Sobieszczuk-Nowicka E, Bruce King S, Milczarek G, Rębiś T, Gajewska J, Jagodzik P et al (2023) Discovery of endogenous nitroxyl as a new redox player in *Arabidopsis thaliana*. Nat Plants 9(1):36–44

Arbona V, Hossain Z, López-Climent MF, Pérez-Clemente RM, Gómez-Cadenas A (2008) Antioxidant enzymatic activity is linked to waterlogging stress tolerance in citrus. Physiol Plant 132:452–466

Arnao MB, Hernández-Ruiz J (2015) Functions of melatonin in plants: a review. J Pineal Res 59:133–150

Arnao MB, Hernández-Ruiz J (2019) Melatonin: a new plant hormone and/or a plant master regulator? Trends Plant Sci 24(1):38–48

Arora D, Bhatla SC (2017) Melatonin and nitric oxide regulate sunflower seedling growth under salt stress accompanying differential expression of cu/Zn SOD and Mn SOD. Free Radic Biol Med 106:315–328

Bidabadi SS, Vander Weide J, Sabbatini P (2020) Exogenous melatonin improves glutathione content, redox state and increases essential oil production in two salvia species under drought stress. Sci Rep 10:6883

Bramley H, Tyerman SD, Turner DW, Turner NC (2011) Root growth of lupins is more sensitive to waterlogging than wheat. Funct Plant Biol 38:910–918

Campos CN, Ávila RG, de Souza KRD, Azevedo LM, Alves JD (2019) Melatonin reduces oxidative stress and promotes drought tolerance in young *Coffeaarabica* L. plants. Agric Water Manag 211:37–47

Can H (2023) Melatonin application at different doses changes the physiological responses in favor of cabbage seedlings (*Brassica oleracea var. capitata*) against flooding stress. GesundePflanzen 75(6):2733–2745

Cao Y, Du P, Shang Y, Ji J, Tan L, Zhang X, Xu J, Liang B (2024) Melatonin and dopamine alleviate waterlogging stress in apples by recruiting beneficial endophytes to enhance physiological resilience. J Integr Agric 23(7):2270–2291

Chen K, Hu Q, Ma X, Zhang X, Qian R, Zheng J (2024) The effect of exogenous melatonin on waterlogging stress in *Clematis*. Front Plant Sci 15:1385165

Cheng XX, Yu M, Zhang N, Zhou ZQ, Xu QT, Mei FZ, Qu LH (2016) Reactive oxygen species regulate programmed cell death progress of endosperm in winter wheat (*Triticumaestivum* L.) under waterlogging. Protoplasma 253:311–327

Cherono S, Ntini C, Wassie M, Mollah MD, Belal MA, Ogutu C, Han Y (2021) Exogenous application of melatonin improves drought tolerance in coffee by regulating photosynthetic efficiency and oxidative damage. J Am Soc Hortic Sci 146:24–32

Cohen I, Zandalinas SI, Huck C, Fritschi FB, Mittler R (2021) Meta-analysis of drought and heat stress combination impact on crop yield and yield components. Physiol Plant 171:66–76

Dickin E, Wright D (2008) The effects of winter waterlogging and summer drought on the growth and yield of winter wheat (*Triticumaestivum* L.). Eur J Agron 28:234–244

Donat MG, Lowry AL, Alexander LV, O'Gorman PA, Maher N (2016) More extreme precipitation in the world's dry and wet regions. Nat Clim Chang 6:508–513

Fernando V, Zheng X, Walia Y, Sharma V, Letson J, Furuta S (2019) S-Nitrosylation: an emerging paradigm of redox signaling. Antioxidants 8(9):404

Gill MB, Zeng F, Shabala L, Zhang G, Yu M, Demidchik V, Shabala S, Zhou M (2019) Identification of QTL related to ROS formation under hypoxia and their association with waterlogging and salt tolerance in Barley. Int J Mol Sci 20:699

Grzesiak MT, Janowiak F, Szczyrek P, Kaczanowska K, Ostrowska A, Rut G, Hura T, Rzepka A, Grzesiak S (2016) Impact of soil compaction stress combined with drought or waterlogging on physiological and biochemical markers in two maize hybrids. Acta Physiol Plant 38:109

Gu X, Xue L, Lu L, Xiao J, Song G, Xie M, Zhang H (2021) Melatonin enhances the waterlogging tolerance of *Prunuspersica* by modulating antioxidant metabolism and anaerobic respiration. J Plant Growth Regul 40:2178–2190

Gupta KJ, Fernie AR, Kaiser WM, van Dongen JT (2011) On the origins of nitric oxide. Trends Plant Sci 16(3):160–168

He H, Yang M, Li S, Zhang G, Ding Z, Zhang L, Shi G, Li Y (2023) Mechanisms and biotechnological applications of transcription factors. Synth Syst Biotechnol 8:565

Huo L, Wang H, Wang Q, Gao Y, Xu K, Sun X (2022) Exogenous treatment with melatonin enhances waterlogging tolerance of kiwifruit plants. Front Plant Sci 13:1081787

Hussain S, Mehmood U, Ashraf U, Naseer MA (2022) Combined salinity and waterlogging stress in plants: limitations and tolerance mechanisms. In Climate change and crop stress. Academic Press, pp 95–112

Hussain A, Faizan M, Ahmed SM (2024) Regulation of photosynthesis by melatonin under optimal and suboptimal conditions. In: Sharma A, Ahammed GJ (eds) Melatonin in plants: role in plant growth, development, and stress response. Springer, pp 35–51

Hwang OJ, Back K (2018) Melatonin is involved in skotomorphogenesis by regulating brassinosteroid biosynthesis in rice plants. J Pineal Res 65(2):e12495

Ismond KP, Dolferus R, De Pauw M, Dennis ES, Good AG (2003) Enhanced low oxygen survival in Arabidopsis through increased metabolic flux in the fermentative pathway. Plant Physiol 132(3):1292–1302

Jahan MS, Shu S, Wang Y, Hasan M, El-Yazied AA, Alabdallah NM, Hajjar D, Altaf MA, Sun J, Guo S (2021) Melatonin pretreatment confers heat tolerance and repression of heat-induced senescence in tomato through the modulation of ABA-and GA-mediated pathways. Front Plant Sci 12:381

Jahan MS, Li G, Xie D, Farag R, Hasan MM, Alabdallah NM, Al-Harbi NA, Al-Qahtani SM, Zeeshan M, Nasar J (2023) Melatonin mitigates salt-induced growth inhibition through the regulation of carbohydrate and nitrogen metabolism in tomato seedlings. J Soil Sci Plant Nutr 23:4290–4308

Jahan MS, Huang Y, Qin NM, Wu HY, Zhou XB (2024) Exogenous application of melatonin attenuates waterlogging stress through adopting quiescence adaptation technique in tomato seedlings. Environ Exp Bot 225:105855

Jalili S, Ehsanpour AA, Javadirad SM (2022) The role of melatonin on caspase-3-like activity and expression of the genes involved in programmed cell death (PCD) induced by in vitro salt stress in alfalfa (*Medicago sativa* L.) roots. Bot Stud 63(1):19

Kaya C, Higgs D, Ashraf M, Alyemeni MAM, Ahmad P (2020) Integrative roles of nitric oxide and hydrogen sulfide in melatonin-induced tolerance of pepper (*Capsicum annuum* L.) plants to iron deficiency and salt stress alone or in combination. Physiol Plant 168:256–277

Khosravi S, Haghighi M, Mottaghipisheh J (2023) Effects of melatonin foliar application on hot pepper growth and stress tolerance. Plant Stress 9:100192

Kim Y, Seo CW, Khan AL, Mun BG, Shahzad R, Ko JW, Yun BW, Park SK Lee IJ (2018) Exoethylene application mitigates waterlogging stress in soybean (*Glycine max* L.). BMC Plant Biol 18:1–16

Komatsu S, Shirasaka N, Sakata K (2013) 'Omics' techniques for identifying flooding-response mechanisms in soybean. J Proteome Res 93:169–178

Krishnan A, Kumar S, Yadav SK (2018) Comparative role of nitric oxide in ethylene-dependent and ethylene-independent signaling events during abiotic stress in plants. Nitric Oxide Biol Chem 73:51–58

Li H, Wang XM, Chen L, Ahammed GJ, Xia XJ, Shi K, Considine MJ, Yu JQ, Zhou YH (2013) Growth temperature-induced changes in biomass accumulation, photosynthesis and glutathione redox homeostasis as influenced by hydrogen peroxide in cucumber. Plant Physiol Biochem 71:1–10

Liang D, Ni Z, Xia H, Xie Y, Lv X, Wang J, Lin L, Deng Q, Luo X (2019) Exogenous melatonin promotes biomass accumulation and photosynthesis of kiwifruit seedlings under drought stress. Sci Hortic 246:34–43

Liu J, Yang J, Zhang H, Cong L, Zhai R, Yang C, Wang Z, Ma F, Xu L (2019) Melatonin inhibits ethylene synthesis via nitric oxide regulation to delay postharvest senescence in pears. J Agric Food Chem 67:2279–2228

Liu J, Wang J, Zhang T, Li M, Yan H, Liu Q, Wei Y, Ji X, Zhao Q (2023) Exogenous melatonin positively regulates rice root growth through promoting the antioxidant system and mediating the auxin signaling under root-zone hypoxia stress. Agronomy 13(2):386

Lu X, Min W, Shi Y, Tian L, Li P, Ma T, Zhang Y, Luo C (2022) Exogenous melatonin alleviates alkaline stress by removing reactive oxygen species and promoting antioxidant defence in rice seedlings. Front Plant Sci 13:849553

Luo X, Xu X, Xu J, Zhao X, Zhang R, Shi Y, Xia M, Xian B, Zhou W, Zheng C, Wei S (2024) Melatonin priming promotes crop seed germination and seedling establishment under flooding stress by mediating ABA, GA, and ROS cascades. J Pineal Res 76(5):e13004

Ma S, Gai P, Geng B, Wang Y, Ullah N, Zhang W, Zhang H, Fan Y, Huang Z (2022) Exogenous melatonin improves waterlogging tolerance in wheat through promoting antioxidant enzymatic activity and carbon assimilation. Agronomy 12(11):2876

Majumder D, Bhowmik P, Das K, Das S, Hossain A, Bhatt R (2024) Versatile role of melatonin in plants. In: Advancement of melatonin research in plants. CRC Press, pp 1–23

Miller G, Shulaev V, Mittler R (2008) Reactive oxygen signaling and abiotic stress. Physiol Plant 133:481–489

Mittler R (2002) Oxidative stress, antioxidants and stress tolerance. Trends Plant Sci 7:405–410

Morales J, Kadota Y, Zipfel C, Molina A, Torres MA (2016) The Arabidopsis NADPH oxidases RbohD and RbohF display differential expression patterns and contributions during plant immunity. J Exp Bot 67(6):1663–1676

Moustafa-Farag M, Mahmoud A, Arnao MB, Sheteiwy MS, Dafea M, Soltan M, Elkelish A, Hasanuzzaman M, Ai S (2020) Melatonin-induced water stress tolerance in plants: recent advances. Antioxidants 9:809

Mukherjee S, Roy S, Arnao MB (2024) Nanovehicles for melatonin: a new journey for agriculture. Trends Plant Sci 29:232–248

Murchie EH, Niyogi KK (2011) Manipulation of photoprotection to improve plant photosynthesis. Plant Physiol 155:86–92

Pan J, Sharif R, Xu X, Chen X (2021) Mechanisms of waterlogging tolerance in plants: research progress and prospects. Front Plant Sci 11:627331

Panozzo A, Dal Cortivo C, Ferrari M, Vicelli B, Varotto S, Vamerali T (2019) Morphological changes and expressions of *AOX1A*, *CYP81D8*, and putative *PFP* genes in a large set of commercial maize hybrids under extreme waterlogging. Front Plant Sci 10:62

Phukan UJ, Mishra S, Shukla RK (2016) Waterlogging and submergence stress: affects and acclimation. Crit Rev Biotechnol 36:956–966

Pucciariello C, Perata P (2017) New insights into reactive oxygen species and nitric oxide signalling under low oxygen in plants. Plant Cell Environ 40:473–482

Quaratiello G, Risoli S, Antichi D, Pellegrini E, Nali C, Lorenzini G, Pampana S, Pisuttu C, Tonelli M, Cotrozzi L (2024) Exogenous melatonin application helps late-sown durum wheat to cope with waterlogging under Mediterranean environmental conditions. Physiol Plant 176(4):e14477

Rai-Kalal P, Tomar RS, Jajoo A (2021) H_2O_2 signaling regulates seed germination in ZnOnanoprimed wheat (*Triticumaestivum* L.) seeds for improving plant performance under drought stress. Environ Exp Bot 189:104561

Razi K, Muneer S (2021) Drought stress-induced physiological mechanisms, signaling pathways and molecular response of chloroplasts in common vegetable crops. Crit Rev Biotechnol 41:669–691

Reddy AR, Chaitanya KV, Vivekanandan M (2004) Drought-induced responses of photosynthesis and antioxidant metabolism in higher plants. J Plant Physiol 161:1189–1202

Reiter RJ, Paredes SD, Manchester LC, Tan DX (2009) Reducing oxidative/nitrosative stress: a newly-discovered genre for melatonin. Crit Rev Biochem Mol Biol 44:175–200

Rzewuski G, Sauter M (2008) Ethylene biosynthesis and signaling in rice. Plant Sci 175:32–42

Sadak MS, Ramadan AAE-M (2021) Impact of melatonin and tryptophan on water stress tolerance in white lupine (*Lupinustermis* L.). Physiol Mol Biol Plants 27:469–481

Saddhe AA, Malvankar MR, Karle SB, Kumar K (2019) Reactive nitrogen species: paradigms of cellular signaling and regulation of salt stress in plants. Environ Exp Bot 161:86–97

Samant SB, Yadav N, Swain J, Joseph J, Kumari A, Praveen A, Sahoo RK, Manjunatha G, Seth CS, Singla-Pareek SL et al (2024) Nitric oxide, energy and redox-dependent responses to hypoxia. J Exp Bot 75:4573–4588

Sandoval Y, Tighe-Neira R, Inostroza-Blancheteau C, Soto-Cerda B, González-Villagra J (2024) Melatonin improves plant water status, photosynthetic performance, and antioxidant defense system in highbush blueberry (*Vacciniumcorymbosum* L.) plants subjected to drought stress. Sci Hortic 323:112528

Seymen M, Çiçek Arı B, Kal Ü, Issı N, Atakul Z, Yavuz D (2023) Mitigation effects of melatonin applied to cauliflower seedlings under different flooding durations. GesundePflanzen 75(4):1031–1045

Shimamura S, Yoshioka T, Yamamoto R, Hiraga S, Nakamura T, Shimada S, Komatsu S (2014) Role of abscisic acid in flood-induced secondary aerenchyma formation in soybean (*Glycine max*) hypocotyls. Plant Prod Sci 17(2):131–137

Shiono K, Ejiri M, Shimizu K, Yamada S (2019) Improved waterlogging tolerance of barley (*Hordeumvulgare*) by pretreatment with ethephon. Plant Prod Sci 22:285–295

Su C, Wang P, Wu J, Wang H, Fan J, Gong W, Hui W, Wang J (2023) Effects of melatonin on the photosynthetic characteristics of *Zanthoxylumarmatum* under waterlogging stress. Russ J Plant Physiol 70(4):82

Szafrańska K, Reiter RJ, Posmyk MM (2017) Melatonin improves the photosynthetic apparatus in pea leaves stressed by paraquat via chlorophyll breakdown regulation and its accelerated de novo synthesis. Front Plant Sci 8:878

Tamang BG, Magliozzi JO, Maroof MA, Fukao T (2014) Physiological and transcriptomic characterization of submergence and reoxygenation responses in soybean seedlings. Plant Cell Environ 37:2350–2365

Tian LX, Zhang YC, Chen PL, Zhang FF, Li J, Yan F, Dong Y, Feng BL (2021) How does the waterlogging regime affect crop yield? A global meta-analysis. Front Plant Sci 12:634898–634906

Verma V, Ravindran P, Kumar PP (2016) Plant hormone-mediated regulation of stress responses. BMC Plant Biol 16:86

Voesenek LACJ, Colmer TD, Pierik R, Millenaar FF, Peeters AJM (2006) How plants cope with complete submergence. New Phytol 170:213–226

Vwioko E, Adinkwu O, El-Esawi MA (2017) Comparative physiological, biochemical, and genetic responses to prolonged waterlogging stress in okra and maize given exogenous ethylene priming. Front Physiol 8:632

Wang Y, Jie K, Chen B, Zhou Z, Meng Y, Zhao W, Liu Z (2014) Waterlogging during flowering and boll forming stages affects sucrose metabolism in the leaves subtending the cotton boll and its relationship with boll weight. Plant Sci 223:79–98

Wang M, Zhang S, Ding F (2020) Melatonin mitigates chilling-induced oxidative stress and photosynthesis inhibition in tomato plants. Antioxidants 9:218

Wang D, Chen Q, Chen W, Guo Q, Xia Y, Wang S, Jing D, Liang G (2021) Physiological and transcription analyses reveal the regulatory mechanism of melatonin in inducing drought resistance in loquat (*Eriobotrya japonica* L.) seedlings. Environ Exp Bot 181:104291

Wurms KV, Reglinski T, Buissink P, Chee AA, Fehlmann C, McDonald S, Cooney J, Jensen D, Hedderley D, McKenzie C, Rikkerink EHA (2023) Effects of drought and flooding on phytohormones and abscisic acid gene expression in kiwifruit. Int J Mol Sci 24:7580

Xia H, Ni Z, Hu R, Lin L, Deng H, Wang J, Tang Y, Sun G, Wang X, Li H, Liao M, Lv X, Liang D (2020) Melatonin alleviates drought stress by a non-enzymatic and enzymatic antioxidative system in kiwifruit seedlings. Int J Mol Sci 21:852

Xu L, Xiang G, Sun Q, Ni Y, Jin Z, Gao S, Yao Y (2019) Melatonin enhances salt tolerance by promoting MYB108A-mediated ethylene biosynthesis in grapevines. Hortic Res 6:114

Xu W, Chen L, Hu X, Zhang L, Huang D, Li J, Xiong R, Huang C, Zhu M (2024) Botanic signal monitor: advanced wearable sensor for plant health analysis. Adv Funct Mater 34(51):2410544

Xu L, Zhao CC, Pang JY, Niu YA, Liu HQ, Zhang WY, Zhou MX (2022) Genome-wide association study reveals quantitative trait loci for waterlogging-triggered adventitious roots and aerenchyma formation in common wheat. Front Plant Sci 13:1066752

Yamane K, Mariyama M, Hirooka Y, Iijima M (2023) Root pruning is effective in alleviating the inhibition of soybean growth caused by anaerobic stress for a short period. J Integr Agric 22:1035–1044

Yamauchi T, Yoshioka M, Fukazawa A, Mori H, Nishizawa NK, Tsutsumi N, Yoshioka H, Nakazono M (2017) An NADPH oxidase RBOH functions in rice roots during lysigenous aerenchyma formation under oxygen-deficient conditions. Plant Cell 29(4):775–790

Yan K, Zhao SJ, Cui MX, Han GX, Wen P (2018) Vulnerability of photosynthesis and photosystem I in Jerusalem artichoke (*Helianthus tuberosus* L.) exposed to waterlogging. Plant Physiol Biochem 125:239–246

Zhang Q, Liu X, Zhang Z, Liu N, Li D, Hu L (2019) Melatonin improved waterlogging tolerance in alfalfa (*Medicago sativa*) by reprogramming polyamine and ethylene metabolism. Front Plant Sci 10:44

Zhang Y, Fan Y, Rui C, Zhang H, Xu N, Dai M, Chen X, Lu X, Wang D, Wang J, Wang J, Wang Q, Wang S, Chen C, Guo L, Zhao L, Ye W (2021) Melatonin improves cotton salt tolerance by regulating ROS scavenging system and Ca^{2+} signal transduction. Front Plant Sci 12:693690

Zhang R, Yue Z, Chen X, Wang Y, Zhou Y, Xu W, Huang R (2022) Foliar applications of urea and melatonin to alleviate waterlogging stress on photosynthesis and antioxidant metabolism in sorghum seedlings. Plant Growth Regul 97:429–438

Zhang Y, Liang T, Dong H (2024) Melatonin enhances waterlogging tolerance of field-grown cotton through quiescence adaptation and compensatory growth strategies. Field Crop Res 306:109217

Zhao G, Zhao Y, Yu X, Kiprotich F, Han H, Guan R, Wang R, Shen W (2018) Nitric oxide is required for melatonin-enhanced tolerance against salinity stress in rapeseed (*Brassica napus* L.) seedlings. Int J Mol Sci 19(7):1912

Zhao T, Pan XJ, Ou ZG, Li Q, Zhang WE (2022) Comprehensive evaluation of waterlogging tolerance of eleven Canna cultivars at flowering stage. Sci Hortic 296:110890

Zheng X, Zhou J, Tan D-X, Wang N, Wang L, Shan D, Kong J (2017) Melatonin improves waterlogging tolerance of *Malusbaccata* (Linn.) Borkh. Seedlings by maintaining aerobic respiration, photosynthesis and ROS migration. Front Plant Sci 8:483–493

Zhou WG, Chen F, Meng YJ, Chandrasekaran U, Luo XF, Yang WY, Shu K (2020) Plant waterlogging/flooding stress responses: from seed germination to maturation. Plant Physiol Biochem 148:228–236

Melatonin as an Emerging Antiviral Agent in Plants

8

Rahul Kumar Tiwari, Milan Kumar Lal, Ravinder Kumar, Sayanti G. Majumdar, Kirtika Swaraj, and Sumit Kumar

Abstract

Melatonin, a multifunctional phytohormone, has shown antiviral properties in animals and plants, offering a new approach to managing viral infections. In the animal system, melatonin has shown conclusive efficacy against multiple pathogens, including severe acute respiratory syndrome coronavirus 2 (SARS-CoV-2), by diminishing cytokine storms and viral replication. This chapter highlights the functions of phytomelatonin in suppressing plant virus infections through biochemical and molecular defence mechanisms. Several studies have shown that melatonin treatment can substantially reduce viral replication, movement, spread and symptom severity in crops such as rice, cucumber, cassava, and tobacco. The major defence mechanisms include reactive species signalling, MAPK signalling activation, and overexpression of pathogen-related proteins. Additionally, melatonin crosstalks with phytohormones and regulates secondary metabolites biosynthesis in the infected plants, which leads to induced defence. Studies have also reported the disruption of melatonin biosynthesis during the interaction of viral coat proteins and host transcription factors such as MeRAV1 and MeRAV2. Furthermore, melatonin influences virus-vector interactions by diminishing vector fecundity and attracting vectors to pretreated plants, and this means reducing

R. K. Tiwari · S. G. Majumdar · K. Swaraj
Division of Crop Physiology and Biochemistry, ICAR-National Rice Research Institute, Cuttack, Odisha, India

M. K. Lal
ICAR-National Rice Research Institute, Cuttack, Odisha, India

R. Kumar (✉)
ICAR-Indian Agricultural Research Institute, New Delhi, India

S. Kumar
Krishi Vigyan Kendra, Mau, UP, India

R. K. Tiwari et al. (eds.), *Melatonin: Signal Transduction Mechanisms and Defense Networks in Plants*, https://doi.org/10.1007/978-981-96-8869-2_8

broad-scale viral transmission. Comparative transcriptomic analysis has also reported that the upregulation of differentially expressed genes is associated with stress mitigation pathways. Although melatonin presents a novel, environment-friendly strategy for plant virus management, more elaborate research is required to standardise its application and evaluate its longstanding impact on crop health.

Keywords
Signalling · Oxidative stress · Coat protein · Cassava · Mosaic

8.1 Introduction

Plant viruses have emerged as a significant challenge to sustainable and productive agriculture globally, leading to annual economic losses accumulating to $30 billion annually (Hilaire et al. 2022; Tatineni and Hein 2023; Kanapiya et al. 2024). Nearly half of all newly emerging and re-emerging plant diseases are caused by viruses (Jones 2021). These pathogens affect cultivated horticultural and field crops and devastate natural vegetation, reducing biodiversity. More severe losses are seen in emerging viral diseases, which have high infection rates and a greater geographical distribution. There are many factors which had contributed to the emergence and spread of viruses. Modern agriculture has a repetitive cropping pattern that relies on sowing the same crops across the seasons, which leads to minimal genetic diversity and vulnerability of crops towards virus epidemics (Scholthof et al. 2011; Parrella et al. 2022). With global trade and trade liberalisation, the exchange of planting materials, quality seeds, cultivated and wild germplasm, and live plants between countries facilitates the unintentional spread of viruses and their insect vectors, exposing new environments to previously minor pathogens. A paradigm shift in rainfall patterns, temperature, humidity and winds also appears as a major factor for the rapid emergence and spread of viruses and vectors in agricultural ecosystems. The majority of plant viruses have an exceptional ability to mutate, evolve and adapt to new hosts and environments, allowing them to spread more efficiently by counteracting host defence. Major symptoms of viral infections in plants include stunted growth, mosaic, mottling, curling, puckering in leaves, reduced seed vigour, reduced photosynthesis and overall deterioration of plant health (Kumar et al. 2019, 2020b, 2021a, 2022, 2023). Virus-infected plants often produce fruits and vegetables of inferior quality, resulting in inferior market value and reduced agricultural income. In many cases, viral outbreaks have led to total crop failure, predominantly in tropical and subtropical zones where a majority of the population depends on agriculture for their livelihood (Jones 2021; Kumar et al. 2021b).

In developing countries such as India, where the majority of the population depends on staple crops for food and nutrition, the viruses are a significant challenge in cultivation (Kumar et al. 2021a, 2023; Lal et al. 2021). In temperate climates, viruses are a major challenge in vegetable and fruit crops and organic cultivation systems are also affected by virus disease outbreaks (Scholthof et al. 2011; Kumar et al. 2020a; Lal et al. 2020).

The diseases of bacteria and fungi can often be managed using antibiotics or chemical fungicides, but there are currently negligible treatments available to mitigate virus-infected plants. Once a plant is infected, the virus continues to spread across the tissue, often without any curable treatment options (Parrella et al. 2022). Therefore, managing viral disease primarily relies on preventive measures. The significant strategies include breeding for virus-resistant varieties, managing insect vector populations, domestic and international quarantine regulations, field sanitation, and weed host management that minimise the risk of initial infection (Soni et al. 2024; Tatineni and Hein 2023; Kanapiya et al. 2024). However, since each virus integrates into its host and rapidly evolves in environments, effective management approaches must be designed specifically for each host-virus pathosystem. The modern methods of boosting plant defence networks through plant-based hormones and other signalling molecules are an intriguing way of managing viral diseases (Santner and Estelle 2009; Yang et al. 2023). Several of these phytoprotectants are actively involved in systemic acquired resistance and the induced systemic resistance mechanism of plant defence against plant pathogens, including viruses (Jameson and Clarke 2002; Collum and Culver 2016; Islam et al. 2019). Almost all reported phytohormones have been documented to play a vital role in growth, development and plant defence against fungal, viral and bacterial pathogens (Fig. 8.1). For example, salicylic acid (SA) regulates basal defence via SAR involving R genes.

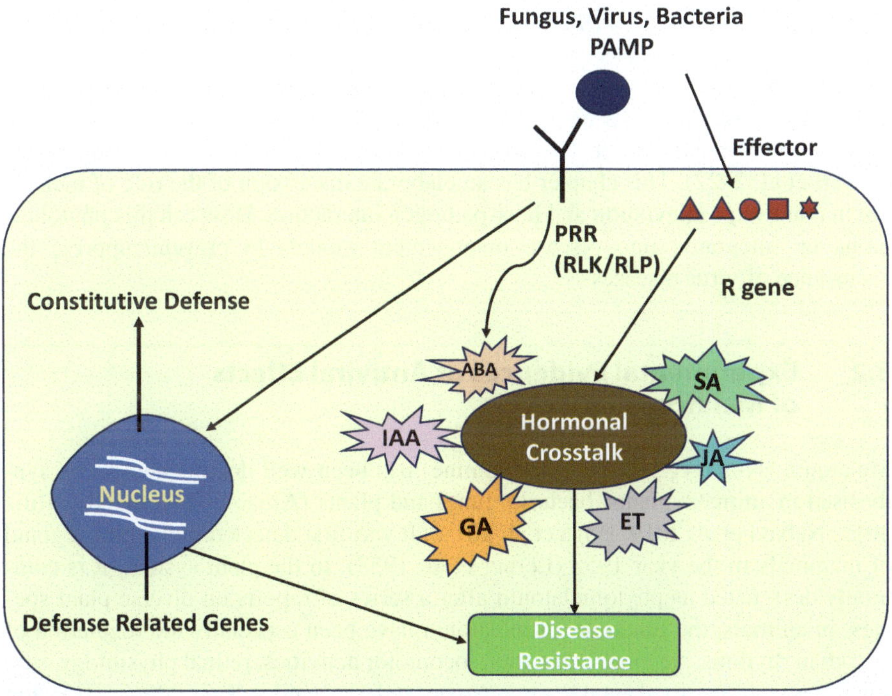

Fig. 8.1 Role of phytohormones in disease resistance

The SA-based defence may include accumulation of reactive oxygen species, callose deposition, activation of pathogenesis-related (PR) proteins, and phenomena of hypersensitive response (Corina Vlot et al. 2009). Jasmonic acid (JA)-SA antagonism is a repeated factor in regulating virus defence processes.

Tobacco mosaic virus movement was enhanced in N gene-resistant tobacco treated with methyl jasmonate. Furthermore, silencing of the JA biosynthetic enzyme, allene oxide synthase, resulted in increased SA accumulation and reduced tobacco mosaic virus (TMV) accumulation (Oka et al. 2013). The crosstalk between phytohormonal pathways and viral components often determines the extent of susceptibility or resistance in host plants. Auxin, gibberellic acid, cytokinin, brassinosteroids, and abscisic acid contribute to defence but primarily regulate plant development and physiological processes (Collum and Culver 2016). The ABA and JA pathways have positive effects on defence against herbivorous insects, which can act as viral vectors. Some viruses compromise the SA-mediated defence responses in plants either by inhibiting the GA and SA pathways or by activating the antagonistic auxin pathway (Nguyen et al. 2016; Pan et al. 2021). Contrarily, other viruses inhibit JA and ABA-mediated defence against their insect vectors either directly or by activating the SA or (Ethylene) ET pathways, which can antagonise ABA. Additionally, SA and ABA are correlated to the general virus defence mechanism of RNA silencing (Yang et al. 2020; Ramesh et al. 2021). This intriguing hormonal crosstalk highlights the balance between plant immunity and virus-induced susceptibility, offering potential targets for imparting resistance through phytohormone modulation (Sharma et al. 2025). Recently, a multifunctional phytoprotectant, 'melatonin', has emerged as a signalling molecule which plays an active role in abiotic and biotic stress modulation (Tiwari et al. 2020, 2021; Altaf et al. 2022). There is a plethora of reports on melatonin-mediated mitigation of biotic stressors, specifically fungi, bacteria and viruses affecting cultivated crop plants (Tiwari et al. 2022). This chapter has an elaborate discussion of the role of melatonin in plant virus alleviation and host-pathogen interaction. How can this phytohormone be integrated into disease management models to prevent/suppress the occurrence of virus diseases?

8.2 Experimental Evidences of Antiviral Effects of Melatonin

Melatonin (N-acetyl-5-methoxytryptamine) has been well documented to be synthesised in animals, plants, bacteria, fungi and plants (Arnao and Hernández-Ruiz 2015; Nabavi et al. 2019; Boga et al. 2019). It was first detected in the pineal gland of mammals in the year 1958 (Lerner et al. 1958). In the plant system, it is commonly designated as phytomelatonin after a series of reports on diverse plant species. In animals, the functions of melatonin have been especially the regulation of circadian rhythms, sleeping behaviour, locomotor activities, retinal physiology, seasonal perceptions, food intake, etc. (Arnao and Hernández-Ruiz 2006, 2015; Tan et al. 2016; Kaya et al. 2020). At the cellular level, melatonin acts as an antioxidant and scavenges reactive oxygen species and reactive nitrogen species via antioxidative enzymes. Melatonin has been reported to be influenced by light signals and plays a vital role in immune system functioning. In many diseases (e.g., Parkinson's, Alzheimer's, dementia, cardiovascular, obesity, and cancer), melatonin functions as an anti-inflammatory, neuroprotective and antioxidant (Jan et al. 2009; Xie et al. 2017; Luo et al. 2019). The antiviral properties of melatonin against Venezuelan equine encephalomyelitis virus and Ebola virus were recorded previously (Bonilla et al. 2004; Tan et al. 2014). But during the global pandemic caused by SARS-CoV-2, melatonin was used in several studies as a vital drug for use at different stages of therapy for COVID-19 (Cho et al. 2021; Reiter et al. 2022). Several studies have shown that melatonin application improved the resistance of animals and humans to viral diseases, namely herpes simplex (Nunes and Pereira 2008), hepatic fibrosis disease in mice (Crespo et al. n.d.) and Venezuelan equine encephalitis (Valero et al. 2015). Computational studies highlighted melatonin as a main candidate against SARS-CoV-2 by mitigating cytokine storm-related respiratory disorders. Laboratory experiments suggested its broad-spectrum antiviral effect against several viruses, including respiratory syncytial virus, dengue virus, transmissible gastroenteritis virus, and encephalomyocarditis virus. Animal studies further confirmed its role in counteracting viral replication and mortality in infections such as Venezuelan equine encephalomyelitis and COVID-19. Although clinical trials have some exploration of their therapeutic potential, confirmatory conclusions on their efficacy and safety remain pending due to the absence of large-scale, double-blind, placebo-controlled studies (Alomari et al. 2023).

The first study that reported an improvement in plant resistance against viruses by the exogenous application of melatonin came in the year 2019 (Zhao et al. 2019). This study investigated the role of melatonin in enhancing plant resistance to tobacco mosaic virus in *Nicotiana glutinosa* (local lesion assay) and tomato (systemic infection). The application of 100 μM phytomelatonin activated antiviral activity by 37.4% in *N. glutinosa*. In *S. lycopersicum*, the similar dose treatment substantially reduced virus load as tested with RNA qRT-PCR and dot-ELISA. The expression of PR1 and PR5 defence-related genes was also enhanced, and this study provided crucial evidence that melatonin plays a critical role in strengthening plant immunity against TMV, offering potential strategies for improving virus resistance in crops. Interestingly, it was observed that the transgenic *Nicotiana tabacum* line expressing SA hydroxylase (nahG) did not reduce the virus concentration in melatonin-treated plants, which suggests the possible crosstalk of melatonin and SA in virus disease resistance in this particular study. When a NO scavenger (cPTIO) was applied to virus-infected plants, the melatonin treatment was less effective in suppressing the resistance, which confirmed the melatonin-mediated NO signalling as a defence mechanism against plant virus. During the same year, melatonin was reported as a potential molecule for propagating virus-free apple plants in tissue culture set-up by enhancing shoot proliferation and mediating virus eradication. In the study on apple (*Malus domestica*) cultivar Gala, the pretreatment of 15 μM melatonin to the tissue culture (shoot proliferation) medium significantly enhanced shoot growth, increasing both shoot number and length (Lu et al. 2019). The shoots cultured on the melatonin-supplemented medium showed the highest levels of endogenous indole-3-acetic acid (IAA), indicating a role in promoting growth. The *Apple stem grooving virus*-infected shoots treated with melatonin under in vitro conditions showed maximum regrowth in 4 weeks as compared to control (non-treated shoots). About 95% of the shoot tips (treated with melatonin) were free of virus particles, while no virus-free shoots were reported in the melatonin-untreated shoots. The virus concentration was quantified using microtissue direct RT-PCR and RT-qPCR, which confirmed the reduction in the concentration of Apple Stem Grooving Virus (ASGV) in melatonin-treated shoots. Virus localisation studies were also performed in shoot tips, which confirmed the increment in the virus-free zones in melatonin-supplemented shoots. This proved that melatonin restricted the spread of the virus in plant cells. The prominent role of melatonin in tissue culture set-up involving virus resistance and growth promotion offered a promising milestone in further studies of this phytoprotectant in virus disease mitigation under in vitro and in vivo conditions.

The prominent role of melatonin in virus disease resistance has also been reported in staple food crops such as rice. A study in China highlighted the significant effect of melatonin in enhancing rice resistance to *Rice stripe virus* (Lu et al. 2019). In this study, the complementary effect of melatonin sodium nitroprusside (SNP), a nitric oxide (NO) donor was observed in the rice cultivar 'Nipponbare' which is susceptible to *Rice stripe virus*. Melatonin in four different concentrations (0.1, 1, 10, and 100 μM) and SNP (10, 50, 100, 500, and 1000 μM) were exogenously applied in 14-day-old seedlings. To express the virus symptoms in plants, the seedlings were kept in dry soil conditions, and darkness was maintained for 12 hours. After inoculation with the rice stripe virus (RSV), the plants were maintained in the greenhouse and observed periodically for symptom expression. Both SNP and melatonin significantly impacted the virus disease occurrence in susceptible plants, and a concentration of 10 μM melatonin and 100 μM SNP showed the strongest positive effects melatonin. Interestingly, exogenous melatonin and NO application upregulated the expression of pathogenesis-related genes (OsPR1b and OsWRKY45), which enhanced the plant defence response. In another study, it was also confirmed that endogenous melatonin level was upregulated in cultivar Zhendao-88 (resistant) after RSV infection as compared to *Nipponbare* (susceptible).

A recent study by Sofy et al. 2021 documented the antiviral effects of melatonin and SA against Alfalfa Mosaic Virus (AMV) in brinjal (*Solanum melongena*) (Sofy et al. 2021). The seeds were grown in a controlled environment in a glasshouse under glasshouse conditions with a 12-hour photoperiod, maintaining temperatures of 27 °C during the day and 23 °C at night. A foliar spray of 100 μM melatonin or SA was done on 14-day-old plants, and AMV was inoculated after 3 days. The virus quantification revealed that the titre was reduced in melatonin and SA-treated plants, and symptom expression was in correlation with concentration. There was improvement in growth parameters, namely, shoot and root length, number of leaves, leaf area, and biomass. In melatonin-sprayed plants, the chlorophyll and carotenoid content was enhanced, and the activity of catalase and peroxidase was upregulated. The expression of defence-related genes glutathione reductase, dehydroascorbate reductase, monodehydroascorbate reductase, pathogenesis-related proteins, and mitogen-activated protein kinase was enhanced compared to untreated, virus-infected controls. The treatments also ameliorated oxidative stress by minimising the accumulation of hydrogen peroxide, superoxide anion, hydroxyl radical, and malondialdehyde. These findings corroborate that melatonin and SA are potential eco-friendly supplements for managing viral infections in cultivated plants, offering a novel approach to enhancing resistance against AMV.

A recent study by Yang et al. 2023 highlighted the multifunctional melatonin signalling mechanism that has shown its effectiveness in managing *Cucumber Green Mottle Mosaic Virus* (CGMMV) infection (Yang et al. 2023). The roots of *N. benthamiana* were saturated with 50 μM concentration melatonin solution for three consecutive days. To examine the impact of melatonin on CGMMV infection in *N. benthamiana*, four concentrations of melatonin were supplemented through root irrigation. Plants were irrigated with 50, 100, 200, and 400 μM melatonin for 3 days followed by CGMMV inoculation. The plant samples were collected 9 days after inoculation, and qRT-PCR analysis was performed to measure the transcript levels of CGMMV coat protein in plant leaves. It was evident that melatonin at 50 and 200 μM significantly reduced CGMMV CP transcription, while the rest of the concentrations were ineffective. Considering the practicality and economy of application, 50 μM melatonin was recommended to be the most appropriate concentration for mitigating CGMMV infection in *N. benthamiana*. A transcriptome analysis was also conducted to understand the underlying molecular mechanisms by comparing mock-inoculated, CGMMV-infected, and melatonin-treated CGMMV-infected tobacco leaves. The major finding included the upregulation of cysteine-rich secretory protein, CRISP1, in response to melatonin. Further experimentation showed that knockdown of CRISP1 activated melatonin's defensive effects against CGMMV; however, CRISP1 suppression itself did not directly influence CGMMV infection. Overall observations suggest that melatonin enhances plant resistance by modulating cysteine-rich secretory protein expression and that targeting these proteins could be a potential melatonin-based antiviral strategy for controlling viral diseases in plants.

In a major viral disease of cassava caused by Cassava common mosaic virus (CsCMV), melatonin played a critical role in regulating essential biosynthetic pathways (Wei et al. 2024). Under virus infection, two of the cassava transcription factors (MeRAV1 and MeRAV2) enhanced disease resistance by activation of melatonin biosynthetic genes (tryptophan decarboxylase, tryptamine 5-hydroxylase, and N-acetylserotonin O-methyltransferase 1) and antioxidative defence-related catalase genes (MeCAT6 and MeCAT7). Additionally, it was observed that the CsCMV coat protein interferes with this defence mechanism by modulating MeRAV1 and MeRAV2, restricting their phosphorylation and restraining their ability to activate these genes. This suppression mechanism weakens the cassava plants' antiviral response, assisting severe CsCMV infection. Understanding this mechanism provides mechanistic insights for genetic improvement strategies to strengthen cassava's resistance to viral infections.

Virus	Host	Mode of application (optimal concentration)	References
Cucumber green mottle mosaic virus (CGMMV)	*N. benthamiana*	Root irrigation (50 μM)	Yang et al. (2023)
Cassava common mosaic virus	Cassava	Not applicable	Zhao et al. (2019)
Tobacco mosaic virus (TMV)	*N. glutinosa* and *S. lycopersicum*	Foliar spray on seedlings 100 μM	Zhao et al. (2019)
Apple stem grooving virus (ASGV)	*Malus domestica*	Incorporation into the media (10 μM)	Chen et al. (2019)
Rice stripe virus (RSV)	*Oryza sativa*	Foliar spray on 14-day-old seedlings (10 μM)	Lu et al. (2019)
Alfalfa mosaic virus (AMV)	*Solanum melongena*	Foliar spray (100 μM)	Sofy et al. (2021)

8.3 Functional Role of Melatonin Role on Vectors of Plant Viruses

In animal science, the potential of melatonin in controlling vector-borne diseases, such as malaria, dengue, and leishmaniasis, has been reported with special reference to its capacity to influence both the human immune response and the behaviour of the transmitting vectors (Zeng et al. 2024). Malaria, a disease caused by Plasmodium parasites transmitted by Anopheles mosquitoes, is a vector-borne disease in which melatonin was observed to regulate the dynamics of both host and vector (Dos Santos et al. 2021). The circadian cycles in mosquitoes affect their feeding and mating behaviours, which may align with human activity patterns (Shetty et al. 2022). Melatonin's capacity to influence these circadian rhythms may disturb these patterns in mosquitoes, so altering their feeding schedules and minimising the probability of human interaction. The transmission of plant viruses is mostly facilitated by insect vectors, which play an essential role in transmission and successful virus infection in host plants. The major insect vectors, such as aphids, play a diversified role in the spread of several plant viral diseases. *Aphis gossypii*, a devastating insect pest in cucumber production, transmits viruses through its feeding mechanisms. Recent studies have shown the possibility of melatonin as a modulator of plant-virus-vector interactions by altering this insect behaviour and enhancing plant resistance. The study elaborated on the effect of exogenous melatonin on *Aphis gossypii* infestation in cucumber plants (Liu et al. 2022). Cucumber seedlings were sprayed with different melatonin concentrations (0, 0.1, 1, and 5 mM) through foliar spray three times at 24-hour intervals. Ten wingless aphids were placed on the second leaf, and developed nymphs were counted after 24, 48, and 96 hours to assess the fecundity levels in aphids feeding on melatonin-treated plants. Additionally, an aphid choice assay (preference assay) was performed by keeping melatonin-treated and control seedlings in a net-covered chamber, jointed by a cardboard bridge. A total of 25 wingless adult aphids were released at the centre. After one day of release, aphid distribution on treated and control seedlings was recorded. The results revealed that melatonin treatment restricted aphid fecundity while enhancing plant attractiveness to aphids, as a result reducing virus transmission by limiting aphid dispersal. The exogenous melatonin application also regulated enzyme activities, specifically methyl-esterified homogalacturonan degradation-associated enzymes and those involved in antioxidant defence and membrane lipid peroxidation. Subsequent transcriptomic analysis demonstrated profound changes in gene expression, with 2360 differentially expressed genes induced before aphid infestation. These genes were primarily associated with hormone signalling, MAPK pathways, and plant-pathogen interactions. In aphid-infested plants, 2397 DEGs were identified, primarily associated with activated jasmonic acid signalling, reactive species signalling, and flavonoid biosynthesis in plants treated with exogenous melatonin. Altogether, these reports suggest that melatonin not only improves plant immunity but also manages aphid-induced adversities and offers novel strategies for managing virus-vector interactions in crop cultivation.

8.4 Conclusion

The management of plant viruses mainly depends on the use of protective measures such as disease resistance, quarantine regulations, healthy planting material and vector management. The emergence of such novel phytoprotectants, which can restrict the accumulation and spread of viruses in the infected plant, paves the way for new strategies for managing these pathogens. Melatonin-mediated virus disease resistance is mainly due to antioxidant defence and activation of systemic resistance-related phytohormones. Studies during the past 5 years indicate that melatonin reduces viral load through the expression of crucial defence-related genes such as PR1, PR5, and transcription factors, ultimately boosting plant immunity. Moreover, melatonin affects virus-vector dynamics by influencing insect behaviour, restricting vector fecundity, and minimising large-scale virus transmission. Plant resistance is attributed to melatonin-induced reactive species homeostasis, flavonoid generation, and sucrose metabolism. Furthermore, melatonin-mediated virus disease amelioration was correlated to the regulation of specific transcription factors, such as MeRAV1 and MeRAV2 in cassava, which regulate melatonin biosynthesis and catalase activity. The antiviral ability of melatonin against the diverse spectrum of viruses, including CGMMV, TMV, AMV, CsCMV, and RSV, suggests its future application in virus disease management strategies. More elaborate field-based studies are also required in this domain for large-scale applications.

References

Alomari T, Al-Abdallat H, Hamamreh R et al (2023) Assessing the antiviral potential of melatonin: a comprehensive systematic review. Rev Med Virol 34(1):e2499. https://doi.org/10.1002/rmv.2499

Altaf MA, Shahid R, Ren MX et al (2022) Melatonin improves drought stress tolerance of tomato by modulation plant growth, root architecture, photosynthesis, and antioxidant defense system. Antioxidants 11:309. https://doi.org/10.3390/ANTIOX11020309/S1

Arnao MB, Hernández-Ruiz J (2006) The physiological function of melatonin in plants. Plant Signal Behav 1:89–95. https://doi.org/10.4161/PSB.1.3.2640

Arnao MB, Hernández-Ruiz J (2015) Functions of melatonin in plants: a review. J Pineal Res 59:133–150. https://doi.org/10.1111/jpi.12253

Boga JA, Caballero B, Potes Y et al (2019) Therapeutic potential of melatonin related to its role as an autophagy regulator: a review. J Pineal Res 66:e12534

Bonilla E, Valero N, Chacín-Bonilla L, Medina-Leendertz S (2004) Melatonin and viral infections. J Pineal Res 36:73. https://doi.org/10.1046/J.1600-079X.2003.00105.X

Chen L, Wang MR, Li JW et al (2019) Exogenous application of melatonin improves eradication of apple stem grooving virus from the infected in vitro shoots by shoot tip culture. Plant Pathol 68:997–1006. https://doi.org/10.1111/ppa.13018

Cho JH, Bhutani S, Kim CH, Irwin MR (2021) Anti-inflammatory effects of melatonin: a systematic review and meta-analysis of clinical trials. Brain Behav Immun 93:245–253. https://doi.org/10.1016/j.bbi.2021.01.034

Collum TD, Culver JN (2016) The impact of phytohormones on virus infection and disease. Curr Opin Virol 17:25–31. https://doi.org/10.1016/J.COVIRO.2015.11.003

Corina Vlot A, Dempsey DA, Klessig DF (2009) Salicylic acid, a multifaceted hormone to combat disease. Annu Rev Phytopathol 47:177–206. https://doi.org/10.1146/ANNUREV.PHYTO.050908.135202

Crespo I, San-Miguel B, Fernández A, et al. Melatonin limits the expression of profibrogenic genes and ameliorates the progression of hepatic fibrosis in mice. Elsevier

Dos Santos BM, Pereira PH, Garcia CR (2021) Molecular basis of synchronous replication of malaria parasites in the blood stage. Curr Opin Microbiol 63:210–215. https://doi.org/10.1016/J.MIB.2021.08.002

Hilaire J, Tindale S, Jones G et al (2022) Risk perception associated with an emerging Agri-food risk in Europe: plant viruses in agriculture. Agric Food Secur 11:21. https://doi.org/10.1186/S40066-022-00366-5

Islam W, Naveed H, Zaynab M et al (2019) Plant defense against virus diseases; growth hormones in highlights. Plant Signal Behav 14. https://doi.org/10.1080/15592324.2019.1596719

Jameson PE, Clarke SF (2002) Hormone-virus interactions in plants. CRC Crit Rev Plant Sci 21:205–228. https://doi.org/10.1080/0735-260291044241

Jan JE, Reiter RJ, Wasdell MB, Bax M (2009) The role of the thalamus in sleep, pineal melatonin production, and circadian rhythm sleep disorders. J Pineal Res 46:1–7. https://doi.org/10.1111/J.1600-079X.2008.00628.X

Jones RAC (2021) Global plant virus disease pandemics and epidemics. Plan Theory 10:233. https://doi.org/10.3390/PLANTS10020233

Kanapiya A, Amanbayeva U, Tulegenova Z et al (2024) Recent advances and challenges in plant viral diagnostics. Front Plant Sci 15. https://doi.org/10.3389/FPLS.2024.1451790/FULL

Kaya C, Higgs D, Ashraf M et al (2020) Integrative roles of nitric oxide and hydrogen sulfide in melatonin-induced tolerance of pepper (*Capsicum annuum* L.) plants to iron deficiency and salt stress alone or in combination. Physiol Plant 168:256–277. https://doi.org/10.1111/ppl.12976

Kumar R, Kumar Tiwari R, Kaundal P et al (2019) Potato viruses and their diagnostic techniques: an overview. phytojournal.com 8

Kumar A, Sahu C, Panda PA et al (2020a) Phytic acid content may affect starch digestibility and glycemic index value of rice (*Oryza sativa* L.). J Sci Food Agric 100:1598–1607. https://doi.org/10.1002/jsfa.10168

Kumar R, Kaundal P, Arjunan J et al (2020b) Development of a visual detection method for Potato virus S by reverse transcription loop-mediated isothermal amplification. 3 Biotech 10. https://doi.org/10.1007/S13205-020-02214-4

Kumar R, Kaundal P, Tiwari RK et al (2021a) Rapid and sensitive detection of potato virus X by one-step reverse transcription-recombinase polymerase amplification method in potato leaves and dormant tubers. Mol Cell Probes 58:101743. https://doi.org/10.1016/J.MCP.2021.101743

Kumar R, Tiwari RK, Jeevalatha A et al (2021b) Potato apical leaf curl disease: current status and perspectives on a disease caused by tomato leaf curl New Delhi virus. J Plant Dis Protect 128:897–911. https://doi.org/10.1007/S41348-021-00463-W

Kumar R, Kaundal P, Tiwari RK et al (2022) Establishment of a one-step reverse transcription recombinase polymerase amplification assay for the detection of potato virus S. J Virol Methods 307:114568. https://doi.org/10.1016/J.JVIROMET.2022.114568

Kumar R, Lal MK, Tiwari RK et al (2023) Investigating the interplay between tomato leaf curl New Delhi virus infection, starch metabolism and antioxidant Defence system in potato (*Solanum tuberosum* L.). Antioxidants (Basel) 12. https://doi.org/10.3390/ANTIOX12071447

Lal MK, Kumar A, Kardile HB et al (2020) Biofortification of vegetables. In: Advances in agri-food biotechnology. Springer, pp 105–129

Lal MK, Tiwari RK, Kumar R et al (2021) Effect of potato apical leaf curl disease on glycemic index and resistant starch of potato (*Solanum tuberosum* L.) tubers. Food Chem 359:129939. https://doi.org/10.1016/j.foodchem.2021.129939

Lerner AB, Case JD, Takahashi Y et al (1958) Isolation of melatonin, the pineal gland factor that lightens melanocytes. J Am Chem Soc 80:2587

Liu X, Du C, Yue C et al (2022) Exogenously applied melatonin alleviates the damage in cucumber plants caused by *Aphis goosypi*i through altering the insect behavior and inducing host plant. Pest Manag Sci 79:140–151. https://doi.org/10.1002/ps.7183

Lu R, Liu Z, Shao Y et al (2019) Melatonin is responsible for rice resistance to rice stripe virus infection through a nitric oxide-dependent pathway. Virol J 16:141. https://doi.org/10.1186/s12985-019-1228-3

Luo C, Yang Q, Liu Y et al (2019) The multiple protective roles and molecular mechanisms of melatonin and its precursor N-acetylserotonin in targeting brain injury and liver damage and in maintaining bone health. Free Radic Biol Med 130:215–233. https://doi.org/10.1016/j.freeradbiomed.2018.10.402

Nabavi SM, Nabavi SF, Sureda A et al (2019) Anti-inflammatory effects of melatonin: a mechanistic review. Crit Rev Food Sci Nutr 59:S4–S16

Nguyen D, Rieu I, Mariani C, van Dam NM (2016) How plants handle multiple stresses: hormonal interactions underlying responses to abiotic stress and insect herbivory. Plant Mol Biol 91:727–740. https://doi.org/10.1007/S11103-016-0481-8

Nunes ODS, Pereira RDS (2008) Regression of herpes viral infection symptoms using melatonin and SB-73: comparison with Acyclovir. J Pineal Res 44(4):373–378

Oka K, Kobayashi M, Mitsuhara I, Seo S (2013) Jasmonic acid negatively regulates resistance to tobacco mosaic virus in tobacco. Plant Cell Physiol 54:1999–2010. https://doi.org/10.1093/PCP/PCT137

Pan LL, Miao H, Wang Q et al (2021) Virus-induced phytohormone dynamics and their effects on plant–insect interactions. New Phytol 230:1305–1320. https://doi.org/10.1111/NPH.17261

Parrella G, Elbeaino T, Guy PL (2022) Editorial: emerging and reemerging plant viruses in a context of global change. Front Plant Sci 13. https://doi.org/10.3389/FPLS.2022.1108211/FULL

Ramesh SV, Yogindran S, Gnanasekaran P et al (2021) Virus and viroid-derived small RNAs as modulators of host gene expression: molecular insights into pathogenesis. Front Microbiol 11. https://doi.org/10.3389/FMICB.2020.614231/FULL

Reiter RJ, Sharma R, Simko F et al (2022) Melatonin: highlighting its use as a potential treatment for SARS-CoV-2 infection. Cell Mol Life Sci 79:143. https://doi.org/10.1007/S00018-021-04102-3

Santner A, Estelle M (2009) Recent advances and emerging trends in plant hormone signalling. Nature 459:1071–1078. https://doi.org/10.1038/NATURE08122

Scholthof KBG, Adkins S, Czosnek H et al (2011) Top 10 plant viruses in molecular plant pathology. Mol Plant Pathol 12:938–954. https://doi.org/10.1111/J.1364-3703.2011.00752.X

Sharma S, Prasad M, Wang W-M (2025) Diverse roles of phytohormonal signaling in modulating plant–virus interactions. J Exp Bot 76:1921. https://doi.org/10.1093/JXB/ERAE468

Shetty V, Meyers JI, Zhang Y et al (2022) Impact of disabled circadian clock on yellow fever mosquito *Aedes aegypti* fitness and behaviors. Sci Rep 12(1):1–14. https://doi.org/10.1038/s41598-022-10825-5

Sofy AR, Sofy MR, Hmed AA et al (2021) (2021) Molecular characterization of the alfalfa mosaic virus infecting *Solanum melongena* in Egypt and the control of its deleterious effects with melatonin and salicylic acid. Plants 10:459. https://doi.org/10.3390/PLANTS10030459

Soni A, Kushvaha R, Biol SS-AJBGM (2024) Undefined current strategies for Management of Plant Viruses and Future Perspectives: enhancing crop health, yield and productivity. researchgate.net

Tan DX, Korkmaz A, Reiter RJ, Manchester LC (2014) Ebola virus disease: potential use of melatonin as a treatment. J Pineal Res 57:381–384. https://doi.org/10.1111/JPI.12186

Tan DX, Hardeland R, Back K et al (2016) On the significance of an alternate pathway of melatonin synthesis via 5-methoxytryptamine: comparisons across species. J Pineal Res 61:27–40. https://doi.org/10.1111/jpi.12336

Tatineni S, Hein GL (2023) Plant viruses of agricultural importance: current and future perspectives of virus disease management strategies. Phytopathology 113:117–141. https://doi.org/10.1094/PHYTO-05-22-0167-RVW

Tiwari RK, Lal MK, Naga KC et al (2020) Emerging roles of melatonin in mitigating abiotic and biotic stresses of horticultural crops. Sci Hortic 272:109592. https://doi.org/10.1016/j.scienta.2020.109592

Tiwari RK, Lal MK, Kumar R et al (2021) Mechanistic insights on melatonin-mediated drought stress mitigation in plants. Physiol Plant 172:1212. https://doi.org/10.1111/ppl.13307

Tiwari RK, Lal MK, Kumar R et al (2022) Insight into melatonin-mediated response and signaling in the regulation of plant defense under biotic stress. Plant Mol Biol 109:385. https://doi.org/10.1007/s11103-021-01202-3

Valero N, Mosquera J, Alcocer S et al (2015) Melatonin, minocycline and ascorbic acid reduce oxidative stress and viral titers and increase survival rate in experimental Venezuelan equine encephalitis. Brain Res 1622:368–376

Wei Y, Xie H, Xu L et al (2024) Coat protein of cassava common mosaic virus targets RAV1 and RAV2 transcription factors to subvert immunity in cassava. Plant Physiol 194:1218–1232. https://doi.org/10.1093/PLPHYS/KIAD569

Xie Z, Chen F, Li WA et al (2017) A review of sleep disorders and melatonin. Neurol Res 39:559–565. https://doi.org/10.1080/01616412.2017.1315864

Yang Z, Huang Y, Yang J et al (2020) Jasmonate signaling enhances RNA silencing and antiviral defense in rice. Cell Host Microbe 28:89–103.e8. https://doi.org/10.1016/j.chom.2020.05.001

Yang L-L, Li Q-L, Han X-Y et al (2023) A cysteine-rich secretory protein involves in phytohormone melatonin mediated plant resistance to CGMMV. BMC Plant Biol 23:215. https://doi.org/10.1186/s12870-023-04226-7

Zeng Y, Xiao X, Yang F, Li T, Huang Y, Shi X, Lai C (2024) Progress towards understanding the effects of artificial light on the transmission of vector-borne diseases. Ecotoxicol Environ Saf 283, p 116780

Zhao L, Chen L, Gu P et al (2019) Exogenous application of melatonin improves plant resistance to virus infection. Plant Pathol 68:1287–1295. https://doi.org/10.1111/ppa.13057

Melatonin Involvement in PAMP-Triggered Immunity (PTI) and Effector-Triggered Immunity (ETI) Against Fungal Diseases

9

Jagmohan Singh, Swagata Thakur, H. M. Akshay Kumar, Vanapalli Lohitha Sai Sree, and Karandeep Singh Chahal

Abstract

Earlier melatonin was considered a multifunctional molecule in animal physiology but has lately been recognized as playing an important role in enhancing plant immunity against fungal pathogens. This chapter now deals with the involvement of melatonin in pathogen-associated molecular pattern (PAMP)-triggered immunity (PTI) and effector-triggered immunity (ETI), two major mechanisms that protect plants against fungal diseases. PTI is a non-specific defense that depends upon the recognition of PAMPs via PRRs, whereas ETI is a specific resistance by the plant to the specific pathogen mediated by an R gene-dependent recognition mechanism of the effectors. It plays a key role in both immunity layers through modulating signaling pathways, activation of defense-related genes, and balancing the production of reactive oxygen species (ROS). Melatonin enhances PTI by influencing PAMP perception and subsequent pattern recognition receptor (PRR) activation, whereas in ETI, it is contributed by amplifying the expression of R genes and promoting systemic acquired resistance (SAR). Melatonin plays a crucial role in strengthening plant defenses by working in synergy with both PTI and ETI, making it a key player in integrating

J. Singh (✉)
Department of Plant Pathology, CCS Haryana Agricultural University, Hisar, Haryana, India
e-mail: jsdhillon@hau.ac.in

S. Thakur · V. L. S. Sree
Division of Plant Pathology, ICAR-Indian Agricultural Research Institute, New Delhi, India

H. M. Akshay Kumar
Division of Crop Protection, ICAR-Indian Agricultural Research Institute, Gogamukh, Assam, India

K. S. Chahal
Department of Plant, Soil and Microbial Sciences, Michigan State University, East Lansing, MI, USA

R. K. Tiwari et al. (eds.), *Melatonin: Signal Transduction Mechanisms and Defense Networks in Plants*, https://doi.org/10.1007/978-981-96-8869-2_9

these immune responses for a more robust defense system. In agriculture, melatonin holds great potential—not only as a natural, phytohormone-like modulator that supports sustainable crop treatments but also as a powerful tool for boosting plant immunity and minimizing damage caused by fungal diseases.

Keywords
Effector · Resistance · Defense · Antioxidants · Catalases · Colonization

9.1 Introduction

Fungal pathogens pose a constant threat to agriculture, leading to significant economic losses and threatening food security worldwide. These pathogens attack crops at different growth stages, significantly reducing yields and compromising the quality of agricultural produce. Diseases such as blights, rusts, powdery mildew, and wilts are caused by a diverse range of fungi, many of which have evolved sophisticated strategies to evade plant defenses. In response, plants have developed intricate immune systems that function like a two-tiered defense: pathogen-associated molecular pattern (PAMP)-triggered immunity (PTI) and effector-triggered immunity (ETI). The constant battle between plants and pathogens has profound implications for sustainable agriculture, highlighting the need for a deeper understanding of plant immunity to develop better crop protection strategies (Zipfel 2008).

PTI serves as the frontline defense, triggered when plants perceive specific molecular patterns, or PAMPs, that are pathogen-associated. Fungal chitin, mannans, and flagellin are some examples of PAMPs. These patterns are identified by pattern recognition receptors (PRRs) on the cell surface of plants, which elicit a set of defensive reactions like the activation of reactive oxygen species (ROS), cell wall reinforcement, and defense gene activation (Jones and Dangl 2006; Boller and Felix 2009). However, some pathogens have evolved to evade this defense by releasing effectors, molecules specifically tailored to suppress PTI and enable infection (Cook et al. 2015). Plants respond to such pathogens by initiating a more specialized and robust defense strategy called as ETI. This second-line response involves immune receptors encoded by resistance (R) genes, which recognize these pathogen effectors. ETI tends to trigger a hypersensitive response (HR), where localized cell death prevents the spread of infection (Dodds and Rathjen 2010; Cui et al. 2015). The integration of PTI and ETI often results in enhanced resistance, collectively creating a dynamic immune network to defend plants against various fungal pathogens (Thomma et al. 2011).

The evolutionary arms race between plants and pathogens has driven sophisticated strategies on both sides. While pathogens continuously evolve new effectors to subvert immunity, plants counterattack by evolving novel R genes or reconfiguring PRRs to recognize the novel pathogenic arsenal. This co-evolution is a driving force behind the complexity of plant immunity and necessitates ongoing research to stay ahead of fungal threats.

However, these natural defense mechanisms are not always effective in protecting crops under intensive agricultural systems or against evolving fungal pathogens. This has led researchers to explore ways to boost the immunity of plants through external interventions, such as the application of molecules like melatonin, which has shown promise as a modulator of plant immune responses. By integrating the knowledge of plant immunity and sustainable agricultural practices, researchers aim to mitigate the impact of fungal pathogens and secure global food production.

The losses caused by fungal pathogens are staggering, with certain diseases like rice blast and wheat rust costing billions annually in yield reduction. For instance, Fusarium head blight in wheat not only reduces yield but also contaminates grain with harmful mycotoxins, rendering it unsafe for consumption. Developing resistant crop varieties is essential, but these interventions tend to trail behind fast-developing pathogen strains (Dean et al. 2012). As climate change exacerbates fungal disease prevalence, identifying novel, sustainable solutions has become critical.

9.1.1 Role of Melatonin in Plant Physiology

Melatonin, a tryptophan-derived compound, was first discovered in plants during the 1990s. Although it was originally known to control circadian rhythms in animals, research has since unveiled its diverse biological roles in plants. It influences growth, enhances stress tolerance, and protects against oxidative damage. Melatonin has been detected in various plant tissues, such as leaves, roots, and seeds, highlighting its widespread importance in plant biology (Arnao and Hernández-Ruiz 2015; Reiter et al. 2015).

Antioxidant Properties of Melatonin

One of the standout features of melatonin is its high antioxidant activity. It not only neutralizes toxic free radicals but also enhances the activities of major antioxidant enzymes such as superoxide dismutase (SOD), catalase (CAT), and peroxidase (POD). This property becomes essential under stressful conditions, such as drought or pathogen attack, where excessive ROS can damage cells (Tan et al. 2012; Hardeland 2016).

Melatonin's ability to regulate ROS levels also plays a pivotal role in signaling pathways. Controlled ROS bursts are crucial in triggering defense responses in plants. Melatonin ensures that ROS production is tightly regulated, balancing defense activation and cellular protection (Zhao et al. 2019).

Hormonal Interplay and Growth Regulation

In addition to its antioxidant function, melatonin also regulates plant hormone signaling pathways, which affect processes such as seed germination, root elongation, and senescence in leaves (Wei et al. 2015; Zhang et al. 2017b). For instance, its interaction with auxins promotes root architecture reprogramming under nutrient-stress conditions, while its interplay with abscisic acid (ABA) enhances drought tolerance (Chen et al. 2019).

Melatonin's impact on ethylene biosynthesis also contributes to its role in delaying leaf senescence and enhancing crop longevity and productivity. Moreover, studies have suggested its potential to influence cytokinins and gibberellins, positioning it as a central hub in plant growth regulation.

Melatonin as a Defense-Priming Agent

Recent studies have turned the spotlight on melatonin's role in plant immunity. It has been observed to condition plants to defend themselves, rendering them more resilient to pathogen attacks. For example, melatonin has been reported to induce defense-related genes, trigger MAPK signaling pathways, and elicit the production of antimicrobial secondary metabolites (Park et al. 2013; Wang et al. 2016). These findings position melatonin as a key signaling molecule in plant immune responses.

Particularly in the battle against fungal pathogens, melatonin's role is multifaceted. It reduces oxidative stress caused by fungal infections and strengthens both PTI and ETI mechanisms. Through its ability to modulate ROS generation and increase PRRs and R gene expression, melatonin affords broad-spectrum and specific immunity. For instance, in rice and wheat, melatonin treatments have been shown to lower fungal infection rates and disease severity (Zhang et al. 2017a, b; Li et al. 2017).

Crosstalk with Other Stress Responses

Melatonin's involvement in abiotic stress responses also further establishes its role as a versatile molecule. Its ability to mitigate drought, salinity, and temperature stresses makes it a unifying component of general plant resilience. The overlap between abiotic and biotic stress signaling pathways suggests that enhancing melatonin pathways might provide multifunctional benefits to crops. For instance, melatonin-treated plants are more tolerant to heavy metal stress, further broadening their potential applications (Kolodziejczyk et al. 2015).

Interaction with Secondary Metabolites

Melatonin also affects the biosynthesis of secondary metabolites such as flavonoids and phenolic compounds, which play critical roles in plant defense. Apart from being antimicrobial agents, these metabolites also serve as antioxidants, reinforcing the plant's immune system. Melatonin's ability to enhance the production of such compounds adds another dimension to its defense-priming capabilities.

9.2 PAMP-Triggered Immunity (PTI) and Role of Melatonin in PTI

9.2.1 Understanding PTI

When we encounter an unfamiliar individual, regardless of the extent of our knowledge of the person's intention, we maintain some level of protection on the first handshake. The same holds true for plants; there exists a basal resistance that operates under perceiving extraneous elements. Such elements might originate from a pathogen or a microbe or from within plants themselves via wounds or damage. A conserved molecular pattern of such foreign entities, called pathogen-associated molecular pattern (PAMP), or microbe-associated molecular pattern (MAMP), or damage-associated molecular pattern (DAMP), respectively, arrives and shakes hands with a specific receptor from the host side, referred to as pattern recognition receptor (PRR). This molecular greeting stimulates the plant's first line of defense, the PAMP-triggered immunity or PTI.

PAMP-triggered immunity (PTI) is the first layer in the immune response, which involves the recognition of various conserved surface molecular patterns of the pathogens or microbes or the altered self-molecules resulting from damage by the host receptors (Ge et al. 2022; Ramlal et al. 2023). PTI is governed by several minor genes, providing a broad and general quantitative defense against numerous pathogens (French et al. 2016). Perception of PAMPs by PRRs triggers transcriptional reprogramming that activates primary immune response genes (Li et al. 2016a). PTI does not produce significant symptoms in most cases; however, extensive molecular changes take place behind the scenes (Szatmári et al. 2014). It acts as an early warning system to detect potential pathogens and triggers a chain of signaling events that leads to the production of antimicrobial compounds, preventing pathogen invasion (Segonzac and Zipfel 2011). It causes the reduction of disease but is not strong enough to provide complete resistance (French et al. 2016).

Pattern recognition receptors (PRRs) are a key part of the innate immune response and are involved in the identification of non-self-molecules. They are extremely sensitive, which is very much essential for the plant's survival. PRRs can often detect even nanomolar concentrations of foreign molecules (Ranf et al. 2016). There are four principal structural subfamilies of PRRs, namely, receptor-like kinases (RLKs), receptor-like proteins (RLPs), extracellular soluble receptors (ESRs), and intracellular soluble receptors (ISRs). RLKs are composed of extracellular leucine-rich repeats acting as ligand binding domain, a transmembrane domain, and a serine/threonine kinase domain. In case of RLPs, they do not possess the kinase domain and, thus, need an adapter molecule for signal transduction. ESRs and ISRs do not possess any transmembrane domain as they are not membrane-bound, and they are found in apoplast or cytoplasm according to their suggestive names (Schwessinger and Ronald 2012). Thus, PRRs constitute the frontline of host defense working in every cell coming in contact with the pathogen. Some examples of PRRs include FLS2, CERK1, CEBiP, LYM1, LYM3, LYP4, and LYP6 (Yamaguchi et al. 2013).

9.2.2 Melatonin's Role in PTI

Several responses start as a part of PTI after the recognition of PAMP by PRR. Based on the immediateness, they can be divided into three types, namely, early responses, i.e., responses in seconds to minutes (e.g., ion fluxes and oxidative burst), intermediate responses, i.e., responses in minutes to hours (e.g., MAPK/CDPK activation, ethylene production, stomatal closure, and transcriptional reprogramming), and late responses, that is, responses in hours to days (e.g., salicylic acid [SA] accumulation and callose deposition) (Zipfel and Robatzek 2010). Melatonin is found to play an important role in almost every step of the activation of PTI, starting from PAMP recognition to signal transmission and defense gene activation.

(a) *How Melatonin Influences PAMP Perception and Signaling*

PTI is strong but not so strong that it cannot be succumbed by the pathogen. And here comes melatonin, which plays a vital role in strengthening the effects of PTI. PAMP perception can lead to a sudden rise of oxidative stress in plants in minutes or seconds as an early response, which may damage plant cells. Production of reactive oxygen species (ROS) is a part of programmed cell death (PCD), but without proper regulation, it can potentially damage plant cells if the oxidative burst becomes too intense. Melatonin functions as an antioxidant, reducing oxidative damage by scavenging free radicals and enhancing the activity of antioxidant enzymes like superoxide dismutase (SOD), ascorbate peroxidase (APX), peroxidase (POD), and catalase (CAT). This helps plants manage oxidative stress induced by pathogen attacks, maintaining cellular integrity during immune responses (Sahu et al. 2022; Sun et al. 2019).

The signaling pathway of melatonin in plants has also been explored gradually. In the year 2016, Lee and Back performed two different research where they proved treatment with 1 μM melatonin activated MAPK3 and MAPK6 via upstream MAPK kinases (MKKs), including MKK4, MKK5, MKK7, and MKK9 (Lee and Back 2016a). The second research found that activation of MAPK3/6 is regulated by an upstream activation of MAPKKK3 and oxidative signal-inducible 1 (OXI1) (Lee and Back 2016b). However, this activation is not associated with G-protein signaling, as the activation levels remain unchanged in the G-protein β knockout mutant (*agb1*) of *Arabidopsis* (Lee and Back 2016a). But in 2018, the first phytomelatonin receptor was identified in *Arabidopsis thaliana*, which was candidate G-protein coupled receptor *2* (*CAND2/PMTR1*), a membrane protein that binds readily with melatonin (Wei et al. 2015). Thus, MAPKKK activation is mediated via melatonin receptors or receptor kinases that recognize pathogen-derived molecular patterns (PAMPs), and followed by channelized downstream activation of MKKs and MAPKs, resulting in the activation of PTI (Zhang et al. 2022).

(b) *Activation of Defense-Related Genes by Melatonin*

Melatonin is found to activate some exclusive genes at specific concentrations; not all genes regulated by low concentrations of melatonin are affected by high concentrations. Transcriptomics studies have proved that many genes involved in plant stress defense were impacted by melatonin. Numerous stress receptors, kinases, and calcium signals linked to stress were upregulated. Many transcription factors also were engaged in plant stress defense. Furthermore, the majority of genes involved in the ABA, ET, SA, and jasmonic acid (JA) pathways were upregulated, whereas the majority of genes linked to cell wall production and modifications, auxin responses and signaling, and peroxidases were down-regulated (Weeda et al. 2014). The transcriptome results were further validated by several research. Melatonin treatment was proved to induce activities of SOD, POD, CAT, APX, phenylalanine deaminase (PAL), and polyphenol oxidase (PPO) in cucumbers against downy mildew. Expression of the *PR1* gene was also found to be upregulated (Sun et al. 2019). Melatonin treatment increased methyl jasmonate (MeJA) content with subsequent upregulation of *SlLoxD, SlAOC* (two genes which are related to the biosynthesis of JAs), and *SlPI II* (proteinase inhibitor II, mainly regulated by JA and MeJA, is a JA-dependent gene); and downregulation of *SlMYC2* and *SlJAZ1* (negative regulators of JA biosynthesis) in tomatoes against *Botrytis cinerea* (Liu et al. 2019). Melatonin increases saponin and flavonoid biosynthesis (Gao et al. 2020; Yang et al. 2021).

(c) *Examples of Fungal Pathogens Where Melatonin-Mediated PTI Has Been Observed*

The melatonin treatment in cucurbits against the foliar pathogen *Podosphaera xanthii* (powdery mildew) and the soil-borne oomycete *Phytophthora capsici* was found to upregulate four genes, namely, Cla021170, Cla002982, Cla012528, and Cla001623 which can directly recognize PAMPs. High-level expression of receptor-like kinases (Cla001254 and Cla015613) was detected, which are orthologs of genes known to be involved in PTI responses in *Arabidopsis* and also function in response to oxidative stress initiated by pathogens (Mandal et al. 2018). Melatonin may also activate callose deposition and other PTI-associated defense mechanisms to promote resistance against *Botrytis cinerea* in tomato (Xia et al. 2024).

9.2.3 Molecular Mechanisms of Melatonin in PTI

Melatonin regulates defense response by interacting with different phytohormones. Melatonin triggers the biosynthesis of phytoalexins and the expression of PR proteins. Melatonin promotes ROS homeostasis by activating the antioxidants (Gupta 2023). The molecular mechanisms involved in the process are discussed below.

(a) *Melatonin-salicylic acid (MT-SA) interaction*: The external application of melatonin activates the *isochorismate synthase (ICS)* gene via mitogen-activated protein kinases (Mkk4/5/7/9). ICS catalyzes the conversion of chorismate to isochorismate, which is then transported to the cytoplasm by enhanced disease susceptibility 5 (EDS5) from the chloroplast. In the cytosol, AVRPPHB SUSCEPTIBLE 3 (PBS3) helps conjugate L-glutamate and isochorismate to produce isochorismate-9-glutamate, which spontaneously decays to form SA. Additionally, ENHANCED PSEUDOMONAS SUSCEPTIBILITY 1 (EPS1) may break down N-pyruvoyl-L-glutamate into SA. WRKY transcription factors interact with NONEXPRESSER OF PR GENES1 (NPR1), recruiting cyclin-dependent kinase 8 (CDK8) to the W-box sequence in the promoter region of NPR1. The MT-induced overexpression of ICS and pathogen-related defense signaling promotes SA accumulation, which triggers redox reactions that reduce NPR1 oligomers to monomers. These monomers move to the nucleus and form an association with transcription factors (TGA), CDK8, EDS1, and SA, leading to the transcription of PR genes. The molecular mechanism of MT-SA interaction has been depicted in Fig. 9.1.

(b) *Melatonin-jasmonic acid (MT-JA) interaction*: Exo- and endogenous melatonin increase JA levels by upregulating lipoxygenase (LOX) genes. PAMPs are detected by PRRs, activating the MEK1/2 cascades. Melatonin also activates MKK4/5/7/9, with MKK4 further activating MPK3/MPK6 and MKS1 (a substrate of MPK4). Camalexin, a phytoalexin, is synthesized by the WRKY33 transcription factor, which MPK3/MPK6 phosphorylates to boost its transactivation activity. The WRKY33-MKS1 complex triggers the transcription of phytoalexin deficient 3 (PAD3), promoting camalexin synthesis. In stimulated cells, SCFcol1 forms through the association of Col1 with S-phase kinase-associated protein 1 (SKP1), Cullin, and JA-Ile protein, leading to the ubiquitination and 26S proteasome-mediated degradation of JAZ repressor proteins. This degradation releases Myelocytomatosis transcription factor MYC2, activating downstream JA responses. MEDIATOR25 (MED25) enhances the transcription of the wound-responsive gene vegetative storage protein 1 (VSP1) by interacting with MYC2.

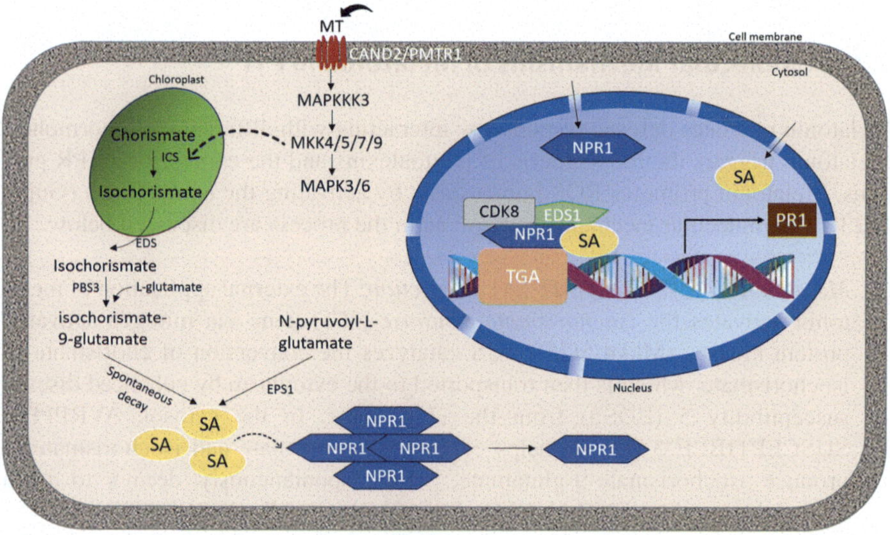

Fig. 9.1 Molecular mechanism of melatonin (MT)-salicylic acid (SA) interaction

(c) *MT-ethylene mutual stimulation*: Melatonin also boosts ethylene production and upregulates ETHYLENE INSENSITIVE LIKE 5 (EIL5). EIL5 binds with HEAT STRESS TRANSCRIPTION FACTOR 20 (HSF20), activating the N-acetylserotonin O-methyltransferase 2 (ASMT2) gene, which produces more melatonin. Under normal conditions, low ethylene levels activate ethylene receptor 1 (ERT1), which binds with constitutive triple response 1 (CTR1) to suppress downstream signaling. However, elevated ethylene levels inactivate CTR1 by binding to its receptors, thereby activating ethylene insensitive 2 (EIN2) and ethylene insensitive 3 (EIN3) transcription factors. EIN3 regulates the expression of ERF1 and ERF94, which interact with the GCC box to activate PDF1.2 and PR3 genes, respectively. Increased EIL5 levels enhance the binding of Hsf20 and PR3, further promoting PR3 synthesis and triggering pathogen resistance genes in plants (Khan et al. 2023).

Interaction with Reactive Oxygen Species (ROS) Production and Defense Enzymes

It has been proposed that there are two different pathways via which the endogenous concentration of melatonin might control ROS levels (Ahammed et al. 2024). Firstly, melatonin can interact directly with ROS, serving as a scavenger and promoting detoxification (Arnao and Hernández-Ruiz 2019a, b; Kaur et al. 2015). Secondly, melatonin has the ability to enhance the activity of antioxidant enzymes and the expression of key genes involved in the biosynthesis of these enzymes (Khan et al. 2020). The most distinctive aspect of melatonin is that it is an excellent antioxidant, having several times the antioxidant capacity of common antioxidants such as vitamin E, ascorbic acid (AsA), and others (Arnao and Hernández-Ruiz 2020). Furthermore, it has been demonstrated that melatonin therapy promotes the formation of several non-enzymatic antioxidant compounds, including carotenoids, flavonoids, phenolic compounds, glutathione (GSH), and AsA (Arnao et al. 2022; Sun et al. 2023; Wang et al. 2022). These molecules are crucial for ROS detoxification. In addition, ROS may enhance the transcription of melatonin biosynthetic genes, which increases the endogenous content of melatonin in plants (Arnao and Hernández-Ruiz 2019a, b). This feedback regulation mechanism directly accounts for the enhancement of plant antioxidant response.

9.3 Effector-Triggered Immunity (ETI) and Role of Melatonin in ETI

Effector-triggered immunity is an important aspect of plant defense mechanisms against pathogens. ETI differs from the first line of defense, which is called pathogen-triggered immunity, elicited by the perception of conserved pathogen-associated molecular patterns by plant pattern recognition receptors (Vannier et al. 2019). In ETI, plants have evolved specialized intracellular immune receptors that can sense pathogen-derived effector molecules directly or indirectly, leading to a robust defense response.

One of the key players in ETI is the resistance (R) genes, which encode proteins that are capable of recognizing specific pathogen effectors. Upon recognition of the effector, the R proteins initiate intricate cascades of signals, such as the activation of phytohormones such as salicylic acid, jasmonic acid, and ethylene, which orchestrate various defense responses (Flors et al. 2009). These responses can include the hypersensitive reaction, which involves the localized programmed cell death at the site of infection, the production of antimicrobial compounds, and the induction of systemic acquired resistance throughout the plant (Aranega-Bou et al. 2014).

The interaction between plant immune receptors and pathogen effectors is a highly dynamic process that has co-evolved over time. Pathogens continually evolve novel effectors to suppress plant immunity, while plants counter by acquiring new resistance genes to recognize these effectors. Understanding the molecular mechanisms of ETI is crucial for developing strategies to enhance crop disease resistance.

9.3.1 Role of Melatonin in Effector-Triggered Immunity (ETI)

Melatonin's role in plant immunity has been a focus area of growing interest over the past few years. Emerging evidence suggests that melatonin plays a pivotal role as a key effector in the plant's effector-triggered immunity response, which is triggered upon perception of pathogen-derived effector molecules by plant resistance (R) proteins (Wiesel et al. 2014). This perception triggers a series of signaling events, which result in the activation of defense responses, including the production of antimicrobial compounds, the hypersensitive response, and the induction of systemic acquired resistance (García-Olmedo et al. 1998; Martin-Rivilla et al. 2020).

Recent studies have demonstrated that melatonin plays a significant role in enhancing the effectiveness of R gene-mediated resistance against fungal pathogens. In addition, melatonin can trigger the production of secondary metabolites like alkaloids, flavonoids, and terpenes, which possess potent antifungal properties. These secondary metabolites can directly suppress the growth and development of fungal pathogens, providing an additional layer of defense for the plant (Tiwari et al. 2021).

Mandal et al. (2018) investigated melatonin's beneficial role in enhancing plant resistance through upregulation of genes related to PAMP-triggered immunity and ETI (effector-triggered immunity) response genes strengthening plant resistance by upregulating genes associated with PTI (PAMP-triggered immunity) and ETI (effector-triggered immunity) responses. Through transcriptomic analysis, they identified 27 genes that were upregulated, contributing to either baseline defense mechanisms or initial priming for melatonin-induced resistance. Additionally, overexpression of the melatonin biosynthetic gene resulted in increased expression of resistance proteins from the TIR-NBS-LRR and CC-NBS-LRR classes. Notably, genes involved included WRKY transcription factors, LRR-receptor-like kinases, a subtilisin-like serine protease, glutathione S-transferase, and serine/threonine kinase.

Melatonin also induces transcription factors (CBF/DREB1s), defense genes like pathogenesis-related 1 (PR1), and antimicrobial peptides like plant defensin 1.2 (PDF1.2) all of which play a regulatory function to protect against fungal infections (Shi et al. 2015b).

9.3.2 Role of Melatonin in Inducing Systemic Resistance

Systemic acquired resistance is an important plant defense system conferring broad-spectrum, long-term protection against a broad array of pathogens. Emerging evidence suggests that melatonin, an endogenously synthesized indolamine, plays a significant role in the activation and regulation of plant defense response (Fig. 9.2). Melatonin has been found to exert varied protective effects, such as strong antioxidant activity through which it can neutralize free radicals, activate antioxidant enzymes, and protect cellular components from oxidative damage (Kopustinskiene and Bernatoniene 2021; Minich et al. 2022).

Melatonin has been found to regulate the levels and signaling of several phytohormones, thereby affecting the defense reactions of the plant. Melatonin can enhance the production of jasmonic acid and salicylic acid, which play a vital role in triggering defense-related gene expression and the hypersensitive response (Altaf et al. 2023). Additionally, melatonin can also interfere with abscisic acid, a phytohormone responsible for abiotic stress responses, to fine-tune the plant's overall stress resilience. The induction of systemic acquired resistance, a key defense mechanism in plants, has been closely correlated with the synthesis of salicylic acid and the subsequent expression of pathogenesis-related proteins. Melatonin has also been found to induce the biosynthesis of salicylic acid and thus activate the plant's defense arsenal against invading pathogens (Shah 2003). In addition to its influence on salicylic acid, melatonin also exhibits a dynamic interplay with other phytohormones, including jasmonic acid and ethylene, which are crucial players in the plant's defense signaling cascade (Hernández-Ruiz and Arnao 2018).

Melatonin acts upstream in the defense gene signaling pathway by enhancing the biosynthesis of salicylic acid (SA), jasmonic acid (JA), nitric oxide (NO), and ethylene (Zhu and Lee 2015). An early oxidative burst in the plant-pathogen interaction appears to increase the endogenous melatonin levels (Arnao and Hernández-Ruiz 2015). Melatonin-induced NO accumulation in mitochondria boosts SA concentration during infection, the enzyme isochorismate synthase-1, crucial for SA synthesis, is upregulated, increasing SA levels and triggering a pathogen-induced response. Additionally, melatonin raises methyl jasmonate (MeJA) levels and activates JA biosynthesis genes AOC and LoxD, while downregulating SlMYC2 and JASMONATE ZIM DOMAIN 1 (a negative regulator in JA signaling), which promote the expression of pathogenesis-related genes. Although the mechanism by which melatonin influences ethylene is not fully understood, melatonin treatment does increase ethylene levels in infected plants, thereby promoting the expression of defense-related genes, facilitated by ACC synthase (ACS6). This further supports pathogen resistance by inducing defense proteins such as NPR1, plant defensin

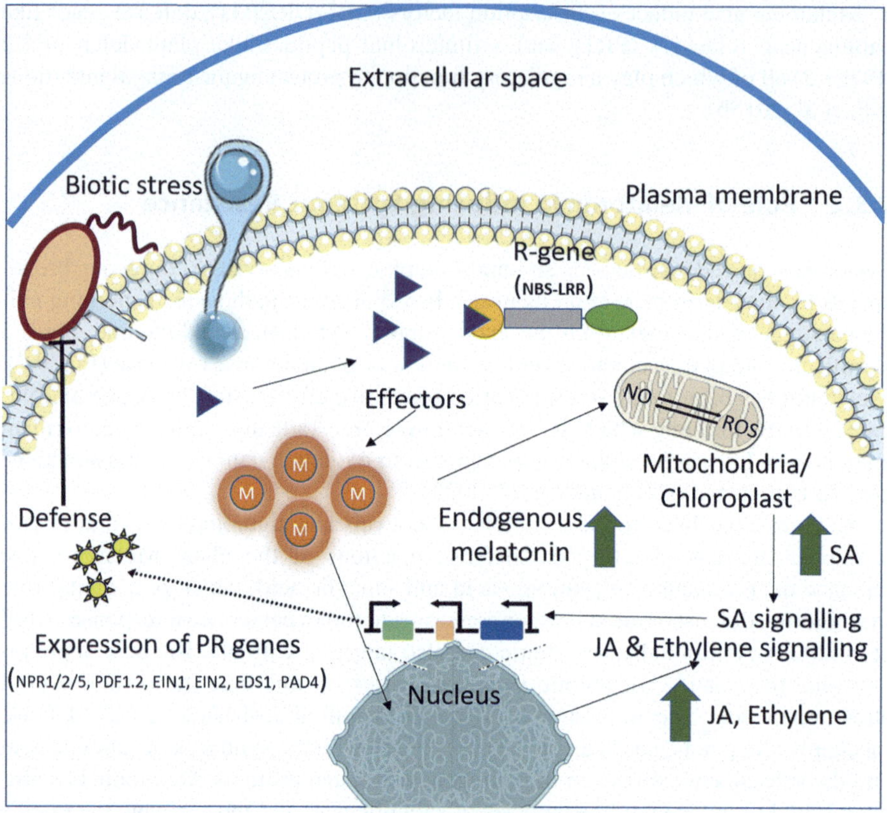

Fig. 9.2 Melatonin-mediated ETI and phyto-hormone signaling in plants under biotic stress. Icons obtained from Biocons (https://bioicons.com/) and Bioart (https://bioart.niaid.nih.gov/)

(PDF), ethylene insensitive (EIN), enhanced disease susceptibility 1 (EDS1), and phytoalexin deficient 4 (PAD4), which are essential in SA- and ethylene-mediated defense signaling. Furthermore, JA and SA independently activate PR genes to reinforce plant defense (Arnao and Hernández-Ruiz 2018) (Tiwari et al. 2021).

Understanding the intricate interplay between melatonin, phytohormones, and ETI is crucial for developing effective strategies to enhance plant disease resistance (Flors et al. 2009; Zhang et al. 2017a, b, 2019; Samota et al. 2017).

9.3.3 The Influence of Melatonin on R Gene-Mediated and Systemic Resistance Against Fungal Pathogens

Studies have highlighted the potential of melatonin in enhancing the expression of R genes. For example, melatonin has been reported to increase the resistance of tomato, against the gray mold pathogen *Botrytis cinerea* (Moo-Koh et al. 2022).

Several plant species show increased resistance to various pathogens through mechanisms that include hormonal regulation, oxidative responses, and induction of defense-related genes and enzymes. For instance, in *Malus prunifolia*, resistance to *Diplocarpon mali* is reinforced by maintaining intracellular H_2O_2 levels and initiating defense enzymes, leading to decreased lesion count (Yin et al. 2013). Similarly, *Musa acuminata* displays increased resistance to *Fusarium oxysporum* by reprogramming hormone-related defense mechanisms through regulation of the MaHSP90s gene (Wei et al. 2017).

In *Fragaria ananassa*, infections by *Botrytis cinerea* and *Rhizopus stolonifer* lead to reduced fruit decay, which is linked to the maintenance of H_2O_2 levels and the enhancement of antioxidant enzyme activity (Aghdam and Fard 2017). Defense mechanisms are also reported in *Citrullus lanatus* against *Podosphaera xanthii* and *Phytophthora capsici*, wherein enhanced resistance is achieved by activating genes associated with pathogen-triggered immunity (PTI) and effector-triggered immunity (ETI) (Mandal et al. 2018). Furthermore, resistance in *Solanum lycopersicum* to *Botrytis cinerea* is enhanced by modulating the level of jasmonic acid (JA) and by inducing the activity of reactive oxygen species (ROS), salicylic acid (SA), chitinase, and β-1,3-glucanase activity (Liu et al. 2019; Li et al. 2019a, b; Xia et al. 2024).

Pathogen resistance in other crops is facilitated through similar mechanisms. For instance, *Cucumis sativus* exhibits decreased disease severity from *Pseudoperonospora cubensis* and *Fusarium oxysporum* due to the activation of antioxidant genes and alleviation of H_2O_2 (Sun et al. 2019; Ahammed et al. 2020). *Capsicum annuum* and *Arabidopsis thaliana* demonstrate enhanced immunity to *Colletotrichum gloeosporioides* through the induction of pathogenesis-related genes, such as chitinase (CaChiIII2) (Ali et al. 2020). Further, in *Brassica rapa*, the resistance against *Sclerotinia sclerotiorum* involves ATP-dependent activation of antioxidants and glucosinolate production to prevent infection (Teng et al. 2021). The above examples highlight the intricate crosstalk among oxidative processes, hormone signaling, and gene activation in plants, which together provide robust protection against diverse pathogens across various crop species.

9.4 Crosstalk Between PTI and ETI Influenced by Melatonin

Melatonin has emerged as a critical molecule influencing the crosstalk between PTI and ETI. It is also reported to regulate ROS levels, a shared signaling component of both PTI and ETI, thus coordinating the two immune layers. Research studies indicate that melatonin treatment in plants enhances the expression of defense-related genes, such as those encoding PRRs for PTI and nucleotide-binding leucine-rich repeat (NLR) proteins for ETI (Tsuda and Katagiri 2010). For instance, Zhang et al. (2015) demonstrated that Arabidopsis plants treated with melatonin showed increased activation of MAPK signaling cascades, the central hub in PTI and ETI responses. This crosstalk ensures an integrated and efficient immune response, reducing redundancy while enhancing pathogen resistance (Shi et al. 2015a, b).

Melatonin acts as a signaling molecule that coordinates PTI and ETI to achieve maximum defense against pathogens. It orchestrates such integration by regulating the production of ROS and nitric oxide (NO), essential signaling molecules involved in plant immunity. Another important feature of melatonin is its regulation of phytohormonal pathways, such as salicylic acid (SA) and jasmonic acid (JA). SA is predominantly associated with systemic acquired resistance (SAR) and ETI, while JA plays a vital role in localized defense mechanisms and PTI. Melatonin ensures a balanced activation of these pathways, preventing trade-offs that could compromise overall plant fitness (Arnao and Hernández-Ruiz 2015; Tan et al. 2012).

In addition, melatonin can interact with transcription factors that govern immune responses, including WRKY and MYB proteins, to elevate defense gene expression, participating in PTI and ETI. For instance, Cao et al. (2018) established that treatment with melatonin triggered WRKY70 expression, a transcription factor that has been implicated in the mediation of SA signaling, thus augmenting both basal and effector-triggered immunity.

9.5 Practical Applications of Melatonin in Crop Protection

9.5.1 Melatonin Use in Enhancing Crop Immunity

The agricultural applications of melatonin are growing rapidly as researchers identify its multifaceted functions in enhancing plant health and immunity. In particular, melatonin's ability to modulate immune responses and enhance resistance against fungal pathogens has drawn significant attention.

(a) *Foliar Application*: Foliar sprays are the most direct and widely used method of applying melatonin in agriculture. This method of application ensures that melatonin reaches the plant tissues that are most susceptible to fungal infection, including leaves and stems. Research demonstrates that melatonin sprays enhance the activity of defense-related enzymes like peroxidase and catalase, inhibiting fungal growth on plant surfaces (Li et al. 2019a, b).
(b) *Soil Amendments*: Soil-based melatonin applications target on enhancing root health and resistance to soil-borne fungal pathogens such as *Fusarium* spp. and *Phytophthora* spp. Studies indicate that melatonin enhances root development and microbial resistance by modifying the biochemistry of the rhizosphere, boosting beneficial microbial populations (Zhang et al. 2021).
(c) *Seed Priming*: Melatonin-treated seeds exhibit improved germination rates and resistance to fungal pathogens. Pre-soaking of seeds in melatonin solutions triggers a systemic immune response, equipping seedlings to combat biotic stresses more efficiently. This technique has particularly proven effective in wheat and maize, where early fungal infections are common (Hernández-Ruiz and Arnao 2018).

9.5.2 Case Studies and Experimental Evidence

Melatonin is gaining attention as a powerful natural compound that enhances plant immunity against fungal pathogens. Research has shown its effectiveness in protecting various crops from devastating diseases. Rice blast, caused by *Magnaporthe oryzae*, poses a serious threat to global rice production. Field trials have demonstrated that applying melatonin reduces lesion formation by 30–40% while boosting the activity of key defense-related genes like *PAL* (phenylalanine ammonia-lyase) and *PR1* (pathogenesis-related protein 1) (Park et al. 2020). Similarly, in tomato plants infected with *Botrytis cinerea* (gray mold), melatonin treatment reduced fungal colonization and stimulated the production of antioxidant enzymes and secondary metabolites essential for immune responses (Wei et al. 2022). Wheat crops treated with melatonin during infection by *Fusarium graminearum* exhibited lower fungal biomass and improved grain quality, suggesting a role in systemic acquired resistance (SAR) (Liu et al. 2021). Melatonin application showed increased resistance to *Pseudomonas syringae*in *Arabidopsis thaliana* by activating genes involved in PTI (e.g., FLS2 encoding a PRR) and ETI (e.g., RPS2 encoding an NLR protein). This dual activation highlights melatonin's role in synchronizing PTI and ETI responses (Zhang et al. 2015; Shi et al. 2015a, b).

The use of melatonin in agriculture has significant advantages:

Reduced Fungicide Dependence: Melatonin enhances plant immunity, reducing the need for chemical fungicides. This approach lowers production costs and minimizes environmental pollution (Wang et al. 2022).

Improved Crop Yield and Quality: By protecting plants from fungal infections, melatonin ensures healthier crops with higher yields, addressing global food security challenges (Hernández-Ruiz and Arnao 2020).

9.6 Future Prospects for Melatonin in Sustainable Agriculture

Melatonin's unique ability to connect basal and specific defense mechanisms makes it a promising tool for sustainable agriculture. By unraveling its molecular role in both pattern-triggered immunity (PTI) and effector-triggered immunity (ETI), researchers can harness its full potential to combat fungal diseases more effectively. As we uncover more about melatonin's interactions with other plant metabolites and signaling pathways, its benefits may extend beyond fungal pathogens to a wider range of agricultural challenges. Furthermore, the future of melatonin in agriculture lies in developing innovative strategies to maximize its effectiveness:

(a) *Genetic Engineering*: Advances in genetic engineering could enable crops to produce higher levels of melatonin naturally, making them inherently resistant to fungal pathogens. Overexpression of genes involved in melatonin biosynthesis has shown promising results in model plants like *Arabidopsis thaliana* (Zhang et al. 2021).

(b) *Bio-fertilizers*: Incorporating melatonin into bio-fertilizers could offer dual benefits: enriching soil fertility while strengthening plant immunity. Such products cater to the growing demand for sustainable and organic farming solutions (Liu et al. 2021).

(c) *Controlled-Release Formulations*: Developing slow-release melatonin formulations ensures its sustained availability in plant tissues, providing long-term protection against fungal pathogens (Wei et al. 2022).

(d) *Development of Melatonin Analogs*: Synthesizing analogs with improved stability and efficacy could overcome current limitations (Wei et al. 2022).

(e) *Molecular Mechanism Studies*: Investigating how melatonin interacts with other plant signaling molecules will help unlock its full role in plant immunity, paving the way for more targeted applications (Zhang et al. 2021).

(f) *Integration with Other Stress Management Strategies*: Combining melatonin treatments with traditional breeding or CRISPR-based gene editing could lead to crops with enhanced resistance to both biotic and abiotic stresses (Liu et al. 2021).

Challenges and Limitations

Despite its potential, there are challenges associated with melatonin's widespread adoption in agriculture, and they must be addressed before it can be widely adopted.

1. *Cost of Production*: The synthesis of melatonin remains more expensive compared to conventional fungicides, posing a barrier to large-scale application (Park et al. 2020).
2. *Stability Issues*: Melatonin is prone to degradation under UV light and extreme temperatures, necessitating the development of more stable formulations (Tan et al. 2019).
3. *Crop-Specific Variability*: The effectiveness of melatonin differs across different crops and pathogens, highlighting the need for crop-specific research and optimization (Zhao et al. 2020).

9.7 Conclusion

Melatonin is emerging as a powerful tool for enhancing crop immunity and managing fungal diseases. Its diverse applications, ranging from seed treatments to foliar sprays, have demonstrated significant benefits in enhancing plant resistance and ensuring better crop yields. However, to fully unlock its potential, challenges related to cost, stability, and variability across crops must be addressed. Future research and technological advancements are expected to unlock new possibilities for using melatonin as a cornerstone of sustainable and eco-friendly agriculture.

References

Aghdam MS, Fard JR (2017) Melatonin treatment attenuates postharvest decay and maintains nutritional quality of strawberry fruits (Fragaria× anannasa cv. Selva) by enhancing GABA shunt activity. Food Chem 221:1650–1657

Ahammed GJ, Mao Q, Yan Y, Wu M, Wang Y, Ren J, Guo P, Liu A, Chen S (2020) Role of melatonin in arbuscular mycorrhizal fungi-induced resistance to Fusarium wilt in cucumber. Phytopathology 110(5):999–1009

Ahammed GJ, Li Z, Chen J, Dong Y, Qu K, Guo T, Wang F, Liu A, Chen S, Li X (2024) Reactive oxygen species signaling in melatonin-mediated plant stress response. Plant Physiol Biochem 207:108398. https://doi.org/10.1016/J.PLAPHY.2024.108398

Ali M, Tumbeh Lamin-Samu A, Muhammad I, Farghal M, Khattak AM, Jan I, ul Haq S, Khan A, Gong ZH, Lu G (2020) Melatonin mitigates the infection of *Colletotrichum gloeosporioides* via modulation of the chitinase gene and antioxidant activity in *Capsicum annuum* L. Antioxidants 10(1):7

Altaf MA, Sharma N, Singh J, Samota MK, Sankhyan P, Singh B, Kumar A, Naz S, Lal MK, Tiwari RK, Kumar R (2023) Mechanistic insights on melatonin-mediated plant growth regulation and hormonal cross-talk process in solanaceous vegetables. Sci Hortic 308:111570

Aranega-Bou P, de la O Leyva M, Finiti I, García-Agustín P, González-Bosch C (2014) Priming of plant resistance by natural compounds. Hexanoic acid as a model. Front Plant Sci 5:488

Arnao MB, Hernández-Ruiz J (2015) Functions of melatonin in plants: a review. J Pineal Res 59(2):133–150

Arnao MB, Hernández-Ruiz J (2018) Melatonin and its relationship to plant hormones. Ann Bot 121(2):195–207

Arnao MB, Hernández-Ruiz J (2019a) Melatonin and its relationship to plant hormones. Front Plant Sci 10:134. https://doi.org/10.3389/fpls.2019.00134

Arnao MB, Hernández-Ruiz J (2019b) Melatonin: a new plant hormone and/or a plant master regulator? Trends Plant Sci 24(1):38–48. https://doi.org/10.1016/J.TPLANTS.2018.10.010/ASSET/4026C84B-136E-496A-BBD7-D68E426426A9/MAIN.ASSETS/GR1.JPG

Arnao MB, Hernández-Ruiz J (2020) Is Phytomelatonin a new plant hormone? Agronomy 10(1):95. https://doi.org/10.3390/AGRONOMY10010095

Arnao MB, Cano A, Hernández-Ruiz J (2022) Phytomelatonin: an unexpected molecule with amazing performances in plants. J Exp Bot 73(17):5779–5800. https://doi.org/10.1093/jxb/erac009

Boller T, Felix G (2009) A renaissance of elicitors: perception of microbe-associated molecular patterns and danger signals by pattern-recognition receptors. Annu Rev Plant Biol 60:379–406. https://doi.org/10.1146/annurev.arplant.57.032905.105346

Cao J, Wang J, Wang M, Wu Y (2018) Melatonin improves cold tolerance in banana seedlings by regulating the antioxidant system and proline metabolism. Plant Physiol Biochem 130:529–539. https://doi.org/10.1016/j.plaphy.2018.07.039

Chen L, Liu Y, Xu C, Bai Q, Liu Y (2019) Roles of melatonin in abiotic stress resistance in plants. J Exp Bot 70(3):647–665. https://doi.org/10.1093/jxb/ery430

Cook DE, Mesarich CH, Thomma BP (2015) Understanding plant immunity against necrotrophic pathogens. Plant J 83(4):662–674. https://doi.org/10.1111/tpj.12945

Cui H, Tsuda K, Parker JE (2015) Effector-triggered immunity: from pathogen perception to robust defense. Annu Rev Plant Biol 66:487–511. https://doi.org/10.1146/annurev-arplant-050213-040012

Dean R, Van Kan JA, Pretorius ZA, Hammond-Kosack KE, Di Pietro A, Spanu PD, Rudd JJ, Dickman M, Kahmann R, Ellis J, Foster GD (2012) The top 10 fungal pathogens in molecular plant pathology. Mol Plant Pathol 13(4):414–430

Dodds PN, Rathjen JP (2010) Plant immunity: towards an integrated view of plant-pathogen interactions. Nat Rev Genet 11(8):539–551. https://doi.org/10.1038/nrg2812

Flors V, Ton J, Mauch-Mani B (2009) Role of abscisic acid in disease resistance. In: Signal crosstalk in plant stress responses. Wiley-Blackwell, pp 1–22

French E, Kim BS, Iyer-Pascuzzi AS (2016) Mechanisms of quantitative disease resistance in plants. Semin Cell Dev Biol 56:201–208. https://doi.org/10.1016/J.SEMCDB.2016.05.015

Gao S, Ma W, Lyu X, Cao X, Yao Y (2020) Melatonin may increase disease resistance and flavonoid biosynthesis through effects on DNA methylation and gene expression in grape berries. BMC Plant Biol 20(1):1–15. https://doi.org/10.1186/S12870-020-02445-W/FIGURES/7

García-Olmedo F, Molina A, Alamillo JM, Rodríguez-Palenzuéla P (1998) Plant defense peptides. Pept Sci 47(6):479–491

Ge D, Yeo IC, Shan L (2022) Knowing me, knowing you: self and non-self recognition in plant immunity. Essays Biochem 66(5):447. https://doi.org/10.1042/EBC20210095

Gupta R (2023) Melatonin: A promising candidate for maintaining food security under the threat of phytopathogens. Plant Physiol Biochem 198:107691. https://doi.org/10.1016/J.PLAPHY.2023.107691

Hardeland R (2016) Melatonin in plants and other phototrophs: advances and gaps concerning the diversity of functions. J Exp Bot 67(13):3877–3891. https://doi.org/10.1093/jxb/erw139

Hernández-Ruiz J, Arnao MB (2018) Relationship of melatonin and salicylic acid in biotic/abiotic plant stress responses. Agronomy 8(4):33

Hernández-Ruiz J, Arnao MB (2020) Melatonin: a growth-stimulating compound for horticultural crops. Horticult Res 7:57. https://doi.org/10.1038/s41438-020-0285-6

Jones JDG, Dangl JL (2006) The plant immune system. Nature 444(7117):323–329. https://doi.org/10.1038/nature05286

Kaur H, Mukherjee S, Baluska F, Bhatla SC (2015) Regulatory roles of serotonin and melatonin in abiotic stress tolerance in plants. Plant Signal Behav 10(11):e1049788. https://doi.org/10.1080/15592324.2015.1049788/ASSET/4BB8B5B2-D2BA-44B9-B30E2DD6EA7F443C/ASSETS/IMAGES/LARGE/KPSB_A_1049788_F0005_OC.JPG

Khan A, Numan M, Khan AL, Lee IJ, Imran M, Asaf S, Al-Harrasi A (2020) Melatonin: awakening the defense mechanisms during plant oxidative stress. Plan Theory 9(4):407. https://doi.org/10.3390/PLANTS9040407

Khan MSS, Ahmed S, ul Ikram A, Hannan F, Yasin MU, Wang J, Zhao B, Islam F, Chen J (2023) Phytomelatonin: a key regulator of redox and phytohormones signaling against biotic/abiotic stresses. Redox Biol 64:102805. https://doi.org/10.1016/J.REDOX.2023.102805

Kolodziejczyk A et al (2015) Melatonin improves tolerance to heavy metals stress. Environ Exp Bot 113:39–49. https://doi.org/10.1016/j.envexpbot.2015.01.005

Kopustinskiene DM, Bernatoniene J (2021) Molecular mechanisms of melatonin-mediated cell protection and signaling in health and disease. Pharmaceutics 13(2):129

Lee HY, Back K (2016a) Mitogen-activated protein kinase pathways are required for melatonin-mediated defense responses in plants. J Pineal Res 60(3):327–335. https://doi.org/10.1111/JPI.12314

Lee HY, Back K (2016b) Melatonin is required for H2O2- and NO-mediated defense signaling through MAPKKK3 and OXI1 in *Arabidopsis thaliana*. J Pineal Res 62(2):e12379. https://doi.org/10.1111/JPI.12379

Li B, Meng X, Shan L, He P (2016a) Transcriptional regulation of pattern-triggered immunity in plants. Cell Host Microbe 19(5):641. https://doi.org/10.1016/J.CHOM.2016.04.011

Li C, Tan DX, Liang D, Chang C, Jia D, Ma F (2016b) Melatonin enhances cold tolerance by promoting antioxidant enzymes and alleviating oxidative stress in apple fruits. J Pineal Res 61(3):241–250. https://doi.org/10.1111/jpi.12354

Li C, Tan DX, Liang D, Chang C, Jia D, Ma F (2017) Melatonin mediates nitric oxide signaling during dark-induced leaf senescence in apple (*Malus domestica Borkh.*). J Pineal Res 63(1):e12427. https://doi.org/10.1111/jpi.12427

Li S, Xu Y, Bi Y, Zhang B, Shen S, Jiang T, Zheng X (2019a) Melatonin treatment inhibits gray mold and induces disease resistance in cherry tomato fruit during postharvest. Postharvest Biol Technol 157:110962

Li M, Zhang S, Wang Y (2019b) Melatonin enhances plant resistance to fungal diseases via ROS-mediated signaling. J Exp Bot 70(9):2551–2563. https://doi.org/10.1093/jxb/erz120

Liu C, Chen L, Zhao R, Li R, Zhang S, Yu W, Sheng J, Shen L (2019) Melatonin induces disease resistance to *Botrytis cinerea* in tomato fruit by activating jasmonic acid signaling pathway. J Agric Food Chem 67(22):6116–6124

Liu J, Wang X, Zhang H (2021) Application of melatonin in wheat disease resistance. Plant Pathol 70(6):1128–1135. https://doi.org/10.1111/ppa.13382

Mandal MK, Suren H, Ward B, Boroujerdi A, Kousik C (2018) Differential roles of melatonin in plant-host resistance and pathogen suppression in cucurbits. J Pineal Res 65(3):e12505

Martin-Rivilla H, Garcia-Villaraco A, Ramos-Solano B, Gutierrez-Mañero FJ, Lucas JA (2020) Bioeffectors as biotechnological tools to boost plant innate immunity: signal transduction pathways involved. Plan Theory 9(12):1731

Minich DM, Henning M, Darley C, Fahoum M, Schuler CB, Frame J (2022) Is melatonin the "next vitamin D"?: a review of emerging science, clinical uses, safety, and dietary supplements. Nutrients 14(19):3934

Moo-Koh FA, Cristóbal-Alejo J, Andrés MF, Martín J, Reyes F, Tun-Suárez JM, Gamboa-Angulo M (2022) In vitro assessment of organic and residual fractions of nematicidal culture filtrates from thirteen tropical Trichoderma strains and metabolic profiles of most-active. J Fungi 8(1):82

Park S, Lee DE, Jang H, Byeon Y, Kim YS, Back K (2013) Melatonin-rich transgenic rice plants exhibit resistance to herbicide-induced oxidative stress. J Pineal Res 54(3):258–263. https://doi.org/10.1111/jpi.12020

Park S, Back K, Lee H (2020) Melatonin biosynthesis and its role in abiotic stress tolerance in rice. J Agric Food Chem 68(15):4218–4226. https://doi.org/10.1021/acs.jafc.9b06672

Ramlal A, Sarma R, Rani A, Nautiyal A, Kumar J, Mishra V (2023) Plant-virus interactions in plant innate immunity. In: Plant RNA viruses: molecular pathogenesis and management. Academic Press, pp 297–310. https://doi.org/10.1016/B978-0-323-95339-9.00002-8

Ranf S, Scheel D, Lee J (2016) Challenges in the identification of microbe-associated molecular patterns in plant and animal innate immunity: a case study with bacterial lipopolysaccharide. Mol Plant Pathol 17(8):1165. https://doi.org/10.1111/MPP.12452

Reiter RJ, Tan DX, Zhou Z, Cruz MHC, Garcia JJ (2015) Physiological functions of melatonin: role of melatonin in antioxidation. Curr Neuropharmacol 13(2):173–183. https://doi.org/10.2174/1570159X13999150224225900

Sahu PK, Jayalakshmi K, Tilgam J, Gupta A, Nagaraju Y, Kumar A, Hamid S, Singh HV, Minkina T, Rajput VD, Rajawat MVS (2022) ROS generated from biotic stress: effects on plants and alleviation by endophytic microbes. Front Plant Sci 13:1042936. https://doi.org/10.3389/FPLS.2022.1042936/BIBTEX

Samota MK, Bhatt L, Garg N, Geat N (2017) Defense induced by jasmonic acid: a review. Int J Curr Microbiol App Sci 6(5):2467

Schwessinger B, Ronald PC (2012) Plant innate immunity: perception of conserved microbial signatures. Annu Rev Plant Biol 63:451–482. https://doi.org/10.1146/ANNUREV-ARPLANT-042811-105518

Segonzac C, Zipfel C (2011) Activation of plant pattern-recognition receptors by bacteria. Curr Opin Microbiol 14(1):54–61. https://doi.org/10.1016/J.MIB.2010.12.005

Shah J (2003) The salicylic acid loop in plant defense. Curr Opin Plant Biol 6(4):365–371

Shi H, Chen Y, Tan DX, Reiter RJ, Chan Z, He C (2015a) Melatonin induces nitric oxide and the mitogen-activated protein kinase pathway to regulate cold responses in grapevine. J Pineal Res 59(2):180–190. https://doi.org/10.1111/jpi.12245

Shi H, Qian Y, Tan DX, Reiter RJ, He C (2015b) Melatonin induces the transcripts of CBF/DREB1s and their involvement in both abiotic and biotic stresses in Arabidopsis. J Pineal Res 59(3):334–342

Sun Y, Liu Z, Lan G, Jiao C, Sun Y (2019) Effect of exogenous melatonin on resistance of cucumber to downy mildew. Sci Hortic 255:231–241. https://doi.org/10.1016/J.SCIENTA.2019.04.057

Sun S, Liu A, Li Z, Guo T, Chen S, Ahammed GJ (2023) Anthocyanin synthesis is critical for melatonin-induced chromium stress tolerance in tomato. J Hazard Mater 453:131456. https://doi.org/10.1016/J.JHAZMAT.2023.131456

Szatmári Á, Zvara Á, Móricz ÁM, Besenyei E, Szabó E, Ott PG, Puskás LG, Bozsó Z (2014) Pattern triggered immunity (PTI) in tobacco: isolation of activated genes suggests role of the Phenylpropanoid pathway in inhibition of bacterial pathogens. PLoS One 9(8):e102869. https://doi.org/10.1371/JOURNAL.PONE.0102869

Tan DX, Manchester LC, Liu X, Rosales-Corral SA, Acuña-Castroviejo D, Reiter RJ (2012) Mitochondria and chloroplasts as the original sites of melatonin synthesis: reactions and consequences. J Pineal Res 52(4):341–358. https://doi.org/10.1111/j.1600-079X.2011.00956.x

Tan DX, Manchester LC, Reiter RJ (2019) Melatonin as a plant growth regulator. Plant Sci 287:110193. https://doi.org/10.1016/j.plantsci.2019.110193

Teng Z, Yu Y, Zhu Z, Hong SB, Yang B, Zang Y (2021) Melatonin elevated *Sclerotinia sclerotiorum* resistance via modulation of ATP and glucosinolate biosynthesis in *Brassica rapa* ssp. pekinensis. J Proteome 243:104264

Thomma BP, Nurnberger T, Joosten MH (2011) Of PAMPs and effectors: the blurred PTI-ETI dichotomy. Plant Cell 23(1):4–15. https://doi.org/10.1105/tpc.110.082602

Tiwari RK, Lal MK, Kumar R, Mangal V, Altaf MA, Sharma S, Singh B, Kumar M (2021) Insight into melatonin-mediated response and signaling in the regulation of plant defense under biotic stress. Plant Mol Biol:1–15

Tsuda K, Katagiri F (2010) Comparing signaling mechanisms engaged in pattern-triggered and effector-triggered immunity. Curr Opin Plant Biol 13(4):459–465. https://doi.org/10.1016/j.pbi.2010.04.006

Vannier N, Agler M, Hacquard S (2019) Microbiota-mediated disease resistance in plants. PLoS Pathog 15(6):e1007740

Wang P, Sun X, Li C, Wei Z, Liang D, Ma F (2016) Long-term exogenous application of melatonin delays drought-induced leaf senescence in apple. J Pineal Res 61(4):411–425. https://doi.org/10.1111/jpi.12359

Wang K, Xing Q, Jalal Ahammed G, Zhou J (2022) Functions and prospects of melatonin in plant growth, yield, and quality. J Exp Bot 73(17):5928–5946. https://academic.oup.com/jxb/article/73/17/5928/6595022

Weeda S, Zhang N, Zhao X, Ndip G, Guo Y, Buck GA, Fu C, Ren S (2014) Arabidopsis transcriptome analysis reveals key roles of melatonin in plant defense systems. PLoS One 9(3):e93462. https://doi.org/10.1371/JOURNAL.PONE.0093462

Wei Y, Hu W, Wang Q, Zeng H, Li X, Yan Y, Reiter RJ, He C, Shi H (2017) Identification, transcriptional and functional analysis of heat-shock protein 90s in banana (*Musaacuminata* L.) highlight their novel role in melatonin-mediated plant response to Fusarium wilt. J Pineal Res 62(1):e12367

Wei W, Li QT, Chu YN, Reiter RJ, Yu XM, Zhu DH, Zhang WK, Ma B, Lin Q, Zhang JS, Chen SY (2015) Melatonin enhances plant growth and abiotic stress tolerance in soybean plants. J Exp Bot 69(8):1837–1848. https://doi.org/10.1093/jxb/ery027

Wei Y, Li X, Zhao Y (2022) Advances in melatonin research for plant immunity. Plant Biol J 24(3):215–228. https://doi.org/10.1016/j.plantsci.2022.110265

Wiesel L, Newton AC, Elliott I, Booty D, Gilroy EM, Birch PR, Hein I (2014) Molecular effects of resistance elicitors from biological origin and their potential for crop protection. Front Plant Sci 5:655

Xia Z, Ye Y, Hu C, Wang H, Zheng L, Hu Y, Sheng L, Xing J, Jia W, Wang Y, Luo Z (2024) Exogenous melatonin orchestrates multiple defense responses against Botrytis cinerea in tomato leaves. Plant Stress 11:100326

Yamaguchi K, Yamada K, Kawasaki T (2013) Receptor-like cytoplasmic kinases are pivotal components in pattern recognition receptor-mediated signaling in plant immunity. Plant Signal Behav 8(10):e25662. https://doi.org/10.4161/PSB.25662

Yang Q, Li J, Ma W, Zhang S, Hou S, Wang Z, Li X, Gao W, Rengel Z, Chen Q, Cui X (2021) Melatonin increases leaf disease resistance and saponin biosynthesis in Panax notogiseng. J Plant Physiol 263:153466. https://doi.org/10.1016/J.JPLPH.2021.153466

Yin L, Wang P, Li M, Ke X, Li C, Liang D, Wu S, Ma X, Li C, Zou Y, Ma F (2013) Exogenous melatonin improves *M alus* resistance to *M arssonina* apple blotch. J Pineal Res 54(4):426–434

Zhang N, Sun Q, Zhang H, Cao Y, Weeda S (2015) Roles of melatonin in abiotic stress resistance in plants. J Exp Bot 66(3):647–656. https://doi.org/10.1093/jxb/eru336

Zhang L, Zhang F, Melotto M, Yao J, He SY (2017a) Jasmonate signaling and manipulation by pathogens and insects. J Exp Bot 68(6):1371–1385

Zhang N, Zhao B, Zhang HJ, Weeda S, Yang C, Yang ZC, Ren S, Guo YD (2017b) Melatonin promotes water-stress tolerance, lateral root formation, and seed germination in cucumber (*Cucumis sativus L.*). J Pineal Res 62(3):e12410. https://doi.org/10.1111/jpi.12410

Zhang H, He Y, Tan X, Xie K, Li L, Hong G, Li J, Cheng Y, Yan F, Chen J, Sun Z (2019) The dual effect of the brassinosteroid pathway on rice black-streaked dwarf virus infection by modulating the peroxidase-mediated oxidative burst and plant defense. Mol Plant-Microbe Interact 32(6):685–696

Zhang N, Sun Q, Zhang H (2021) Melatonin's role in plant-microbe interactions and soil health. Soil Biol Biochem 152:108057. https://doi.org/10.1016/j.soilbio.2020.108057

Zhang M, Gao C, Xu L, Niu H, Liu Q, Huang Y, Lv G, Yang H, Li M (2022) Melatonin and Indole-3-acetic acid synergistically regulate plant growth and stress resistance. Cells 11(20):3250. https://doi.org/10.3390/CELLS11203250

Zhao H, Zhang K, Zhou X, Xi L, Wang Y, Xu H (2019) Melatonin-mediated regulation of plant defense against pathogen attack. Plant Signal Behav 14(10):e1659700. https://doi.org/10.1080/15592324.2019.1659700

Zhao Y, Tan DX, Reiter RJ (2020) Regulatory roles of melatonin in plant-pathogen interactions. Plant Mol Biol Report 38(5):529–545. https://doi.org/10.1007/s11105-020-01246-w

Zhu Z, Lee B (2015) Friends or foes: new insights in jasmonate and ethylene co-actions. Plant Cell Physiol 56(3):414–420

Zipfel C (2008) Pattern-recognition receptors in plant innate immunity. Curr Opin Immunol 20(1):10–16. https://doi.org/10.1016/j.coi.2007.11.003

Zipfel C, Robatzek S (2010) Pathogen-associated molecular pattern-triggered immunity: Veni, Vidi...? Plant Physiol 154(2):551. https://doi.org/10.1104/PP.110.161547

Melatonin's Role in Protease Activity, Biofilm Formation, and mRNA Expression to Combat Bacterial Diseases in Plants

10

Jagmohan Singh, Shiv Singla, M. S. Gambo, Asharani Patel, Kuleshwar Parsad Sahu, and Nikita Gambhir

Abstract

The huge economic losses caused by bacterial phytopathogens and an increase in bactericide resistance underscore the need for novel and sustainable strategies to manage bacterial plant diseases. Melatonin is a phytochemical that plays a significant role in enhancing plant defense against bacterial pathogens. In this chapter, the impact of melatonin on protease activity, biofilm formation, and mRNA expression, and the practical application of melatonin to manage bacterial plant diseases are discussed. Melatonin enhances plant immunity by regulating protease activity and stress responses to bacterial pathogen infections. It also directly suppresses bacterial pathogens by disrupting biofilm formation and reducing bacterial virulence through the inhibition of quorum sensing. Though melatonin is a high-potential ecofriendly alternative to chemical pesticides, practical appli-

J. Singh
Department of Plant Pathology, CCS Haryana Agricultural University, Hisar, Haryana, India

S. Singla
Department of Plant and Soil Sciences, University of Delaware, Newark, DE, USA

M. S. Gambo
Department of Plant Pathology, CCS Haryana Agricultural University, Hisar, Haryana, India

Aliko Dangote University of Science and Technology, Wudil, Kano State, Nigeria

A. Patel
Division of Plant Pathology, ICAR-Indian Agricultural Research Institute, Hazaribag, Jharkhand, India

K. P. Sahu
College of Agriculture, Rani Lakshmi Bai Central Agricultural University, Jhansi, India

N. Gambhir (✉)
FMC Stine Research Center, Newark, DE, USA

R. K. Tiwari et al. (eds.), *Melatonin: Signal Transduction Mechanisms and Defense Networks in Plants*, https://doi.org/10.1007/978-981-96-8869-2_10

cation challenges still persist. Future research should optimize procedures to standardize and enhance melatonin production for practical application.

Keywords
Effector · Signal transduction · Catalase · Reactive species · Signaling

10.1 Introduction

The natural synthesis of melatonin was first reported in α-proteobacteria and photosynthetic cyanobacteria (Vivien-Roels and Pévet 1993). As per the classical endosymbiotic theory, the early eukaryotes engulfed these bacteria by phagocytosis, and led to a symbiotic relationship between the host and the engulfed α-proteobacteria and photosynthetic cyanobacteria (Zhao et al. 2019). This evolutionary process led to the development of mitochondria and chloroplasts from α-proteobacteria and cyanobacteria, respectively. Therefore, both these organelles inherited the ability for melatonin biosynthesis from their respective ancestors (Zhao et al. 2021). The eukaryotic life forms such as plants and animals possess the biosynthetic pathways for melatonin synthesis, which alters with life stages (Back et al. 2016; Hardeland 2019; Zhao et al. 2019). The universal precursor for melatonin biosynthesis is tryptophan in all organisms. Tryptophan is synthesized by a shikimic acid pathway from substrates such as D-erythrose-4-phosphate, phosphoenolpyruvate, or carbon dioxide in plants, fungi, and bacteria, whereas for animals, tryptophan is a dietary supplement. Melatonin production is more sensitive in animals as compared to plants when tryptophan levels are low. The biosynthesis of tryptophan in plants guarantees optimal melatonin levels, thereby enhancing protection against oxidative stress.

TJ Burril (1839–1916) is renowned for his discovery of the first bacterial disease in plants, specifically bacterial blight of the pear tree. He isolated the pathogen and designated it as *Micrococcus amylovorus*, which was subsequently renamed *Erwinia amylovora* (Tanner 1937). Since then, bacterial plant pathology has developed rapidly. Phytopathogenic bacteria constitute a significant category of pathogens responsible for various critical diseases that impact crop productivity. Of the approximately 7100 diverse species within Kingdom Monera, around 200 species are identified as pathogens capable of causing disease in plants (Buttimer et al. 2017). Despite the diversity of phytopathogenic bacteria across various genera, only a few genera are significant in agricultural production including: *Xanthomonas, Ralstonia, Xylella, Pseudomonas, Erwinia, Dickeya, Clavibacter, Pectobacterium,* and *Agrobacterium* (Mansfield et al. 2012). Bacterial plant diseases provide serious threats to economic stability and ecological balance. They result in significant agricultural losses, elevated management expenses, and socio-economic difficulties, especially in developing countries. They disrupt plant microbiomes and alter interactions within ecosystems, impacting biodiversity and ecosystem services. Mitigating these effects necessitates innovative management strategies and a comprehensive understanding of plant-pathogen interactions. For instance, Xanthomonas Wilt of Banana has led to around 50% yield losses in Uganda, threatening the food security of millions

(Tripathi et al. 2009). The efficacy of melatonin is reported against multiple antibiotic-resistant bacteria such as *Staphylococcus aureus*, *Acinetobacter baumannii*, and *Pseudomonas aeruginosa* (Chen et al. 2018; Wei et al. 2018). Melatonin inhibits the bacterial activity in the host by modulating cAMP and Ca^{2+} regulation, decreasing the intracellular substrates and destabilization of the bacterial cell wall. The bacterial pathogens of paddy, i.e., *X. oryzae* pv. *oryzae* and *X. oryzae* pv. *oryzicola* inciting leaf blight and leaf streak, respectively, is inhibited by the application of melatonin at 200 μg mL^{-1} concentration (Chen et al. 2018). Melatonin is documented to upregulate transcription activators, e.g., MeRAV1 & MeRAV2 genes in cassava to provide resistance against *Xanthomonas axonopodis* pv. *manihotis* (Wei et al. 2018). Melatonin-induced plant cell wall fortification and callose deposition in *Arabidopsis thaliana* restrict the infection by *Pseudomonas syringae* pv. tomato *(Pst)* DC3000 (Shi et al. 2015). Ghosh and coworkers reported the role of melatonin in mitigating the severe infection of *Candidatus* Liberibacter asiaticus in citrus plants (Ghosh et al. 2018).

Melatonin application has been documented to directly suppress fungal plant pathogens. The exogenous application of melatonin in potato deforms the spore structure of *Phytophthora infestans* and inhibits mycelial growth in potato (Zhang et al. 2017). The RNA sequencing data indicated that the genes related to pathogenicity and virulence were downregulated in *P. infestans* (Zhang et al. 2017). Another study in *Capsicum annum* revealed that the mycelial growth of *Colletotrichum gloeosporioides* and *Colletotrichum acutatum* was reduced by 76% and 71%, respectively by 100 μM of phytomelatonin (Ali et al. 2020). The effectiveness of melatonin is documented in some vegetable crops such as tomato against *Botrytis cinerea* (Liu et al. 2019) and in cucurbits against *Podosphaera xanthii* (Mandal et al. 2018). Among fruit crops, the effectiveness of melatonin is reported likewise against Fusarium wilt in banana by *Fusarium oxysporum* (Wei et al. 2017), marssonia blotch by *Diplocarpon mali* (Yin et al. 2013), and gray mold in strawberries by *Botrytis cinerea* (Liu et al. 2019; Tiwari et al. 2020).

10.2 Melatonin: A Phytochemical Defense Molecule

Melatonin is a crucial biomolecule that is present in several organisms including bacteria, fungi, plants, and animals (Zhao et al. 2019). In plants, melatonin serves as a central signaling molecule that activates further defense signaling pathways to provide resistance against phytopathogens and tolerance to various biotic stresses (Lee and Back 2016).

10.2.1 Structure and Biosynthesis of Melatonin in Plants

Melatonin has a molecular weight of 232.283 g/mol with the chemical formula $C_{13}H_{16}N_2O_2$. A 3-amide group and a 5-alkoxygroup of indole scaffold gives the

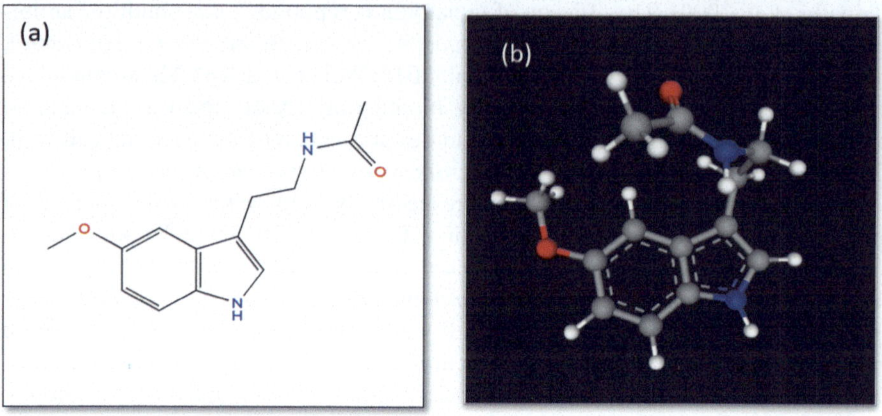

Fig. 10.1 Structure of melatonin. (**a**) 2D molecular structure; (**b**) 3D molecular structure

amphiphilicity property to melatonin molecule and plays crucial role to pass through biological membranes of cell and cell organelles (Fig. 10.1) (Omer et al. 2021).

10.2.2 Biosynthesis of Melatonin in Plants

In plants, the melatonin precursor-tryptophan is converted to tryptomine by the enzyme tryptophan decarboxylase, which is further converted to serotonin by tryptomine hydrolase. Later, serotonin undergoes N-acetylation followed by methylation, leading to the synthesis of the final product, i.e., melatonin by N-acetyltransferase and hydroxy indole-O-methyltransferase enzymes, respectively (Khan et al. 2023).

Pathways of Melatonin Synthesis

Basically, there are four pathways of melatonin biosynthesis classified on the basis of the presence of intermediates and enzymes, but serotonin is a common intermediate present in all the biosynthesis pathways in plants.

Pathway I The biosynthesis of melatonin begins in the cytoplasm, where the enzyme tryptophan decarboxylase (TDC) catalyzes the conversion of tryptophan into tryptamine, which is transported to the endoplasmic reticulum, where tryptamine 5-hydroxylase (T5H) catalyze the reaction to convert tryptamine to serotonin. Serotonin then undergoes acetylation by serotonin-N-acetyltransferase (SNAT), producing N-acetylserotonin. The final step involves methylation, which is carried out in the cytoplasm by either acetyl serotonin methyltransferase (ASMT) or caffeic acid O-methyltransferase (COMT), to yield final product as melatonin (Fig. 10.2).

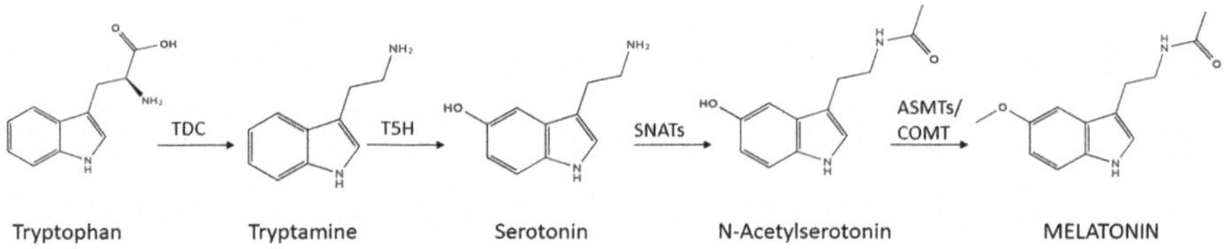

Fig. 10.2 Pathway I of melatonin biosynthesis in plants. TDC, tryptophan decarboxylase; T5H, tryptamine 5-hydroxylase; SNAT, serotonin N-acetyltransferase; ASMT, N-acetylserotonin methyltransferase; COMT, caffeic acid O-methyltransferase

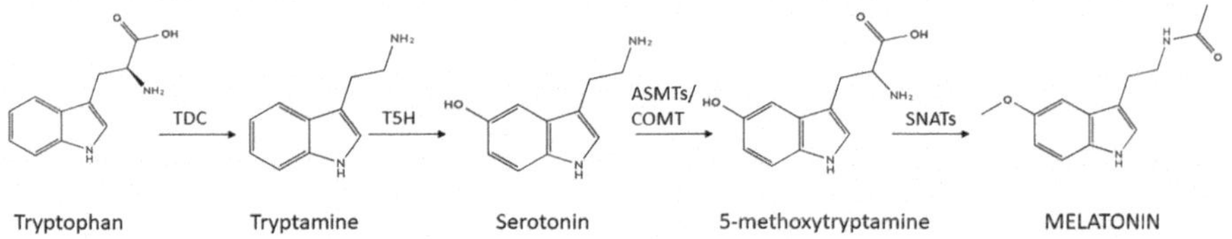

Fig. 10.3 Pathway II of melatonin biosynthesis in plants. *TDC*, tryptophan decarboxylase; *T5H, tryptamine 5-hydroxylase; ASMT,* N-acetylse*rotonin methyltransferase; COMT,* caffeic acid O-methyltransferase*; SNAT,* sero*tonin N-acetyltransferase*

Pathway II In this pathway, the production of serotonin takes place by same procedure as in pathway I, but serotonin further undergoes methylation, instead of acetylation by ASMT or COMT to form the intermediate compound 5-methoxytryptamine. Finally, 5-methoxytryptamine is converted to melatonin in chloroplast by serotonin N-acetyltransferase (SNAT) (Fig. 10.3).

Pathway III In this pathway, tryptophan at first step undergoes hydrolysis by tryptophan 5-hydroxylase (TPH) to yield 5-hydroxytryptophan in the cytoplasm, later that is transformed to serotonin TDC. Serotonin undergoes N-acetylation by SNAT and further methylation by ASMT/COMT to yield melatonin in the cytoplasm of plant cell (Fig. 10.4).

Pathway IV In this pathway, initially TPH introduces a hydroxyl group to tryptophan, creating 5-hydroxytryptophan as the first metabolic intermediate. The carboxyl group is removed from 5-hydroxytryptophan by tryptophan decarboxylase to produce serotonin in cytoplasm. Then, serotonin undergoes enzymatic modification through methylation by either ASMT or COMT enzymes to produces 5-methoxytryptamine, the immediate precursor to melatonin. The 5-methoxytrpamine is finally transformed into melatonin in chloroplast by SNAT enzyme (Fig. 10.5).

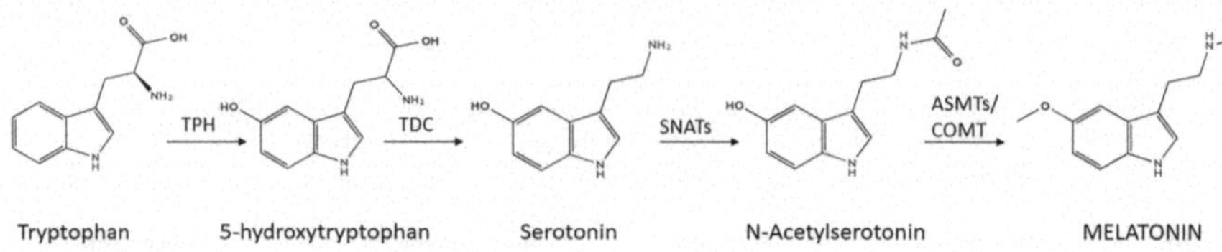

Fig. 10.4 Pathway III of melatonin biosynthesis in plants. TPH, tryptophan hydroxylase; TDC, tryptophan decarboxylase; SNAT, serotonin N-acetyltransferase; ASMT, N-acetylserotonin methyltransferase; COMT, caffeic acid O-methyltransferase

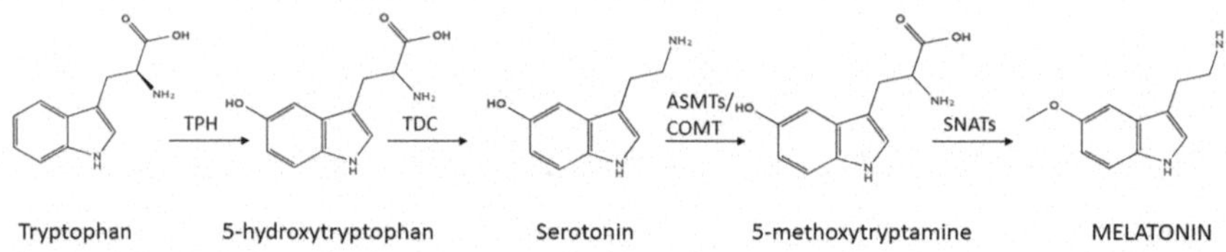

Fig. 10.5 Pathway IV of melatonin biosynthesis in plants. TPH, tryptophan hydroxylase; TDC, tryptophan decarboxylase; ASMT, N-acetylserotonin methyltransferase; COMT, caffeic acid O-methyltransferase; SNAT, serotonin N-acetyltransferase

10.2.3 Melatonin as an Antioxidant and Stress-Reliever

Plant growth and yield are adversely affected by abiotic stresses such as drought, flooding, heat, cold, and salinity. Melatonin is a central regulator phytohormone that helps plant to perform better under stress conditions. Several studies have revealed that under abiotic stress conditions, the endogenous levels of melatonin increase in plants. It is also reported that the exogenous application of melatonin mitigates abiotic stress in plants by regulating the gene expressions in plants under stress. Melatonin act as a regulator for combating abiotic stress by activating defense signaling pathways and scavenging reactive oxygen species in plants.

10.3 Melatonin and Protease Activities

10.3.1 Role of Proteases in Plant Defense

Proteases are enzymes that partially or completely degrade proteins to achieve several end goals including disposal of misfolded, damaged, or excess proteins, maturation of proteins, and regulation of different processes. The MEROPS database employs a comprehensive framework for protease classification, organizing these hydrolytic enzymes based on their fundamental biochemical characteristics (https://www.ebi.ac.uk/merops/). Specifically, proteases are grouped into nine major categories determined by their catalytic properties: aspartic proteases (A), cysteine proteases (C), glutamic proteases (G), metalloproteases (M), asparagine proteases (N), serine proteases (S), threonine proteases (T), mixed-type proteases (P), and proteases with undetermined mechanisms (U) (Rawlings and Barrett 1999). Of these, the major plant protease classes involved in pathogen resistance are aspartic, cysteine, serine, and metalloproteases (Thomas and Van der Hoorn 2018). Protease classes are subdivided into clans \, which are further subdivided into families (denoted by the letter representing the catalytic residue and a number, e.g., A1 is a family in the aspartic protease class).

In plants, serine proteases are the largest class (9 clans and 14 families), followed by aspartic proteases (3 clans and 4 families), cysteine proteases (5 clans and 15 families), and metalloproteases (11 clans and 19 families). The most prominent plant proteases involved in resistance against bacterial pathogens are summarized in Table 10.1.

Proteases are a key regulator of several facets of plant immunity including pathogen recognition, disease resistance signaling, priming of plant immunity, and hypersensitive response (HR) or programmed cell death (PCD) (Van der Hoorn 2008). The constitutive disease resistance 1 (CDR1), an aspartic protease, is involved in disease resistance signaling in *Arabidopsis thaliana* (Van der Hoorn 2008; Xia et al. 2004). The overexpression of *CDR1* induces resistance and pathogenesis-related (PR) gene expression in an SA-dependent manner when the plants are challenged by *Pseudomonas syringae.* Xia et al. (2004) demonstrated that the protease activity of CDR1 is required for its biological function in regulating plant immunity. CDR1 protein is released in the apoplast during pathogen attack, and it is proposed that its protease activity releases peptides, which systemically induce PR gene expression.

The Subtilase 3.3 (SBT3.3) protease is an extracellular serine protease that positively regulates defense priming in *A. thaliana* via an SA-dependent mechanism (Ramírez et al. 2013; Thomas and Van der Hoorn 2018). Overexpression of *SBT3.3* elicits an amplified resistance response to *P. syringae*. The mutant *sbt3.3* plants are susceptible to *P. syringae* owing to a severe reduction in SA-related defense gene expression and inhibition of activation of mitogen-associated protein kinases (MAPKs).

Cathepsin B (CathB) is an apoplastic cysteine protease, which positively regulates HR in *A. thaliana* and *Nicotiana benthamiana* (Gilroy et al. 2007; McLellan et al. 2009; Thomas and Van der Hoorn 2018). In *A. thaliana, CathB* genes provide basal resistance to *P. syringae* in a functionally redundant manner and are not needed for *R* gene-mediated resistance to this pathogen (McLellan et al. 2009). In *N. benthamiana*, virus-induced gene silencing and protease inhibition showed that CathB prevented PCD when challenged with the virulent *Erwinia amylovora* and avirulent *P. syringae* (Gilroy et al. 2007).

10.3.2 Impact of Melatonin on Proteases

Melatonin has been shown to regulate protein degradation in several organisms including the human malarial parasite, *Plasmodium falciparum* (Koyama et al. 2014) and the Chinese crab apple, *Malus hupehensis* (Wang et al. 2013). Although

Table 10.1 Plant proteases involved in resistance against bacterial pathogens

Protease catalytic type	*MEROPS* family	Protease	Function in defense	Plant host	Pathogen	Reference
Aspartic	A01	CDR1	Signaling	*A. thaliana*	*Pseudomonas syringae*	Xia et al. (2004)
Cysteine	C01	CathB	Hypersensitive response (HR)	*N. benthamiana*	*Erwinia amylovora*, *P. syringae*	Gilroy et al. (2007)
				A. thaliana	*P. syringae*	McLellan et al. (2009)
		C14	HR	Tomato	*P. syringae*	Shindo et al. (2016)
	C13	VPE	Membrane fusion	*A. thaliana*	*P. syringae*	Rojo et al. (2004)
	C14	AtMC1	HR	*A. thaliana*	*P. syringae*	Coll et al. (2010)
Metallo	M10	MMP2	Unknown	*Glycine max*	*P. syringae pv. glycinea*	Liu et al. (2001)
Serine	S08	SBT3.3	Priming	*A. thaliana*	*P. syringae*	Ramírez et al. (2013)
		S1P	Signaling, hormone release	*A. thaliana*	*P. syringae*	Stegmann et al. (2017)
Threonine	T01	PBA1	HR, membrane fusion	*A. thaliana*	*P. syringae*	Hatsugai et al. (2009)

there is no direct evidence of melatonin-mediated regulation of proteases involved in plant defense, Wang et al. (2013) found that long-term melatonin application regulates protein degradation and delays leaf senescence in *M. hupehensis*. Melatonin can also interact with PR proteins. *PR1* has been shown to enhance disease resistance upon expression in transgenic tobacco (Alexander et al. 1993; Li et al. 2011). The mechanism of action of PR1 is proteolytic cleavage to release a C-terminal CAPE1 peptide, which induces the plant defense response (Chen et al. 2014). Guo et al. (2022) found that the melatonin biosynthetic enzyme, *N*-acetylserotonin *O*-methyltransferase 2 (ASMT2) physically interacted with PR1 in vitro and in vivo and enhanced cassava's defense response against *Xanthomonas axonopodis* pv. *manihotis* (*Xam*). Together, these studies point toward the interaction of melatonin and proteases to modulate plant processes including plant defense.

Extracellular proteases are important virulence factors of pathogenic bacteria (Engel et al. 1998; Huang et al. 2005; Lee et al. 2002). Melatonin treatment has been shown to reduce the virulence of *Xanthomonas oryzae* pv. *oryzicola (Xoc)* on rice (Chen et al. 2019). This reduced virulence was attributed to the impact of melatonin on several growth and virulence aspects of *Xoc* including decreased extracellular protease activities of the pathogen. Thus, the effect of plant melatonin on pathogen proteases could be a plausible defense mechanism.

10.4 Biofilm Formation and Melatonin Intervention

Biofilms are either monolayer or multilayers consisting of either homogeneous or heterogeneous bacterial populations embedded in a matrix of extracellular polymeric molecules secreted by the constituent populations of the biofilm. Costerton defined a biofilm as a structured assemblage of bacterial cells encased in a self-generated polymeric matrix and attached to either living or non-living surface (Costerton et al. 1999). In the majority of biofilm formations, unicellular organisms aggregate to establish a community that envelopes a solid substrate by exopolysaccharide matrix embedded with cells. The biofilm consists of microorganisms <10%, while the matrix comprises >90% of total dry mass. Diverse mechanisms enable various microbial species to establish proximity to a surface, adhere securely, facilitate cell–cell interactions, and develop into a complex structure (Verstraeten et al. 2008). Currently, five basic generalized stages are illustrated for biofilm formation (Fig. 10.1). Step 1: Planktonic cells adhere to the substrate via an adhesion mechanism. Step 2: Cells initiate adsorption and proliferation. Step 3: Early biofilm architecture develops, accompanied by the production of cell–cell signaling molecules. Finally, Step 4: A robust mature biofilm architecture is established, characterized by extracellular polymeric substances (EPS), followed by Step 5: Dispersion of individual cells from the biofilm. Is reported that genetic and environmental factors both play a crucial role in bacterial biofilm formation (Marić and Vraneš 2007). EPS are referred to as "the dark matter of biofilms" due to the vast range of matrix biopolymers and the challenges associated with their analysis (Flemming et al. 2007). EPS mainly consists of polysaccharides and biomolecules, such as proteins, lipids, and nucleic acids, along with polymers like glycopeptides, lipids, and lipopolysaccharides, which form a framework to retain integrity of biofilm (Cortés et al. 2011). The complex structure and metabolism of biofilms have led to analogies being drawn with the tissues of more multicellular organisms (Costerton et al. 1995).

Multiple mechanisms are believed to contribute to biofilm tolerance and resistance, including the gradual penetration of the antimicrobial agent through the biofilm and alterations in the chemical microenvironment within the biofilm. A mechanism by which biofilms resist antimicrobial agents is the inability of the agent to penetrate the entire depth of the biofilm. Polymeric substances constituting the biofilm matrix are recognized for impeding the diffusion of antibiotics (Nickel et al. 1985). The mathematical models indicate that a significant penetration barrier will form if the antimicrobial agent is inactivated in the outer layers of the biofilm more rapidly than it diffuses (Vrany et al. 1997). Occasionally, if the antibiotic degrades during its penetration of the biofilm, its efficacy diminishes swiftly. Antibiotics may adsorb onto the extracellular polymeric surfaces of the biofilm, potentially reducing the penetration of aminoglycosides (Shigeta et al. 1997). Occasionally, positively charged antibiotics can adhere to the negatively charged molecules within the biofilm matrix. This interaction impedes the penetration of the antibiotic into the biofilm's depth (Nichols et al. 1988). Another mechanism is the rapid alteration of the biofilm's microenvironment, which results in antibiotic ineffectiveness. In the subterranean strata of the biofilm, oxygen is absent, resulting in an anaerobic environment (De Beer et al. 1994), illustrating the environmental heterogeneity produced within biofilms. Aminoglycosides, a class of antibiotics, have been demonstrated to be ineffective in anaerobic environments (Tack and Sabath 1985). Bacteria can accumulate elevated levels of beta-lactamases in response to antibiotics.

Studies using confocal laser microscopy have revealed that *Pseudomonas aeruginosa* biofilms exhibit *ampC* promoter activity, as indicated by green fluorescent protein (GFP) expression, following exposure to ceftazidime at 100 mg/ml for 4 hours (Bagge et al. 2004). Another research study suggests that BrlR, a DNA-binding regulatory protein, contributes significantly to biofilm-mediated antibiotic resistance. By downregulating *phoPQ* expression, BrlR enhances bacterial susceptibility to colistin—likely due to membrane alterations, while simultaneously increasing resistance to tobramycin (Chambers and Sauer 2013). Another critical factor in antibiotic tolerance is the sigma factor σ^{22} (encoded by *algT/algU*), which is normally inhibited by MucA but activates in response to cell wall stress (Wood and Ohman 2012). The literature review revealed that melatonin influences pathogenic bacteria and mitigates their detrimental effects on host organisms. Key findings include the following:

1. In vitro effects: Melatonin modulates bacterial activity by altering free radical production and neutralization. It disrupts bacterial replication by targeting cell wall integrity and reducing intracellular nutrient availability (e.g., iron). Additionally, it inhibits glucose and glutamate metabolism, which are critical for bacterial growth and division.

2. In planta: Melatonin enhances host defenses by activating genes linked to phytohormone-mediated immune responses and MAPK signaling pathways.
3. In animal systems: Melatonin interacts with multiple signaling cascades, including NF-κB, STAT1, and Nrf2, which play roles in reducing inflammation, enhancing antioxidant activity, and protecting mitochondrial function.

Another study investigated the function of melatonin receptor agonists, specifically melatonin, in suppressing biofilm formation and the viability of *Porphyromonas gingivalis*, a principal pathogen in periodontal diseases. Melatonin effectively suppressed biofilm formation, diminished established biofilms, and lowered biofilm viability. At sub-minimum inhibitory concentrations (sub-MIC), melatonin inhibited the activities of proteinases, hemolysis, and the expression of virulence genes associated with biofilm formation, while enhancing the expression of iron/heme utilization genes. The disruption of biofilm architecture by melatonin treatment can be visualized by Confocal laser scanning microscopy (CLSM). Further research demonstrated melatonin's ability to inhibit the proliferation of food-borne *Bacillus* species, including *B. cereus*, *B. licheniformis*, and *B. subtilis*, on cherry tomatoes. The mechanism involved two synergistic actions: first, melatonin directly inhibited *B. subtilis* growth by impairing cell division, oxidative phosphorylation, and motility, while also reducing biofilm formation. Second, it enhanced the antioxidant activity of cherry tomatoes, increased phenolic and ethylene production, and upregulated the expression of defense-related genes (*PT16* and *PR1b1*). Notably, this protective response only occurred when both *B. subtilis* and melatonin were present, suggesting a targeted interaction.

10.5 Regulation of mRNA Expression by Melatonin

10.5.1 Importance of mRNA Expression in Plant Defense

Messenger RNA (mRNA) expression is a key indicator of immune activation in plants as it mediates several aspects of plant defense. The first line of plant defense begins with recognition of pathogen-associated molecular patterns (PAMPs) of pathogen by pattern recognition receptors (PRRs) present on the plant cell surface, which inducing pattern-triggered immunity (PTI). To combat PTI, virulent pathogens use effector proteins. Plants may have corresponding resistance (R) genes encoding R proteins, which comprise nucleotide-binding site leucine-rich repeat regions (NBS-LRR or NLR) that recognize the specific effectors that lead to effector-triggered immunity (ETI). It is important to note that effector recognition by R proteins initiates ETI in biotrophs but usually leads to effector-triggered susceptibility (ETS) in necrotrophs (Ghozlan et al. 2024). The activation of the receptors mentioned above may lead to rapid transcriptional reprogramming, which is reflected by modified mRNA expression.

mRNA expression mediates the production of defense proteins like PR proteins and transcriptional reprogramming via transcription factors such as WRKY and

MYB (Aerts et al. 2022). Expression of genes encoding enzymes involved in cell wall reinforcement enhance plant defense (Bethke et al. 2014). Changes in mRNA expression of genes involved in the biosynthesis of defense phytohormones such as salicylic acid (SA), jasmonic acid (JA), and ethylene (ET) are important for local and systemic defense signaling (Yang et al. 2015).

10.5.2 Melatonin Influence on Defense-Related Gene Expression

Melatonin affects defense-related gene expression in plants through several mechanisms. Exogenous melatonin treatment has been shown to alter genome-wide expression in plants, indicating broad transcriptional reprogramming in response to pathogen attack (Lee et al. 2014). It activates the expression of reactive oxygen species (ROS) related genes, transcription factors involved in plant defense and modulates hormone signaling pathways, including SA, JA, and ET, which in turn affect transcription of defense genes (Wei et al. 2016; Xie et al. 2022; Yang et al. 2015). Melatonin has also been shown to act upstream of MAPK cascades to activate defense-related genes (Lee and Back 2016; Tiwari et al. 2021). Additionally, melatonin enhances the expression of genes involved in secondary metabolism such as phenylalanine metabolism, phenylpropanoid biosynthesis, flavonoids, and flavonol biosynthesis (Gan et al. 2024; Xie et al. 2022). On the pathogen side, exogenous melatonin treatment results in the downregulation of pathogenicity, metabolism, and cell division-related genes (Chen et al. 2019).

The number of genes transcribed in response to melatonin application increases in direct proportion to the melatonin dose (Weeda et al. 2014). Interestingly, not all genes expressed under low melatonin treatment are affected by high melatonin treatment. This suggests that plants may utilize melatonin in a dose-dependent manner to regulate different processes such as growth, development, and defense.

10.5.3 Molecular Insights and Transcriptomic Studies

A few studies have examined the plant host transcriptome in response to infection by bacterial pathogens. Wei et al. (2016) found that challenging cassava plants with *Xanthomonas axonopodis* pv. *manihotis* (Xam) and the bacterial flagellin protein, flg22 resulted in significant transcriptional changes of melatonin synthesis genes, *MeTDC1, MeTDC2, MeSNAT, MeASMT1, MeASMT2 MeASMT3, and MeT5H* (Wei et al. 2016). After infiltration in *Nicotiana benthamiana*, the transient expression of these genes resulted in cell death and HR, callose deposition, and significantly increased the expression of ROS-related genes (superoxide dismutase, catalase, and ascorbate peroxidase) and defense-related genes (RbohA, RbohB, PR1, PR2, and PR5).

Lee et al. (2014) found that melatonin treatment of Arabidopsis and tobacco challenged with *P. syringae* induced the expression of PR genes. In Arabidopsis mutants (npr1, ein2, and mpk6), these PR genes were completely or partially suppressed, which suggested the role of SA and ET dependent pathways in melatonin-induced plant defense responses. In a follow-up study conducted by the authors, they generated Arabidopsis knockout mutant lines for SNAT, which had reduced levels of endogenous melatonin. Challenging the *snat* mutant lines with avirulent *P. syringae* led to enhanced susceptibility and reduced induction of several defense genes including PR1, ICS1, and PDF1.2 (Lee et al. 2015). Further, Lee and Back (2016) demonstrated that MKK4/5/7/9-MPK3/6 cascades trigger melatonin-mediated innate immunity and suppression of MPK3 and MPK6 reduces the induction of defense-related gene expression and pathogen resistance in transgenic Arabidopsis mutants.

In another study conducted on *Xanthomonas oryzae* pv. *oryzae* (*Xoo*) infecting rice, exogenous melatonin significantly downregulated cell division and metabolism of *Xoo* (Chen et al. 2018). Quantitative polymerase chain reaction (qPCR) and RNA sequencing experiments showed that the expression of genes related to cell division, catalytic and metal-binding activity, and carbohydrate and amino acid metabolism were downregulated.

Shi et al. (2015) demonstrated that upon challenging Arabidopsis with *P. syringae,* melatonin levels rise, which in turn elevate nitric oxide levels and when a nitric oxide scavenger is used, it compromises the expression of SA-related genes and disease resistance against bacterial pathogen infection were compromised, suggesting the importance of melatonin-induced NO production in plant innate immune response.

10.6 Practical Implications in Agriculture

Recent studies indicated melatonin's role as a potent antimicrobial agent in plant-pathogen systems. The *Xoc* pathogen inciting bacterial streak in paddy is sensitive to melatonin application (Chen et al. 2018). Similarly, Nehela and coworkers discovered that melatonin-supplemented feed for psyllids, the insect vectors of *Candidatus* Liberibacter asiaticus, enhances the life of both healthy and infected vectors, as well as reduces bacterial load in vectors (Nehela and Killiny 2018). Another study by Wei and coworkers indicated that upregulation of the transcriptional activators MeRAV2 and MeRAV1 boosts the biosynthesis of melatonin in cassava that imparts disease resistance against bacterial blight (Wei et al. 2018). Melatonin plays a key role in this defense system by activating several defense genes—notably PDF1.2, PR1, and PR5—through different signaling pathways. Research by Shi et al. (2015) showed that melatonin and nitric oxide work together to boost the expression of important defense-related genes like PAD4, EDS1, and various PR proteins. Further, in Arabidopsis plants infected with Pst-DC3000 bacteria, melatonin triggers MAPK signaling cascades. This activation leads to increased production of ICS1, a crucial enzyme for making salicylic acid (Lee et al. 2014–2017). Another fascinating finding comes from Zhao et al. (2015), who observed that melatonin treatment strengthens plant cell walls. This happens through increased activity of cell wall invertase (CWI), which promotes the buildup of protective callose deposits containing cellulose, xylose, and galactose. Essentially, melatonin helps plants build better physical barriers against bacterial invaders.

Sugar and glycerol participate in melatonin-mediated defense against *Pst* DC3000 through salicylic acid- and nitric oxide-dependent pathways in Arabidopsis. Melatonin therapy is said to stimulate 1-aminocyclopropane-1-carboxylate synthase 6 (ACS6), the principal enzyme in ethylene biosynthesis, subsequently leading to the expression of PDF1.2 (Lee et al. 2014). Given that JA can also stimulate PDF1.2 expression, JA's possible role alongside melatonin in pathogen resistance cannot be dismissed (Arnao and Hernández-Ruiz 2018). Studied reported that sugar and glycerol contribute to melatonin-mediated protection against *Pst*DC3000 via salicylic acid- and nitric oxide-dependent pathways in Arabidopsis (Qian et al. 2015). Recent transcriptomic data revealed that exogenous melatonin treatment activates ETI- and PTI-related genes in watermelon and Arabidopsis (Weeda et al. 2014).

Plant viruses are major persisting threat to agriculture systems due to the absence of direct chemical solution against viruses (Lal et al. 2021). An alternative approach to plant viral disease management involves enhancing the host defense and management of insect vector. In this context, melatonin has emerged as an important phytoprotectant against viral diseases management in plants such as melatonin (100 μM) applied through irrigation decreased the viral load of tobacco mosaic virus in *Nicotiana glutinosa* and *Solanum lycopersicum* (Zhao et al. 2019). The expression analysis of pathogenesis-related genes indicated enhanced expression of PR1 and PR5 genes on application of melatonin and this coincides with reduction viral load. Furthermore, the application of melatonin effectively eliminated apple stem grooving virus (ASGV) from in vitro virus-infected samples. The method described by Chen et al. (2019) demonstrates a mechanism effective in producing virus-free apple shoots of the "Gala" variety. Insect pests are the most prominent biotic stressors in the global agricultural production system. They inflict both direct and indirect damage to the plant through feeding and the transmission of viral and bacterial pathogens (Dhaliwal et al. 2015). A considerable number of studies document melatonin biosynthesis and regulation in insect pests, influencing their circadian rhythms, feeding behavior, reproductive capacity, photoperiodism, molting, and reproduction. A study by Saremba et al. (2017) illustrated the function of melatonin as a signaling compound in resistant American elm (*Ulmus americana* Planch) infested by bark beetle (*Scolytus multistriatus*). Nonetheless, the molecular and regulatory mechanisms in this regard remain unexamined and require further investigation. The psyllid vector, *Daphrina citri* associated with the transmission of the destructive citrus greening disease (Huanglongbing) has demonstrated reduced bacterial titer following melatonin supplementation (Nehela and Killiny 2018). This underscores the beneficial role of melatonin in reducing the vector-mediated transmission of citrus greening in sweet oranges. No direct evidence exists regarding the impact of melatonin on insect pest management in plants. It can be hypothesized that, similar to other forms of stress, antioxidative defense may also be evident against insect pests and viral vectors. A multitude of unanswered questions and unexamined mechanisms exist regarding melatonin's role in mitigating viral, fungal, and bacterial plant pathogens. The research on the regulation of viral vectors via melatonin is entirely deficient. In the following sections, we aim to summarize and elucidate the defense mechanisms related to this singular regulatory chemical and their interconnections.

Plants have evolved sophisticated defense systems, and melatonin appears to be a key player in this natural protection. Recent research reveals that this remarkable molecule boosts production of various defensive chemicals that either repel or harm plant-eating insects and animals. These include important secondary metabolites like phenolics, alkaloids, and terpenoids—nature's own pest deterrents (Arnao et al. 2022). What makes melatonin particularly interesting is its ability to influence multiple aspects of plant defense. It does not just stimulate chemical defenses; it actually modifies crucial hormone pathways involving JA and ETH that plants use to trigger their anti-herbivore responses (Liu et al. 2019; Khan et al. 2022). This multipronged approach means melatonin helps plants both detect threats faster and respond more effectively. The effects go beyond biochemistry. Researchers have observed real changes in insect behavior—from altered feeding patterns to modified egg-laying habits—when plants contain elevated melatonin levels (Li et al. 2021). Some studies suggest melatonin might even help recruit the herbivores' natural predators, creating a clever biological control system (Maathuis et al. 2018).

10.7 Future Perspectives and Research Directions

10.7.1 Role of Melatonin in Enhancing IPM Strategies

Recent studies have revealed melatonin's surprising potential in agricultural pest control. This naturally occurring molecule, long studied for its roles in animal physiology, is now showing remarkable benefits for crop protection. Researchers have found that melatonin can actually interfere with bacterial infections in plants through multiple mechanisms—from altering protein-digesting enzymes to preventing the formation of biofilms of bacteria (Reiter et al. 2017). Moreover, melatonin can serve as one of the important components in integrated disease management of bacterial diseases. The research on combined application of melatonin with various biocontrol agents is need of the hour to determine whether the synergetic effect exists in disease management (Tan et al. 2015). This synergy could be game-changing for sustainable agriculture. As melatonin enhances plants' natural defenses, farmers may be able to significantly reduce their dependence on traditional chemical pesticides.

10.7.2 Need for In-Depth Studies on Long-Term Impacts

The long-term impacts of widespread application of melatonin in agriculture need to be studied specifically impact on environment and ecology. Research on impact of continuous use of melatonin on soil microbiome should be initiated. The microbial niche of rhizosphere and phyllo-sphere is very important constituents of a plant, so effect of melatonin on off-targets beneficial microbes must be assessed. Melatonin is known to have adverse effect on the insect vectors of various phytopathogens; therefore, we have to explore the impact on beneficial insect and existing biodiversity.

Another promising area for future research is to increase the endogenous melatonin biosynthesis in crop plants by biotechnological approaches such as genome editing or genetic engineering, which will enhance the ability of plants to resist the pathogen infection. Moreover, genome editing by modern tools like CRISPR/cas9 could develop resilient crop varieties that require fewer external inputs for pest control.

10.8 Conclusion

Melatonin's role to mitigate bacterial diseases in plants is well explored, along with its widely known mechanisms like inhibition of biofilm formation and regulation of mRNA expression. Among sustainable agriculture inputs and practices, melatonin has a potential use that can reduce dependence on synthetic chemical pesticides by strengthening plant defense system. This could lead to farming practices that will enhance crop production in a sustainable manner.

References

Aerts N, Chhillar H, Ding P, Van Wees SC (2022) Transcriptional regulation of plant innate immunity. Essays Biochem 66(5):607–620

Alexander D, Goodman RM, Gut-Rella M, Glascock C, Weymann K, Friedrich L, Maddox D, Ahl-Goy P, Luntz T, Ward E (1993) Increased tolerance to two oomycete pathogens in transgenic tobacco expressing pathogenesis-related protein 1a. Proc Natl Acad Sci 90(15):7327–7331

Ali M, Tumbeh Lamin-Samu A, Muhammad I, Farghal M, Khattak AM, Jan I et al (2020) Melatonin mitigates the infection of *Colletotrichum gloeosporioides* via modulation of the chitinase gene and antioxidant activity in *Capsicum annuum* L. Antioxidants 10(1):7

Arnao MB, Hernández-Ruiz J (2018) Melatonin and its relationship to plant hormones. Ann Bot 121(2):195–207

Arnao MB, Cano A, Hernández-Ruiz J (2022) Phytomelatonin: an unexpected molecule with amazing performances in plants. J Exp Bot 73(17):5779–5800

Back K, Tan DX, Reiter RJ (2016) Melatonin biosynthesis in plants: multiple pathways catalyze tryptophan to melatonin in the cytoplasm or chloroplasts. J Pineal Res 61(4):426–437

Bagge N, Hentzer M, Andersen JB, Ciofu O, Givskov M, Høiby N (2004) Dynamics and spatial distribution of β-lactamase expression in *Pseudomonas aeruginosa* biofilms. Antimicrob Agents Chemother 48(4):1168–1174

Bethke G, Grundman RE, Sreekanta S, Truman W, Katagiri F, Glazebrook J (2014) Arabidopsis PECTIN METHYLESTERASEs contribute to immunity against *Pseudomonas syringae*. Plant Physiol 164(2):1093–1107

Buttimer C, McAuliffe O, Ross RP, Hill C, O'Mahony J, Coffey A (2017) Bacteriophages and bacterial plant diseases. Front Microbiol 8:34

Chambers JR, Sauer K (2013) The MerR-like regulator BrlR impairs *Pseudomonas aeruginosa* biofilm tolerance to colistin by repressing PhoPQ. J Bacteriol 195(20):4678–4688

Chen YL, Lee CY, Cheng KT, Chang WH, Huang RN, Nam HG, Chen YR (2014) Quantitative peptidomics study reveals that a wound-induced peptide from PR-1 regulates immune signaling in tomato. Plant Cell 26(10):4135–4148

Chen X, Sun C, Laborda P, Zhao Y, Palmer I, Fu ZQ, Qiu J, Liu F (2018) Melatonin treatment inhibits the growth of *Xanthomonas oryzae* pv. oryzae. Front Microbiol 9:2280

Chen X, Sun C, Laborda P, He Y, Zhao Y, Li C, Liu FJPP (2019) Melatonin treatments reduce the pathogenicity and inhibit the growth of *Xanthomonas oryzae* pv. *oryzicola*. Plant Pathol 68(2):288–296

Coll NS, Vercammen D, Smidler A, Clover C, Van Breusegem F, Dangl JL, Epple P (2010) Arabidopsis type I metacaspases control cell death. Science 330(6009):1393–1397

Cortés ME, Bonilla JC, Sinisterra RD (2011) Biofilm formation, control and novel strategies for eradication. Sci Against Microbial Pathog Commun Curr Res Technol Adv 2:896–905

Costerton JW, Lewandowski Z, Caldwell DE, Korber DR, Lappin-Scott HM (1995) Microbial biofilms. Annu Rev Microbiol 49:711–745

Costerton JW, Stewart PS, Greenberg EP (1999) Bacterial biofilms: a common cause of persistent infections. Science 284(5418):1318–1322

De Beer D, Stoodley P, Roe F, Lewandowski Z (1994) Effects of biofilm structures on oxygen distribution and mass transport. Biotechnol Bioeng 43(11):1131–1138

Dhaliwal GS, Jindal V, Mohindru B (2015) Crop losses due to insect pests: global and Indian scenario. Indian J Entomol 77(2):165–168

Engel LS, Hill JM, Caballero AR, Green LC, O'Callaghan RJ (1998) Protease IV, a unique extracellular protease and virulence factor from *Pseudomonas aeruginosa*. J Biol Chem 273(27):16792–16797

Flemming HC, Neu TR, Wozniak DJ (2007) The EPS matrix: the "house of biofilm cells". J Bacteriol 189(22):7945–7947

Gan Y, Tu Z, Yang Y, Cheng L, Wang N, Fan S, Wu C (2024) Enhancing cowpea wilt resistance: insights from gene coexpression network analysis with exogenous melatonin treatment. BMC Plant Biol 24(1):599

Ghosh DK, Kokane SB, Kokane AD, Warghane AJ, Motghare MR, Bhose S et al (2018) Development of a recombinase polymerase based isothermal amplification combined with lateral flow assay (HLB-RPA-LFA) for rapid detection of "*Candidatus Liberibacter asiaticus*". PLoS One 13(12):e0208530

Ghozlan MH, Gambhir N, Eman EA, Tokgöz S, Lakshman DK, Mitra A (2024) Erratum to "plant defense against necrotrophic pathogens". Am J Plant Sci 15(2):122–138

Gilroy EM, Hein I, Van Der Hoorn R, Boevink PC, Venter E, McLellan H, Kaffarnik F, Hrubikova K, Shaw J, Holeva M, López EC (2007) Involvement of cathepsin B in the plant disease resistance hypersensitive response. Plant J 52(1):1–13

Guo J, Bai Y, Wei Y, Dong Y, Zeng H, Reiter RJ, Shi H (2022) Fine-tuning of pathogenesis-related protein 1 (PR1) activity by the melatonin biosynthetic enzyme ASMT2 in defense response to cassava bacterial blight. J Pineal Res 72(2):e12784

Hardeland R (2019) Melatonin in the evolution of plants and other phototrophs. Melatonin Res 2(3):10–36

Hatsugai N, Iwasaki S, Tamura K, Kondo M, Fuji K, Ogasawara K, Nishimura M, Hara-Nishimura I (2009) A novel membrane fusion-mediated plant immunity against bacterial pathogens. Genes Dev 23(21):2496–2506

Huang X, Tian B, Niu Q, Yang J, Zhang L, Zhang K (2005) An extracellular protease from *Brevibacillus laterosporus* G4 without parasporal crystals can serve as a pathogenic factor in infection of nematodes. Res Microbiol 156(5–6):719–727

Khan M, Ali S, Manghwar H, Saqib S, Ullah F, Ayaz A, Zaman W (2022) Melatonin function and crosstalk with other phytohormones under normal and stressful conditions. Genes 13(10):1699

Khan D, Cai N, Zhu W, Li L, Guan M, Pu X, Chen Q (2023) The role of hytomelatonin receptor 1-mediated signaling in plant growth and stress response. Front Plant Sci 14:1142753

Koyama FC, Azevedo MF, Budu A, Chakrabarti D, Garcia CR (2014) Melatonin-induced temporal up-regulation of gene expression related to ubiquitin/proteasome system (UPS) in the human malaria parasite *Plasmodium falciparum*. Int J Mol Sci 15(12):22320–22330

Lal MK, Tiwari RK, Kumar R, Naga KC, Kumar A, Singh B, Raigond P, Dutt S, Chourasia KN, Kumar D, Parmar V (2021) Effect of potato apical leaf curl disease on glycemic index and resistant starch of potato (*Solanum tuberosum* L.) tubers. Food Chem 359:129939

Lee HY, Back K (2016) Mitogen-activated protein kinase pathways are required for melatonin-mediated defense responses in plants. J Pineal Res 60(3):327–335

Lee CY, Cheng MF, Yu MS, Pan MJ (2002) Purification and characterization of a putative virulence factor, serine protease, from *Vibrio parahaemolyticus*. FEMS Microbiol Lett 209(1):31–37

Lee HY, Byeon Y, Back K (2014) Melatonin as a signal molecule triggering defense responses against pathogen attack in Arabidopsis and tobacco. J Pineal Res 57(3):262–268

Lee HY, Byeon Y, Tan DX, Reiter RJ, Back K (2015) Arabidopsis serotonin N-acetyltransferase knockout mutant plants exhibit decreased melatonin and salicylic acid levels resulting in susceptibility to an avirulent pathogen. J Pineal Res 58(3):291–299

Li ZT, Dhekney SA, Gray DJ (2011) PR-1 gene family of grapevine: a uniquely duplicated PR-1 gene from a Vitis interspecific hybrid confers high level resistance to bacterial disease in transgenic tobacco. Plant Cell Rep 30:1–11

Li Y, Cheah BH, Fang YF, Kuang YH, Lin SC, Liao CT et al (2021) Transcriptomics identifies key defense mechanisms in rice resistant to both leaf-feeding and phloem feeding herbivores. BMC Plant Biol 21:1–18

Liu Y, Dammann C, Bhattacharyya MK (2001) The matrix metalloproteinase gene GmMMP2 is activated in response to pathogenic infections in soybean. Plant Physiol 127(4):1788–1797

Liu C, Chen L, Zhao R, Li R, Zhang S, Yu W et al (2019) Melatonin induces disease resistance to Botrytis cinerea in tomato fruit by activating jasmonic acid signaling pathway. J Agric Food Chem 67(22):6116–6124

Maathuis F, Voitsekhovskaja O, Shabala S, Tyutereva EV, Voitsekhovskaja OV, Tsyganova AV et al (2018) In: Demidchik V (ed) Plant Signalling: from molecules to behaviour. CSIRO Publishing

Mandal MK, Suren H, Ward B, Boroujerdi A, Kousik C (2018) Differential roles of melatonin in plant-host resistance and pathogen suppression in cucurbits. J Pineal Res 65(3):e12505

Mansfield J, Genin S, Magori S, Citovsky V, Sriariyanum M, Ronald P et al (2012) Top 10 plant pathogenic bacteria in molecular plant pathology. Mol Plant Pathol 13(6):614–629

Marić S, Vraneš J (2007) Characteristics and significance of microbial biofilm formation. Period Biol 109:115–121

McLellan H, Gilroy EM, Yun BW, Birch PR, Loake GJ (2009) Functional redundancy in the Arabidopsis Cathepsin B gene family contributes to basal defence, the hypersensitive response and senescence. New Phytol 183(2):408–418

Nehela Y, Killiny N (2018) Infection with phytopathogenic bacterium inhibits melatonin biosynthesis, decreases longevity of its vector, and suppresses the free radical-defense. J Pineal Res 65(3):e12511

Nichols WW, Dorrington SM, Slack MP, Walmsley HL (1988) Inhibition of tobramycin diffusion by binding to alginate. Antimicrob Agents Chemother 32(4):518–523

Nickel JC, Ruseska I, Wright JB, Costerton JW (1985) Tobramycin resistance of *Pseudomonas aeruginosa* cells growing as a biofilm on urinary catheter material. Antimicrob Agents Chemother 27(4):619–624

Omer RA, Koparir P, Ahmed L, Koparir M (2021) Computational and spectroscopy study of melatonin. Indian J Chem B 60:732–741

Qian Y, Tan DX, Reiter RJ, Shi H (2015) Comparative metabolomic analysis highlights the involvement of sugars and glycerol in melatonin-mediated innate immunity against bacterial pathogen in Arabidopsis. Sci Rep 5(1):15815

Ramírez V, López A, Mauch-Mani B, Gil MJ, Vera P (2013) An extracellular subtilase switch for immune priming in Arabidopsis. PLoS Pathog 9(6):e1003445

Rawlings ND, Barrett AJ (1999) MEROPS: the peptidase database. Nucleic Acids Res 27(1):325–331

Reiter RJ, Tan DX, Zhou Z, Cruz MH, Fuentes-Broto L (2017) The value of melatonin in reducing oxidative stress and inflammation: a review of the current evidence. Cell Biochem Funct 35(2):95–116

Rojo E, Martın R, Carter C, Zouhar J, Pan S, Plotnikova J, Jin H, Paneque M, Sánchez-Serrano JJ, Baker B, Ausubel FM (2004) VPEγ exhibits a caspase-like activity that contributes to defense against pathogens. Curr Biol 14(21):1897–1906

Saremba BM, Tymm FJ, Baethke K, Rheault MR, Sherif SM, Saxena PK, Murch SJ (2017) Plant signals during beetle (*Scolytus multistriatus*) feeding in American elm (*Ulmus americana* Planch). Plant Signal Behav 12(5):e1296997

Shi H, Chen Y, Tan DX, Reiter RJ, Chan Z, He C (2015) Melatonin induces nitric oxide and the potential mechanisms relate to innate immunity against bacterial pathogen infection in Arabidopsis. J Pineal Res 59(1):102–108

Shigeta M, Tanaka G, Komatsuzawa H, Sugai M, Suginaka H, Usui T (1997) Permeation of antimicrobial agents through *Pseudomonas aeruginosa* biofilms: a simple method. Chemotherapy 43(5):340–345

Shindo T, Kaschani F, Yang F, Kovács J, Tian F, Kourelis J, Hong TN, Colby T, Shabab M, Chawla R, Kumari S (2016) Screen of non-annotated small secreted proteins of *Pseudomonas syringae* reveals a virulence factor that inhibits tomato immune proteases. PLoS Pathog 12(9):e1005874

Stegmann M, Monaghan J, Smakowska-Luzan E, Rovenich H, Lehner A, Holton N, Belkhadir Y, Zipfel C (2017) The receptor kinase FER is a RALF-regulated scaffold controlling plant immune signaling. Science 355(6322):287–289

Tack KJ, Sabath LD (1985) Increased minimum inhibitory concentrations with an aerobiasis for tobramycin, gentamicin, and amikacin, compared to latamoxef, piperacillin, chloramphenicol, and clindamycin. Chemotherapy 31(3):204–210

Tan DX, Hardeland R, Manchester LC, Paredes SD, Korkmaz A, Sainz RM, Reiter RJ (2015) The changing biological roles of melatonin during evolution: from an antioxidant to signals of darkness, sexual selection and fitness. Biol Rev 90(4):1115–1151

Tanner FW (1937) Bacteriology: a text-book of microorganisms. Wiley

Thomas EL, Van der Hoorn RA (2018) Ten prominent host proteases in plant-pathogen interactions. Int J Mol Sci 19(2):639

Tiwari RK, Lal MK, Naga KC, Kumar R, Chourasia KN, Subhash S et al (2020) Emerging roles of melatonin in mitigating abiotic and biotic stresses of horticultural crops. Sci Hortic 272:109592

Tiwari RK, Lal MK, Kumar R, Mangal V, Altaf MA, Sharma S, Singh B, Kumar M (2021) Insight into melatonin-mediated response and signaling in the regulation of plant defense under biotic stress. Plant Mol Biol:1–15

Tripathi L, Mwangi M, Abele S, Aritua V, Tushemereirwe WK, Bandyopadhyay R (2009) Xanthomonas wilt: a threat to banana production in east and Central Africa. Plant Dis 93(5):440–451

Van der Hoorn RA (2008) Plant proteases: from phenotypes to molecular mechanisms. Annu Rev Plant Biol 59(1):191–223

Verstraeten N, Braeken K, Debkumari B, Fauvart M, Fransaer J, Vermant J, Michiels J (2008) Living on a surface: swarming and biofilm formation. Trends Microbiol 16(10):496–506

Vivien-Roels B, Pévet P (1993) Melatonin: presence and formation in invertebrates. Experientia 49:642–647

Vrany JD, Stewart PS, Suci PA (1997) Comparison of recalcitrance to ciprofloxacin and levofloxacin exhibited by *Pseudomonas aeruginosa* bofilms displaying rapid-transport characteristics. Antimicrob Agents Chemother 41(6):1352–1358

Wang P, Sun X, Chang C, Feng F, Liang D, Cheng L, Ma F (2013) Delay in leaf senescence of *Malus hupehensis* by long-term melatonin application is associated with its regulation of metabolic status and protein degradation. J Pineal Res 55(4):424–434

Weeda S, Zhang N, Zhao X, Ndip G, Guo Y, Buck GA, Fu C, Ren S (2014) Arabidopsis transcriptome analysis reveals key roles of melatonin in plant defense systems. PLoS One 9(3):e93462

Wei Y, Hu W, Wang Q, Liu W, Wu C, Zeng H, Yan Y, Li X, He C, Shi H (2016) Comprehensive transcriptional and functional analyses of melatonin synthesis genes in cassava reveal their novel role in hypersensitive-like cell death. Sci Rep 6(1):35029

Wei Y, Hu W, Wang Q, Zeng H, Li X, Yan Y et al (2017) Identification, transcriptional and functional analysis of heat-shock protein 90s in banana (*Musa acuminata* L.) highlight their novel role in melatonin-mediated plant response to Fusarium wilt. J Pineal Res 62(1):e12367

Wei Y, Chang Y, Zeng H, Liu G, He C, Shi H (2018) RAV transcription factors are essential for disease resistance against cassava bacterial blight via activation of melatonin biosynthesis genes. J Pineal Res 64(1):e12454

Wood LF, Ohman DE (2012) Identification of genes in the σ22 regulon of *Pseudomonas aeruginosa* required for cell envelope homeostasis in either the planktonic or the sessile mode of growth. MBio 3(3):10–1128

Xia Y, Suzuki H, Borevitz J, Blount J, Guo Z, Patel K, Dixon RA, Lamb C (2004) An extracellular aspartic protease functions in Arabidopsis disease resistance signaling. EMBO J 23(4):980–988

Xie X, Han Y, Yuan X, Zhang M, Li P, Ding A, Wang J, Cheng T, Zhang Q (2022) Transcriptome analysis reveals that exogenous melatonin confers lilium disease resistance to *Botrytis elliptica*. Front Genet 13:892674

Yang YX, Ahammed GJ, Wu C, Fan SY, Zhou YH (2015) Crosstalk among jasmonate, salicylate and ethylene signaling pathways in plant disease and immune responses. Curr Protein Pept Sci 16(5):450–461

Yin L, Wang P, Li M, Ke X, Li C, Liang D et al (2013) Exogenous melatonin improves Malus resistance to *M. arssonina* apple blotch. J Pineal Res 54(4):426–434

Zhang S, Zheng X, Reiter RJ, Feng S, Wang Y, Liu S et al (2017) Melatonin attenuates potato late blight by disrupting cell growth, stress tolerance, fungicide susceptibility and homeostasis of gene expression in *Phytophthora infestans*. Front Plant Sci 8:1993

Zhao H, Xu L, Su T, Jiang Y, Hu L, Ma F (2015) Melatonin regulates carbohydrate metabolism and defenses against *Pseudomonas syringae* pv. *tomato* DC 3000 infection in *Arabidopsis thaliana*. J Pineal Res 59(1):109–119

Zhao D, Yu Y, Zhao Z, Sharma R, Reiter RJ (2019) Melatonin synthesis and function: evolutionary history in animals and plants. Front Endocrinol 10:441357

Zhao D, Wang H, Chen S, Yu D, Reiter RJ (2021) Phytomelatonin: an emerging regulator of plant biotic stress resistance. Trends Plant Sci 26(1):70–82

Enhancing Biotic Stress Defense: Melatonin and Its Synergy with Other Compounds

11

Sucharita Mohapatra, Lellapalli Rithesh, Mehjebin Rahman, and Debashis Mahapatra

Abstract

Melatonin is an indolamine bioactive molecule that regulates various developmental processes in plants and strengthens their resistance to biotic stress, making it a multifunctional compound in plant biology. There is a considerable increase in studies with regard to melatonin in a plant in recent times, and its significance in a plant has been reported. It has been shown to promote root elongation, lateral root development, photosynthesis, and overall plant growth and productivity. It shows antioxidant properties and mitigates oxidative stress by scavenging reactive oxygen species. Additionally, melatonin plays a pivotal role in regulating the plant's circadian rhythm, influencing essential physiological processes that shape growth patterns. Melatonin helps plants to defend against diseases and pests in response to biotic stress. It also activates defense pathways by producing compounds, such as phytoalexins, pathogen-related proteins, and secondary metabolites with antimicrobial activity. It also acts as an essential regulator of gene expression linked with plant hormones, influencing the metabolism of indole-3-acetic acid, cytokinin, gibberellins, ethylene, abscisic acid, and

S. Mohapatra (✉)
Department of Plant Pathology, College of Agriculture, Odisha University of Agriculture and Technology, Bhubaneswar, Odisha, India

L. Rithesh
Department of Plant Pathology, College of Agriculture, Thiruvananthapuram, Kerala, India

M. Rahman
Department of Plant Pathology, Assam Agricultural University, Assam, India

D. Mahapatra
Department of Plant Physiology, College of Agriculture, Odisha University of Agriculture and Technology, Bhubaneswar, Odisha, India

R. K. Tiwari et al. (eds.), *Melatonin: Signal Transduction Mechanisms and Defense Networks in Plants*, https://doi.org/10.1007/978-981-96-8869-2_11

auxin carrier proteins. All these processes involving melatonin highlight its significant role in crop improvement and regulation of biotic stress.

Keywords
Biotic stress · Defense mechanism · Melatonin · Phytohormone · Plant growth · Biosynthesis

11.1 Introduction

Plants come across a wide range of environmental stresses, both abiotic and biotic, which include flooding, drought, heat, salinity, heavy metals, and pathogenic microbes. These stresses considerably affect the improvement and yield of agricultural crops, ultimately threatening food security worldwide. Such adverse conditions tend to alter the fundamental metabolism of affected plants. Particularly biotic stress, poses a critical challenge, requiring plants to adapt to environmental constraints in order to successfully complete their life cycle. Various management methods including cultural, chemical, biological, and integrated pest management have been incorporated to mitigate biotic stresses. To counteract these stresses, plants especially synthesize and regulate various biomolecules, which help them to survive under unfavorable conditions (Naeem et al. 2022; Saqib et al. 2022). Among these biomolecules, melatonin serves as a key protective agent against these different stresses. Melatonin is an indoleamine chemically known as N-acetyl-5-methoxytryptamine; it was first identified in 1958 from the bovine pineal gland (Lerner et al. 1958) and plays a pivotal role in regulating circadian rhythms, seasonal reproduction, antioxidant activity, and immune enhancement (Jiang et al. 2016; Reiter et al. 2018).

The discovery of melatonin in plants has expanded our understanding of its significance, revealing that it is a multifunctional metabolite. Plants produce melatonin in different plant parts including seeds, roots, fruits, and leaves (Fan et al. 2018), and produced within different plant cellular organelles such as mitochondria and chloroplasts. Furthermore, enzymes involved in melatonin production have been identified in the cytoplasm and endoplasmic reticulum (ER), suggesting that these organelles also contribute to its biosynthesis (Meenu et al. 2024). Melatonin is recognized as a pleiotropic hormone and has beneficial effects on plant physiology, which enhances water retention, seed and fruit yield, osmoregulation, and regulates various metabolic pathways, including sulfur, phosphorus, carbohydrate, and lipid assimilation pathways (Altaf et al. 2022). Melatonin is a master plant regulator and helps in regulating other plant hormones, such as auxins (AUXs), gibberellins (GAs), cytokinins (CKs), and abscisic acid (ABA) thereby affecting different stages of plant growth and development (Arnao and Hernandez-Ruiz 2021).

In relation to secondary metabolism, melatonin also promotes the biosynthesis of several essential compounds, which includes terpenoids, flavonoids, anthocyanins, carotenoids, and phenolic compounds. It helps in the production of defense-related chemicals such as phytoalexins, pathogenesis-related proteins, and antimicrobial secondary metabolites (Tiwari et al. 2021a, b; Kaur et al. 2022). It also influences stomatal conductance, osmoregulation, immunomodulation, and photorespiration. Manipulation of melatonin levels or its signaling pathways presents a promising strategy for improving plant growth, enhancing stress resistance, and reducing plant diseases. The melatonin concentration in plants varies depending on species, growth stage, and environmental conditions. Understanding different mechanisms of melatonins in plants can provide valuable insights for increasing pioneering approaches to sustainable agriculture.

11.2 Melatonin Biosynthesis

The biosynthesis of melatonin in plants is much more complex in character. Its synthesis is regulated through many factors, one of them being a key one, and that is light. In plants, the cellular organelles, such as mitochondrial and chloroplasts, serve as principal locations for melatonin biosynthesis. Several groups of enzymes are involved in producing melatonin through biosynthetic processes in these organelles. A series of sequential processes of enzymes is involved in melatonin biosynthesis, and serotonin acts as an intermediate. Several enzymes such as tryptophan decarboxylase (TDC), aromatic-L-amino-acid decarboxylase (AADC), tryptophan hydroxylase (TPH), tryptophan 5-hydroxylase (T_5H), serotonin-N-acetyltransferase (SNAT), arylalkyl amine N-acetyltransferase (AANAT), N-acetyl serotonin methyltransferase (ASMT), and caffeic acid O-methyltransferase (COMT) are involved in its synthesis,.

Tryptophan acts as precursor for producing melatonin in different plants. Two different processes are required for tryptophan in producing melatonin for its biosynthesis which include (i) decarboxylation and (ii) hydroxylation. Synthesis is initiated through tryptophan (Trp) decarboxylation, initiated through TDC, and forms tryptamine (Zhou et al. 2020a, b), and then tryptamine is modified to form serotonin through an enzyme, T_5H. SNAT/AANAT then converts serotonin to N-acetyl serotonin. In addition, SNAT can convert tryptamine to N-acetyl-tryptamine. N-acetyl-serotonin is then catalyzed to melatonin through ASMT/COMT.

Under certain environmental stresses, in plants, melatonin is produced through an alternative mechanism through 5-hydroxytryptophan or 5-methoxytryptophan or both intermediate(s). In which its biosynthesis is initiated through tryptophan, and it is catalyzed into 5-hydroxytryptophan through TPH. 5-hydroxytryptophan is converted to serotonin through TDC/ AADC. For example, during senescence, a lot of serotonin is synthesized in plants, and its synthetic path involves its transformation from serotonin to 5-methoxytryptamine through ASMT/COMT. Subsequently, 5-methoxytryptamine is then acetylated to yield melatonin through SNAT (Back et al. 2016; Choi et al. 2017; Colombage et al. 2023). Tryptophan is a precursor molecule for IAA and melatonin, and for that reason, melatonin has a range of functions in plants (Fig. 11.1).

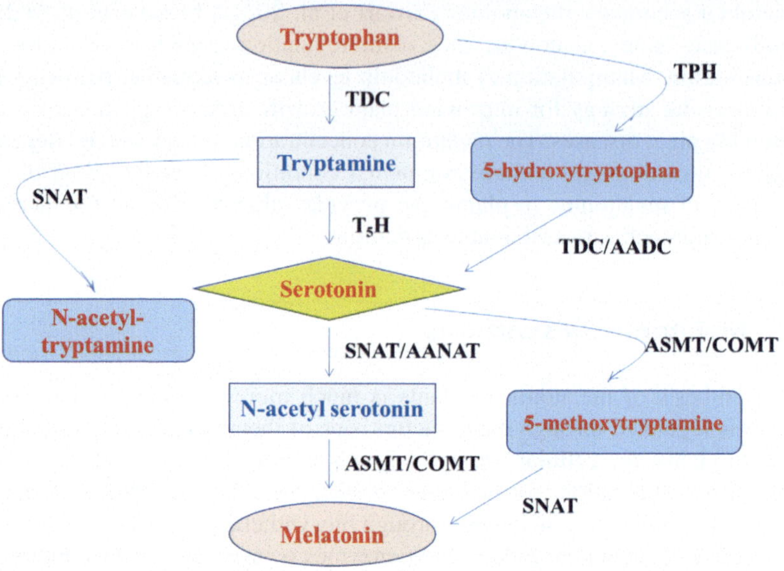

Fig. 11.1 The biosynthetic pathway of melatonin in plant is depicted, with solid-line arrows indicating the melatonin biosynthetic routes. Tryptophan decarboxylase (TDC), aromatic-L-amino-acid decarboxylase (AADC), tryptophan hydroxylase (TPH), tryptophan 5-hydroxylase (T_5H), serotonin-N-acetyltransferase (SNAT), arylalkyl amine N-acetyltransferase (AANAT), N-acetyl serotonin methyltransferase (ASMT), and caffeic acid O-methyltransferase (COMT)

11.3 Role of Melatonin in Plant Growth and Development

Numerous studies have highlighted the essential role of melatonin in various plant processes, including germination of seeds, development process of roots, circadian rhythms, photosynthesis, flowering, senescence, and adaptation to environmental stresses (Fig. 11.2). Recent research indicates that melatonin may function as a novel growth regulator or phytohormone in plants. Due to its structural resemblance to indole-3-acetic acid, it has gained the interest of plant scientists, leading to further exploration of its impact on growth and development in plants.

11.3.1 Role of Melatonin in Vegetative Growth

Melatonin enhances germination of seeds under both optimal and stressful conditions by increasing water uptake and activating antioxidant defense mechanisms. It plays a role in modulating shoot initiation, cell division, and seed dormancy in a wide range of plant species, including Arabidopsis, sweet corn, cucumber, cotton, bermudagrass, and pepper. To hasten germination, melatonin interacts with key phytohormones such as indole-3-acetic acid, giberellins, abscisic acid, and cytokinin (Yin et al. 2022; Chen et al. 2021). One of the important role of melatonin in

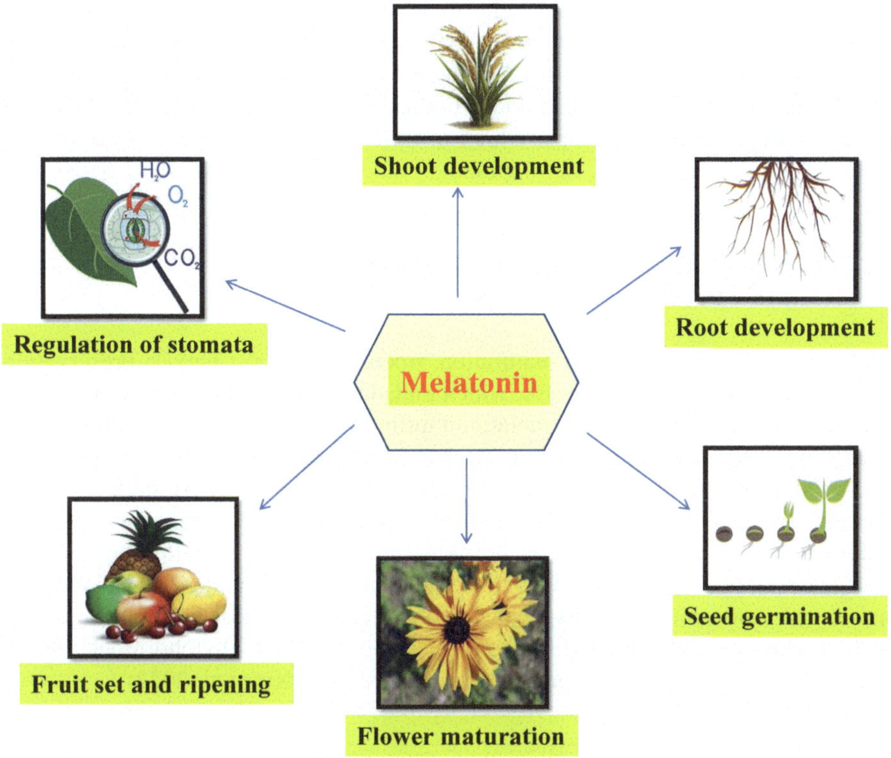

Fig. 11.2 Role of melatonin in enhancing plant growth and development

vegetative growth is to enhance root architecture, promote root elongation, and formation of lateral roots. Additionally, melatonin improves photosynthetic efficiency by promoting chlorophyll synthesis and regulating stomatal function, optimizing CO_2 assimilation. Melatonin also plays a crucial role in root system development, enhancing root growth in crops such as tomato, rice, and apple. It has also been found to increase plant height, biomass, leaf dimensions, stem diameter, water content, branching, chlorophyll levels, and photosynthetic performance. Melatonin is also responsible for establishing beneficial symbiotic relationships with microorganisms, thereby improving plant growth, productivity, and survival.

11.3.2 Role of Melatonin in Reproductive Growth

Melatonin is essential for reproductive processes, including flowering, fruit formation, and seed production. It helps to regulate the flowering induction by interacting with hormones such as gibberellins (GAs) and auxins. Melatonin promotes flowering and is involved in multiple mechanisms, including the activation of flowering locus C, stabilization of DELLA proteins (transcription regulators), and modulation of volatile compounds. Melatonin helps to protect reproductive tissues and counteract stress-induced reductions in pollen viability and germination, which facilitates carbon transport from leaves to reproductive organs and regulates carbohydrate metabolism in both male and female tissues (Hu et al. 2020). Furthermore, it increases male fertility in crops, which in turn improves autophagy-related gene expression and promotes the development of autophagosomes (Qi et al. 2018).

11.3.3 Role of Melatonin in Fruit Maturation and Post-harvest Management

Melatonin is essential for fruit maturation, ripening, and post-harvest preservation by regulating the ripening process, which enhances fruit quality and extends shelf life. As a powerful antioxidant, melatonin mitigates oxidative stress, delaying fruit senescence and preventing post-harvest deterioration (Lal et al. 2023; Tiwari et al. 2024). Moreover, melatonin helps to maintain cellular integrity by stabilizing cell membranes and reducing lipid peroxidation, thereby preventing premature spoilage. Additionally, melatonin influences hormonal metabolism, escalating the levels of abscisic acid, hydrogen peroxide, and ethylene, and also regulates fruit texture and nutritional composition. Fruits treated with melatonin exhibit improved quality, increased number, weight, and size, underscoring its role in phytohormone regulation. Furthermore, melatonin enhances fruit storability (Xie et al. 2022).

11.4 Role of Melatonin in Biotic Stress

Biotic stress refers to the negative impacts of pathogens, which include fungi, bacteria, viruses, and other microorganisms on plants. These pathogens can infect and damage several types of plant tissues and reduce both the quality and productivity of different crops. To combat the negative impact of biotic stresses, plants have developed an immune system. Physical barriers like wax layers, thick cuticles, and specialized trichomes act as the first line of defense, inhibiting disease and insect attachment. Plants detect pathogens through two different mechanisms to activate defense responses. In the first pathway, pattern recognition receptors (PRRs) detect pathogen-associated molecular patterns (PAMPs) like flagellin and elicit PAMP-triggered immunity (PTI). However, the second pathway relies on various plant resistance (R) proteins, which detect specific pathogen effectors (avirulent proteins) and activate defense mechanisms via effector-triggered immunity (ETI) (Shahid et al. 2019; Khan et al. 2021).

Melatonin is primarily responsible for plant growth and development, and several recent researches reveal that it also plays an important role in regulating plant defense responses against different biotic stresses such as pathogens and herbivores (Fig. 11.3). Additionally, exogenous application of melatonin or genetic makeup of melatonin-related genes may differ depending on several plant species and

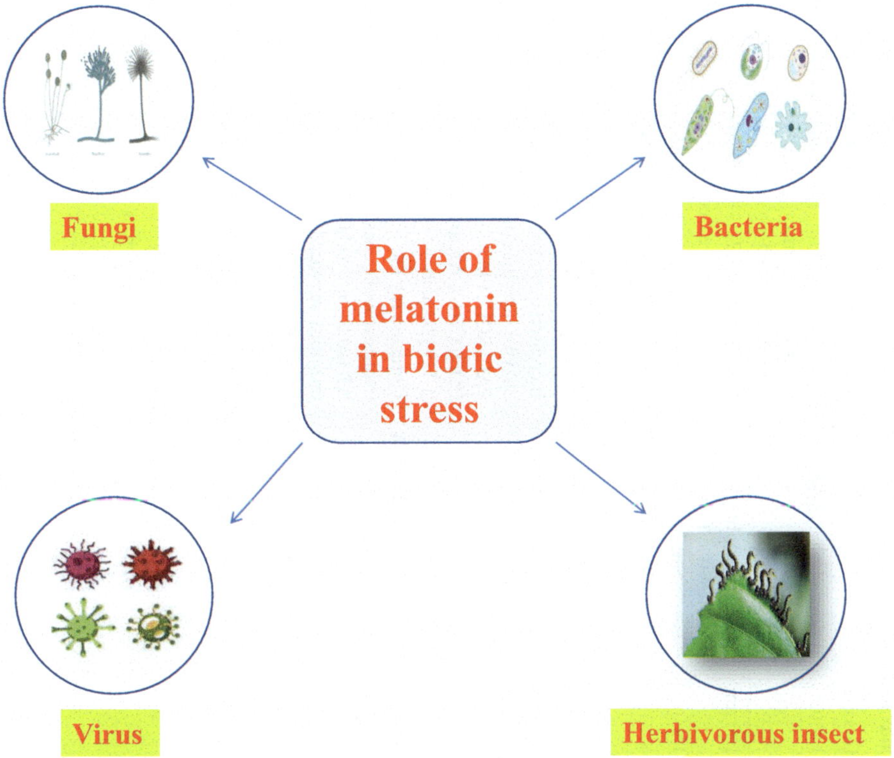

Fig. 11.3 Role of melatonin in mitigation of biotic stress which includes fungi, bacteria, virus, and herbivorous insects

experimental conditions (Table 11.1). Melatonin's positive effects on plant-pathogen interactions have been highlighted in recent studies.

11.4.1 Antifungal Effects of Melatonin

Melatonin improves plant biotic stress resistance by enhancing the tolerance to fungal pathogen infections. Antifungal activities of melatonin have been established in research on several plant species, indicating its ability to increase the plant defenses against fungal infections. Melatonin promotes defense-related chemical production, such as pathogenesis-related proteins, phytoalexins, and antimicrobial peptides, which serve as natural barriers to fungal diseases by reducing their growth and spread. Melatonin also affects the immunological signaling pathways, improving plant defense responses against harmful fungus.

Melatonin has also been shown in plant-fungal interaction models to increase resistance. Plants generally utilize programmed cell death (PCD) as a key defense mechanism against fungal infections. Melatonin helps to balance cell death and

Table. 11.1 The role of exogenously applied melatonin in plant–microbe interactions

Sl	Plant	Pathogen/Disease	Dosage	Delivery	Mechanism	Reference
1.	Watermelon	*Podosphaera xanthii, Phytophthora capsici*	1 mol/L	Foliar spray	27 genes related to constitutive defense were upregulated	Mandal et al. (2018)
2.	Tomato	Tobacco mosaic virus (TMV)	100 μm	Seedling treatment	Expression of *PR1* and *PR5*was increased	Zhao et al. (2019)
3.	Rice	Rice bacterial blight	100 μg/ml	Submerge in hydroponic solutions	Expression of three pathogenesis-related genes *PR1b,PR8a,* PR9. And NPR1 and PDF1.2 pathways	Chen et al. (2020)
4.	Grapes	*Botrytis cinerea*	2 mol/L	Fruit immersion	Increased defensive enzyme activities of PAL, SOD, PPO POD, CHI, CAT, and β-1,3-glucanase	Li et al. (2022a, b)
5.	Apple	Apple replant disease	200 μM	Soil application	Upregulating the expression of genes related to antioxidant enzymes increased ROS scavenging enzyme activity. Increased expression of nitrogen absorption genes and the activity of nitrogen metabolism enzymes	He et al. (2023)

Sl	Plant	Pathogen/Disease	Dosage	Delivery	Mechanism	Reference
6.	Maize	*Fusarium graminearum* PH-1	400 µM	Stalks and corn were pretreated	Upregulated the defense-associated gene expression of *PR1, LOX1, OXR, serPIN,* and *WIPI* and regulated antioxidant enzymes	Kong et al. (2023)
7	Rice	Blast	100 µmol/L	Seedling treatment	Increase superoxide dismutase (SOD) activity, peroxidase (POD) activity, catalase (CAT) activity, malondialdehyde (MDA) content, hydrogen peroxide (H_2O_2) content, and soluble protein content	Yuan et al. (2025)
8.	Ginger	*Fusarium oxysporum* and *Penicillium brevicompactum*	0.1 mM	Rhizome immersion	Increased expression of defense-related genes, including β-1,3-glucanase (GLU), phenylalanine ammonia-lyase (PAL), and coiled-coil nucleotide-binding site leucine-rich repeat (CC-NBS-LRR)	Huang et al. (2021)

(continued)

Table. 11.1 (continued)

Sl	Plant	Pathogen/Disease	Dosage	Delivery	Mechanism	Reference
9.	Pear	Ring rot (*Botryosphaeria dothidea*)	100 μmol/L	Fruit soaking	Increased the activity of antioxidant enzymes, including peroxidase (POD), superoxide dismutase (SOD), and polyphenol oxidase (PPO), activated the phenylpropanoid metabolic pathway, increased the contents of jasmonic acid and phlorizin	Xu et al. (2024)
10.	Tea	Sooty mold	10 mM	Spray	Enhanced CAT, SOD, PPO, and PAL activities	Thangaraj et al. (2022)

survival pathways, which allows plants to destroy contaminated cells while retaining healthy tissue. This regulated cell death mechanism makes the plants more resistant to fungal diseases (Wang et al. 2015).

Exogenous application of melatonin treatment demonstrated some promising effects on various plant-pathogen interactions. Spraying melatonin on potato leaves and tuber slices dramatically reduced late blight disease by modifying cellular ultrastructure, preventing mycelial growth, and reducing the pathogenic mold's stress tolerance (Zhang et al. 2017). Chang et al. (2018) found that melatonin treatment in *Arabidopsis thaliana* decreased the number of sporangia and the effect of *Plasmodiophora brassicae*. It was also associated with the overexpression of JA-responsive PR3 and PR4 genes.

11.4.2 Antiviral Effects of Melatonin

Viral infections damage the photosynthetic system, resulting in decreased plant growth and productivity. Melatonin, a plant growth regulator, has antiviral characteristics, which protect plants from harmful effects of different biotic stressors. It also protects the photosynthetic system against viral damage by reducing the extent of oxidative stress (Sofy et al. 2021). Hernandez-ruiz et al. (2023) found that melatonin shows an antiviral effect against specific plant viruses, by reducing viral reproduction and limiting infection spread in plant tissues. It also explains melatonin's ability to increase the production of antiviral proteins and other defense-related molecules.

Zhao et al. (2019) observed that melatonin (100 μM, twice) effectively reduced TMV RNA levels and viral concentration in highly infected tobacco and tomato seedlings. This was highly associated with an increase in salicylic acid (SA) concentrations through the nitric oxide (NO)-dependent pathway. Additionally, it efficiently removed Apple Stem Grooving Virus (ASGV) infected apple shoots of the 'Gala' variety under in vitro conditions, indicating a feasible strategy for producing virus-free plants (Chen et al. 2019a, b). However, the precise methods through which melatonin gives resistance against plant viral infections are still unknown.

11.4.3 Antibacterial Effects of Melatonin

Melatonin has also antibacterial properties, which regulate gene expression in plants by increasing the expression of genes that encode pathogenesis-related (PR) proteins, antimicrobial peptides, and other defense-related compounds. These gene products are essential for plant defense against bacterial infections. Melatonin also interacts with phytohormones such as salicylic acid (SA), jasmonic acid (JA), and ethylene (ETH) to increase plant immunity to bacterial infections by regulating their concentrations and activity (Gao et al. 2020). Melatonin affects bacterial cell membrane integrity, inhibits bacterial cell division, reduces bacterial virulence, inhibits quorum sensing, and causes oxidative stress, which results in bacterial cells to die. It also enhances plant immunological responses to harmful bacteria by activating the innate immune system, increasing reactive oxygen species (ROS) production, strengthening cell walls, and inducing programmed cell death (PCD) in infected tissues. These responses reduce the infection caused by bacteria and prevent the spread of disease. Melatonin exhibits a significant antibacterial effect against bacterial infections during plant-bacteria interactions. It modulates immune functions by stimulating antimicrobial peptide production and enhances phagocytic activity in host cells. For example, applying melatonin treatment exogenously reduced the incidence of bacterial leaf streak (BLS) in rice caused by *Xanthomonas oryzae* pv. *oryzicola* (Xoo) by 17% when applied to leaves (Chen et al. 2019a, b).

11.4.4 Role of Melatonin in Resistance Against Herbivores

Melatonin has a substantial influence on herbivore behavior. It increases the production of numerous defense-related chemicals which are poisonous or repellant to herbivores. For example, melatonin increases the synthesis of various secondary metabolites such as alkaloids, phenolics, and terpenoids, which repel or injure herbivores upon intake or contact (Arnao et al. 2022).

Melatonin regulates essential signaling pathways, which are associated with plant defense against herbivores, and regulates the activity of phytohormones like jasmonic acid (JA) and ethylene (ETH), both of which are required to activate herbivore-induced defenses. By controlling these pathways, melatonin improves the ability of the plant to recognize and counter to herbivory (Khan et al. 2022). Melatonin also activates the genes which are associated with herbivore tolerance. The susceptibility of rice plants to leaf folder and brown planthopper was greatly improved by enhancing the synthesis of JA, salicylic acid (SA), melatonin, flavonoids, and lignin. Additionally, reactive oxygen species (ROS), which are produced by herbivore grazing, induce cellular damage in plant tissues and are reduced by melatonin. However, antioxidant qualities mitigate these negative effects, boosting plant resilience (Tiwari et al. 2020).

Similarly, higher levels of melatonin, serotonin, and JA in American elm increased resistance to insect feeding. Plants that have naturally high melatonin levels, such as walnuts and herbivore consumption, can elevate melatonin levels in their bodies, potentially disrupting their physiological processes (Saremba et al. 2017). These findings indicate that melatonin can be very effectively utilized to enhance plant resistance against herbivores.

11.5 Synergistic Interactions of Melatonin

Melatonin is a bioactive molecule, known for its different roles in plant growth, development, and stress responses. While its independent functions have been extensively studied, its synergistic interactions with other plant hormones and signaling molecules become an evolving area of research. Interaction between melatonin and other plant growth hormones like auxins, gibberellins, cytokinins, abscisic acid, salicylic acid, jasmonic acid, ethylene, brassinosteroids, nitric oxide, and other bioactive compounds provide insights into their overall information on different physiological activities in plants.

11.5.1 Melatonin and Auxins

Auxin, mainly indole-3-acetic acid (IAA), is a vital phytohormone involved in cell elongation, differentiation, and organogenesis. Melatonin, a multifunctional signaling molecule in plants, shares structural similarities with different types of auxins and can influence their biosynthesis, transport, and signaling pathways. Recent studies highlight the synergistic relations between melatonin and auxin, contributing to plant development and stress alterations (Zhang et al. 2022; Samanta and Roychoudhury 2023; Jindal et al. 2024). Application of melatonin upregulated TRYPTOPHAN AMINOTRANSFERASE OF ARABIDOPSIS (TAA) and YUCCA family genes, which are critical for IAA production (Zhang et al. 2022). In tomatoes, melatonin treatment elevated YUCCA gene expression significantly, leading to an increase in the biosynthesis of IAA and improved root architecture under different stress circumstances (Tian et al. 2024). Similarly, Arabidopsis demonstrated that melatonin supplementation boosted endogenous IAA levels and facilitated lateral root formation and shoot elongation (Castro-Camba et al. 2022). Furthermore, exogenous application of melatonin elevated IAA quantity in both root and shoot tissues, leading to enhanced root elongation and lateral root development (Yang et al. 2021).

The directional transport of auxin is the fundamental key to maintaining hormonal balance and guiding plant growth responses. It influences auxin-related gene expression, enhancing auxin transporters, which regulate auxin distribution. Melatonin modulates auxin transport by influencing the expression of PIN-FORMED (PIN) and ATP-BINDING CASSETTE B (ABCB) transporters, which are necessary for auxin distribution across plant tissues. Studies in Arabidopsis revealed that treating plants with melatonin upregulated PIN1 and PIN2 genes, leading to improved auxin flow from the shoot apical meristem to root tips (Wang et al. 2016), which increased adventitious root formation, improved shoot growth, and ultimately, increased stress resilience. Similarly, melatonin-auxin interactions played a significant role in optimizing root system architecture by increasing lateral root density and modulating PIN-mediated auxin gradients in soybean (Wei et al. 2022). These studies recommend that melatonin plays as a key coordinator of auxin transport, which promote efficient distribution of auxin and allowing plants to adapt to various environmental challenges.

Melatonin influences auxin signaling by modulating activities of AUXIN RESPONSE FACTORS (ARFs) and AUXIN/INDOLE-3-ACETIC ACID (AUX/IAA) proteins (Zhang et al. 2022). Generally, ARFs control auxin-receptive gene activity, while AUX/IAA proteins function as inhibitors of auxin signaling. Research has shown that melatonin changes the degeneration dynamics of AUX/IAA proteins, thereby influencing ARF-mediated gene activation. For example, in rice, melatonin-induced auxin signaling was shown to improve seedling growth by upregulating OsARF genes and downregulating AUX/IAA repressor proteins (Liang et al. 2017).

In the development of roots, melatonin increases lateral and adventitious roots by modulating auxin biosynthesis. Studies in tomatoes have shown that melatonin application upregulated YUCCA gene activity, which leads to increased IAA levels and improved root system architecture under different stresses (Zhang et al. 2022). Similarly, in Arabidopsis, melatonin-mediated upregulation of PIN1 and PIN2 transporters facilitates auxin redistribution, which increases lateral root density (Wang et al. 2016). In maize, exogenous melatonin treatment enhanced ZmPIN-mediated auxin flow and improved drought tolerance by optimizing root elongation and water productivity (Li et al. 2023). Melatonin also contributes to biomass accumulation and shoot elongation by influencing ARFs and PIN proteins. Studies in wheat demonstrated that melatonin treatment increased shoot height and leaf area, positively correlating with increased auxin signaling (Fahad et al. 2025). Zhang et al. (2022) stated that in rice, melatonin-auxin crosstalk promoted coleoptile elongation by upregulating OsARF genes while downregulating AUX/IAA repressors.

Melatonin also plays a key role in stress adaptation by regulating auxin-mediated responses under various adverse environmental conditions (Yu et al. 2024). Tropic responses, including gravitropism and phototropism, are also regulated by melatonin-auxin interactions. In Arabidopsis, melatonin treatment altered auxin gradients, leading to enhance hypocotyl bending under asymmetric light conditions, thereby improving phototropic efficiency (Arnao and Hernández-Ruiz 2017). In maize, melatonin improved root gravitropic responses by modulating the asymmetric distribution of IAA in root cap tissues, ensuring proper root orientation during the early establishment of seedlings (Gao et al. 2024).

11.5.2 Melatonin and Cytokinin

Cytokinin is an important factor in modulating cell division and shoot formation in different parts of plants. Different researches associated with cytokinin suggest that melatonin interacts with different cytokinins to modulate shoot-to-root ratio and stress responses. Exogenous application of melatonin enhances cytokinin biosynthesis by upregulating IPT (isopentenyl transferase) genes, thereby promoting delayed senescence and increased photosynthetic effectiveness under various stresses (Xu et al. 2020; Samanta and Roychoudhury 2023). Research have shown that melatonin treatment increases endogenous cytokinin levels in Arabidopsis and rice, leading to increased shoot development and delayed leaf senescence (Zhao et al. 2021). It modulates cytokinin signaling by interacting with the HISTIDINE KINASE (HK) and RESPONSE REGULATOR (RR) proteins, which regulate the perception and signal transduction mechanisms of cytokinin.

Melatonin increases cytokinin-induced shoot proliferation and biomass accumulation by regulating cytokinin biosynthesis and signaling pathways. In Arabidopsis, melatonin treatment upregulates the activity of IPT genes, which are essential for cytokinin biosynthesis, which leads to increased shoot branching and leaf expansion (Samanta and Roychoudhury 2023). Similarly, in rice, exogenous melatonin application increases cytokinin activity, resulting in higher tiller numbers and improved grain yield under stress conditions (Yan et al. 2024). Additionally, in tobacco, the combined outcome of melatonin and cytokinin stimulates cytokinesis in shoot apical meristems, leading to increased shoot regeneration efficiency in tissue culture systems (Arnao and Hernández-Ruiz 2018). Research on wheat have suggested that melatonin-cytokinin interaction reduces leaf yellowing and extends the photosynthetic lifespan by suppressing senescence-associated genes (SAGs) while upregulating STAY-GREEN (SGR) genes (Li et al. 2022a, b). In Arabidopsis, melatonin application prevents cytokinin degradation by inhibiting CYTOKININ OXIDASE/DEHYDROGENASE (CKX) activity, thereby sustaining cytokinin levels and delaying age-induced decline in chloroplast function (Hussain et al. 2025).

11.5.3 Melatonin and Gibberellins

Gibberellins (GAs) are essential for stem elongation and seed germination. Melatonin has been reported to synergize with gibberellins to promote seed germination and stem growth under suboptimal conditions. In *Arabidopsis*, melatonin upregulated GA biosynthesis genes (GA3ox and GA20ox), while downregulating DELLA repressors, and facilitating enhanced gibberellin signaling (Husen et al. 2023). It enhances gibberellin biosynthesis by promoting the expression of GA3ox and GA20ox genes. It also interacts with gibberellin signaling by modulating DELLA proteins, which act as negative regulators of GA responses. In Arabidopsis, treating plants with melatonin led to the accumulation of a very lesser amount of DELLA protein, which promotes GA-mediated stem elongation and flowering (Jindal et al. 2024). Melatonin synergizes with GAs and regulates stem elongation by stimulating internode growth and increasing plant height, as reported in soybeans and wheat (Wu et al. 2023).

11.5.4 Melatonin and Abscisic Acid

Abscisic acid (ABA) plays a major role in regulating different kinds of stress responses. The interaction between melatonin and ABA is primarily antagonistic, where melatonin helps to mitigate ABA-induced stress responses. Melatonin counteracts ABA-induced stomatal closure, allowing plants to maintain better gas exchange under drought conditions (Ahmad et al. 2021). Additionally, melatonin reduces ABA biosynthesis by enhancing ABA catabolism, which helps in maintaining a balanced stress response. Beyond abiotic stress mitigation, they also influence seed germination and dormancy (Lv et al. 2021). While ABA generally suppresses germination, melatonin can alleviate this inhibition and promote faster and more uniform seedlings under unfavorable conditions. It is particularly beneficial in agricultural settings where rapid germination is desirable under stress conditions.

Melatonin reduces ABA-induced stomatal closure by downregulating ABA signaling genes such as PYRABACTIN RESISTANCE1 (*PYR1*) and SNF1-RELATED PROTEIN KINASE2 (*SnRK2*) and improves photosynthesis and water-use efficiency in Arabidopsis and rice (Parwez et al. 2022). This regulatory effect strengthens plant resilience by maintaining cellular redox balance, improving overall stress tolerance, and ensuring better growth and survival under adverse environmental conditions.

11.5.5 Melatonin and Salicylic Acid

Salicylic acid (SA) is an important player in plant immunity and systemic acquired resistance. Melatonin enhances SA-mediated defense responses against pathogens in different plants. The crosstalk between melatonin and SA was implicated in the activation of several pathogenesis-related (PR) genes, leading to enhance resistance against biotrophic pathogens (Tiwari et al. 2022). It also accumulates a higher concentration of reactive oxygen species (ROS), which act as signaling molecules to strengthen SA-driven immunity. They also enhance the activity of antioxidant enzymes, reducing oxidative damage and maintaining cellular homeostasis. This cooperative action of melatonin and salicylic acid suggests a dual role under biotic stress conditions.

Melatonin induces systemic resistance against biotic stress and enhances SA-mediated defense pathways, as observed in Arabidopsis and rice. It upregulates PATHOGENESIS-RELATED (*PR*) genes, boosts callose deposition, and improves resistance against fungal and bacterial pathogens (Tiwari et al. 2022). Additionally, melatonin synergizes with SA to activate systemic acquired resistance (SAR) and reinforce plant immunity under pathogen attack. Melatonin also enhances different antioxidant enzyme activity, and reduces oxidative stress during biotic challenges. By stimulating SOD, CAT, and POD activity, melatonin-SA crosstalk maintains cellular redox homeostasis, as demonstrated in wheat and tomato exposed to pathogen-induced oxidative bursts (Tiwari et al. 2022). Furthermore, melatonin regulates SA-responsive defense gene expression, including NONEXPRESSOR OF PATHOGENESIS-RELATED GENES1 (*NPR1*) and WRKY transcription factors, strengthening plant resilience against environmental stressors (Melotto et al. 2024).

11.5.6 Melatonin and Jasmonic Acid

Jasmonic acid (JA) is involved in plant defense against herbivores and pathogens. Melatonin increases JA biosynthesis and signaling through different signal transduction mechanisms. Some studies indicate that melatonin upregulates JA-responsive genes, amplifying defense mechanisms against necrotrophic fungi and insect herbivores (Rithesh et al. 2023). This interaction increases plant immunity, particularly in such cases where JA alone may be insufficient to confer resistance. Additionally, melatonin and JA co-regulate secondary metabolite production, which contributes to plant adaptation under stress conditions. Phenolic compounds and flavonoid build-up are often amplified by the combined effects of these two signaling molecules, further boosting plant defense.

Melatonin increases JA-mediated resistance against necrotrophic pathogens and herbivores by upregulating key defense-related genes such as JASMONATE ZIM-DOMAIN (*JAZ*) and LIPOXYGENASE (*LOX*), observed in Arabidopsis and tomato, where plants exhibited higher resistance to *Botrytis cinerea* and *Spodoptera litura* (Azeem and Hakeem 2024). Furthermore, melatonin-JA crosstalk increases the deposition of callose and accumulation of phytoalexin, which reinforces plant immunity under pathogen attack (Kumar et al. 2023, 2024). It also increases the production of secondary metabolites by regulating the biosynthesis of alkaloids, flavonoids, and phenolics, which contributes to plant defense and stress tolerance (Jan et al. 2021). In rice and soybean, melatonin-JA synergy increased anthocyanin and lignin content, strengthening structural barriers against pests and environmental stressors(Zou et al. 2021).

11.5.7 Melatonin and Ethylene

Ethylene regulates fruit ripening, senescence, and different stress responses. Melatonin influences ethylene biosynthesis and signaling through different signal transduction mechanisms. The interaction between melatonin and ethylene is particularly significant in delaying fruit ripening and leaf senescence. Melatonin downregulates key ethylene biosynthetic genes, reducing ethylene production and prolonging shelf life in harvested fruits (Ze et al. 2021).In different stress responses, melatonin modulated ethylene signaling pathways to enhance plant adaptation.

Melatonin delays ethylene-induced ripening and senescence by downregulating ETHYLENE RESPONSE FACTOR (*ERF*) genes and decreasing ACC OXIDASE (*ACO*) activity, as observed in tomato and banana, where melatonin treatment prolonged fruit shelf life and maintained chlorophyll content in aging leaves (Tipu and Sherif 2024).Furthermore, melatonin modulates ethylene biosynthesis to slow down ripening processes, which ensures better post-harvest quality and extended vegetative growth. Moreover, melatonin regulates ethylene-responsive transcription factors such as ERFs and WRKYs, by optimizing stress-adaptive responses like antioxidant enzyme activation and osmoprotectant accumulation, thereby improving plant survival under adverse environmental conditions.

11.5.8 Melatonin and Brassinosteroids

Brassinosteroids (BRs) are involved in cell elongation, photomorphogenesis, and vaious kind of stress responses (Manghwar et al. 2022). Melatonin synergizes with BRs, which enhance growth and resilience of plants. The interplay between these two molecules was shown to promote cell expansion and increase biomass accumulation in plants. Additionally, melatonin modulates the expression of BR-responsive genes, which leads to an improved photomorphogenic responses and shows greater adaptability to light conditions.

It enhances BR-mediated cell expansion and differentiation by upregulating BRASSINOSTEROID-INSENSITIVE1 (BRI1) and BRASSINAZOLE-RESISTANT1 (BZR1) genes, as seen in Arabidopsis and rice, where application of melatonin led to improved root elongation and leaf development (Ahammed et al. 2020). Additionally, melatonin-BR synergy regulates plant architecture and biomass accumulation by regulating BR signaling pathways, which results in increased shoot growth and enhanced productivity in wheat and soybean (Zou et al. 2021).

11.5.9 Melatonin and Nitric Oxide (NO)

Nitric oxide (NO) is a gaseous signaling molecule, which is responsible for different stress responses and hormone signaling. It modulates plant defense responses and adapt to various stress conditions in presence of melatonin. Their interaction primarily enhances antioxidant defense system, regulates stomatal movement, and coordinates different hormonal crosstalk. Melatonin enhances NO-mediated antioxidant defense system by activating key enzymes such as superoxide dismutase, catalase, and peroxidase, thereby enhancing stress tolerance in crops like maize and Arabidopsis under stress conditions (Pardo-Hernández et al. 2020). Yin et al. (2023) also noticed similar type of results where synergistic effect of NO and melatonin could improve ROS scavenging and protect cellular structures from oxidative damage in wheat and soybean.

Furthermore, crosstalk between both of them serves an important role in hormonal regulation by influencing jasmonic acid, abscisic acid, and auxin signaling networks. This interaction fine-tunes stress-responsive gene expression, including WRKY and NAC transcription factors, thereby enhancing plant adaptability to environmental fluctuations in tomato and barley (Sun et al. 2021).

11.5.10 Other Synergistic Interactions

Melatonin interacts with other different biochemical compounds, which include polyamines, flavonoids, and sugars, and increases plant stress tolerance and metabolic regulation (Samanta and Roychoudhury 2023). Melatonin enhances the production of putrescine and spermidine by upregulating different key biosynthetic enzymes such as ornithine decarboxylase (ODC) and spermidine synthase (SPDS) (Zhou et al. 2020a, b). It also plays an important role in sugar metabolism, which is crucial for adapting to different kinds of stresses and maintaining energy homeostasis in the cell. Cao et al. (2021) found that melatonin modulates sucrose metabolism by upregulating the expression of sucrose synthase (SUS) and invertase (INV) genes by optimizing carbohydrate distribution in maize and cucumber under osmotic stress conditions. By coordinating sugar signaling with antioxidant defense system, melatonin ensures optimum plant growth and resilience against various types of adverse environmental conditions.

11.6 Conclusions and Future Prospects

Melatonin is a bioactive molecule that plays an essential role in plant growth and defense, particularly in response to biotic stress. It enhances the growth of plants by regulating different physiological processes and interacting with plant growth regulators. It is also known to affect numerous gene expressions and regulatory factors, thereby reducing or reversing the negative effects of stress on plant growth and development. Its involvement is important in preventing oxidative damage and activating different stress-responsive genes. While numerous aspects of the functions of melatonin remain unknown, it has been speculated as a major regulator under stressful situations, indicating its potential as a common effector in plant defense and growth control. Future research should explore on extensive study on melatonin including gene expression, enzymes involved in melatonin production, and metabolism. Overall, multiple functions of melatonin suggest potential for enhancing plant performance and stress tolerance, but further studies are essential to fully understand its mechanisms and practical applications.

References

Ahammed GJ, Li X, Liu A, Chen S (2020) Brassinosteroids in plant tolerance to abiotic stress. J Plant Growth Regul 39(4):1451–1464. https://doi.org/10.1007/s00344-020-10098-0

Ahmad S, Muhammad I, Wang GY, Zeeshan M, Yang L, Ali I, Zhou XB (2021) Ameliorative effect of melatonin improves drought tolerance by regulating growth, photosynthetic traits and leaf ultrastructure of maize seedlings. BMC Plant Biol 21(1):368. https://doi.org/10.1186/s12870-021-03160-w

Altaf MA, Shahid R, Ren MX, Naz S, Altaf MM, Khan LU, Tiwari RK, Lal MK, Shahid MA, Kumar R, Nawaz MA (2022) Melatonin improves drought stress tolerance of tomato by modulating plant growth, root architecture, photosynthesis, and antioxidant defense system. Antioxidants 11(2):309

Arnao MB, Hernández-Ruiz J (2017) Growth activity, rooting capacity, and tropism: three auxinic precepts fulfilled by melatonin. Acta Physiol Plant 39(6):127. https://doi.org/10.1007/s11738-017-2428-3

Arnao MB, Hernández-Ruiz J (2018) Melatonin and its relationship to plant hormones. Ann Bot 121(2):195–207. https://doi.org/10.1093/aob/mcx114

Arnao MB, Hernandez-Ruiz J (2021) Regulatory role of melatonin in the redox network of plants and plant hormone relationship in stress. In: Hormones and plant response. Springer, Cham, pp 235–272

Arnao MB, Cano A, Hernández-Ruiz J (2022) Phytomelatonin: an unexpected molecule with amazing performances in plants. J Exp Bot 73:5779–5800

Azeem U, Hakeem KR (2024) Jasmonates and plant defense an eco-friendly and sustainable approach for stress tolerance. Apple Academic Press, Incorporated, Milton

Back K, Tan DX, Reiter RJ (2016) Melatonin biosynthesis in plants: multiple pathways catalyze tryptophan to melatonin in the cytoplasm or chloroplasts. J Pineal Res 61(4):426–437

Cao L, Qin B, Zhang YX (2021) Exogenous application of melatonin may contribute to enhancement of soybean drought tolerance via its effects on glucose metabolism. Biotechnol Biotechnol Equip 35(1):964–976. https://doi.org/10.1080/13102818.2021.1941254

Castro-Camba R, Sanchez C, Vidal N, Vielba J (2022) Interactions of gibberellins with phytohormones and their role in stress responses. Horticulturae 8(3):241. https://doi.org/10.3390/horticulturae8030241

Chang F, Cao X, Peng J, Tan X, Chen T, Zeng J, Cao Y, Qiao D (2018) Induced resistance to *Plasmodiophora brassicae* in Arabidopsis by melatonin. Chin J Appl Environ Bio 24:75–80

Chen X, Sun C, Laborda P, He Y, Zhao Y, Li C, Liu F (2019a) Melatonin treatments reduce the pathogenicity and inhibit the growth of *Xanthomonas oryzae* pv. oryzicola. Plant Patho 68:288–296

Chen L, Wang M, Li J, Feng C, Cui Z, Zhao L, Wang Q (2019b) Exogenous application of melatonin improves eradication of apple stem grooving virus from the infected in vitro shoots by shoot tip culture. Plant Pathol 68:997–1006

Chen X, Laborda P, Liu F (2020) Exogenous melatonin enhances rice plant resistance against *Xanthomonas oryzae* pv. oryzae. Plant Dis 104(6):1701–1708

Chen L, Lu B, Liu L, Duan W, Jiang D, Li J, Zhang K, Sun H, Zhang Y, Li C (2021) Melatonin promotes seed germination under salt stress by regulating ABA and GA3 in cotton (*Gossypium hirsutum* L.). Plant Physio Biochem 162:506–516

Choi GH, Lee HY, Back K (2017) Chloroplast overexpression of rice caffeic acid O-methyltransferase increases melatonin production in chloroplasts via the 5-methoxytryptamine pathway in transgenic rice plants. J Pineal Res 63:e12412

Colombage R, Singh MB, Bhalla PL (2023) Melatonin and abiotic stress tolerance in crop plants. Int J MolSci 24:7447. https://doi.org/10.3390/ijms24087447

Fahad S, Muhammad I, Zhang S, Alwahibi MS, Elshikh MS, Wang J (2025) Impact of melatonin application on wheat agronomic traits under abiotic stress: a meta-analysis. J Agron Crop Sci 211(1):e70016. https://doi.org/10.1111/jac.70016

Fan J, Xie Y, Zhang Z, Chen L (2018) Melatonin: a multifunctional factor in plants. Int J MolSci 19:152

Gao S, Ma W, Lyu X, Cao X, Yao Y (2020) Melatonin may increase disease resistance and flavonoid biosynthesis through effects on DNA methylation and gene expression in grape berries. BMC Plant Biol 20:231

Gao J, Zhuang S, Zhang W (2024) Advances in plant auxin biology: synthesis, metabolism, signaling, interaction with other hormones, and roles under abiotic stress. Plan Theory 13(17):2523. https://doi.org/10.3390/plants13172523

He X, Yin B, Zhang J, Zhou S, Li Z, Zhang X, Liang B (2023) Exogenous melatonin alleviates apple replant disease by regulating rhizosphere soil microbial community structure and nitrogen metabolism. Sci Total Environ 884:163830

Hernandez-Ruiz J, Giraldo-Acosta M, El Mihyaoui A, Cano A, Arnao MB (2023) Melatonin as a possible natural anti-viral compound in plant biocontrol. Plan Theory 12:781

Hu W, Cao Y, Loka DA, Harris-Shultz KR, Reiter RJ, Ali S, Liu Y, Zhou Z (2020) Exogenous melatonin improves cotton (*Gossypium hirsutum* L.) pollen fertility under drought by regulating carbohydrate metabolism in male tissues. Plant Physiol Biochem 151:579–588

Huang K, Sui Y, Miao C, Chang C, Wang L, Cao S, Wang Z (2021) Melatonin enhances the resistance of ginger rhizomes to postharvest fungal decay. Postharvest Biol Technol 182:111706

Husen A, Zhang W, EBSCOhost (eds) (2023) Hormonal cross-talk, plant defense and development. Academic Press, an imprint of Elsevier, London/San Diego, CA

Hussain S, Chang J, Li J, Chen L, Ahmad S, Song Z, Zhang B, Chen X (2025) Multifunctional role of Cytokinin in horticultural crops. IJMS 26(3):1037. https://doi.org/10.3390/ijms26031037

Jan R, Asaf S, Numan M, Lubna, Kim K-M (2021) Plant secondary metabolite biosynthesis and transcriptional regulation in response to biotic and abiotic stress conditions. Agronomy 11(5):968. https://doi.org/10.3390/agronomy11050968

Jiang C, Cui Q, Feng K, Xu D, Li C, Zheng Q (2016) Melatonin improves antioxidant capacity and ion homeostasis and enhances salt tolerance in maize seedlings. Acta Physiologiae Plantarum 38:82

Jindal P, Kant K, Kaur N, Gupta S, Ali A, Naeem M (2024) Melatonin: discovery, biosynthesis, phytohormones crosstalk, and roles in agricultural crops under abiotic stress conditions. Environ Exp Bot 226:105942. https://doi.org/10.1016/j.envexpbot.2024.105942

Kaur S, Samota MK, Choudhary M, Choudhary M, Pandey AK, Sharma A, Thakur J (2022) How do plants defend themselves against pathogens—biochemical mechanisms and genetic interventions. Physiol Mol Biol Plants 28:485–504

Khan M, Nazar T, Pande A, Mun BG, Lee D, Hussain A, Lee BH, Yun BW (2021) The role of nitric oxide-induced ATILL6 in growth and disease resistance in *Arabidopsis thaliana*. Front Plant Sci 12:1314

Khan M, Ali S, Manghwar H, Saqib S, Ullah F, Ayaz A, Zaman W (2022) Melatonin function and crosstalk with other phytohormones under normal and stressful conditions. Genes 13:1699

Kong M, Ali Q, Jing H, Hussain A, Wang F, Liu X, Xu HL (2023) Exogenous melatonin regulates plant-disease interaction by inducing maize resistance and decreasing the pathogenicity of *Fusarium graminearum*. Physiologia Plantarum 175(6):e14108

Kumar R, Altaf MA, Lal MK, Tiwari RK (eds) (2023) Melatonin in plants: a regulator for plant growth and development. Springer Nature Singapore

Kumar R, Altaf MA, Lal MK, Tiwari RK (eds) (2024) Melatonin in plants: a pleiotropic molecule for abiotic stresses and pathogen infection. Springer

Lal MK, Tiwari RK, Lal P, Kumar A, Kumar R (2023) Regulatory role of melatonin in post-harvest management of vegetables and fruits. In: Melatonin in plants: a regulator for plant growth and development. Springer, pp 219–244

Lerner AB, Case JD, Takahashi Y, Lee TH, Mori W (1958) Isolation of melatonin, the pineal gland factor that lightens melanocyteS1. J Am Chem Soc 80:2587–2587

Li Q, Yang H, Guo J, Huang Q, Zhong S, Tan F, Ren T, Li Z, Chen C, Luo P (2022a) Comparative transcriptome analysis revealed differential gene expression involved in wheat leaf senescence between stay-green and non-stay-green cultivars. Front Plant Sci 13:971927. https://doi.org/10.3389/fpls.2022.971927

Li Z, Zhang S, Xue J, Mu B, Song H, Liu Y (2022b) Exogenous melatonin treatment induces disease resistance against Botrytis cinerea on post-harvest grapes by activating defence responses. Food Secur 11(15):2231

Li J, Zhu Q, Jiao F, Yan Z, Zhang H, Zhang Y, Ding Z, Mu C, Liu X, Li Y, Chen J, Wang M (2023) Research Progress on the mechanism of salt tolerance in maize: a classic field that needs new efforts. Plan Theory 12(12):2356. https://doi.org/10.3390/plants12122356

Liang C, Li A, Yu H, Li W, Liang C, Guo S, Zhang R, Chu C (2017) Melatonin regulates root architecture by modulating auxin response in rice. Front Plant Sci 8. https://doi.org/10.3389/fpls.2017.00134

Lv Y, Pan J, Wang H, Reiter RJ, Li X, Mou Z, Zhang J, Yao Z, Zhao D, Yu D (2021) Melatonin inhibits seed germination by crosstalk with abscisic acid, gibberellin, and auxin in Arabidopsis. J Pineal Res 70(4):e12736. https://doi.org/10.1111/jpi.12736

Mandal MK, Suren H, Ward B, Boroujerdi A, Kousik C (2018) Differential roles of melatonin in plant-host resistance and pathogen suppression in cucurbits. J Pineal Res 65(3):e12505

Manghwar H, Hussain A, Ali Q, Liu F (2022) Brassinosteroids (BRs) role in plant development and coping with different stresses. IJMS 23(3):1012. https://doi.org/10.3390/ijms23031012

Meenu M, Faizan M, Sharma KP, Giri L, Sobti AK, Bansal V (2024) Phytomelatonin: history, biosynthesis and functions. In: Sharma A, Ahammed GJ (eds) Melatonin in plants: role in plant growth, development, and stress response. Springer, Singapore

Melotto M, Fochs B, Jaramillo Z, Rodrigues O (2024) Fighting for survival at the stomatal gate. Annu Rev Plant Biol 75(1):551–577. https://doi.org/10.1146/annurev-arplant-070623-091552

Naeem M, Shahzad K, Saqib S, Shahzad A, Younas M, Afridi MI (2022) The Solanummelongena COP1LIKE manipulates fruit ripening and flowering time in tomato (*Solanum lycopersicum*). Plant Growth Regul 96:369–382

Pardo-Hernández M, López-Delacalle M, Rivero RM (2020) ROS and NO regulation by melatonin under abiotic stress in plants. Antioxidants 9(11):1078. https://doi.org/10.3390/antiox9111078

Parwez R, Aftab T, Gill SS, Naeem M (2022) Abscisic acid signaling and crosstalk with phytohormones in regulation of environmental stress responses. Environ Exp Bot 199:104885. https://doi.org/10.1016/j.envexpbot.2022.104885

Qi ZY, Wang KX, Yan MY, Kanwar MK, Li DY, Wijaya L, Alyemeni MN, Ahmad P, Zhou J (2018) Melatonin alleviates high temperature-induced pollen abortion in *Solanum lycopersicum*. Molecules 23:386

Reiter RJ, Tan DX, Sharma R (2018) Historical perspective and evaluation of the mechanisms by which melatonin mediates seasonal reproduction in mammals. Melatonin Res 1:59–77. https://doi.org/10.32794/mr11250004

Rithesh L, Krishna GV, Rao SS, Shiva B (2023) Emerging role of melatonin in integrated management of crop pathogens. In: Kumar R, Altaf MA, Lal MK, Tiwari RK (eds) Melatonin in plants: a pleiotropic molecule for abiotic stresses and pathogen infection. Springer Nature Singapore, Singapore, pp 195–221

Samanta S, Roychoudhury A (2023) Crosstalk of melatonin with major phytohormones and growth regulators in mediating abiotic stress tolerance in plants. S Afr J Bot 163:201–216. https://doi.org/10.1016/j.sajb.2023.10.040

Saqib S, Nazeer A, Ali M, Zaman W, Younas M, Shahzad A, Nisar M (2022) Catalytic potential of endophytes facilitates synthesis of biometallic zinc oxide nanoparticles for agricultural application. Biometals 35:1–19

Saremba BM, Tymm FJ, Baethke K, Rheault MR, Sherif SM, Saxena PK, Murch SJ (2017) Plant signals during beetle (*Scolytus multistriatus*) feeding in American elm (*Ulmus americana* Planch). Plant Signal Behav 12(5):e1296997

Shahid M, Imran QM, Hussai A, Khan M, Lee SU, Mun BG, Yun BW (2019) Comprehensive analyses of nitric oxide-induced plant stem cell-related genes in *Arabidopsis thaliana*. Genes 10:190

Sofy AR, Sofy MR, Hmed AA, Dawoud RA, Refaey EE, Mohamed HI, El-Dougdoug NK (2021) Molecular characterization of the Alfalfa mosaic virus infecting Solanummelongena in Egypt and the control of its deleterious effects with melatonin and salicylic acid. Plan Theory 10:459

Sun C, Liu L, Wang L, Li B, Jin C, Lin X (2021) Melatonin: a master regulator of plant development and stress responses. JIPB 63(1):126–145. https://doi.org/10.1111/jipb.12993

Thangaraj K, Liu S, Li J, Zhao Z, Han R, Mei H, Li X (2022) Exogenous melatonin alleviates sooty mould on tea plants (*Camellia sinensis* L.). Scientia Horticulturae 299:111056

Tian Q, Wang G, Dou J, Niu Y, Li R, An W, Tang Z, Yu J (2024) Melatonin modulates tomato root morphology by regulating key genes and endogenous hormones. Plan Theory 13(3):383. https://doi.org/10.3390/plants13030383

Tipu MMH, Sherif SM (2024) Ethylene and its crosstalk with hormonal pathways in fruit ripening: mechanisms, modulation, and commercial exploitation. Front Plant Sci 15:1475496. https://doi.org/10.3389/fpls.2024.1475496

Tiwari RK, Lal MK, Naga KC, Kumar R, Chourasia KN, Subhash S, Kumar D, Sharma S (2020) Emerging roles of melatonin in mitigating abiotic and biotic stresses of horticultural crops. Sci Hortic 272:109592

Tiwari RK, Lal MK, Kumar R, Kumar R (2022) Insight into melatonin-mediated response and signaling in the regulation of plant defense under biotic stress. Plant Mol Biol 109:385–399. https://doi.org/10.1007/s11103-021-01202-34

Tiwari RK, Altaf MA, Kumar R, Lal MK (2024) Evolution of melatonin as an oxidative stress mitigator in plant. In Melatonin in plants: a pleiotropic molecule for abiotic stresses and pathogen infection. Singapore, Springer Nature Singapore, pp 1–20

Tiwari RK, Lal MK, Kumar R, Chourasia KN, Naga KC, Kumar D, Das SK, Zinta G (2021a) Mechanistic insights on melatonin-mediated drought stress mitigation in plants. Physiol Plant 172(2):1212–1226. https://doi.org/10.1111/ppl.13307

Tiwari RK, Lal MK, Kumar R, Mangal V, Altaf MA, Sharma S, Singh B, Kumar M (2021b) Insight into melatonin-mediated response and signaling in the regulation of plant defense under biotic stress. Plant Mol Biol 109:385–399

Wang P, Sun X, Wang N, Tan DX, Ma F (2015) Melatonin enhances the occurrence of autophagy induced by oxidative stress in Arabidopsis seedlings. J Pineal Res 58:479–489

Wang Q, An B, Wei Y, Reiter RJ, Shi H, Luo H, He C (2016) Melatonin regulates root meristem by repressing auxin synthesis and polar auxin transport in Arabidopsis. Front Plant Sci 07. https://doi.org/10.3389/fpls.2016.01882

Wei W, Tao J-J, Yin C-C, Chen S-Y, Zhang J-S, Zhang W-K (2022) Melatonin regulates gene expressions through activating auxin synthesis and signaling pathways. Front Plant Sci 13:1057993. https://doi.org/10.3389/fpls.2022.1057993

Wu Y, Liu J, Wu H, Zhu Y, Ahmad I, Zhou G (2023) The roles of Mepiquate chloride and melatonin in the Morpho-physiological activity of cotton under abiotic stress. IJMS 25(1):235. https://doi.org/10.3390/ijms25010235

Xie Q, Zhang Y, Cheng Y, Tian Y, Luo J, Hu Z, Chen G (2022) The role of melatonin in tomato stress response, growth and development. Plant Cell Rep 41:1631–1650

Xu Y, Li K, Zhu K, Tian Y, Yu Q, Zhang W, Wang Z (2020) Effect of exogenous plant hormones on agronomic and physiological performance of a leaf early-senescent rice mutant osled. Plant Growth Regul 92(3):517–533. https://doi.org/10.1007/s10725-020-00653-w

Xu H, Zhang S, Liang C, Li M, Wang R, Song J, Li D (2024) Melatonin enhances resistance to *Botryosphaeria dothidea* in pear by promoting jasmonic acid and phlorizin biosynthesis. BMC Plant Biol 24(1):470

Yan F, Zhang G, Zhao H, Huang Z, Niu Y, Zhu M (2024) Foliar application of melatonin improve the number of secondary branches and secondary branch grains quality of rice. PLoS One 19(8):e0307368. https://doi.org/10.1371/journal.pone.0307368

Yang L, You J, Li J, Wang Y, Chan Z (2021) Melatonin promotes Arabidopsis primary root growth in an IAA-dependent manner. J Exp Bot 72(15):5599–5611. https://doi.org/10.1093/jxb/erab196

Yin X, Bai YL, Gong C, Song W, Wu Y, Ye T, Feng YQ (2022) The phytomelatonin receptor PMTR1 regulates seed development and germination by modulating abscisic acid homeostasis in *Arabidopsis thaliana*. J Pineal Res 72:e12797

Yin Y, Zhang R, Tian X, Yang Z, Fang W (2023) Nitric oxide affects melatonin mediates enrichment of isoflavones and physiological biochemistry in germinated soybeans under ultraviolet-B stress. Plant Growth Regul 100(3):657–666. https://doi.org/10.1007/s10725-022-00958-y

Yuan H, Qian J, Wang C, Shi W, Chang H, Yin H, Li Q (2025) Exogenous melatonin enhances Rice blast disease resistance by promoting seedling growth and antioxidant defense in rice. Int J Mol Sci 26(3):1171

Yu J, Luo X, Zhou Q, Li Z, Chen S (2024) Exogenous melatonin alleviates osmotic stress by enhancing antioxidant metabolism, photosynthetic maintenance, and hormone homeostasis in forage oat (*Avena sativa*) seedlings. Grasses 3:190–204. https://doi.org/10.3390/grasses3030014

Ze Y, Gao H, Li T, Yang B, Jiang Y (2021) Insights into the roles of melatonin in maintaining quality and extending shelf life of postharvest fruits. Trends Food Sci Technol 109:569–578. https://doi.org/10.1016/j.tifs.2021.01.051

Zhang S, Zheng X, Reiter RJ, Feng S, Wang Y, Liu S (2017) Melatonin attenuates potato late blight by disrupting cell growth, stress tolerance, fungicide susceptibility and homeostasis of gene expression in *Phytophthora infestans*. Front Plant Sci 8. https://doi.org/10.3389/fpls.2017.01993

Zhang M, Gao C, Xu L, Niu H, Liu Q, Huang Y, Lv G, Yang H, Li M (2022) Melatonin and Indole-3-acetic acid synergistically regulate plant growth and stress resistance. Cells 11(20):3250. https://doi.org/10.3390/cells11203250

Zhao L, Chen L, Gu P, Zhan X, Zhang Y, Hou C, Wang QC (2019) Exogenous application of melatonin improves plant resistance to virus infection. Plant Pathol 68(7):1287–1295

Zhao Y-Q, Zhang Z-W, Chen Y-E, Ding C-B, Yuan S, Reiter RJ, Yuan M (2021) Melatonin: a potential agent in delaying leaf senescence. Crit Rev Plant Sci 40(1):1–22. https://doi.org/10.1080/07352689.2020.1865637

Zhou R, Hu Q, Pu Q, Chen M, Zhu X, Gao C, Zhou G, Liu L, Wang Z, Yang J, Zhang J, Cao Y (2020a) Spermidine enhanced free polyamine levels and expression of polyamine biosynthesis enzyme gene in Rice Spikelets under heat tolerance before heading. Sci Rep 10(1):8976. https://doi.org/10.1038/s41598-020-64978-2

Zhou Y, Liao L, Liu X, Liu B, Chen X, Guo Y (2020b) Crystal structure of Oryza sativa TDC reveals the substrate specificity for TDC-mediated melatonin biosynthesis. J Adv Res 24:501–511. https://doi.org/10.1016/j.jare.2020.06.004

Zou J, Yu H, Yu Q, Jin X, Cao L, Wang M, Wang M, Ren C, Zhang Y (2021) Physiological and UPLC-MS/MS widely targeted metabolites mechanisms of alleviation of drought stress-induced soybean growth inhibition by melatonin. Ind Crop Prod 163:113323. https://doi.org/10.1016/j.indcrop.2021.113323

The Emerging Role of Melatonin in Insect Behavior, Parasitology, and Plant–Insect Interactions: Ecological Implications and Applications

12

Deepak Kumar Mahanta, Ipsita Samal,
Tanmaya Kumar Bhoi, and J. Komal

Abstract

The established circadian rhythm regulator known as melatonin demonstrates growing importance in ecosystem-wide effects which extend past its biological functions. The importance of melatonin for controlling insect behavior and parasitology and plant–insect interactions continues to grow according to recent scientific investigations. Insect organisms use melatonin to control their circadian cycles and stress reactions and feeding patterns which impacts how different species connect with each other in food chains and ecosystem functions. Research in parasitology reveals that melatonin exhibits potential effects on host–parasite relationships and disease parasite life cycles as well as immunological responses, thus providing new possibilities for disease control. The growing interest in plant defense biology focuses on melatonin because this molecule demonstrates the ability to reduce insect feeding and strengthen plant resistance against insect attacks. This chapter examines how melatonin functions in various ecological roles to reveal its potential applications in sustainable farming and pest control and biodiversity preservation. This research study combines botanical, entomo-

D. K. Mahanta
Forest Entomology Discipline, Forest Protection Division, Indian Council of Forestry Research and Education (ICFRE)-Forest Research Institute (ICFRE-FRI), Dehradun, Uttarakhand, India

I. Samal (✉)
ICAR-National Research Centre on Litchi, Mushahari, Ramna, Muzaffarpur, Bihar, India

T. K. Bhoi
Forest Protection Division, ICFRE-Arid Forest Research Institute (ICFRE-AFRI), Jodhpur, Rajasthan, India

J. Komal
Basic Seed Multiplication and Training Centre, Central Silk Board, Kharsawan, Jharkhand, India

R. K. Tiwari et al. (eds.), *Melatonin: Signal Transduction Mechanisms and Defense Networks in Plants*, https://doi.org/10.1007/978-981-96-8869-2_12

logical, and parasitic findings about melatonin to demonstrate its ecological significance in boosting ecosystem resilience against environmental threats.

Keywords
Melatonin · Insect behavior · Parasitology · Plant–insect interactions

12.1 Introduction

Melatonin, or *N-acetyl-5-methoxytryptamine*, is a multifaceted, ancient molecule characterized by a largely conserved structure present in several creatures, including mammals, plants, fungi, and some bacteria (Rajotia et al. 2023). Melatonin is produced through enzymatic processes that convert tryptophan into the substance (Arnao and Hernández-Ruiz 2014). The vertebrate pineal gland functions as the primary melatonin-producing structure which responds to the dark periods of the light–dark cycle (Maugars et al. 2020). The traditional role of melatonin encompasses regulation of circadian rhythms through its night-time signaling mechanism which maintains internal biological clocks in harmony with external environmental conditions (Pandi-Perumal et al. 2022). The system controls vital operations including sleep patterns and seasonal reproduction and daily metabolic processes (Pache 2020). The antioxidant properties of melatonin enable it to eliminate harmful reactive oxygen and nitrogen species that result from cellular metabolic processes (Ain et al. 2024). The antioxidant properties of melatonin include direct free radical elimination while simultaneously activating antioxidant enzymes like superoxide dismutase and glutathione peroxidase to enhance overall body protection against oxidative stress (Tumilaar et al. 2024; Altaf et al. 2022; Tiwari et al. 2024). The special characteristic of melatonin enables it to cross cellular membranes and blood–brain barrier protection which safeguards multiple tissues including the brain from damage (Mineiro et al. 2024). Studies have proven that melatonin functions as a fundamental chemical to reduce age-related and stress-induced degenerative damage (Verma et al. 2023a, b). The proven roles of melatonin include immunomodulatory effects which control inflammatory reactions through their impact on immune cells (Koul et al. 2024). The compound supports cardiovascular health while functioning in mitochondria and signaling pathways that control cell death and cell reproduction (Reiter et al. 2024; Kumar et al. 2023, 2024). Research on melatonin in plants and microbial systems has elevated its importance because these findings show its role in growth control, stress response, and environmental adaptation (Erland 2024). The multiple functions of this important molecule make it a subject of intense research interest because of its potential therapeutic applications in treating sleep disorders and neurological diseases and conditions related to oxidative stress.

The study of melatonin within insect behavior and parasitology and plant–insect interactions helps explain its broader biological significance and ecological roles (Kim et al. 2024). Figure 12.1 shows the scientific productivity regarding insect research and melatonin function among different countries. Studies on melatonin

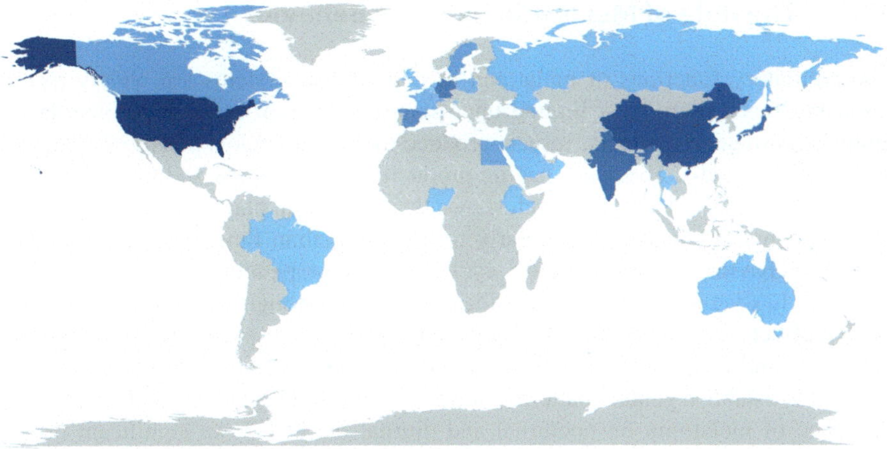

Fig. 12.1 Country-wise scientific production on the role of melatonin in insect research. The darker shades represent the higher level of scientific work in melatonin research on insect pest

regulatory functions in mammals have received attention but scientists are currently exploring its presence and effects in non-mammalian organisms including insects and plants and parasites (Liu et al. 2023). Research demonstrates that melatonin regulates circadian cycles and stress responses and reproductive behaviors in insects, yet scientists have not fully explained its specific mechanisms or wider behavioral effects (Bloch et al. 2013). Research into how melatonin affects insect foraging along with navigation and communication could reveal fundamental principles of insect behavioral biology because of their ecological and diverse nature (Takeda and Suzuki 2022).

The scientific field of parasitology identifies melatonin as a key element in host–parasite relationships that modifies parasite infection cycles and their ability to infect hosts (Bagnaresi et al. 2012). Studies of melatonin regulation of immunological responses and oxidative stress provide understanding about how parasites evade or control host defense mechanisms (Calvo et al. 2013). The obtained results can help develop new treatment methods that target melatonin pathways in parasitic infections specifically in vector-borne diseases including malaria and leishmaniasis (Daryani et al. 2018). The research field of plant physiology and herbivorous–insect interactions with melatonin shows strong growth potential according to Subala et al. (2017) and Barberà et al. (2020). Plants use melatonin as a signaling chemical to control growth and stress resilience as well as defense systems which affects feeding behavior of herbivorous insects and plant–insect interactions (Sun et al. 2021). Exploration of these relationships could generate new agricultural applications for pest control that include both tolerance enhancement in crops and sustainable pest management systems. Researchers can improve their understanding of melatonin's diverse applications by filling the knowledge gaps in these scientific fields.

12.2 The Role of Melatonin in Insect Behavior

The regulatory functions of melatonin in mammalian systems are widely recognized but researchers have limited understanding of its roles in non-mammalian animals, particularly insects. Insects produce melatonin endogenously, yet experts still lack knowledge about its specific production mechanisms and biological outcomes (Zhao et al. 2019). Research shows that melatonin plays a role in regulating circadian rhythms in insects similarly to its mammalian functions which modify feeding behavior and reproductive cycles and flight patterns (Fan et al. 2021). Research has yet to identify the chemical pathways through which melatonin triggers its effects in insect bodies. The large number of insect species with their diverse behaviors and ecological requirements and physiological characteristics makes it difficult to establish general patterns for melatonin activities (Wang et al. 2015). The influence of melatonin on nocturnal and diurnal insects shows significant differences but researchers have not thoroughly studied these variations (Subala and Shivakumar 2018). Research on how melatonin fights stress and protects against oxidative damage in insects during adverse environmental situations remains a poorly understood field. The lack of understanding about melatonin studies prevents their future application in entomology and ecology research (Schwarzenberger and Wacker 2015). Scientific insights about melatonin effects on insect behavior could lead to new pest management practices while improving pollinator health (Li et al. 2022a, b). The gap requires multiple research disciplines to understand both molecular mechanisms and environmental effects and evolutionary roles of melatonin in insects. The complete understanding of its functions combined with its potential applications for ecological and agricultural development can be achieved through such research. Figure 12.2 depicts research trends on the role of melatonin in insect research.

Research should concentrate on the specific ways melatonin affects insect behavior by studying its effects on circadian rhythm and stress response and behavioral modulation. The biological processes of insects heavily depend on circadian rhythms which control their foraging behavior and reproductive cycles and migration patterns (Yamano et al. 2001). The signaling molecule role of melatonin helps regulate these rhythms to explain how insects match their behavioral patterns with natural light–dark transitions (Kamruzzaman et al. 2021). Research on the relationship between insect circadian genes and neurochemical pathways and melatonin could help explain how this hormone controls night-time moth flying behavior and day-time bee foraging (Lampel et al. 2005). Research on multiple species has proven melatonin functions as an effective antioxidant, thus establishing its role in insect stress response mechanisms (Wang et al. 2013). The natural environment produces various stresses that affect insects through ultraviolet light-induced oxidative damage and harsh temperature conditions and pollution exposure (Ali et al. 2017). Investigating how melatonin fights stress through free radical defense mechanisms and enhanced antioxidant enzymes could help researchers understand insect survival strategies in hostile conditions. The study of how melatonin affects behavioral responses will help scientists understand its ecological importance for insects.

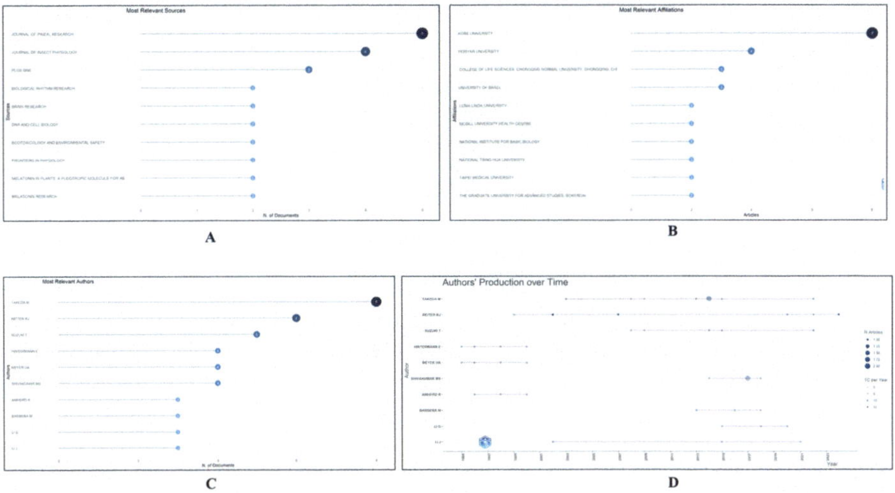

Fig. 12.2 Research information on the role of melatonin in insect research from bibliometric data (20th December 2024). (**a**) Most relevant sources. (**b**) Most relevant affiliations. (**c**) Most relevant authors. (**d**) Authors' production over time

The survival and reproduction processes including navigation and mating and pheromone signaling appear to be influenced by melatonin according to Shi et al. 2021. Multiple insect species will benefit from experimental research using transcriptome analysis and behavioral testing and melatonin pathway modifications to reveal its distinctive activities. The study would improve our understanding of insect biology while delivering practical outcomes. Research on melatonin regulatory functions will help establish methods for preserving pollinators and controlling pests while improving ecosystem stability. Research in the key domains of insects will create new possibilities for agricultural along with environmental applications within melatonin studies. Figure 12.3 represents co-occurrence of keywords related to the role of melatonin in insect research.

The scientific community recognizes honeybees because of their complex social organization and precise behavioral patterns such as foraging and navigation (Even et al. 2022). Research demonstrates that honeybees use melatonin to regulate their circadian rhythms which helps them match their internal clock mechanisms with external light–dark patterns (Yang et al. 2007). The timing of feeding operations and hive maintenance tasks which sustain colony survival seems to be influenced by melatonin in these insects. The scientific community has demonstrated that melatonin affects how honeybees remember and learn information which could affect their flower recognition abilities and sun navigation (Lehmann et al. 2011). The circadian cycle regulation through melatonin appears to control the feeding and mating behaviors of mosquitoes (Rund et al. 2016). The timing of biological rhythms depends on how melatonin affects them and happens most frequently during dusk or morning hours (Dias et al. 2024). The external stress response of mosquitoes is

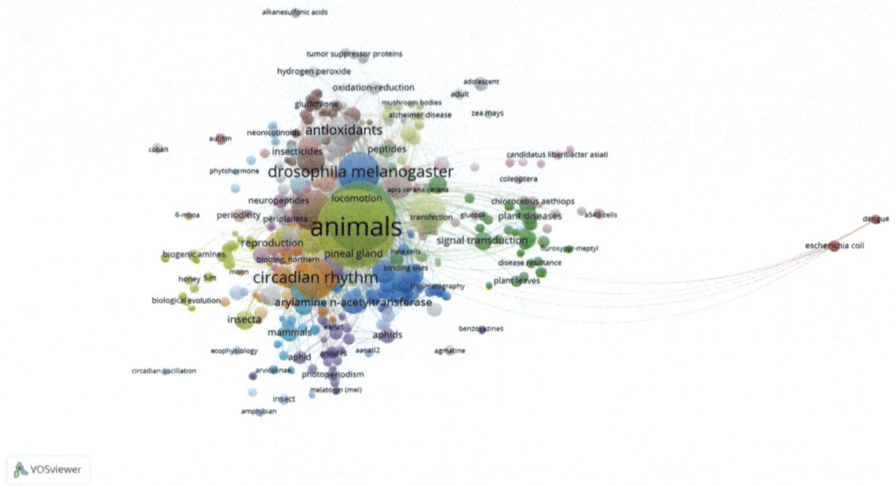

Fig. 12.3 Co-occurrence of keywords related to the role of melatonin in insect research

altered by melatonin which affects their ability to survive and reproduce effectively. Understanding how melatonin affects mosquito behavior could help explain vector-borne disease transmission patterns since the timing of feeding and mating activities determines disease spread. Two nocturnal moth species *Spodoptera* and *Manduca sexta* display behaviors that are linked to the presence of melatonin according to Karthi and Shivakumar (2015) and Maxwell and Hildebrand (1981). The daily rhythms of these moths control their nighttime activities for flying and sexual mating. Through melatonin regulation, the insects become better able to match their mating and foraging activities to the night time (Linn et al. 1995). The antioxidant properties of melatonin protect moths from nocturnal flying damage caused by low-light conditions which increases their chance of survival (Desouhant et al. 2019). Research shows that *Drosophila melanogaster* exhibits melatonin-related behaviors as described in Ortega-Arellano et al. 2021. The hormone melatonin controls natural body clocks which in turn affect the timing of eating habits together with sleep cycles and movement patterns. Scientific research shows that melatonin helps fruit flies adjust to environmental changes including different light conditions and food availability. The protective mechanism of melatonin against oxidative stress plays a vital role in both extending life span and improving environmental resistance for fruit flies because they regularly experience high reactive oxygen species (ROS) levels during their active time (Wong et al. 2021). Cicadas display behaviors that link to melatonin even though they are best known for their specific vocalizations and cyclic life cycles. The timing regulation of their emergence depends on melatonin because it helps synchronize mating while providing protection from predators.

12.3 Melatonin in Parasitology

The study of melatonin in parasitology faces challenges because of intricate host–parasite relationships which depend on both host immune responses and parasite evasion abilities of immunological defenses. Malaria-causing *Plasmodium* spp. and trypanosomiasis-causing *Trypanosoma* spp. along with leishmaniasis-causing *Leishmania* spp. employ complex strategies to benefit from their host's physiological environment (Singh et al. 2020). The parasite achieves this through three different strategies which include immune system adjustment and detection evasion and circadian rhythm manipulation to match its life cycle (Franco et al. 2024). Treatment discovery becomes more difficult because of these parasitic techniques. The immune regulatory and antioxidant properties of melatonin have been identified but scientists need to study how this compound interacts with parasites and their hosts (Huang et al. 2023). Studies show that melatonin controls immune system function through cytokine synthesis regulation and other immune substances which affect infection resistance in hosts (Maestroni 2024). Scientists have yet to understand how melatonin affects parasite survival together with host defense mechanisms (Calvo et al. 2013). Parasites use melatonin to modify host immune cell signaling which helps them evade immune monitoring, allowing better parasite multiplication and survival within the host (Klimczak et al. 2024). Circadian rhythm regulation through melatonin use creates additional complexities in the system. The host biological clock serves parasites by helping their replication cycles match up with periods of weakened immune defenses or host vulnerability (Reece et al. 2017). Studies show that specific parasitic diseases follow daily cycles which melatonin potentially controls in both host organisms and parasites (Carvalho Cabral et al. 2019). Research needs to expand regarding how melatonin affects parasitic conditions because it influences both parasite conduct and host immune system reactions (Vielma et al. 2014). The lack of understanding about these interactions makes it difficult to develop treatment methods based on melatonin. The therapeutic application of melatonin faces challenges as a treatment or supporting agent for parasitic diseases because of its complex interactions within host systems and parasites (Szewczyk-Golec et al. 2021). Future research needs to explore melatonin's potential uses for parasitic disorder treatment through behavioral modification of parasites or immune system support or free radical protection. The immunomodulatory and life-cycle regulating properties of melatonin enable researchers to create new treatment strategies for parasitic diseases in the field of parasitology (Singh et al. 2019). The established immune regulatory properties of melatonin enable scientists to strengthen host immune responses which boosts resistance against parasite diseases (Bagnaresi et al. 2012). The administration of melatonin enables hosts to better detect and eliminate parasite intruders through its effect on essential immune mediator synthesis including cytokines and interferons (Kotb Elmahallawy et al. 2015). The inflammatory response in malaria patients can be better controlled by melatonin which enhances immune cell activation and reduces parasite numbers. The ability of melatonin to control oxidative stress serves as a vital mechanism that medical professionals can use for parasite treatment. The infection process by parasites leads to oxidative stress in the host which can damage tissues and impair immune function (Pawłowska et al. 2023). Through its ability to boost antioxidant defenses, melatonin reduces oxidative damage and strengthens overall health which results in improved immune system infection-fighting capabilities. Research shows that melatonin defends immune system cells by protecting them from reactive oxygen species (ROS) which form during parasite invasion (Maldonado et al. 2020). The regulatory mechanism of melatonin for circadian rhythms presents a possible approach to disrupt parasite life cycles (Srinivasan et al. 2012). The activity patterns of parasitic organisms such as Plasmodium (malaria) and Trypanosoma (sleeping sickness) follow the circadian rhythms of their host during day or night periods. The administration of melatonin as a biological clock regulator disrupts the natural alignment between the immune system of the host and the reproductive cycle of the parasite (Barik 2019). The parasite's ability to reproduce and thrive is reduced during essential life cycle stages which leads to increased chances for the host to recover from infection. The combination of conventional treatments with melatonin-based therapies shows promise in reducing medication resistance in parasitic diseases according to Chitimus et al. (2020). The combination of melatonin-directed parasitic life-cycle stage monitoring with host–parasite interaction research strengthens antimalarial and antiparasitic drugs and immunomodulatory treatments while reducing drug resistance development potential. The complete potential of melatonin in parasitology needs additional research to identify how it influences both parasite biology and host immune system processes. Research needs to evaluate how melatonin affects immune cell signaling and oxidative stress responses and life cycle patterns of various parasitic species. Researchers can develop new therapeutic medicines by studying these domains because melatonin regulates parasite life cycles and modulates immune responses.

The immunomodulatory and life-cycle regulating properties of melatonin enable researchers to create new treatment strategies for parasitic diseases in the field of parasitology (Singh et al. 2019). The established immune regulatory properties of melatonin enable scientists to strengthen host immune responses which boosts resistance against parasite diseases (Bagnaresi et al. 2012). The administration of melatonin enables hosts to better detect and eliminate parasite intruders through its effect on essential immune mediator synthesis including cytokines and interferons (Kotb Elmahallawy et al. 2015). The inflammatory response in malaria patients can be better controlled by melatonin which enhances immune cell activation and reduces parasite numbers. The ability of melatonin to control oxidative stress serves as a vital mechanism that medical professionals can use for parasite treatment. The infection process by parasites leads to oxidative stress in the host which can damage tissues and impair immune function (Pawłowska et al. 2023). Through its ability to boost antioxidant defenses, melatonin reduces oxidative damage and strengthens overall health which results in improved immune system infection-fighting capabilities. Research shows that melatonin defends immune system cells by protecting them from reactive oxygen species (ROS) which form during parasite invasion (Maldonado et al. 2020). The regulatory mechanism of melatonin for circadian rhythms presents a possible approach to disrupt parasite life cycles (Srinivasan et al.

2012). The activity patterns of parasitic organisms such as Plasmodium (malaria) and Trypanosoma (sleeping sickness) follow the circadian rhythms of their host during day or night periods. The administration of melatonin as a biological clock regulator disrupts the natural alignment between the immune system of the host and the reproductive cycle of the parasite (Barik 2019). The parasite's ability to reproduce and thrive is reduced during essential life cycle stages which leads to increased chances for the host to recover from infection. The combination of conventional treatments with melatonin-based therapies shows promise in reducing medication resistance in parasitic diseases according to Chitimus et al. 2020. The combination of melatonin-directed parasitic life-cycle stage monitoring with host–parasite interaction research strengthens antimalarial and antiparasitic drugs and immunomodulatory treatments while reducing drug resistance development potential. The complete potential of melatonin in parasitology needs additional research to identify how it influences both parasite biology and host immune system processes. Research needs to evaluate how melatonin affects immune cell signaling and oxidative stress responses and life cycle patterns of various parasitic species. Researchers can develop new therapeutic medicines by studying these domains because melatonin regulates parasite life cycles and modulates immune responses.

12.4 Melatonin's Influence on Plant Feeding Behavior

Modern agriculture faces herbivore and pest management as a primary challenge according to Horgan (2020), Samal et al. (2024a, b), and Bhatnagar et al. (2024). Herbivorous insects such as aphids caterpillars and beetles damage crop output quality by eating plant tissues which reduces productivity and makes plants more susceptible to diseases and creates economic losses (Kumar and Rathor 2020; Mahanta et al. 2022, 2023a, b; Bhoi et al. 2022; Majhi et al. 2023). The control of these pests through conventional methods primarily depends on chemical pesticides (Zhou et al. 2024). These pest control methods lead to various negative outcomes which include pesticide resistance alongside damage to beneficial insects and environmental pollution as well as health risks for both humans and animals (Hashimi et al. 2020). Modern agriculture faces a dual challenge because sustainable and environmentally friendly agricultural methods are becoming more essential (Tiwari et al. 2020, 2021a, b). Soil and water pollution, combined with ecosystem destruction, occur when chemical pesticides are used excessively (Alengebawy et al. 2021). The resistance development of pests against these chemicals through time causes their effectiveness to decline, thereby requiring the use of increasingly toxic substances (Qu et al. 2022; Samal et al. 2024c, d). Agricultural ecosystems face a risk of natural predator–prey disruption which leads to unstable pest control outcomes. The identification of alternative sustainable pest management solutions has gained attention because of current problems (Mahanta et al. 2023c, 2024; Siddiqui et al. 2024; Bhoi et al. 2023). Research into plant signaling chemicals such as melatonin can reveal their effects on both herbivore behaviors and plant defense mechanisms (Samal et al. 2024c). The antioxidant chemical known as melatonin shows ability to modify plant stress reactions while potentially influencing insect feeding behavior, thus enabling plants to better fight against insect damage (Bhoi et al. 2024a). The agricultural sector experiences significant losses because insects both assist crops and inflict substantial harm to plant crops (Komal et al. 2023; Samal et al. 2023a, b, 2024b; Bhoi et al. 2024b;Uma et al. 2024). Research on how melatonin affects plant–insect relationships remains unclear because of multiple uncertainties (Chowdhury et al. 2024; Mahanta and Komal 2022; Bhoi et al. 2024c). Research has established that plants contain melatonin which enhances their resistance to abiotic stressors but scientists lack understanding about how this compound regulates biotic stress from herbivory. Research into the effects of melatonin on herbivore feeding patterns and plant defense signaling and insect attractiveness will help develop efficient sustainable pest management strategies (Vyas et al. 2024; Samal et al. 2023a, b). The complex nature of these relationships creates a research opportunity to understand how melatonin could be used to reduce herbivory while enhancing plant defensive capabilities and developing environmentally friendly pest control solutions that do not depend on harmful chemicals. Figure 12.4 represents signaling mechanism and stress management (biotic) by melatonin.

The defensive properties of melatonin combined with its insect-feeding inhibiting ability create a potential solution for herbivory control in agricultural pest management. The understanding of how melatonin affects plant–insect relationships is still in its early stages but multiple studies suggest that this natural compound could be an effective method for protecting plants against herbivores without requiring artificial pesticides. Plants use melatonin to activate antioxidant systems while modifying jasmonic acid and salicylic acid pathways and enhancing their resistance to oxidative stress according to Luo et al. (2023) and Li et al. (2022a, b). Plants need these defense systems to defend themselves from herbivore-caused damage. The defense systems of plants become stronger through melatonin application which leads to better insect predation survival rates (Serag et al. 2023). Melatonin facilitates phytoalexin synthesis together with protein synthesis and the production of detrimental or repellant secondary metabolites like alkaloids and flavonoids which

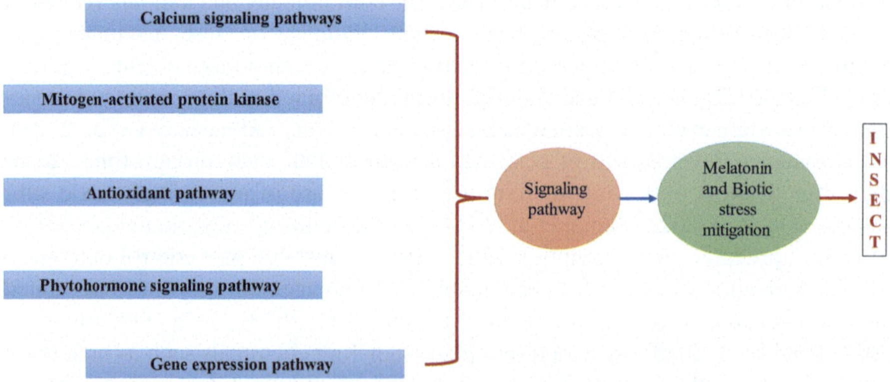

Fig. 12.4 Signaling mechanism and biotic stress management (insect pests) by melatonin

enhances plant resistance. The implementation of melatonin in agricultural operations leads to healthier crops that possess stronger resistance against insects (Khan et al. 2024). The compound melatonin functions as an insect repellent against herbivorous pests. Scientific studies prove that melatonin modifies the feeding patterns of numerous insect species. Melatonin disrupts the neurophysiological processes that lead to feeding and modifies both gustatory and olfactory responses of insects to prevent them from recognizing plants as food. The insect's circadian cycles are disrupted by melatonin which reduces its feeding activity during essential times or the plant becomes less appetizing due to melatonin-modified VOC emissions that attract herbivores (Zhou et al. 2021). The high concentration of melatonin in plants leads insects to avoid them because they become unpalatable or unfit for eating (Liu et al. 2023). The physiological effects of melatonin on insects disrupt their growth processes while hindering their reproductive abilities. The reproductive cycles of multiple insect species show sensitivity to melatonin exposure which leads to reduced pest populations, according to Kumari et al. (2023). The timing of insect maturation and oviposition regulation mechanisms become disrupted by melatonin exposure which leads to reduced pest population levels through time (Jiang et al. 2023). The approach serves dual purposes by creating a management solution for herbivores while reducing pesticide usage through life cycle-focused strategies. The implementation of melatonin in sustainable pest management enables farmers to use an environmentally friendly alternative to traditional chemical pest control methods. Crops receive melatonin-based treatments through three methods: foliar spraying, soil application, and genetically modified plants which produce melatonin. The applied treatments lead plants to develop stronger defense systems while blocking herbivorous damage without causing harm to beneficial insects or the surrounding ecosystem. The combination of oxidative stress reduction and increased plant vitality through melatonin treatment enables farmers to achieve higher yields in addition to stronger plants which becomes particularly beneficial for organic farming operations. The effectiveness of melatonin as a plant defense mechanism becomes stronger through its combination with natural predators and parasitoids or plant growth-promoting rhizobacteria (PGPR) (Asif et al. 2023). The combination of melatonin with plant defense mechanisms and herbivore consumption inhibition leads to comprehensive pest control solutions which eliminate the environmental problems caused by chemical pesticides. Figure 12.5 *represents* melatonin action model on biotic stress responses that result in resistance development towards insects.

The role of melatonin as a defense booster against herbivores in plants creates various potential applications for sustainable agricultural practices (Selwal et al. 2023). The growing demand for organic agricultural techniques combined with sustainable practices makes methods based on melatonin hold great potential to reduce chemical pesticide use while producing healthier and more resilient crops. The agricultural application of melatonin includes its use as a natural insect deterrent (Tiwari et al. 2020). Melatonin regulates insect behavior because these organisms respond strongly to environmental cues including light and temperature and chemical indications. Plants treated with melatonin tend to produce lower amounts of volatile

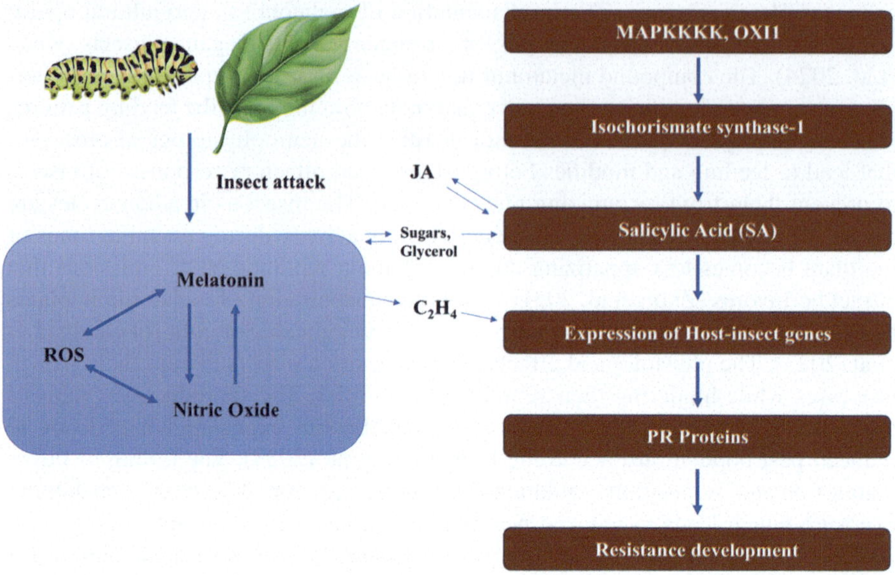

Fig. 12.5 Melatonin action model on biotic stress responses that result in resistance development towards insects

compounds that attract herbivorous pests or the melatonin itself makes plants unattractive to pests (Golubkina et al. 2022). The reduction of chemical pesticide usage would become possible through this method because it eliminates risks to beneficial insects and pollinators. The leaves of cabbage and tomato crops can take up melatonin to block caterpillars and aphids from damaging them (Tan et al. 2020; Masoumi et al. 2024). The research shows that melatonin activates plant defense mechanisms through phytoalexin and secondary compound production which deters herbivorous insects (Yu et al. 2018). The chemicals produced by plants make them unattractive and dangerous to eating insects. Organic agricultural systems without chemical pesticides would gain specific benefits from this approach. Plants that receive melatonin treatment demonstrate better pest resistance to aphids and weevils and grasshoppers because their natural defense mechanisms become more robust. The addition of melatonin to wheat and rice crops triggers chemical production which leads to aphid repulsion and root-feeding larva inhibition, thus creating healthier thriving plants (Muhammad et al. 2024). Research shows that melatonin treatment leads to better stress tolerance in plants alongside its herbivore protection benefits. Oxidative stress from herbivore attacks creates tissue damage in plants. Plants that receive melatonin treatment benefit from its antioxidant properties which counteract free radicals and lower oxidative stress levels. The application of melatonin as a stress resistance booster in plants allows farmers to improve both crop quantity and quality and minimize pesticide requirements. The application of melatonin in tomato and strawberry crops helps plants withstand herbivore damage, thus preserving their health and productivity (Stürtz et al. 2011). Sustainable agriculture commonly uses various biocontrol methods to achieve population reduction of insects. The application of melatonin extends to combine with additional biocontrol strategies through natural predator implementation such as using ladybirds for aphid control or parasitoids. Plants protected by melatonin defense mechanisms develop increased resistance against pest attacks which allows their natural predators to better control herbivore populations. The behavioral effects of melatonin create possibilities to steer pest and predator relationships. The feeding patterns of herbivores become shorter and their reproductive cycles change when exposed to melatonin which makes pests more susceptible to biological control agents. The advantages of melatonin could be achieved through genetic engineering as a sustainable agricultural solution (Nawaz et al. 2020). Plants can be genetically modified to create melatonin production in response to stress or herbivore attacks which results in permanent defense enhancement. The technique demonstrates particular value for protecting valuable crops and regions where insect threats are severe. Genetically modified maize plants producing melatonin can resist maize borers and other pests through an environmentally friendly pest control solution which eliminates the need for external pesticides. Herbivorous insects damage both aerial plant parts and underground structures which leads to serious crop destruction. The root defense mechanisms against soil-based pests like root aphids, nematodes, and beetles enhance through the application of melatonin according to Darshan et al. (2024). The antioxidant and defense mechanisms of plant roots become stronger through melatonin application which helps protect roots from pest damage and boosts plant health. The treatment shows particular usefulness for potato and carrot crops together with other root vegetables that are vulnerable to underground insect damage. The implementation of melatonin-based strategies allows farmers to reduce their reliance on chemical pesticides used in conventional agriculture. The application of melatonin decreases dependence on synthetic pesticides because it strengthens plants against herbivores while activating natural defense mechanisms that stop feeding behavior. The adoption of these methods would establish environmentally friendly farming practices which protect helpful species while reducing pesticide levels in food products. The adoption of such systems would create safer food production methods that benefit environmental sustainability together with human health.

12.5 Towards Ecosystem-Level Understanding

Research on melatonin in mammals has advanced significantly but scientists still lack sufficient knowledge about its ecological functions across ecosystems involving plants and insects and bacteria (Grubisic et al. 2019). The insufficient knowledge about melatonin creates substantial barriers for scientists to fully understand its effects on ecosystem dynamics (Tan et al. 2014). Ecosystem interactions show melatonin affects species-to-species relationships between plants and insects and predators and their prey as well as between hosts and parasites. The integration of melatonin's ecological functions remains limited because researchers do not recognize it sufficiently across various species. Scientists recognize melatonin as a regulator of circadian rhythms and sleep–wake cycles and stress responses in mammals but its plant physiological and insect behavioral functions are still being studied (Jones et al. 2015). Research demonstrates that plant melatonin enhances both stress resistance and protects against insect herbivores, yet scientists have not yet integrated these findings into an extensive ecological theory. The complete understanding of how melatonin affects plant–insect relations inside shared environments will provide useful pest management solutions yet requires thorough investigation of both plants and their insect prey. The research on how melatonin affects microbial populations remains insufficiently studied according to He et al. (2023). Soil health and plant development as well as ecosystem nutrient cycling depend on microorganisms including bacteria and fungus (Madigan et al. 2019). Research shows that melatonin occurs in specific microorganisms, although scientists have yet to study how this compound impacts their behavioral patterns or growth or their plant symbiotic relationships. Ecosystem models currently lack melatonin integration which limits our ability to understand its ecological equilibrium mediation potential. The behavioral influence of melatonin on particular species has been confirmed but its effects on ecological processes such as species interactions and community dynamics and trophic cascades remain unclear. The regulatory role of melatonin in plant flowering time and insect feeding behavior shapes food availability synchrony in ecosystems which results in altered populations of herbivores and predators (Zhang et al. 2019). The wide-ranging ecological effects of changing times are not well understood by scientists. Environmental research and ecological studies will reach their maximum potential through integrated investigations of melatonin across multiple biological organizational levels beyond studying individual organisms. The development of ecosystem knowledge regarding melatonin effects requires multidisciplinary research which combines plant biology with insect ecology and parasitology and microbial interaction studies. Research ability to forecast ecosystem responses to changes in melatonin levels becomes stronger when scientists understand the extensive ecological functions of melatonin.

Insects are essential to ecosystems, serving as herbivores, predators, pollinators, and decomposers, hence influencing plant–animal interactions and the overall functioning of ecosystems (Verma et al. 2023a, b). Examining the effects of melatonin on insect behavior at various trophic levels is crucial for comprehending its wider ecological implications. Investigating the role of melatonin in mediating interactions among plants, herbivores, predators, and other organisms enables researchers to comprehend the cascading impacts of this chemical on ecosystem dynamics. A crucial area of investigation is the impact of plant-derived melatonin on herbivore behavior. Melatonin has been demonstrated to regulate plant defense systems, encompassing the synthesis of volatile organic compounds (VOCs) and secondary metabolites that can repel herbivores or render plants poisonous. Alterations in plant chemistry may affect insect feeding habits and behavior. Melatonin-induced alterations in plant defenses may influence the appeal of plants to herbivores, hence diminishing herbivore impact on crops or indigenous flora. Insects may be discouraged from feeding or exhibit diminished fitness, thereby affecting the quantity and distribution of herbivores in an ecosystem. This alteration in herbivore behavior may have considerable consequences for agricultural pest control and natural plant–insect interactions. Besides plant–insect interactions, melatonin may influence predator–prey dynamics by modifying herbivore behavior. If melatonin affects herbivores' feeding behavior, it may indirectly impact predators that depend on these herbivores for sustenance. Alterations in herbivore activity may modify predator hunting behaviors, perhaps resulting in fluctuations in predator populations (Pettorelli et al. 2011). The study of melatonin's influence on herbivore and predator interactions shows how disturbances in one part of the food web spread throughout the complete system. The acquired knowledge enables researchers to create sustainable pest management systems when predators are used in agricultural settings to control pests. The influence of melatonin extends to pollinators in addition to herbivores because these pollinators play a vital role in plant reproduction. Bees and butterflies alongside moths serve as pollinators through their vital activities which are influenced by circadian rhythms that melatonin regulates (Singhal et al. 2021). The timing and effectiveness of pollinators visiting plants are altered by melatonin exposure which results in modified plant reproductive outcomes (Lehmann et al. 2011). Plants that produce melatonin during specific periods draw more active pollinators thus leading to enhanced pollination success. Plants that produce melatonin create a chain reaction which affects their reproductive processes as well as the availability of herbivores and ecological equilibrium. Light together with temperature and seasonal changes substantially influence how insects produce melatonin (Niva and Takeda 2003). Environmental signals which undergo changes due to climate change can affect the melatonin levels in insects while altering their behavioral patterns. Higher temperatures have the ability to modify insect circadian cycles which leads to changes in their feeding behavior and reproductive patterns as well as migration patterns. The analysis of melatonin–environment interactions will reveal insect adaptation mechanisms for changing temperatures while showing their influence on ecosystem operations. The comprehension of these behavioral changes holds essential value in current environmental challenges because they can substantially affect ecosystem diversity and food systems. Soil microbial populations function under the influence of insects which leads to enhanced nitrogen cycling and decomposition processes (Beare et al. 1992). The behavior modifications caused by melatonin can affect insect conduct which subsequently affects microbial activities either positively or negatively. The presence of melatonin in insects that move organic matter into soils including dung beetles and termites might modify decomposition rates of microorganisms. Soil health together with nutrient cycling becomes affected which are vital elements for maintaining ecosystem stability. Analyzing the role of melatonin in insect–microbe relationships can provide essential knowledge about how this compound affects ecosystem nutrient cycles and overall system health. The knowledge of how melatonin affects insect behaviors offers potential for agricultural scientists to develop innovative sustainable pest management systems. Melatonin shows potential to restrict herbivores or adapt their feeding patterns which in turn reduces the need for chemical pesticides. The application of melatonin-based pest control techniques by farmers would enable them to manage pests effectively without disrupting the natural insect population dynamics between predators and pollinators. The defensive mechanisms of plants that include melatonin can be developed into strategies to improve crop resistance which will help protect yields and sustain agricultural production. The approach demonstrates particular value within integrated pest management (IPM) systems which unite biological and chemical and cultural pest control methods.

12.6 Conclusion

The hormone melatonin has traditionally controlled circadian rhythms and sleep–wake cycles in mammals, yet scientists now understand its broader ecological importance. The study of melatonin in insects and plant–insect relationships and parasitology allows researchers to better understand ecosystem dynamics. Studies into melatonin's influence on plant defenses and insect feeding and host–parasite relationships enable better management of ecosystems and sustainable agriculture and conservation practices. A comprehensive understanding of how melatonin affects herbivore behavior and predator–prey relationships and pollinator dynamics enables better understanding of its regulatory role in trophic interactions and community structures. The findings from this research will enable scientists to create better sustainable pest control strategies and improve biodiversity preservation in changing environmental conditions. The influence of melatonin on microbial communities and soil health creates potential impacts which will affect how ecosystems cycle nutrients and operate. The promising study outcomes exist despite the ongoing need to understand melatonin's ecological functions in all species beyond mammals. The study faces two main challenges because scientists fail to recognize melatonin's significance to species biology and require better interdisciplinary research between plant biology and insect ecology and parasitology and environmental sciences. Research must focus on explaining how melatonin affects insect behaviors and plant reactions and ecosystem species interactions. Future research should investigate how melatonin functions within ecosystem-level models. Understanding how various environmental elements such as light and temperature and climate change affect melatonin production in multiple species will reveal critical insights about ecosystem responses to global environmental changes. The study of melatonin inside ecological restoration and agroecology could lead to innovative approaches for developing sustainable agricultural practices and enhancing ecosystem resilience. The mounting scientific evidence demonstrates how melatonin functions as a vital mechanism which influences both ecological processes and evolutionary change. Strengthened understanding of how melatonin affects species relationships and ecosystem functions will enable the development of innovative strategies for managing ecosystems and conserving biodiversity and addressing environmental change challenges. The extensive research potential of melatonin in ecology shows promise for practical implementations in environmental and ecological management which will benefit both natural ecosystems and agricultural systems.

References

Ain QU, Mir HA, Mobasher MA, Guru SA (2024) Melatonin in oxidative stress management. In: Melatonin. Academic Press, pp 259–272

Alengebawy A, Abdelkhalek ST, Qureshi SR, Wang MQ (2021) Heavy metals and pesticides toxicity in agricultural soil and plants: ecological risks and human health implications. Toxics 9(3):42

Ali A, Rashid MA, Huang QY, Lei CL (2017) Influence of UV-A radiation on oxidative stress and antioxidant enzymes in Mythimna separata (Lepidoptera: Noctuidae). Environ Sci Pollut Res 24:8392–8398

Altaf MA, Shahid R, Ren MX, Naz S, Altaf MM, Khan LU, Tiwari RK, Lal MK, Shahid MA, Kumar R, Nawaz MA (2022) Melatonin improves drought stress tolerance of tomato by modulating plant growth, root architecture, photosynthesis, and antioxidant defense system. Antioxidants 11(2):309

Arnao MB, Hernández-Ruiz J (2014) Melatonin: plant growth regulator and/or biostimulator during stress? Trends Plant Sci 19(12):789–797

Asif M, Ahmad R, Pervez A, Al Farraj DA, Elshikh MS, Shahzad M et al (2023) Combination of melatonin and plant growth promoting rhizobacteria improved the growth of *Spinacia oleracea* L. under the arsenic and cadmium stresses. Physiol Mol Plant Pathol 127:102097

Bagnaresi P, Nakabashi M, Thomas AP, Reiter RJ, Garcia CR (2012) The role of melatonin in parasite biology. Mol Biochem Parasitol 181(1):1–6

Barberà M, Escrivá L, Collantes-Alegre JM, Meca G, Rosato E, Martínez-Torres D (2020) Melatonin in the seasonal response of the aphid *Acyrthosiphon pisum*. Insect Sci 27(2):224–238

Barik S (2019) Molecular interactions between pathogens and the circadian clock. Int J Mol Sci 20(23):5824

Beare MH, Parmelee RW, Hendrix PF, Cheng W, Coleman DC, Crossley DA Jr (1992) Microbial and faunal interactions and effects on litter nitrogen and decomposition in agroecosystems. Ecol Monogr 62(4):569–591

Bhatnagar S, Mahanta DK, Vyas V, Samal I, Komal J, Bhoi TK (2024) Storage pest management with nanopesticides incorporating silicon nanoparticles: a novel approach for sustainable crop preservation and food security. Silicon 16(2):471–483

Bhoi TK, Samal I, Majhi PK, Komal J, Mahanta DK, Pradhan AK et al (2022) Insight into aphid mediated potato virus Y transmission: A molecular to bioinformatics prospective. Front Microbiol 13:1001454

Bhoi TK, Samal I, Mahanta DK, Komal J, Jinger D, Sahoo MR et al (2023) Understanding how silicon fertilization impacts chemical ecology and multitrophic interactions among plants, insects and beneficial arthropods. Silicon 15(6):2529–2549

Bhoi TK, Samal I, Mahanta DK, Komal J, Majhi PK, Ankur. (2024a) Exploring melatonin's potential as an alternative strategy for protecting plants from biotic stresses. In: Melatonin in plants: a pleiotropic molecule for abiotic stresses and pathogen infection. Springer Nature Singapore, Singapore, pp 223–242

Bhoi TK, Samal I, Mahanta DK, Komal J, Dinesh Kumar PV, Desai HR et al (2024b) Advancements in double-stranded RNA (dsRNA) technology for effective management of forest insect pests. South For J For Sci 86:1–19

Bhoi TK, Samal I, Mahanta DK, Komal J, Majhi PK, Bhatnagar S (2024c) Genetic engineering and genome editing approaches for identification and sustainable management of insect pests under forest dynamics. In: Biotechnological approaches for sustaining forest trees and their products. Springer Nature Singapore, Singapore, pp 385–423

Bloch G, Hazan E, Rafaeli A (2013) Circadian rhythms and endocrine functions in adult insects. J Insect Physiol 59(1):56–69

Calvo JR, Gonzalez-Yanes C, Maldonado MD (2013) The role of melatonin in the cells of the innate immunity: a review. J Pineal Res 55(2):103–120

Carvalho Cabral P, Olivier M, Cermakian N (2019) The complex interplay of parasites, their hosts, and circadian clocks. Front Cell Infect Microbiol 9:425

Chitimus DM, Popescu MR, Voiculescu SE, Panaitescu AM, Pavel B, Zagrean L, Zagrean AM (2020) Melatonin's impact on antioxidative and anti-inflammatory reprogramming in homeostasis and disease. Biomol Ther 10(9):1211

Chowdhury S, Adhikari A, Gaddanakeri S, Sunil V, Singh S, Samal I et al (2024) An in-depth analysis of artificial intelligence-based crop Pest management and water supply regulation. In: Artificial Intelligence techniques in smart agriculture. Springer Nature Singapore, Singapore, pp 151–171

Darshan K, Harshitha KN, Shreedevasena S, Tailor A, Bhoi TK, Nigam S, Kulkarni N (2024) Role of melatonin in management of stress tolerance of forest tree species. In: Melatonin in plants: a pleiotropic molecule for abiotic stresses and pathogen infection. Singapore, Springer Nature Singapore, pp 177–194

Daryani A, Montazeri M, Pagheh AS, Sharif M, Sarvi S, Hosseinzadeh A et al (2018) The potential use of melatonin to treat protozoan parasitic infections: a review. Biomed Pharmacother 97:948–957

Desouhant E, Gomes E, Mondy N, Amat I (2019) Mechanistic, ecological, and evolutionary consequences of artificial light at night for insects: review and prospective. Entomol Exp Appl 167(1):37–58

Dias BK, Mohanty A, Garcia CR (2024) Melatonin as a circadian marker for plasmodium rhythms. Int J Mol Sci 25(14):7815

Erland LA (2024) Views and perspectives on the indoleamines serotonin and melatonin in plants: past, present and future. Plant Signal Behav 19(1):2366545

Even N, Bertrand O, Lihoreau M (2022) Navigation by honey bees. In: Encyclopedia of animal cognition and behavior. Springer International Publishing, Cham, pp 4565–4573

Fan W, Li G, Zhang X, Wang Y, Wang C, Xu B et al (2021) The role of melatonin and Tryptophan-5-hydroxylase-1 in different abiotic stressors in Apis cerana cerana. J Insect Physiol 128:104180

Franco PIR, do Carmo Neto JR, Guerra RO, da Silva PEF, Braga YLL, Celes MRN et al (2024) Melatonin: a look at protozoal and helminths. Biochimie 219:96–109

Golubkina N, Zayachkovsky V, Sheshnitsan S, Skrypnik L, Antoshkina M, Smirnova A et al (2022) Prospects of the application of garlic extracts and selenium and silicon compounds for plant protection against herbivorous pests: a review. Agriculture 12(1):64

Grubisic M, Haim A, Bhusal P, Dominoni DM, Gabriel KM, Jechow A et al (2019) Light pollution, circadian photoreception, and melatonin in vertebrates. Sustain For 11(22):6400

Hashimi MH, Hashimi R, Ryan Q (2020) Toxic effects of pesticides on humans, plants, animals, pollinators and beneficial organisms. Asian Plant Res J 5(4):37–47

He X, Yin B, Zhang J, Zhou S, Li Z, Zhang X et al (2023) Exogenous melatonin alleviates apple replant disease by regulating rhizosphere soil microbial community structure and nitrogen metabolism. Sci Total Environ 884:163830

Horgan FG (2020) Potential for an impact of global climate change on insect herbivory in cereal crops. In: Crop protection under changing climate. Springer, pp 101–144

Huang H, Mehta A, Kalmanovich J, Anand A, Bejarano MC, Garg T et al (2023) Immunological and inflammatory effects of infectious diseases in circadian rhythm disruption and future therapeutic directions. Mol Biol Rep 50(4):3739–3753

Jiang XL, Ren Z, Hai XX, Zhang L, Wang ZG, Lyu F (2023) Exposure to artificial light at night mediates the locomotion activity and oviposition capacity of *Dastarcus helophoroides* (Fairmaire). Front Physiol 14:1063601

Jones TM, Durrant J, Michaelides EB, Green MP (2015) Melatonin: a possible link between the presence of artificial light at night and reductions in biological fitness. Philos Trans R Soc Lond B Biol Sci 370(1667):20140122

Kamruzzaman ASM, Hiragaki S, Watari Y, Natsukawa T, Yasuhara A, Ichihara N et al (2021) Clock-controlled arylalkylamine N-acetyltransferase (aaNAT) regulates circadian rhythms of locomotor activity in the American cockroach, Periplaneta americana, via melatonin/MT2-like receptor. J Pineal Res 71(2):e12751

Karthi S, Shivakumar MS (2015) The protective effect of melatonin against cypermethrin-induced oxidative stress damage in *Spodoptera litura* (Lepidoptera: Noctuidae). Biol Rhythm Res 46(1):1–12

Khan M, Hussain A, Yun BW, Mun BG (2024) Melatonin: the multifaceted molecule in plant growth and defense. Int J Mol Sci 25(12):6799

Kim TK, Slominski RM, Pyza E, Kleszczynski K, Tuckey RC, Reiter RJ et al (2024) Evolutionary formation of melatonin and vitamin D in early life forms: insects take centre stage. Biol Rev 99:1772

Klimczak S, Packi K, Rudek A, Wenclewska S, Kurowski M, Kurczabińska D, Śliwińska A (2024) The influence of the protozoan *Giardia lamblia* on the modulation of the immune system and alterations in host glucose and lipid metabolism. Int J Mol Sci 25(16):8627

Komal J, Desai HR, Samal I, Mastinu A, Patel RD, Kumar PD et al (2023) Unveiling the genetic symphony: harnessing CRISPR-Cas genome editing for effective insect pest management. Plan Theory 12(23):3961

Kotb Elmahallawy E, Ortega Luque J, Saleh Aloweidi A, Gutiérrez-Fernández J, Sampedro-Martínez A, Rodriguez-Granger J et al (2015) Potential relevance of melatonin against some infectious agents: a review and assessment of recent research. Curr Med Chem 22(33):3848–3861

Koul AM, Shafi T, Anwar I, Banday M, Iqra S, Gull A et al (2024) Melatonin and immune modulation. In: Melatonin. Elsevier, pp 163–185

Kumar R, Rathor VS (2020) Nature and types of damage by insect pests. J Entomol Res 44(4):639–646

Kumar R, Altaf MA, Lal MK, Tiwari RK (eds) (2023) Melatonin in plants: a regulator for plant growth and development. Springer Nature Singapore

Kumar R, Altaf MA, Lal MK, Tiwari RK (eds) (2024) Melatonin in plants: a pleiotropic molecule for abiotic stresses and pathogen infection. Springer

Kumari A, Singh SK, Mathpal B, Verma KK, Garg VK, Bhattacharyya M, Bhatt R (2023) The biosynthesis, mechanism of action, and physiological functions of melatonin in horticultural plants: a review. Horticulturae 9(8):913

Lampel J, Briscoe AD, Wasserthal LT (2005) Expression of UV-, blue-, long-wavelength-sensitive opsins and melatonin in extraretinal photoreceptors of the optic lobes of hawkmoths. Cell Tissue Res 321(3):443–458

Lehmann M, Gustav D, Galizia CG (2011) The early bee catches the flower-circadian rhythmicity influences learning performance in honey bees, *Apis mellifera*. Behav Ecol Sociobiol 65:205–215

Li S, Huan C, Liu Y, Zheng X, Bi Y (2022a) Melatonin induces improved protection against *Botrytis cinerea* in cherry tomato fruit by activating salicylic acid signaling pathway. Sci Hortic 304:111299

Li Z, Duan J, Chen L, Wang Y, Qin Q, Dang X, Zhou Z (2022b) Melatonin enhances the antioxidant capacity to rescue the honey bee *Apis mellifera* from the ecotoxicological effects caused by environmental imidacloprid. Ecotoxicol Environ Saf 239:113622

Linn CE, Poole KR, Roelofs WL, Wu WQ (1995) Circadian changes in melatonin in the nervous system and hemolymph of the cabbage looper moth, Trichoplusia ni. J Comp Physiol A 176:761–771

Liu X, Du C, Yue C, Tan Y, Fan H (2023) Exogenously applied melatonin alleviates the damage in cucumber plants caused by *Aphis goosypii* through altering the insect behavior and inducing host plant resistance. Pest Manag Sci 79(1):140–151

Luo M, Wang D, Delaplace P, Pan Y, Zhou Y, Tang W et al (2023) Melatonin enhances drought tolerance by affecting jasmonic acid and lignin biosynthesis in wheat (*Triticum aestivum* L.). Plant Physiol Biochem 202:107974

Madigan AP, Egidi E, Bedon F, Franks AE, Plummer KM (2019) Bacterial and fungal communities are differentially modified by melatonin in agricultural soils under abiotic stress. Front Microbiol 10:2616

Maestroni GJ (2024) Role of melatonin in viral, bacterial and parasitic infections. Biomolecules 14(3):356

Mahanta DK, Komal J (2022) Sericulture and its salient features. Vigyan Varta 3:105–110

Mahanta DK, Jangra S, Priti, Ghosh A, Sharma PK, Iquebal MA et al (2022) Groundnut bud necrosis virus modulates the expression of innate immune, endocytosis, and cuticle development-associated genes to circulate and propagate in its vector, Thrips palmi. Front Microbiol 13:773238

Mahanta DK, Bhoi TK, Komal J, Samal I, Nikhil RM, Paschapur AU et al (2023a) Insect-pathogen crosstalk and the cellular-molecular mechanisms of insect immunity: uncovering the underlying signaling pathways and immune regulatory function of non-coding RNAs. Front Immunol 14:1169152

Mahanta DK, Samal I, Komal J, Bhoi TK, Majhi PK, Ahmad MA (2023b) Understanding anthropogenic climate change, its consequences on insect pests, and techniques in forecasting and monitoring pest dynamics: a contemporary scenario. In: Climate change and insect biodiversity. CRC Press, pp 44–67

Mahanta DK, Komal J, Samal I, Bhoi TK, Dubey VK, Pradhan K et al (2023c) Nutritional aspects and dietary benefits of "silkworms": current scenario and future outlook. Front Nutr 10:1121508

Mahanta DK, Bhoi TK, Komal J, Samal I, Mastinu A (2024) Spatial, spectral and temporal insights: harnessing high-resolution satellite remote sensing and artificial intelligence for early monitoring of wood boring pests in forests. Plant Stress 11:100381

Majhi PK, Samal I, Bhoi TK, Pattnaik P, Pradhan C, Gupta A et al (2023) Climate-smart agriculture: an integrated approach to address climate change and food security. In: Climate change and insect biodiversity. CRC Press, pp 221–245

Maldonado E, Rojas DA, Morales S, Miralles V, Solari A (2020) Dual and opposite roles of reactive oxygen species (ROS) in Chagas disease: beneficial on the pathogen and harmful on the host. Oxidative Med Cell Longev 2020(1):8867701

Masoumi Z, Haghighi M, Mozafarian M (2024) Effects of foliar spraying with melatonin and chitosan Nano-encapsulated melatonin on tomato (*Lycopersicon esculentum* L. cv. Falcato) plants under salinity stress. BMC Plant Biol 24(1):961

Maugars G, Nourizadeh-Lillabadi R, Weltzien FA (2020) New insights into the evolutionary history of melatonin receptors in vertebrates, with particular focus on teleosts. Front Endocrinol 11:538196

Maxwell GD, Hildebrand JG (1981) Anatomical and neurochemical consequences of deafferentation in the development of the visual system of the moth Manduca sexta. J Comp Neurol 195(4):667–680

Mineiro R, Cardoso MR, Duarte AC, Santos C, Cipolla-Neto J, do Amaral FG et al (2024) Melatonin and brain barriers: the protection conferred by melatonin to the blood-brain barrier and blood-cerebrospinal fluid barrier. Front Neuroendocrinol 75:101158

Muhammad I, Ahmad S, Shen W (2024) Melatonin-mediated molecular responses in plants: enhancing stress tolerance and mitigating environmental challenges in cereal crop production. Int J Mol Sci 25(8):4551

Nawaz K, Chaudhary R, Sarwar A, Ahmad B, Gul A, Hano C et al (2020) Melatonin as master regulator in plant growth, development and stress alleviator for sustainable agricultural production: current status and future perspectives. Sustain For 13(1):294

Niva CC, Takeda M (2003) Effects of photoperiod, temperature and melatonin on nymphal development, polyphenism and reproduction in Halyomorpha halys (Heteroptera: Pentatomidae). Zool Sci 20(8):963–970

Ortega-Arellano HF, Jimenez-Del-Rio M, Velez-Pardo C (2021) Melatonin increases life span, restores the locomotor activity, and reduces lipid peroxidation (LPO) in transgenic knockdown parkin drosophila melanogaster exposed to paraquat or paraquat/iron. Neurotox Res 39(5):1551–1563

Pache D (2020) Melatonin-its role in your life part 1: biosynthesis, regulation, physiology and pharmacology. Aust J Pharm 101(1200):44–51

Pandi-Perumal SR, Cardinali DP, Zaki NF, Karthikeyan R, Spence DW, Reiter RJ, Brown GM (2022) Timing is everything: circadian rhythms and their role in the control of sleep. Front Neuroendocrinol 66:100978
Pawłowska M, Mila-Kierzenkowska C, Szczegielniak J, Woźniak A (2023) Oxidative stress in parasitic diseases—reactive oxygen species as mediators of interactions between the host and the parasites. Antioxidants 13(1):38
Pettorelli N, Coulson T, Durant SM, Gaillard JM (2011) Predation, individual variability and vertebrate population dynamics. Oecologia 167:305–314
Qu M, Merzendorfer H, Moussian B, Yang Q (2022) Bioinsecticides as future mainstream pest control agents: opportunities and challenges. Front Agric Sci Eng 9(1):82–97
Rajotia MS, Sharma A, Bhatt A, Mangal V, Sood S, Thakur AK et al (2023) Melatonin detection and quantification techniques. In: Melatonin in plants: a regulator for plant growth and development. Springer Nature Singapore, Singapore, pp 19–38
Reece SE, Prior KF, Mideo N (2017) The life and times of parasites: rhythms in strategies for within-host survival and between-host transmission. J Biol Rhythm 32(6):516–533
Reiter RJ, Sharma R, Chuffa LGDA, Simko F, Dominguez-Rodriguez A (2024) Mitochondrial melatonin: beneficial effects in protecting against heart failure. Life 14(1):88
Rund SS, O'Donnell AJ, Gentile JE, Reece SE (2016) Daily rhythms in mosquitoes and their consequences for malaria transmission. Insects 7(2):14
Samal I, Bhoi TK, Majhi PK, Murmu S, Pradhan AK, Kumar D et al (2023a) Combatting insects mediated biotic stress through plant associated endophytic entomopathogenic fungi in horticultural crops. Front Plant Sci 13:1098673
Samal I, Bhoi TK, Pradhan AK, Mahanta DK (2023b) Plantix app: a success story of artificial intelligence in plant protection. Van Sangyan 24
Samal I, Bhoi TK, Mahanta DK, Komal J (2024a) Molecular basis of Entomopathogenesis and the way forward. In: Entomopathogenic fungi: prospects and challenges. Springer Nature Singapore, Singapore, pp 481–505
Samal I, Bhoi TK, Mahanta DK, Komal J (2024b) Establishing the role of silicon (Si) in plant resistance to insects: a bibliometric approach. SILICON 16(5):2119–2128
Samal I, Bhoi TK, Vyas V, Majhi PK, Mahanta DK, Komal J et al (2024c) Resistance to fungicides in entomopathogenic fungi: underlying mechanisms, consequences, and opportunities for progress. Trop Plant Pathol 49(1):5–17
Samal I, Bhoi TK, Mahanta DK, Komal J, Singh S (2024d) Biorational pest management: potentials, unintended consequences, and future concerns. In: Biorationals and biopesticides: pest management. Walter de Gruyter GmbH, p 47
Schwarzenberger A, Wacker A (2015) Melatonin synthesis follows a daily cycle in daphnia. J Plankton Res 37(3):636–644
Selwal N, Rahayu F, Herwati A, Latifah E, Suhara C, Suastika IBK et al (2023) Enhancing secondary metabolite production in plants: exploring traditional and modern strategies. J Agric Food Res 14:100702
Serag A, Salem MA, Gong S, Wu JL, Farag MA (2023) Decoding metabolic reprogramming in plants under pathogen attacks, a comprehensive review of emerging metabolomics technologies to maximize their applications. Meta 13(3):424
Shi L, Zhang J, Qiu L, Jiang Z, Xie Z, Zhan Z (2021) Behavioral changes in the brown planthopper, Nilaparvata lugens, mediated by melatonin. CABI Agric Biosci 2:1–11
Siddiqui SA, Mahanta DK, Bhoi TK, Ahmad A, Fernando I (2024) Mopane worm (Gonimbrasia belina)-an exclusive African edible insect as human food-a comprehensive review. Fut Foods 10:100438
Singh B, Varikuti S, Halsey G, Volpedo G, Hamza OM, Satoskar AR (2019) Host-directed therapies for parasitic diseases. Future Med Chem 11(16):1999–2018
Singh MK, Dias BKDM, Garcia CR (2020) Role of melatonin in the synchronization of asexual forms in the parasite *Plasmodium falciparum*. Biomol Ther 10(9):1243
Singhal RK, Chauhan J, Jatav HS, Rajput VD, Singh GS, Bose B (2021) Artificial night light alters ecosystem services provided by biotic components. Biologia Futura 72(2):169–185

Srinivasan V, Ahmad HA, Mohamed M, Zakaria R (2012) Melatonin effects on plasmodium life cycle: new avenues for therapeutic approach. Recent Pat Endocr Metab Immune Drug Discov 6(2):139–147

Stürtz M, Cerezo AB, Cantos-Villar E, Garcia-Parrilla MC (2011) Determination of the melatonin content of different varieties of tomatoes (*Lycopersicon esculentum*) and strawberries (*Fragaria ananassa*). Food Chem 127(3):1329–1334

Subala SPRR, Shivakumar MS (2018) Changes in light and dark periods affect the arylalkylamine N-acetyl transferase, melatonin activities and redox status in the head and hemolymph of nocturnal insect Spodoptera litura. Biol Rhythm Res 49(1):13–28

Subala SP, Zubero EE, Alatorre-Jimenez MA, Shivakumar MS (2017) Pre-treatment with melatonin decreases abamectin induced toxicity in a nocturnal insect *Spodoptera litura* (Lepidoptera: Noctuidae). Environ Toxicol Pharmacol 56:76–85

Sun C, Liu L, Wang L, Li B, Jin C, Lin X (2021) Melatonin: A master regulator of plant development and stress responses. J Integr Plant Biol 63(1):126–145

Szewczyk-Golec K, Pawłowska M, Wesołowski R, Wróblewski M, Mila-Kierzenkowska C (2021) Oxidative stress as a possible target in the treatment of toxoplasmosis: perspectives and ambiguities. Int J Mol Sci 22(11):5705

Takeda M, Suzuki T (2022) Circadian and neuroendocrine basis of photoperiodism controlling diapause in insects and mites: a review. Front Physiol 13:867621

Tan DX, Zheng X, Kong J, Manchester LC, Hardeland R, Kim SJ et al (2014) Fundamental issues related to the origin of melatonin and melatonin isomers during evolution: relation to their biological functions. Int J Mol Sci 15(9):15858–15890

Tan XL, Zhao YT, Shan W, Kuang JF, Lu WJ, Su XG et al (2020) Melatonin delays leaf senescence of postharvest Chinese flowering cabbage through ROS homeostasis. Food Res Int 138:109790

Tiwari RK, Lal MK, Naga KC, Kumar R, Chourasia KN, Subhash S et al (2020) Emerging roles of melatonin in mitigating abiotic and biotic stresses of horticultural crops. Sci Hortic 272:109592

Tiwari RK, Lal MK, Kumar R, Chourasia KN, Naga KC, Kumar D, Das SK, Zinta G (2021a) Mechanistic insights on melatonin-mediated drought stress mitigation in plants. Physiol Plant 172(2):1212–1226

Tiwari RK, Lal MK, Kumar R, Mangal V, Altaf MA, Sharma S, Singh B, Kumar M (2021b) Insight into melatonin-mediated response and signaling in the regulation of plant defense under biotic stress. Plant Mol Biol:1–15

Tiwari RK, Altaf MA, Kumar R, Lal MK (2024) Evolution of melatonin as an oxidative stress mitigator in plant. In: Melatonin in plants: a pleiotropic molecule for abiotic stresses and pathogen infection. Springer Nature Singapore, Singapore, pp 1–20

Tumilaar SG, Hardianto A, Dohi H, Kurnia D (2024) A comprehensive review of free radicals, oxidative stress, and antioxidants: overview, clinical applications, global perspectives, future directions, and mechanisms of antioxidant activity of flavonoid compounds. J Chem 2024(1):5594386

Uma GS, Mahanta DK, Sharma L (2024) Advancing Forest insect pest management: a focus on biotechnological approaches. In: Biotechnological approaches for sustaining Forest trees and their products. Springer Nature Singapore, Singapore, pp 347–383

Verma AK, Singh S, Rizvi SI (2023a) Therapeutic potential of melatonin and its derivatives in aging and neurodegenerative diseases. Biogerontology 24(2):183–206

Verma RC, Waseem MA, Sharma N, Bharathi K, Singh S, Anto Rashwin A et al (2023b) The role of insects in ecosystems, an in-depth review of entomological research. Int J Environ Climate Change 13(10):4340–4348

Vielma JR, Bonilla E, Chacín-Bonilla L, Mora M, Medina-Leendertz S, Bravo Y (2014) Effects of melatonin on oxidative stress, and resistance to bacterial, parasitic, and viral infections: a review. Acta Trop 137:31–38

Vyas V, Bhoi TK, Samal I, Singh S, Mahanta DK (2024) Revitalizing degraded soils with agroforestry interventions: opportunities, challenges, and future direction. In: Agroforestry to combat global challenges: current prospects and future challenges. Springer, pp 529–549

Wang Q, Mohamed AA, Takeda M (2013) Serotonin receptor B may lock the gate of PTTH release/synthesis in the Chinese silk moth, *Antheraea pernyi*; a diapause initiation/maintenance mechanism? PLoS One 8(11):e79381

Wang Q, Egi Y, Takeda M, Oishi K, Sakamoto K (2015) Melatonin pathway transmits information to terminate pupal diapause in the C hinese oak silkmoth A ntheraea pernyi and through reciprocated inhibition of dopamine pathway functions as a photoperiodic counter. Entomolog Sci 18(1):74–84

Wong KC, Sankaran S, Jayapalan JJ, Subramanian P, Abdul-Rahman PS (2021) Melatonin improves cognitive behavior, oxidative stress, and metabolism in tumor-prone lethal giant larvae mutant of Drosophila melanogaster. Arch Insect Biochem Physiol 107(1):e21785

Yamano H, Watari Y, Arai T, Takeda M (2001) Melatonin in drinking water influences a circadian rhythm of locomotor activity in the house cricket, *Acheta domesticus*. J Insect Physiol 47(8):943–949

Yang L, Qin Y, Li X, Song D, Qi M (2007) Brain melatonin content and polyethism in adult workers of *Apis mellifera* and *Apis cerana* (Hym., Apidae). J Appl Entomol 131(9–10):734–739

Yu Y, Lv Y, Shi Y, Li T, Chen Y, Zhao D, Zhao Z (2018) The role of phyto-melatonin and related metabolites in response to stress. Molecules 23(8):1887

Zhang H, Wang L, Shi K, Shan D, Zhu Y, Wang C et al (2019) Apple tree flowering is mediated by low level of melatonin under the regulation of seasonal light signal. J Pineal Res 66(2):e12551

Zhao D, Yu Y, Zhao Z, Sharma R, Reiter RJ (2019) Melatonin synthesis and function: evolutionary history in animals and plants. Front Endocrinol 10:441357

Zhou C, Li D, Shi X, Zhang J, An Q, Wu Y et al (2021) Nanoselenium enhanced wheat resistance to aphids by regulating biosynthesis of DIMBOA and volatile components. J Agric Food Chem 69(47):14103–14114

Zhou W, Arcot Y, Medina RF, Bernal J, Cisneros-Zevallos L, Akbulut ME (2024) Integrated pest management: an update on the sustainability approach to crop protection. ACS Omega 9(40):41130–41147

Mechanistic Understanding of the Crucial Role of Melatonin in Enhancing the Shelf Life of Perishable Food Products

13

Ansheef Ali, Abeer Ali, H. R. Raghavendra, and Rajeev Kumar

Abstract

Melatonin, a natural plant hormone, has recently gained notable attention for its multifunctional roles in animals and plants, particularly in extending the shelf life of perishable food products. As an endogenous antioxidant and regulator of physiological processes, melatonin is potentially responsible for delaying senescence, mitigating oxidative stress, and enhancing the overall quality of harvested products. This chapter explores the mechanistic understanding of melatonin's crucial role in extending the shelf life of perishable food products, focusing on its biochemical properties and interaction with reactive oxygen species (ROS), which are primary factors of food spoilage. This chapter explores the role of melatonin's influence on post-harvest physiology, especially in modulating cellular processes such as gene expression, membrane integrity, and ethylene production, which are crucial for delaying ripening and spoilage. The discussion will also review recent research demonstrating melatonin's effectiveness in various food systems, including fruits and vegetables. This chapter also highlights the probable integration of melatonin with other preservation technologies and the potential for genetically engineered crops with enhanced endogenous melatonin levels. This comprehensive overview of the molecular/biochemical mechanism and practical melatonin application in different food systems may aid

A. Ali (✉) · H. R. Raghavendra
School of Crop Improvement, ICAR-Indian Agricultural Research Institute, Gogamukh, Dhemaji, Assam, India

A. Ali
Division of Fruit Science, ICAR-Indian Agricultural Research Institute, New Delhi, India

R. Kumar
Division of Plant Physiology and Biochemistry, ICAR-Indian Institute of Sugarcane Research, Lucknow, Uttar Pradesh, India

R. K. Tiwari et al. (eds.), *Melatonin: Signal Transduction Mechanisms and Defense Networks in Plants*, https://doi.org/10.1007/978-981-96-8869-2_13

researchers and industries in developing food products with minimal post-harvest losses and ensuring food security.

Keywords
Antioxidants · Melatonin · Phytohormones · Post-harvest · Shelf-life

13.1 Introduction

The discovery of melatonin in plants marked a crucial advancement in plant physiology and biochemistry, opening new areas in our understanding of plant regulatory mechanisms (Altaf et al. 2022; Kumar et al. 2023, 2024). Initially identified as a neurohormone in animals, melatonin (N-acetyl-5-methoxytryptamine) is synthesised from the amino acid tryptophan and plays critical roles in circadian rhythm, photo-response regulation, seed germination, immune response, and antioxidative protection of the plant system under different stress conditions etc. (Lerner et al. 1958; Lal et al. 2023). It is possibly synthesised in the chloroplasts and mitochondria of leaves and roots and then transferred to the remaining plant parts (Tan et al. 2013). Although many of its roles in plant systems are carried out independently, it has a cross-talk with other phytohormones like gibberellin, auxin, cytokinin, ethylene, abscisic acid, salicylic acid, etc., in regulating plant growth and development (Yang et al. 2021; Sun et al. 2023).

Research has later shown that melatonin may act as a natural preservative for fruits, vegetables, and other perishable products, effectively improving shelf life. Unlike artificial preservatives, melatonin does not harm the human body (Feng et al. 2022). Increasing shelf life without compromising nutritional quality and safety is a long-term challenge for researchers, regardless of the agricultural commodity (Ali et al. 2023a). So, in this regard, melatonin became a promising candidate with wide application in various food systems. The initial studies identified it as an excellent molecule with antioxidant properties in plants (Tiwari et al. 2021a, b, 2024). Melatonin not only reduces the reactive oxygen species (ROS) directly, but it can also act as a signalling molecule that regulates the gene expression of antioxidant enzymes, the membrane permeability, and the expression of ripening-related genes through which the quality of the food commodity can be enhanced (Wang et al. 2020a). So, in this chapter, we will discuss the biochemical basis of melatonin regulation in plants through which preservation or shelf-life enhancement can be altered. The chapter will also review the recent progress of melatonin research in various fruits, vegetables, and other perishable food products along with the possible application of melatonin with other treatments/bioactive compounds to enhance shelf-life aspects of food products. Lastly, a recent update on the interventions in biotechnological aspects in enhancing innate levels of melatonin in plants is also being addressed.

13.2 Biochemical Basis of Melatonin Action in Enhancing Shelf Life of Perishable Food Products

Soon after the discovery of melatonin in plants, its various roles in growth and development were studied. It emerged as a crucial signalling molecule in plants, where it regulates various physiological processes. In plants, melatonin is synthesised primarily through the tryptophan pathway and acts by modulating antioxidant systems, promoting stress tolerance. Biochemically, melatonin scavenges reactive oxygen species (ROS) directly and indirectly by enhancing the activities of antioxidant enzymes such as superoxide dismutase, catalase, and peroxidase. This helps maintain cellular redox balance, which is crucial during environmental stresses like drought, salinity, and extreme temperatures. Melatonin also regulates gene expression by activating transcription factors that respond to stress, facilitating the upregulation of stress-responsive genes. Additionally, it modulates hormone signalling pathways, particularly ethylene (ET) and abscisic acid (ABA), which are involved in stomatal regulation and stress response and the shelf-life quality of several perishable food products. In this section, we will discuss the biochemical basis of melatonin regulation and the post-harvest shelf life of perishable food products.

13.2.1 Melatonin Biosynthesis in Plants

Melatonin enhances shelf life and is a preservative in perishable food products primarily due to its antioxidant, anti-browning, and antimicrobial properties. In plants, at least four different biosynthetic pathways for melatonin that occur from tryptophan and involve several enzymatic reactions have been proposed. The most common pathway of its biosynthesis is shown in Fig. 13.1. The precursor tryptophan is first converted into tryptamine by the enzyme tryptophan decarboxylase. Next, tryptamine 5-hydroxylase (T5H) hydroxylates tryptamine at the C5 position, resulting in the formation of serotonin (5-hydroxytryptamine), which is then transformed into melatonin through a series of reactions involving N-acetylation, catalysed by arylalkylamine N-acetyltransferase, followed by O-methylation (Khan et al. 2020). Apart from this general pathway, there are other minor pathways which are generally activated under stressful conditions in plants (Tan et al. 2016; Ye et al. 2019). In addition, it has been studied that melatonin biosynthesis is primarily located in the chloroplast, endoplasmic reticulum and cytoplasm of plants, as the enzymes that convert tryptophan to melatonin are located in different cell organelles (Back et al. 2016). If the chloroplast pathway is inhibited for any reason, the mitochondrial pathway can activate to sustain melatonin biosynthesis. Mitochondria are recognised as primary sources of reactive oxygen species (ROS) and reactive nitrogen species (RNS), and melatonin synthesis within mitochondria enhances antioxidant activity, offering greater protection for organelles. This demonstrates that melatonin possesses multiple biosynthetic pathways (Wang et al. 2017). Although the presence and quantification of melatonin in fruits and vegetables after harvest is well established, it's interesting that most of the post-harvested fruits and some

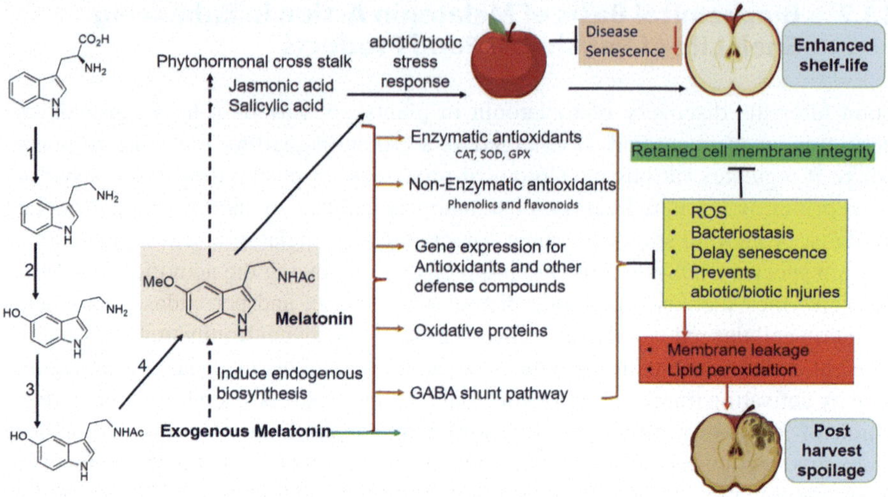

Fig. 13.1 Biosynthetic pathway of melatonin in plants and its role in shelf-life enhancement of perishable food products. Abbreviation: 1. Tryptophan decarboxylase. 2. Tryptamine 5-hydroxylase. 3. Arylalkylamine N-acetyltransferase. 4. N-acetylserotonin methyltransferase

vegetables don't have chloroplasts, especially in seeds and pulp, in which melatonin biosynthesis needs further investigation.

13.2.2 Antioxidant Potential of Melatonin

The scavenging capacity of melatonin was first recognised about three decades ago in animal systems, and soon after its discovery in vascular plants, its antioxidant properties were well-studied (Fuhrberg et al. 1997). The small molecular weight and amphipathic properties of melatonin allow it to readily traverse through cell membranes to reach the cytoplasm, showing its potential role as an antioxidant in cells. Later, it was found that the scavenging potential of melatonin is 5 and 15 times more than glutathione and mannitol, respectively (Khan et al. 2020). Melatonin also upregulates several genes associated with antioxidant enzymes, which in turn improves its efficiency as an antioxidant. Besides the activation of the main antioxidative enzymes, such as peroxidase and superoxide dismutase, and the contents of other antioxidant molecules, such as glutathione, ascorbic acid, phenolic and flavonoid compounds, are also indirectly modified by melatonin (Arnao and Hernández-Ruiz 2021). On the other hand, melatonin acts as a mediator in several antioxidant pathways such as the glutathione-ascorbate cycle, through different mechanisms (Debnath et al. 2019). It uses both direct and indirect mechanisms to scavenge the ROS (Tiwari et al. 2020). In direct, melatonin can neutralise various ROS, including hydroxyl radicals, hydrogen peroxide, singlet oxygen, nitric oxide, and peroxynitrite anion, etc. In addition, it can also help in modulating mitochondrial homeostasis, which protects cells against oxidative damage (Florido et al. 2022). Several metabolites that are produced during the biosynthesis of melatonin, such as 2-hydroxy melatonin and cyclic 3-hydroxy melatonin, are also influential antioxidants, promoting the antioxidant capabilities of this biomolecule (Tan et al. 2007). So, it is well clear by now that melatonin can regulate the antioxidant potential of cells in various ways.

13.2.3 Melatonin Cross-Talk with Phytohormones and Regulation of Post-Harvest Quality

Apart from contributing to the natural antioxidant potential of cells, melatonin can be a signalling molecule, through which it can modulate hormone levels significantly during the post-harvest period of perishable food products. It is well known that ethylene (ET) and abscisic acid ABA are the major hormones involved in the senescence and ripening of most crops. Melatonin was found to be effective in inhibiting ET and ABA in some products, by which the senescence and shelf life of products can be enhanced. For example, when melatonin is applied externally during storage, the ethylene biosynthesis genes (ACSs and ACOs) are inhibited, reducing its formation and extending colour changes in tomatoes and bananas (Sun et al. 2020). Such results are also observed in other products like kiwi (Dou et al. 2022), apple (Onik et al. 2021), pear (Liu et al. 2019), mango (Liu et al. 2020), etc. Zhang et al. (2020) observed that the application of melatonin along with nitric oxide can further improve the post-harvest quality of fruits. On the other hand, melatonin stimulates the ethylene biosynthesis in apples by upregulating the genes involved (ACSs and ACOs), hence making apples more colour, flavour, and firm for consumer appeal (Verde et al. 2023). It is also found that melatonin can enhance ethylene production by upregulating the transcription factors involved in ethylene biosynthetic genes (Ma et al. 2021). In grape berries, exogenous melatonin has caused increased accumulation of ABA, ethylene and H_2O_2, promoting ripening (Xu et al. 2018). Later, Li et al. (2021) observed that the endogenous content of ethylene can be improved by the application of combined red and blue light during storage periods of fruits, which in turn produces ethylene, hence accelerating fruit ripening. Shan et al. (2022) recently observed that melatonin changes the DNA methylation pattern and regulates the ethylene biosynthesis in fruit crops. So based on the literature, melatonin can act as a potential candidate that regulates the post-harvest shelf life of perishable food crops. However, as there are a significant number of studies that state the dual role of melatonin in promoting as well as suppressing product quality, further investigation is required in this area.

13.2.4 Melatonin Cross-Talk with Metabolites and Regulation of Post-Harvest Quality

Melatonin is also reported to interfere with metabolites like GABA (gamma-aminobutyric acid) and ascorbic acid, which regulate the final product's quality.

GABA is produced by the GABA shunt from the citric acid cycle and regulates the cell's ROS accumulation. It's reported that the exogenous application of melatonin can increase the synthesis of GABA through upregulating biosynthetic enzymes involved, thus increasing the ATP production in fruits and vegetables, helping them to remove excess H_2O_2 and reactive oxygen species (ROS), which helps in prolonging their post-harvest shelf life (Palma et al. 2015). Wu et al. (2023) observed that post-harvest exogenous application of melatonin has increased the GABA content in peaches by increasing the expression of its biosynthesis genes as well as suppressing of expression of its degradative genes. Such findings are also observed in other agricultural products like okra (Shi et al. 2024b), tomato (Sharafi et al. 2019), etc. Apart from GABA, other metabolites like proline (Ali et al. 2023b), polyamines (Bhardwaj et al. 2022), glycine betaine (Khalid et al. 2022), ascorbic acid (Hu et al. 2021; Li et al. 2022a), etc., are also accumulated in a plant cell in response to melatonin application, which provides stress resistance through their antioxidant activities and osmotic balances, hence extending the shelf life of perishable food products.

13.3 Role in Extending the Shelf Life of Perishable Food Products

Melatonin is known for its role in regulating circadian rhythms and has gained significant attention in plant science due to its involvement in post-harvest physiology, particularly in the ripening and senescence of fruits, vegetables, and other perishable food products. In the last decade, intensive research advancement has been observed in applying melatonin in perishable products in terms of their shelf-life enhancement, quality attributes and resistance to biotic and abiotic stress during storage periods. So, in this section, the effect of melatonin on the post-harvest quality aspects of perishable food products is briefly discussed.

13.3.1 Role of Melatonin in Ripening and Senescence of Post-Harvest Perishable Food Products

Melatonin can act as a signalling molecule regulating various physiological metabolism of post-harvested food products. As listed in Table 13.1, melatonin reduces post-harvest decay/senescence of several food products. It is well known that climacteric fruits undergo biochemical and physiological changes after harvesting before reaching their full ripening stage. This ripening process helps to form quality attributes, including nutrition, texture, appearance, aroma, and flavour of the respective food products. Recently, a vast number of studies have shown the involvement of melatonin in the ripening process. After the exogenous application of melatonin in optimum concentration, pericarp colour development was significantly delayed in fruits like bananas (Hu et al. 2017), strawberries (Mansouri et al. 2023), and mango (Liu et al. 2020) due to the arrest of starch and cell wall degrading enzymes. In vegetables, like okra (Shi et al. 2024a) and pepper (Li et al. 2024), softening of

Table 13.1 Effect of exogenous melatonin treatment on quality attributes of horticultural crops

Crop	Treatment	Effect	Physiological response/ remarks	References
Fruits				
Apple	Sprayed 1 mM	Senescence ↓	Weight loss, MDA, ET ↓	Onik et al. (2021)
Banana	10 mM 3 min	Decay ↓	Disease spots, CO_2 ↓ Firmness, hemicellulose, esters, waxes ↑	Li et al. (2019b)
Grape	50 μM 5 s	Decay ↓	Flavonoids ↑	Gao et al. (2020)
Litchi	0.4 mM 5 min	Senescence ↓	Browning index, leakage rate, MDA ↓ TPC, Anthocyanins ↑	Wang et al. (2020b)
Mango	0.5 mM 1 h	Ripening ↓	Firmness ↑ b*, CO_2, ethylene, β-carotene, TSS ↓	Liu et al. (2020)
Sweet cherry	Sprayed 100 μM	Senescence ↓	Fruit weight, TSS, Total phenolics ↑	Xia et al. (2020)
Papaya	1.5 mM sprayed	Senescence ↓	Ethylene, CO_2 ↓ AoA, firmness ↑	Borthakur et al. (2024)
Vegetables				
Spinach	0.20 mg/mL	Decay ↓	MDA ↓ TSS, soluble solid, chlorophyll, AoA ↑	Wang et al. (2023)
Tomato	50 μM 1 h	Ripening ↑ Decay ↓	TPC, TFC, lignin ↑	Li et al. (2019a)
Broccoli	100 μM 5 min	Storage life ↑	–	Zhu et al. (2018)
Brinjal	100 μM	Decay ↓	Firmness ↑ MDA, PME ↓	Song et al. (2022)
Cucumber	100, 500 μM	Decay ↓	Electrolyte leakage MDA ↓ Proline ↑	Liu et al. (2022)
Chinese flowering cabbage	100 μM	Decay ↓	Weight loss, respiration ↓ Soluble sugar, protein ↑	Tan et al. (2021)
Cherry tomato	0.1 mM	Senescence ↓	Weight loss, titratable acid ↓ Fruit firmness, total soluble solids ↑	Yan et al. (2022)

Abbreviations: *mM* millimolar, *μM* micromolar, *MDA* malondialdehyde, *ET* ethylene, *min* minutes, *s* seconds, *TPC* total phenolic content, *TSS* total soluble sugar, *TFC* total flavonoid content, *AoA* antioxidant activities, *PME* pectin methyl esterase, *b** hue colour values

cell walls was reduced, and hence, the ripening was arrested in response to melatonin. Apart from this, the exogenous application of melatonin affects several quality attributes such as carotene content, titrable acidity, respiratory rate, soluble sugar, volatile aroma, and organic acid and weight loss in fruits and vegetables, ultimately improving its shelf-life quality (Arabia et al. 2023). But on the other hand, external melatonin application has enhanced the ripening of tomatoes. Here, the accumulation of primary and secondary metabolites was observed. The study observed that melatonin has enhanced soluble sugar by upregulating the enzymes involved in free sugar synthesis, such as sucrose synthase and sucrose phosphate synthase. It also observed a significant reduction in the organic acid content of fruits during storage, along with an accumulation of different phenolics, flavonoids, and volatile compounds that, in turn, enhance the final fruit quality (Dou et al. 2022).

Mostly the melatonin-mediated delayed ripening is regulated through its interaction with other phytohormones like ethylene, abscisic acid, jasmonic acid, salicylic acid, etc. (Hardeland 2016). Additionally, melatonin treatment can increase endogenous melatonin levels and stimulate nitric oxide (NO) production during the ripening of fruits such as bananas (Hu et al. 2017) and pears. NO treatment not only decreases ROS accumulation in fruit but also inhibits the expression of ethylene biosynthetic genes, helping to delay or reduce the ethylene peak in various climacteric fruits (Zhang et al. 2020). Additionally, melatonin improves the efficacy of pesticides in delaying fruit senescence, as indicated by reduced weight loss, lower decay rates, and maintained firmness, while preserving peel structure (Deng et al. 2021). It also boosts ascorbic acid levels, total phenolic content, and overall antioxidant capacity in jujube fruit. Melatonin mitigates colour changes and helps maintain titratable acidity in jujubes. It also enhances the activities of enzymes like ascorbate peroxidase, glutathione reductase, and monodehydroascorbate reductase, while increasing the levels of ascorbate, dehydroascorbate, and reduced glutathione. Moreover, melatonin has been shown to suppress the activities of cellulase, pectin methylesterase, β-glucosidase, and polygalacturonase, reducing water-soluble pectin production and preserving water-insoluble pectin, thereby slowing down fruit softening (Tang et al. 2020).

Although terms like ripening and senescence are often used interchangeably, ripening is an active, genetically regulated process in which the fruits and vegetables undergo molecular and biochemical changes necessary to complete the final stages of development. In senescence, quality starts to gradually deteriorate. Antioxidants, such as phenolic compounds, organic acids, and ethylene, mainly regulate senescence (Wang et al. 2020a). Although the response of melatonin to ripening and senescence is overlapping, it was reported that during senescence, the exogenous melatonin has caused a significant reduction in acetaldehyde, ethanol, volatile aroma, etc., along with a significant increase in contents of anthocyanins, fatty acids, amino acids, phenolics, and flavonoids (Wang et al. 2019; Ze et al. 2021). Considering the fact that both auxin and melatonin are formed from a single precursor, tryptophan (Li et al. 2019b), it is observed that melatonin application has induced auxin biosynthesis, and hence, the banana fruit shelf life has extended. Additionally, exogenous melatonin has reduced the ABA content in strawberries during storage, thereby increasing the post-harvest quality (Liu et al. 2018).

Exogenous melatonin retains chlorophyll content, decreases ethylene production, and boosts the activity of antioxidant enzymes such as catalase, superoxide dismutase, and glutathione reductase. It also lowers free radical levels and reduces the activity of cell wall degrading enzymes. Together, these effects enhance antioxidant capacity, delay fruit softening, and effectively slow the ageing process in fruits and vegetables. Nevertheless, there is limited research on the effects of pre-harvest melatonin treatment on post-harvest fruits and vegetables, highlighting the need for further investigation.

13.3.2 Role of Melatonin in Disease Resistance and Chilling Injuries of Post-Harvest Perishable Food Products

One of the major challenges during perishable food storage is the occurrence of disease. Most of these diseases are caused by fungal infections during storage, hence accelerating the senescence process. In such cases, the useful effects of exogenous melatonin have been proven in fruits like grapes (Li et al. 2022c), citrus (Lin et al. 2019), blueberries (Qu et al. 2022), and vegetable tomato (Li et al. 2019a), potato (Zhang et al. 2017), etc. Conversely, negative consequences of melatonin application in post-harvest citrus are also reported (Lin et al. 2019). Furthermore, no significant effect of melatonin on *Colletotrichum musae* (Li et al. 2019b) and *Penicillium digitatum* (Lin et al. 2019) has also been studied. The biochemical/molecular-based innate defence system is critical for disease resistance during the storage period. Exogenous melatonin has significantly increased ROS such as H_2O_2, O_2^-, and NO at the early stages of fruits and vegetables, improving resistance to stored diseases (Zhang et al. 2020). Melatonin may act as a signalling molecule and may induce other phytohormones, MAPK (mitogen-activated protein kinase), CDPK (calcium-dependent protein kinase), NO (nitric oxide) biosynthesis, etc., contributing altogether in defence resistance towards diseases (Palma et al. 2019; Li et al. 2022b). As a result, melatonin induces several defence-related proteins such as pathogenesis-related (PR) proteins, peroxidase, catalase, lipid-transfer proteins, chitinases, and β-1,3-glucanases (Gao et al. 2020). Apart from this, non-enzymatic antioxidants such as ascorbic acid, glutathione, phenolics, flavonoids, and anthocyanins also contributed to the melatonin-induced post-harvest disease resistance in perishable food products (Bal 2019). Exogenous melatonin could also decrease fungal infection in banana fruit by enhancing the production of cell wall, wax, and lipids on the fruit peel (Li et al. 2019b).

Since most horticultural products are stored at low temperatures, applying exogenous melatonin could notably enhance cold resistance, thereby reducing the incidence of chilling injury in post-harvest fruits and vegetables such as peach (Gao et al. 2018), pomegranates (Azadshahraki et al. 2018), and sapota (Mirshekari et al. 2020). Table 13.2 provides recent studies on melatonin in various fruits and vegetables to prevent chilling injury during storage.

Exogenous melatonin treatment reduced electrolyte leakage, MDA content, and lipoxygenase activity of perishable horticultural crops during low-temperature storage. As in disease resistance, exogenous melatonin application enhanced enzymatic and non-enzymatic antioxidants to scavenge ROS of post-harvest horticultural

Table 13.2 Effect of exogenous melatonin on disease resistance and chilling injuries of perishable crops

Crop	Treatment	Physiological response/remarks	References
Potato	10 mM 12 h sprayed	Potato late blight ↓	Zhang et al. (2017)
Strawberry	100 μM 15 min	Grey mould (*B. cinerea*) ↓	Promyou et al. (2023)
Banana	–	Anthracnose (*Colletotrichum musae*) ↓	Li et al. (2019b)
Cherry tomato	0.1 mM 1 h	Grey mould (*B. cinerea*) ↓	Li et al. (2019a)
Mango	0.2 mmol L^{-1}	Anthracnose (*Colletotrichum gloeosporioides*) ↓	Lu et al. (2024)
Mango	0.2 mmol L^{-1} 30 min	Chilling injury ↓ Proline ↑	Kebbeh et al. (2023)
Kiwifruit	0.05 mM	Chilling injury ↓ Salicylic acid ↑	Guo et al. (2023)
Brinjal	100 μmol L^{-1}	Chilling injury ↓ Polyamines ↑	Song et al. (2023)
Potato	0.1 μM L^{-1}	Chilling resistance ↑	Golovatskaya et al. (2024)

Abbreviations: *mM* millimolar, *μM* micromolar, *min* minute, *h* hour

products during low-temperature storage, such as ascorbate peroxidase, catalase, Glutathione reductase, phenolics, flavonoids, and ascorbic acid (Mirshekari et al. 2020). Browning of grains and horticultural products such as apples and potatoes is one of the major challenges that affect consumer acceptance (Ali et al. 2022; Sidhu et al. 2023). Melatonin exogenous application has proven to inhibit the activity of polyphenol oxidase (PPO), the major culprit of browning, hence reducing the browning significantly (Xiao et al. 2021) in crops such as pear (Zheng et al. 2019), sweet potato (Santos e Silva et al. 2024). It also noted that, during cold storage, melatonin treatment could alleviate the nitric oxide (NO) level in tissues, thus scavenging the formation of ROS (Aghdam et al. 2019). Exogenous melatonin also helped to maintain a higher ratio of unsaturated fatty acid to saturated fatty acid in the membrane during the cold stage, hence helping to reduce membrane leakage and lipid peroxidation (Gao et al. 2018). Increased level of nitrogenous compounds such as GABA, polyamine, and proline is also reported to accumulate in response to melatonin application during cold storage (Cao et al. 2016). In addition, melatonin application has also induced biosynthesis of phytohormones such as salicylic acid during cold storage to induce cold resistance in horticultural products (Gao et al. 2018). It is fascinating that melatonin's response to disease and chilling stress overlaps in most studies. Externally applied melatonin can enhance defence responses by regulating hormone synthesis and various other metabolic pathways, ultimately increasing the shelf life of perishable food products.

13.4 Synergistic Role of Melatonin with Other Treatments in Perishable Food Products Quality

So far, we have discussed the potential role of melatonin in regulating the post-harvested shelf-life quality of perishable food products. It is now well known that the combined action of melatonin with other established treatments could further improve the quality of food products. In the initial studies, Lin et al. (2020) studied the combined effect of melatonin and ethanol on the post-harvest quality of bitter melon fruit. During the optimum treatment of melatonin (120 μmol L^{-1}), with ethanol (6%) and an immersing time of 10 min, there was a reduction in chilling and a reduction in the accumulation of MDA and ion leakage. The study also observed a significantly higher accumulation of increased content of total soluble solids, soluble sugar, chlorophyll content, ascorbic acid, total phenolic, and flavonoids. Later, Sarkar et al. (2021) observed in green plantains (*Musa paradisiaca*) that a lower dose of gamma irradiation (0.04 kGy) caused a rise in melatonin along with serotonin content, hence improving shelf-life quality. Combining melatonin (1 mM) with 1-methylcyclopropene (2400 ppb) has improved chilling tolerance in zucchini fruit. This study observed a reduction in respiratory rate, MDA content and ethylene production and an increase in chlorophyll and antioxidant potential in response to treatments. The simultaneous effect of melatonin (1 mM) and ascorbic acid (20 mM) has improved the shelf life of fresh-cut avocado fruits. The study observed a reduction in post-harvest weight loss, membrane leakage, and respiratory rate along with increased enzymatic (POD, SOD) and non-enzymatic antioxidant activities (total phenolic and flavonoid content), etc. Later, Mwelase et al. (2022) showed that the chitosan-based melatonin composite coating has quality and delays the senescence of minimally processed pomegranate aril-sacs during cold storage. An optimum concentration of chitosan and 0.8–1 mM melatonin treatment shows the best results with retained higher aril chroma, anthocyanin content, and ascorbic acid, along with improved antioxidant activities (FRAP, DPPH) and reduced surface browning. Zang et al. (2022) showed the synergistic effect of melatonin with selenium in post-harvest resistance towards grey mould disease caused by the fungal pathogen *Botrytis cinerea* in tomato fruits. An optimal sodium selenite (20 mg/L) and melatonin (10 mmol/L) dose has improved the post-harvest shelf life. The combined treatment has increased the expression of genes involved in the pathogen-related protein, along with improved antioxidant enzyme activities, such as POD, SOD, and CAT.

13.5 Engineered Crops with Optimum Melatonin Levels for Improved Shelf Life

While the over-expression of melatonin in plants through transgenic approaches holds promise for enhancing the post-harvest preservation of fruits, vegetables, and other perishable food products, efforts to boost endogenous melatonin levels in plants remain limited (Xu et al. 2019). Engineering transgenic lines overexpressing genes involved in melatonin biosynthesis, such as serotonin N-acetyltransferase (SNAT) or N-acetylserotonin methyltransferase (ASMT), researchers have achieved enhanced endogenous melatonin production (Zuo et al. 2014). Although such transgenic approaches are associated with plants having higher melatonin content for salt tolerance (Wu et al. 2021), drought tolerance (Yang et al. 2019), cold resistance (Kang et al. 2010), resistance to herbicide-induced oxidative stress (Park et al. 2013), improved water use efficiency (Zhou et al. 2022), etc. In one study, it was found that over-expression of melatonin in rice caused pleiotropic effects such as increased seedling growth, delayed flowering, and reduced grain yield (Byeon and Back 2014). A transgenic approach for high-melatonin plants for improved shelf-life properties is now needed. Such approaches may improve the plants' resistance to biotic and abiotic stress, reduce ethylene-induced ripening, and limit lipid peroxidation in harvested produce. Enhanced melatonin levels also promote cellular homeostasis and upregulate antioxidant defence systems, further contributing to extended freshness and reduced spoilage during storage. These advances could significantly reduce food wastage, increase market value, and support global food security efforts, particularly in regions with limited cold storage facilities.

13.6 Conclusion

Melatonin (N-acetyl-5-methoxytryptamine), originally identified as a neurohormone in animals, plays a vital role in regulating plant growth, development, and stress responses. Synthesised from tryptophan, melatonin acts as a powerful antioxidant, interacting with phytohormones like gibberellins, ethylene, and abscisic acid to modulate plant physiology. It is increasingly recognised for its potential as a natural preservative to enhance the shelf life of perishable food products by reducing oxidative damage and improving post-harvest quality. Melatonin synthesis involves the conversion of tryptophan to serotonin and then melatonin through enzymatic reactions occurring mainly in chloroplasts, mitochondria, and the endoplasmic reticulum. Under stress, alternative pathways ensure continuous production. Melatonin's amphipathic nature allows it to cross cell membranes, effectively scavenging reactive oxygen species (ROS) and boosting antioxidant enzyme activities such as superoxide dismutase, catalase, and peroxidase. Its metabolite, 2-hydroxy melatonin, further enhances this antioxidative capacity. Melatonin influences senescence and ripening by modulating ethylene and abscisic acid pathways. It can delay ripening in fruits like bananas and mangoes by inhibiting cell wall degrading enzymes while accelerating ripening in others like tomatoes. It also reduces oxidative stress, preserves chlorophyll, and boosts antioxidant enzyme activity, extending the shelf life of various fruits and vegetables. In addition to improving oxidative stress tolerance, melatonin enhances disease resistance by stimulating defence proteins and promoting physical barriers such as wax layers. It also mitigates chilling injuries by reducing electrolyte leakage, lipid peroxidation, and ROS accumulation. Combined treatments with melatonin, such as with ethanol, gamma irradiation, chitosan coating, or selenium, further enhance its benefits, improving shelf life and nutritional quality. Genetically engineered crops with higher melatonin production show promise for improved post-harvest quality, stress tolerance, and reduced spoilage, though challenges regarding growth and yield trade-offs remain. Overall, melatonin holds great potential to reduce food waste and support global food security.

References

Aghdam MS, Luo Z, Jannatizadeh A et al (2019) Employing exogenous melatonin applying confers chilling tolerance in tomato fruits by upregulating ZAT2/6/12 giving rise to promoting endogenous polyamines, proline, and nitric oxide accumulation by triggering arginine pathway activity. Food Chem 275:549–556. https://doi.org/10.1016/j.foodchem.2018.09.157

Ali A, Kumar RR, Vinutha T et al (2022) Grain phenolics: critical role in quality, storage stability and effects of processing in major grain crops—a concise review. Eur Food Res Technol 248(8):2197–2213. https://doi.org/10.1007/S00217-022-04026-7

Ali A, Kumar RR, Vinutha T et al (2023a) Characterization of biochemical indicators and metabolites linked with rancidity and browning of pearl millet flour during storage. J Plant Biochem Biotechnol 32:121–131. https://doi.org/10.1007/s13562-022-00787-0

Ali S, Nawaz A, Naz S et al (2023b) Exogenous melatonin mitigates chilling injury in zucchini fruit by enhancing antioxidant system activity, promoting endogenous proline and GABA accumulation, and preserving cell wall stability. Postharvest Biol Technol 204:112445. https://doi.org/10.1016/j.postharvbio.2023.112445

Altaf MA, Shahid R, Ren MX, Naz S, Altaf MM, Khan LU, Tiwari RK, Lal MK, Shahid MA, Kumar R, Nawaz MA (2022) Melatonin improves drought stress tolerance of tomato by modulating plant growth, root architecture, photosynthesis, and antioxidant defense system. Antioxidants 11(2):309

Arabia A, Muñoz P, Pallarés N, Munné-Bosch S (2023) Experimental approaches in studying active biomolecules modulating fruit ripening: melatonin as a case study. Plant Physiol 192:1747–1767. https://doi.org/10.1093/plphys/kiad106

Arnao MB, Hernández-Ruiz J (2021) Melatonin as a plant biostimulant in crops and during postharvest: a new approach is needed. J Sci Food Agric 101:5297–5304. https://doi.org/10.1002/JSFA.11318

Azadshahraki F, Jamshidi B, Mohebbi S (2018) Postharvest melatonin treatment reduces chilling injury and enhances antioxidant capacity of tomato fruit during cold storage. Adv Hortic Sci 32:299–309. https://doi.org/10.13128/ahs-22260

Back K, Tan DX, Reiter RJ (2016) Melatonin biosynthesis in plants: multiple pathways catalyze tryptophan to melatonin in the cytoplasm or chloroplasts. J Pineal Res 61:426–437. https://doi.org/10.1111/jpi.12364

Bal E (2019) Physicochemical changes in "Santa Rosa" plum fruit treated with melatonin during cold storage. J Food Meas Charact 13:1713. https://doi.org/10.1007/s11694-019-00088-6

Bhardwaj R, Pareek S, Saravanan C, Yahia EM (2022) Contribution of pre-storage melatonin application to chilling tolerance of some mango fruit cultivars and relationship with polyamines metabolism and γ-aminobutyric acid shunt pathway. Environ Exp Bot 194:104691. https://doi.org/10.1016/j.envexpbot.2021.104691

Borthakur P, Chinnasamy K, Paramasivam SK et al (2024) Exogenous melatonin as pre- and postharvest application on quality attributes, antioxidant capacity, and extension of shelf life of papaya. Horticulturae 10:1099. https://doi.org/10.3390/horticulturae10101099

Byeon Y, Back K (2014) An increase in melatonin in transgenic rice causes pleiotropic phenotypes, including enhanced seedling growth, delayed flowering, and low grain yield. J Pineal Res 56:408–414. https://doi.org/10.1111/jpi.12129

Cao S, Song C, Shao J et al (2016) Exogenous melatonin treatment increases chilling tolerance and induces defense response in harvested peach fruit during cold storage. J Agric Food Chem 64:5215–5222. https://doi.org/10.1021/acs.jafc.6b01118
Debnath B, Islam W, Li M et al (2019) Melatonin mediates enhancement of stress tolerance in plants. Int J Mol Sci 20:1040. https://doi.org/10.3390/IJMS20051040
Deng B, Xia C, Tian S, Shi H (2021) Melatonin reduces pesticide residue, delays senescence, and improves antioxidant nutrient accumulation in postharvest jujube fruit. Postharvest Biol Technol 173:111419. https://doi.org/10.1016/j.postharvbio.2020.111419
Dou J, Wang J, Tang Z et al (2022) Application of exogenous melatonin improves tomato fruit quality by promoting the accumulation of primary and secondary metabolites. Foods 11:4097. https://doi.org/10.3390/foods11244097
Feng BS, Kang DC, Sun J et al (2022) Research on melatonin in fruits and vegetables and the mechanism of exogenous melatonin on postharvest preservation. Food Biosci 50:102196. https://doi.org/10.1016/j.fbio.2022.102196
Florido J, Rodriguez-Santana C, Martinez-Ruiz L et al (2022) Understanding the mechanism of action of melatonin, which induces ROS production in cancer cells. Antioxidants 11:1621. https://doi.org/10.3390/antiox11081621
Fuhrberg B, Hardeland R, Poeggeler B, Behrmann G (1997) Dramatic rises of melatonin and 5-methoxytryptamine in Gonyaulax exposed to decreased temperature. Biol Rhythm Res 28:144–150. https://doi.org/10.1076/brhm.28.1.144.12978
Gao H, Lu ZM, Yang Y et al (2018) Melatonin treatment reduces chilling injury in peach fruit through its regulation of membrane fatty acid contents and phenolic metabolism. Food Chem 245:659–666. https://doi.org/10.1016/j.foodchem.2017.10.008
Gao S, Ma W, Lyu X et al (2020) Melatonin may increase disease resistance and flavonoid biosynthesis through effects on DNA methylation and gene expression in grape berries. BMC Plant Biol 20:1–15. https://doi.org/10.1186/s12870-020-02445-w
Golovatskaya IF, Kadyrbaev MK, Boyko EV (2024) Protective role of melatonin and IAA in the regulation of resistance of potato regenerants to cold stress. Potato Res 67:421–449. https://doi.org/10.1007/s11540-023-09642-8
Guo W, Zhang C, Yang R et al (2023) Endogenous salicylic acid mediates melatonin-induced chilling-and oxidative-stress tolerance in harvested kiwifruit. Postharvest Biol Technol 201:112341. https://doi.org/10.1016/j.postharvbio.2023.112341
Hardeland R (2016) Melatonin in plants – diversity of levels and multiplicity of functions. Front Plant Sci 7:198. https://doi.org/10.3389/fpls.2016.00198
Hu W, Yang H, Tie W et al (2017) Natural variation in banana varieties highlights the role of melatonin in postharvest ripening and quality. J Agric Food Chem 65:9987–9994. https://doi.org/10.1021/acs.jafc.7b03354
Hu E, Liu M, Zhou R et al (2021) Relationship between melatonin and abscisic acid in response to salt stress of tomato. Sci Hortic 285:110176. https://doi.org/10.1016/j.scienta.2021.110176
Kang K, Lee K, Park S et al (2010) Enhanced production of melatonin by ectopic overexpression of human serotonin N-acetyltransferase plays a role in cold resistance in transgenic rice seedlings. J Pineal Res 49:176–182. https://doi.org/10.1111/J.1600-079X.2010.00783.X
Kebbeh M, Dong JX, Huan C et al (2023) Melatonin treatment alleviates chilling injury in mango fruit "Keitt" by modulating proline metabolism under chilling stress. J Integr Agric 22:935–944. https://doi.org/10.1016/j.jia.2023.02.008
Khalid M, Rehman HM, Ahmed N et al (2022) Using exogenous melatonin, glutathione, proline, and glycine betaine treatments to combat abiotic stresses in crops. Int J Mol Sci 23:12913. https://doi.org/10.3390/IJMS232112913
Khan A, Numan M, Khan AL et al (2020) Melatonin: awakening the defense mechanisms during plant oxidative stress. Plants 9:407. https://doi.org/10.3390/PLANTS9040407
Kumar R, Altaf MA, Lal MK, Tiwari RK (eds) (2023) Melatonin in plants: a regulator for plant growth and development. Springer, Singapore
Kumar R, Altaf MA, Lal MK, Tiwari RK (eds) (2024) Melatonin in plants: a pleiotropic molecule for abiotic stresses and pathogen infection. Springer, Singapore

Lal MK, Tiwari RK, Lal P et al (2023) Regulatory role of melatonin in post-harvest management of vegetables and fruits. In: Melatonin in plants: a regulator for plant growth and development. Springer, Singapore, pp 219–244. https://doi.org/10.1007/978-981-99-6745-2_10

Lerner AB, Case JD, Takahashi Y et al (1958) Isolation of melatonin, the pineal gland factor that lightens melanocytes. J Am Chem Soc 80:2587. https://doi.org/10.1021/JA01543A060/ASSET/JA01543A060.FP.PNG_V03

Li S, Xu Y, Bi Y et al (2019a) Melatonin treatment inhibits gray mold and induces disease resistance in cherry tomato fruit during postharvest. Postharvest Biol Technol 157:110962. https://doi.org/10.1016/j.postharvbio.2019.110962

Li T, Wu Q, Zhu H et al (2019b) Comparative transcriptomic and metabolic analysis reveals the effect of melatonin on delaying anthracnose incidence upon postharvest banana fruit peel. BMC Plant Biol 19:289. https://doi.org/10.1186/s12870-019-1855-2

Li Y, Liu C, Shi Q et al (2021) Mixed red and blue light promotes ripening and improves quality of tomato fruit by influencing melatonin content. Environ Exp Bot 185:104407. https://doi.org/10.1016/j.envexpbot.2021.104407

Li GZ, Wang YY, Liu J et al (2022a) Exogenous melatonin mitigates cadmium toxicity through ascorbic acid and glutathione pathway in wheat. Ecotoxicol Environ Saf 237:113533. https://doi.org/10.1016/j.ecoenv.2022.113533

Li S, Cheng Y, Yan R et al (2022b) Preharvest spray with melatonin improves postharvest disease resistance in cherry tomato fruit. Postharvest Biol Technol 193:112055. https://doi.org/10.1016/j.postharvbio.2022.112055

Li Z, Zhang S, Xue J et al (2022c) Exogenous melatonin treatment induces disease resistance against Botrytis cinerea on post-harvest grapes by activating defence responses. Foods 11:2231. https://doi.org/10.3390/foods11152231

Li P, Zhang R, Zhou H et al (2024) Melatonin delays softening of postharvest pepper fruits (*Capsicum annuum* L.) by regulating cell wall degradation, membrane stability and antioxidant systems. Postharvest Biol Technol 212:112852. https://doi.org/10.1016/j.postharvbio.2024.112852

Lin Y, Fan L, Xia X et al (2019) Melatonin decreases resistance to postharvest green mold on citrus fruit by scavenging defense-related reactive oxygen species. Postharvest Biol Technol 153:21–30. https://doi.org/10.1016/j.postharvbio.2019.03.016

Lin X, Wang L, Hou Y et al (2020) A combination of melatonin and ethanol treatment improves postharvest quality in bitter melon fruit. Foods 9:1376. https://doi.org/10.3390/foods9101376

Liu C, Zheng H, Sheng K, Liu W (2018) Effects of melatonin treatment on the postharvest quality of strawberry fruit. Postharvest Biol Technol 139:47–55. https://doi.org/10.1016/j.postharvbio.2018.01.016

Liu J, Yang J, Zhang H et al (2019) Melatonin inhibits ethylene synthesis via nitric oxide regulation to delay postharvest senescence in pears. J Agric Food Chem 67:2279–2288. https://doi.org/10.1021/acs.jafc.8b06580

Liu S, Huang H, Huber DJ et al (2020) Delay of ripening and softening in "Guifei" mango fruit by postharvest application of melatonin. Postharvest Biol Technol 163:111136. https://doi.org/10.1016/j.postharvbio.2020.111136

Liu Q, Xin D, Xi L et al (2022) Novel applications of exogenous melatonin on cold stress mitigation in postharvest cucumbers. J Agric Food Res 10:100459. https://doi.org/10.1016/j.jafr.2022.100459

Lu D, Ren Y, Yan T et al (2024) Melatonin improves the postharvest anthracnose resistance of mango fruit by regulating antioxidant activity, the phenylpropane pathway and cell wall metabolism. Eur J Plant Pathol 171:17. https://doi.org/10.1007/s10658-024-02930-0

Ma W, Xu L, Gao S et al (2021) Melatonin alters the secondary metabolite profile of grape berry skin by promoting VvMYB14-mediated ethylene biosynthesis. Hortic Res 8:43. https://doi.org/10.1038/s41438-021-00478-2

Mansouri S, Koushesh Saba M, Sarikhani H (2023) Exogenous melatonin delays strawberry fruit ripening by suppressing endogenous ABA signaling. Sci Rep 13:14209. https://doi.org/10.1038/s41598-023-41311-1

Mirshekari A, Madani B, Yahia EM et al (2020) Postharvest melatonin treatment reduces chilling injury in sapota fruit. J Sci Food Agric 100:1897–1903. https://doi.org/10.1002/jsfa.10198

Mwelase S, Opara UL, Fawole OA (2022) Effect of chitosan-based melatonin composite coating on the quality of minimally processed pomegranate aril-sacs during cold storage. J Food Process Preserv 46:1–16. https://doi.org/10.1111/jfpp.17096

Onik JC, Wai SC, Li A et al (2021) Melatonin treatment reduces ethylene production and maintains fruit quality in apple during postharvest storage. Food Chem 337:127753. https://doi.org/10.1016/j.foodchem.2020.127753

Palma F, Carvajal F, Ramos JM et al (2015) Effect of putrescine application on maintenance of zucchini fruit quality during cold storage: contribution of GABA shunt and other related nitrogen metabolites. Postharvest Biol Technol 99:131–140. https://doi.org/10.1016/j.postharvbio.2014.08.010

Palma JM, Freschi L, Rodríguez-Ruiz M et al (2019) Nitric oxide in the physiology and quality of fleshy fruits. J Exp Bot 70:4405–4417. https://doi.org/10.1093/jxb/erz350

Park S, Lee DE, Jang H et al (2013) Melatonin-rich transgenic rice plants exhibit resistance to herbicide-induced oxidative stress. J Pineal Res 54:258–263. https://doi.org/10.1111/j.1600-079X.2012.01029.x

Promyou S, Raruang Y, Chen ZY (2023) Melatonin treatment of strawberry fruit during storage extends its post-harvest quality and reduces infection caused by *Botrytis cinerea*. Foods 12:1445. https://doi.org/10.3390/foods12071445

Qu G, Wu W, Ba L et al (2022) Melatonin enhances the postharvest disease resistance of blueberries fruit by modulating the jasmonic acid signaling pathway and phenylpropanoid metabolites. Front Chem 10:1–10. https://doi.org/10.3389/fchem.2022.957581

Santos e Silva VN, da Silva Santos I, dos Santos GL et al (2024) Anti-browning action of melatonin is dependent on the type of tissue in minimally processed sweet potatoes. Sci Hortic 337:113572. https://doi.org/10.1016/j.scienta.2024.113572

Sarkar P, Tamili D, Bhattacharjee P (2021) Low dose gamma-irradiation enhances shelf-life and contents of serotonin and melatonin in green plantains (*Musa paradisiaca*): a study involving antioxidant synergy. J Food Process Preserv 45:e15934. https://doi.org/10.1111/jfpp.15934

Shan S, Wang Z, Pu H et al (2022) DNA methylation mediated by melatonin was involved in ethylene signal transmission and ripening of tomato fruit. Sci Hortic 291:110566. https://doi.org/10.1016/j.scienta.2021.110566

Sharafi Y, Aghdam MS, Luo Z et al (2019) Melatonin treatment promotes endogenous melatonin accumulation and triggers GABA shunt pathway activity in tomato fruits during cold storage. Sci Hortic 254:222–227. https://doi.org/10.1016/j.scienta.2019.04.056

Shi L, Cao M, Lu X et al (2024a) Melatonin extends shelf life in postharvest okra via delaying fruit softening and reducing weight loss. J Sci Food Agric 104:9506. https://doi.org/10.1002/jsfa.13773

Shi L, Chen Y, Dong W et al (2024b) Melatonin delayed senescence by modulating the contents of plant signalling molecules in postharvest okras. Front Plant Sci 15:1–11. https://doi.org/10.3389/fpls.2024.1304913

Sidhu RS, Bound SA, Swarts ND (2023) Internal flesh browning in apple and its predisposing factors—a review. Physiologia 3:145–172. https://doi.org/10.3390/physiologia3020012

Song L, Zhang W, Li Q et al (2022) Melatonin alleviates chilling injury and maintains postharvest quality by enhancing antioxidant capacity and inhibiting cell wall degradation in cold-stored eggplant fruit. Postharvest Biol Technol 194:112092. https://doi.org/10.1016/j.postharvbio.2022.112092

Song L, Tan Z, Zhang W et al (2023) Exogenous melatonin improves the chilling tolerance and preharvest fruit shelf life in eggplant by affecting ROS- and senescence-related processes. Hortic Plant J 9:523–540. https://doi.org/10.1016/j.hpj.2022.08.002

Sun Q, Liu L, Zhang L et al (2020) Melatonin promotes carotenoid biosynthesis in an ethylene-dependent manner in tomato fruits. Plant Sci 298:110580. https://doi.org/10.1016/j.plantsci.2020.110580

Sun H, Jia M, Wang Y et al (2023) The complexity of melatonin and other phytohormones cross-talk with other signaling molecules for drought tolerance in horticultural crops. Sci Hortic 321:112348. https://doi.org/10.1016/J.SCIENTA.2023.112348

Tan DX, Manchester LC, Terron MP et al (2007) One molecule, many derivatives: a never-ending interaction of melatonin with reactive oxygen and nitrogen species? J Pineal Res 42:28–42. https://doi.org/10.1111/J.1600-079X.2006.00407.X

Tan DX, Manchester LC, Liu X et al (2013) Mitochondria and chloroplasts as the original sites of melatonin synthesis: a hypothesis related to melatonin's primary function and evolution in eukaryotes. J Pineal Res 54:127–138. https://doi.org/10.1111/JPI.12026

Tan DX, Hardeland R, Back K et al (2016) On the significance of an alternate pathway of melatonin synthesis via 5-methoxytryptamine: comparisons across species. J Pineal Res 61:27–40. https://doi.org/10.1111/JPI.12336

Tan X-l, Fan Z-q, Zeng Z-x et al (2021) Exogenous melatonin maintains leaf quality of postharvest Chinese flowering cabbage by modulating respiratory metabolism and energy status. Postharvest Biol Technol 177:111524. https://doi.org/10.1016/j.postharvbio.2021.111524

Tang Q, Li C, Ge Y et al (2020) Exogenous application of melatonin maintains storage quality of jujubes by enhancing anti-oxidative ability and suppressing the activity of cell wall-degrading enzymes. LWT 127:109431. https://doi.org/10.1016/j.lwt.2020.109431

Tiwari RK, Lal MK, Naga KC, Kumar R, Chourasia KN, Kumar D, Sharma S (2020) Emerging roles of melatonin in mitigating abiotic and biotic stresses of horticultural crops. Sci Hortic 272:109592

Tiwari RK, Lal MK, Kumar R, Chourasia KN, Naga KC, Kumar D, Das SK, Zinta G (2021a) Mechanistic insights on melatonin-mediated drought stress mitigation in plants. Physiol Plant 172(2):1212–1226

Tiwari RK, Lal MK, Kumar R, Mangal V, Altaf MA, Sharma S, Singh B, Kumar M (2021b) Insight into melatonin-mediated response and signaling in the regulation of plant defense under biotic stress. Plant Mol Biol 109:385–399

Tiwari RK, Altaf MA, Kumar R, Lal MK (2024) Evolution of melatonin as an oxidative stress mitigator in plant. In: Melatonin in plants: a pleiotropic molecule for abiotic stresses and pathogen infection. Springer, Singapore, pp 1–20

Verde A, Míguez JM, Gallardo M (2023) Melatonin stimulates postharvest ripening of apples by up-regulating gene expression of ethylene synthesis enzymes. Postharvest Biol Technol 195:112133. https://doi.org/10.1016/j.postharvbio.2022.112133

Wang L, Feng C, Zheng X et al (2017) Plant mitochondria synthesize melatonin and enhance the tolerance of plants to drought stress. J Pineal Res 63:e12429. https://doi.org/10.1111/JPI.12429

Wang X, Liang D, Xie Y et al (2019) Melatonin application increases accumulation of phenol substances in kiwifruit during storage. Emir J Food Agric 31:361–367. https://doi.org/10.9755/ejfa.2019.v31.i5.1954

Wang SY, Shi XC, Wang R et al (2020a) Melatonin in fruit production and postharvest preservation: a review. Food Chem 320:126642. https://doi.org/10.1016/j.foodchem.2020.126642

Wang T, Hu M, Yuan D et al (2020b) Melatonin alleviates pericarp browning in litchi fruit by regulating membrane lipid and energy metabolisms. Postharvest Biol Technol 160:111066. https://doi.org/10.1016/j.postharvbio.2019.111066

Wang M, Xu J, Ding Z, Xie J (2023) Prolong the postharvest shelf life of spinach through the antioxidative ability of melatonin. Food Chem X 19:100769. https://doi.org/10.1016/j.fochx.2023.100769

Wu Y, Fan X, Zhang Y et al (2021) VvSNAT1 overexpression enhances melatonin production and salt tolerance in transgenic Arabidopsis. Plant Physiol Biochem 166:485–494. https://doi.org/10.1016/J.PLAPHY.2021.06.025

Wu C, Hao W, Yan L et al (2023) Postharvest melatonin treatment enhanced antioxidant activity and promoted GABA biosynthesis in yellow-flesh peach. Food Chem 419:136088. https://doi.org/10.1016/j.foodchem.2023.136088

Xia H, Shen Y, Shen T et al (2020) Melatonin accumulation in sweet cherry and its influence on fruit quality and antioxidant properties. Molecules 25:753. https://doi.org/10.3390/molecules25030753

Xiao Y, Xie J, Wu C et al (2021) Effects of melatonin treatment on browning alleviation of fresh-cut foods. J Food Biochem 45:e13798. https://doi.org/10.1111/jfbc.13798

Xu L, Yue Q, Xiang G et al (2018) Melatonin promotes ripening of grape berry via increasing the levels of ABA, H2O2, and particularly ethylene. Hortic Res 5:41. https://doi.org/10.1038/s41438-018-0045-y

Xu T, Chen Y, Kang H (2019) Melatonin is a potential target for improving post-harvest preservation of fruits and vegetables. Front Plant Sci 10:488368. https://doi.org/10.3389/FPLS.2019.01388/BIBTEX

Yan R, Li S, Cheng Y et al (2022) Melatonin treatment maintains the quality of cherry tomato by regulating endogenous melatonin and ascorbate-glutathione cycle during room temperature. J Food Biochem 46:e14285. https://doi.org/10.1111/jfbc.14285

Yang WJ, Du YT, Zhou YB et al (2019) Overexpression of TaCOMT improves melatonin production and enhances drought tolerance in transgenic Arabidopsis. Int J Mol Sci 20:652. https://doi.org/10.3390/ijms20030652

Yang L, You J, Li J et al (2021) Melatonin promotes Arabidopsis primary root growth in an IAA-dependent manner. J Exp Bot 72:5599–5611. https://doi.org/10.1093/JXB/ERAB196

Ye T, Yin X, Yu L et al (2019) Metabolic analysis of the melatonin biosynthesis pathway using chemical labeling coupled with liquid chromatography-mass spectrometry. J Pineal Res 66:e12531. https://doi.org/10.1111/jpi.12531

Zang H, Ma J, Wu Z et al (2022) Synergistic effect of melatonin and selenium improves resistance to postharvest gray mold disease of tomato fruit. Front Plant Sci 13:1–10. https://doi.org/10.3389/fpls.2022.903936

Ze Y, Gao H, Li T et al (2021) Insights into the roles of melatonin in maintaining quality and extending shelf life of postharvest fruits. Trends Food Sci Technol 109:569–578. https://doi.org/10.1016/j.tifs.2021.01.051

Zhang S, Zheng X, Reiter RJ et al (2017) Melatonin attenuates potato late blight by disrupting cell growth, stress tolerance, fungicide susceptibility and homeostasis of gene expression in *Phytophthora infestans*. Front Plant Sci 8:1993. https://doi.org/10.3389/fpls.2017.01993

Zhang W, Cao J, Fan X, Jiang W (2020) Applications of nitric oxide and melatonin in improving postharvest fruit quality and the separate and crosstalk biochemical mechanisms. Trends Food Sci Technol 99:531–541. https://doi.org/10.1016/j.tifs.2020.03.024

Zheng H, Liu W, Liu S et al (2019) Effects of melatonin treatment on the enzymatic browning and nutritional quality of fresh-cut pear fruit. Food Chem 299:125116. https://doi.org/10.1016/j.foodchem.2019.125116

Zhou K, Li Y, Hu L et al (2022) Overexpression of MdASMT9, an N-acetylserotonin methyltransferase gene, increases melatonin biosynthesis and improves water-use efficiency in transgenic apple. Tree Physiol 42:1114–1126. https://doi.org/10.1093/treephys/tpab157

Zhu L, Hu H, Luo S et al (2018) Melatonin delaying senescence of postharvest broccoli by regulating respiratory metabolism and antioxidant activity. Nongye Gongcheng Xuebao/Trans Chin Soc Agric Eng 34:300–308. https://doi.org/10.11975/j.issn.1002-6819.2018.03.040

Zuo B, Zheng X, He P et al (2014) Overexpression of MzASMT improves melatonin production and enhances drought tolerance in transgenic *Arabidopsis thaliana* plants. J Pineal Res 57:408–417. https://doi.org/10.1111/JPI.12180

How Melatonin Emerges as a Prominent Strategy to Tackle Recent Episodes of Combined Stress?

14

Arti Kumari, Mala Kumari, Jyoti Kumari, Manish Kumar, and Rajeev Kumar

Abstract

Melatonin is not only found in animals but also in plants, where it is synthesized from the amino acid tryptophan. Plants, unlike animals, are sessile and susceptible to various biotic and abiotic stressors. These factors can occur alone in the environment or in conjunction with other stresses, such as heat and drought, drought and salt, heavy metal toxicity and senescence, and other abiotic and biotic stresses. These circumstances influence plant growth, development, and yield. Melatonin is synthesized in plants and has been suggested as an antioxidant, photoperiod regulator, growth stimulator, etc. Exogenous melatonin treatment protects against biotic and abiotic stresses such as high temperature, drought, cold, salt, oxidative stress, senescence, and so on. Melatonin's role involves an independent effect or a synergistic effect with phytohormones or secondary signalling mediators such as NO, Ca^{2+}, and H_2S. Exogenous melatonin administration has previously been described for single stressors. However, more research is needed to determine the mechanism by which exogenous melatonin reduces combined stress. Therefore, this chapter discusses the role and mechanisms involved in the exogenous application of melatonin to alleviate combined stress in plants.

Keywords

Melatonin · Combined stress · Biotic stress · Abiotic stress

A. Kumari (✉) · M. Kumari · J. Kumari
Nalanda College of Horticulture, Noorsarai, Nalanda, Bihar, India

M. Kumar
Dr. Rajendra Prasad Central Agricultural University, Pusa, Samastipur, Bihar, India

R. Kumar
ICAR-Indian Institute of Sugarcane Research, Lucknow, Uttar Pradesh, India

R. K. Tiwari et al. (eds.), *Melatonin: Signal Transduction Mechanisms and Defense Networks in Plants*, https://doi.org/10.1007/978-981-96-8869-2_14

14.1 Introduction

N-acetyl-5-methoxy-tryptamine is another name for melatonin, an intrinsic compound available in an extensive group of species (Reiter et al. 2024). Van Tassel et al. (2001) first described the existence of melatonin in plants. As a pleiotropic molecule, it has a different role in plants. It acts as a plant growth regulator, mitigation of stress, signal transduction, and physio-biochemical functions (Sun et al. 2021). It has an important role in ROS scavenger and many cellular functions to protect plants from stresses (Mauriz et al. 2013; Kumar et al. 2023b, 2024). Despite this, it also has a role in functional activity from seed germination to flowering and fruiting as well as in the regulation of hormone pathways and cellular function (Tiwari et al. 2020, 2021; Tiwari 2021). It enhances productivity and the shelf life of crops by conserving chlorophyll, increasing photosynthetic efficiency, and reducing senescence. It also improves the defence mechanism and quality by stimulating secondary metabolites. Overall, melatonin has a crucial role in the growth, development and productivity of plants. Melatonin's effect in field crops has been studied under a variety of stress situations, including drought, heat, salinity, and autotoxicity (Liu et al. 2023; Fu et al. 2024; Martinez et al. 2018). However, in field conditions, abiotic stressors always exist in combination. Therefore, this chapter emphasizes the role of melatonin in combined stress tolerance.

14.2 Melatonin's Role in Plants

14.2.1 Effect of Melatonin on Plant Physiological and Developmental Processes

(i) *Seed sprouting and growth initiation*

The role of melatonin in the initial stage of growth is very critical because it is involved in many important physio-biochemical processes. Its role is to activate all the essential enzymes which are responsible for germination of seed, water uptake, dormancy breaking, and mobilization of nutrients (Ahsan et al. 2025; Tiwari et al. 2024). Furthermore, it is also responsible for the protection of seeds from oxidative stresses caused due to different biotic and abiotic stresses by scavenging reactive oxygen species at the time of germination. During abiotic stress conditions, melatonin maintains a good balance at the initial stage of growth like germination and development of seedlings by interacting with different plant hormones like GA and ABA. Gemination is a crucial and complicated stage in the plant life cycle, and it involves many balanced metabolic and biological processes. Castañares and Bouzo (2019) identified the positive role of melatonin in *Cucumis melo*, at the time of germination. Most importantly, abiotic stress like high temperatures had a bad impact on cotton seed germination, which led to reduced crop growth (Chen et al. 2020). To reduce this negative effect of stress, seed priming with melatonin is helpful to enhance the process of germination (Rajora et al. 2022).

(ii) *Growth of root and shoot*

Melatonin plays a role in the development of roots and shoots by increasing the scavenging of oxidative stress, i.e. reactive oxygen and nitrogen species (Shafi et al. 2021). Sultana and Barthakur (2023) in wheat and Ahmad et al. (2021) in maize reported that seed priming with melatonin is responsible for better growth and development of roots, including length, area, and volume. During the condition of drought stress, hormones like ABA increase, which downregulate the enzyme responsible for loosening of the cell wall and enhance the pH of apoplastic cells from acidic to alkaline, which leads to the restriction of normal root and shoots growth in the plant (Tatar et al. 2023). Melatonin can decrease the pH and upregulate the enzyme responsible for cell wall loosening (Yang et al. 2021). MT along with nitrogen is beneficial for seedling and shoot growth in Maize (Ahmad et al. 2022). MT can enhance soluble sugar and protein for osmotic adjustment during drought conditions in cotton (Duan et al. 2022).

(iii) *Melatonin on gas exchange attributes*

Chlorophyll degradation is reduced by melatonin by downregulating the pheophorbide-a-oxygenase (PAO) enzyme transcript (Liu et al. 2022). Endogenous ABA amount is reduced by upregulation of ABA catabolism and downregulation of ABA anabolism which is responsible for stomatal opening modulated by melatonin (Kuppusamy et al. 2023). Green plants' primary process, photosynthesis, is susceptible to abiotic stressors (Morales et al. 2020), which destroy the electron transport system and chloroplast grana, resulting in decreased ATP and NADPH synthesis and rubisco activity (Abdulbaki et al. 2022). Melatonin is responsible for photosynthetic pigment concentration in soybean suggested by Wang et al. (2022). Melatonin is responsible for negative regulation of chlorophyll-degrading genes like Chlase, PPH, and Chl-PRX in *Agrostis stolonifera* (Ma et al. 2018).

(iv) *Oxidative stress mitigation*

ROS produced under stress conditions have harmful effects on cell membranes and physio-biochemical processes (Kumar et al. 2023a, b). When the plant is treated with melatonin, the concentration of metabolites is enhanced, which increases amino acid and overall physiological activities (Jiang et al. 2022). Free radicals like NO, H_2O_2, O_2^-, and OḤ are hazardous reactive species, and melatonin can fight and scavenge these free radicals and protect against various abiotic stresses (Lal et al. 2023). Melatonin enhances the nitrogen content and chlorophyll content, which enhances the photosynthetic rate and amount of protein and rubisco.

14.2.2 Melatonin as a Key Modulator of Plant Hormone Signalling

Melatonin and indole-3-acetic acid are structurally similar, so the role of melatonin in growth and development is important (Arnao and Hernández-Ruiz 2007). Phytomelatonin has different roles in plants, i.e. stimulator, protector, and regulator. Stimulators help in the germination of seeds, the development of roots, and the overall development of plants, while protectors help in protection from harsh environmental conditions, and regulators are for growth regulation.

(i) *Growth promotion and root formation*

Hernandez-Ruiz et al. (2004) first identified that melatonin influences plant growth in cereals and canary grass. Melatonin influences growth promotion by affecting different physiological changes. It promotes division, elongation, and differentiation of cells, which overall leads to the growth of different parts of the plant like root and shoot. Asif et al. (2019) suggested that melatonin is the main component which is responsible for root and microorganism interaction in the rhizosphere. Root organogenesis is started by the external application of melatonin. Adventitious and lateral root formation is increased by two- to threefold, respectively, by the application of melatonin in Arabidopsis (Pelagio-Flores et al. 2012).

(ii) *Flowering initiation*

Flowering time is also controlled by melatonin, which acts as a growth regulator and was recently identified (Arnao and Hernández-Ruiz 2020). The role of melatonin in circadian regulation was first identified in *C. rubrum* (Wolf et al. 2001). Melatonin is involved in some important phases of floral transition, but it does not affect the flowering phase, w*hile it is responsible for the transition of the crop from the vegetative to the reproductive stage in Arabidopsis* (Shi et al. 2016). Strigolactones, a carotenoid-derived phytohormone and melatonin, are directly linked and responsible for Arabidopsis floral transition (Zhang et al. 2019).

(iii) *Delaying senescence and improving chlorophyll retention*

Melatonin acts as a signal that helps in the activation of antioxidant enzymes and other antioxidants. ROS scavenging enzymes like SOD, CAT, and PPO gene are expressed and maintain ROS homeostasis by CAND2/PMTR1 in Arabidopsis (Li et al. 2020). Melatonin reduces the photo-inhibited senescence in barley leaves (Arnao and Hernández-Ruiz 2009). Melatonin acts as an anti-senescence because it has antioxidant properties. It shows that melatonin decreases the ROS level and increases the amount of ascorbate, glutathione, and antioxidant enzymes, which act as anti-senescence agent.

14.2.3 Role of Melatonin in the Expression of Secondary Metabolites

Exogenous application of melatonin enhances the plant metabolite, and its concentration was altered at the cellular level, which was reduced by the abiotic stresses. Thereby, it is both directly and indirectly involved in tolerance against drought stress in green gram, and the metabolites are involved in another metabolic pathway (Udhayabharathi 2017). Different pathways were upregulated by MT like oxPPP, glycolysis, and TCA cycle and revealed the direct connection between melatonin and carbon metabolic pathway in rice (Xie et al. 2021). Cadmium stress in plants leads to the accumulation of proline enhances osmotic adjustment and maintains membrane integrity. Foliar application of melatonin improves the metabolic group and its pathway in the tomato (Umapathi et al. 2022). Sheikhalipour et al. (2022) revealed that the protein structure which is denatured under drought stress is stabilized by enhanced proline content by the application of melatonin.

14.2.4 Effect of Melatonin on Quality and Productivity of Agricultural Crop

Global warming has been a growing threat over the past three decades, it has significant impact on crop yield, and the problem has been further extended by the quick growth of civilization and technological improvements (Kumari et al. 2024). Melatonin is responsible for high yield by improving the factors responsible for growth and photosynthetic pigment which in turn produce yield. Yield, quality, and number of fruits in tomato plants are increased when plants are treated with melatonin (Khan et al. 2023). Ball formation and seed yield are increased in cotton after melatonin treatment (Hu et al. 2021). Various physio-biochemical processes are regulated by melatonin responsible for the higher productivity of crops (Hu et al. 2017). The yield of pomegranate is enhanced after the application of melatonin (Medina-Santamarina et al. 2021). Melatonin directly improves the yield and quality of rice by reducing ROS production and improving photosynthesis (Jiang et al. 2022). Rapeseed cultivar's oil quality is improved by priming of seed with melatonin, which increases unsaturated fatty acids and decreases saturated fatty acids (Mohamed et al. 2020).

14.2.5 Melatonin Regulating Fruit Maturity and Shelf Life

Melatonin has an important role in modifying different physio-biochemical processes to regulate fruit ripening. It scavenges ROS to reduce oxidative stress which is responsible for ripening and decaying (Kumar et al. 2023a, b, 2024). It helps to delay the ripening process and maintain the fruit quality. Melatonin levels increased in ripened fruit in tomato (Feng et al. 2014), and external application of melatonin hastens the ripening and enhancement in nutritional quality in tomatoes (Sun et al. 2015). Melatonin decreases ethylene biosynthesis and delays the ripening process in bananas (Hu et al. 2017).

14.3 Melatonin as a Prominent Strategy to Tackle Recent Episodes of Combined Stress

14.3.1 Combine Stress

Plants face many challenges like abiotic and biotic stresses which lead reduction in productivity and overall performance in the prevailing environment (Da Costa et al. 2021; Priya et al. 2023). Recognizing the complexity of these stresses and interaction (abiotic and biotic) and the effect of combined stresses is the need for time (Choudhary and Senthil-Kumar 2024; Pandey et al. 2017). This chapter emphasizes how these combined stresses influence plant responses and adaptations and how melatonin works in the case of combined stress. Abiotic stresses can reduce yield by more than 50% of total crop production. There are more severe losses when two abiotic or biotic stresses show a combined effect. *Drought and high*-temperature stress together can reduce crop yield drastically. Mumithrakamatchi et al. (2024) studied the impacts of drought and heat stress together on the growth, physiology, and yield of crops. They concluded that this combined stress reduced water uptake, a rise in canopy temperature, and reactive oxygen species accumulated in cells, which led to membrane damage and also affected photosynthesis in plants. Mostly, reproductive processes in plants were negatively impacted.

A review by Ranjan et al. (2023) studied the effect of abiotic stresses such as drought, salinity, and heat interacting with *Macrophomina phaseolina* in a changing climate. These stresses may worsen the pathogen's impact on plants, causing significant crop yield losses. A review by Siddiqui and Aziz (2024) emphasized the interactions between plant parasitic nematodes (PPN) and fungi, both of which are major contributors to crop yield losses. These two factors mutually influence one another, resulting in the formation of disease complexes. The stresses induced by their combined effects include synergistic and additive interactions, the creation of wounds, physiological changes in the host, and modifications to the rhizosphere, microbial community, which impacts the plant economically. The interaction between various stresses is influenced by several factors, including the age and type of the plant, the timing of stress occurrence, the severity and duration of each stress, the order in which the plants perceive the stresses, and the specific types of stressors involved. Under conditions of combined stress, plants demonstrate shared and unique responses that are specifically tailored to the particular stress combinations. This complexity in plant responses poses challenges in accurately predicting outcomes based solely on studies focused on single stressors.

14.3.2 Role of Melatonin in Tackling Combined Stresses

Melatonin (N-acetyl-5-methoxytryptamine) is a hormone that plays a crucial role in metabolic functions, including biological rhythms, immunomodulation, and the reduction of oxidative stress. Melatonin also has anti-ageing properties, so due to its multiple functionalities, it received great attention from researchers. Melatonin plays an important role in germination, plant growth, and defence against biotic and abiotic stresses. It is a growth regulator so it enhances drought resistance in the plant (Sharma and Zheng 2019). Melatonin has garnered significant interest among researchers as an effective strategy for enhancing crop tolerance to drought, salinity, heavy metal exposure, extreme temperatures, nutritional deficiencies, and various diseases (Tiwari et al. 2021). It plays a crucial role in stimulating the plant antioxidant system, improving photosynthesis, enhancing ion homeostasis, and regulating hormone metabolism under conditions of salt stress (Arnao and Hernández-Ruiz 2014). Pre-treating rice with melatonin has been shown to increase tolerance to salt stress by boosting the plant's fresh and dry weight while reducing plasma membrane damage (Yan et al. 2021). Additionally, exogenous melatonin has been found to enhance the activities of catalase (CAT), peroxidase (POD), superoxide dismutase (SOD), and ascorbate peroxidase (APX) in maize (*Zea mays*) (Alharby and Fahad 2020).

A recent study on cotton (*Gossypium hirsutum*) shows that combined salt and drought stress significantly reduces plant growth, chlorophyll content, and photosynthesis in cotton (Ibrahim et al. 2019). This combined stress also decreases the activity of antioxidant enzymes, ultimately affecting the physiological performance of sunflower plants. Salinity and drought alter osmotic and ionic signalling pathways in various crops (Anwar et al. 2021). The Salt Overly Sensitive (SOS) pathway is essential for maintaining cellular ion balance during salinity stress (Kumar et al. 2022). Both salt and drought stress cause oxidative damage by generating reactive oxygen species (ROS) and damaging membranes through lipid peroxidation, which leads to reduced growth and development in rice plants. Additionally, applying exogenous melatonin can effectively mitigate both salt and drought stress, either alone or in combination (Khan et al. 2024). Melatonin increased the fresh and dry weight of rice under these stress conditions. Moreover, melatonin treatment significantly reduced ROS accumulation and boosted antioxidant activity. It also upregulated the expression of genes involved in ion homeostasis.

Future studies should focus on understanding melatonin's mechanisms, optimizing application strategies, and validating its effectiveness in field trials for enhancing crop resilience to salt and drought stresses. Drought, heat, and the combination of both stresses lead to higher canopy temperatures, which result in reduced fruit volume and weight in tomatoes. Among these stressors, the combination of drought and heat is more harmful than either stress alone. However, a foliar application of 100 μM melatonin effectively increased chlorophyll content, relative water content, and proline levels compared to stressed control plants, indicating better osmotic adjustment (Ramakrishna et al. 2012). This treatment improved water use efficiency, fruit yield, and fruit quality. Overall, exogenous melatonin proves to be an effective strategy for alleviating the negative effects of drought, heat, and their combination by promoting osmolyte accumulation and facilitating ABA-independent stomatal closure. This research on melatonin illustrates how plants adapt to withstand stress.

Jensen et al. (2023) showed that exogenous melatonin modulates stomatal behaviour during heat and drought stress, without the hypothesized mediation of ABA regulation. Instead, it may affect reactive oxygen species (ROS) metabolism. The discovery of melatonin in the alga *Gonyaulax polyhedra* by Balzer and Hardeland in 1991 initiated interest in endogenous melatonin in plants. Research on melatonin's role in plant biology has expanded, starting with its identification in tomato and morning glory in 1993 (Van Tassel 1997). It is now known that melatonin regulates processes like metabolism, osmoregulation, germination, photosynthesis, anti-senescence, and hormone regulation (Arnao and Hernández-Ruiz 2019). Recognized as a ROS scavenger, melatonin serves as a photo-protectant against stressors such as heat, drought, cold, salinity, and pathogens, which generate ROS that can damage cells and disrupt homeostasis. During a pathogen attack, this melatonin facilitates the upregulation of PAMP-triggered immunity (PTI), effector-triggered immunity (ETI), nitric oxide, and salicylic acid-related genes which protect the plants from severe attack after initial infection (Shi et al. 2015).

14.3.3 The Mechanism of Melatonin to Overcome Combined Stresses

14.3.3.1 Improved Growth and Germination

Abiotic stress affects plant growth. The combined heat and salt stress reduced root, stem, and leaf biomass (55%, 60%, and 45%, respectively). However, the application of melatonin during the combined heat and drought stress increased biomass output, comparable to the control plants (Martinez et al. 2018). Exogenous melatonin improves seed germination potential and rate, as demonstrated in a variety of crops (Wang et al. 2024; Tian et al. 2024; Zhang et al. 2020). Melatonin's impact in enhancing germination under combined stress conditions such as autotoxicity and saline-alkali combined stress was examined in melon seeds (Liu et al. 2023). The two biggest challenges to melon production are autotoxicity and soil salinization. Melon autotoxins, such as cinnamic acid, hinder normal plant growth and development (Zhang et al. 2020). However, $NaHCO_3$ and NaCl are the primary causes of soil salinization, which impact plant germination, growth, development, photosynthesis, and water usage efficiency. Seed vigour is the most important indicator of how combined stress affects uniform germination. Combining autotoxicity and soil salinization causes poor seed germination, ROS metabolism, and altered growth and development. This combined stress reduced germination-related measures such as seed vigour index, germination index, germination rate, and germination potential. Melatonin's relieving effect was dose-dependent, with the highest efficacy found at a concentration of 16 μM.

14.3.3.2 Improved Root Length and Root Hydraulic Conductivity

At the germination stage, melon plants' root growth was influenced by a combination of stressors, including autotoxicity and soil salinization. Melatonin supplementation led to increased radical growth, root length, root surface area, and root volume in plants subjected to these combined stresses (Liu et al. 2023). Melatonin treatment of wheat varieties under combined salt and drought stress resulted in increased root length and hydraulic conductivity determines the plant's nutrient and water absorption ability. The root hydraulic conductivity is dependent on the root system of the plant. Drought and salt stress impair root hydraulic conductivity in wheat, tomato, maize, and rice. Melatonin treatment increased wheat root hydraulic conductivity during both individual and *combination drought and salinity stress (*Fu et al. 2024*).*

14.3.3.3 Enhanced Antioxidant Defence System

The activity of antioxidant enzymes namely catalase, superoxide dismutase (SOD) and peroxidase (POD) was increased in the melatonin-treated group under the combined autotoxicity and soil salinization effect in the melon plant at germination stage (Liu et al. 2023). Drought and salinity stress enhanced OMT1 gene expression and melatonin content, which were associated with improved antioxidant ability, greater osmolyte concentration, and reactive oxygen species neutralization capacity (Shamloo-Dashtpagerdi et al. 2022). Free radicals (H_2O_2 and O_2), malondialdehyde, and electrolyte leakage were detected in tomato plants under drought and heat stress. Reduced reactive oxygen species generation led to lower malondialdehyde levels and electrolyte leakage, resulting in better membrane stability and integrity in melatonin-treated tomato plants under drought and heat stress (Annadurai et al. 2023). The foliar spray of melatonin increased antioxidant enzyme activity in the plants. Melatonin priming treatments in wheat boosted SOD and CAT activities during drought and salt stress (Fu et al. 2024). Heat and salinity stress reduced total antioxidant capacity (TAC) in control plants, but melatonin treatment increased TAC (Martinez et al. 2018). Additionally, melatonin-treated plants also displayed decreased levels of H_2O_2, ROS formation, and protein oxidation when exposed to combined heat and salinity stress. The plant's MDA (malondialdehyde) content is used to measure lipid peroxidation. Melatonin treatment for tomato plants under low temperature and salinity stress reduced MDA levels while enhancing SOD and peroxidase activity (Fu et al. 2023).

14.3.3.4 Modulating the Photosynthesis-Related Parameters

Drought and heat are more likely to occur simultaneously due to less precipitation. Drought affects photosynthesis, stomatal conductance, and assimilate partitioning in plants. High-temperature stress, on the other hand, affects the light reaction of photosynthesis, denatures photosynthetic pigments, and destroys the thylakoid membrane. Tomato plants are more sensitive to drought or heat stress during their reproductive stage, which includes pre- and post-fertilization. Annadurai et al. (2023) investigated melatonin's protective effect during drought and heat stress in tomatoes. Melatonin administration enhanced the chlorophyll index while reducing thylakoid membrane damage in plants subjected to a combination of drought and heat treatment. The chlorophyll index represents the chlorophyll content and determines the photosynthetic efficiency of plants. The relative tolerance index (RTI) compares plants' tolerance to drought stress based on stomatal conductance. Melatonin-treated plants demonstrated enhanced RTI under combined drought and heat stress, indicating that foliar melatonin treatment could be a viable management method for improving plant stress tolerance (Annadurai et al. 2023). Photosynthesis-related metrics, such as CO_2 assimilation rate (Pn), transpiration rate (E), and stomatal conductance (gs), are plant stress markers. Melatonin treatment resulted in improved Pn, E, and gs, which were comparable to the control condition under combined salt and heat stress. The photosynthetic pigments chlorophyll (Chla, chlorophyll-a) and carotenoids play a role in light harvesting and CO_2 assimilation. Melatonin-treated plants have greater chlorophyll and carotenoid content when exposed to both heat and salinity stress (Martinez et al. 2018). Exogenous melatonin treatment increased chlorophyll and Fv/Fm in pepper plants subjected to iron deficit and salt stress (Kaya et al. 2020). Melatonin's influence on tomato plants was studied under drought and cold stress conditions in the presence of increased CO_2 levels (Zhou et al. 2020). Melatonin increased photosynthesis-related measures such as net photosynthetic rate, chlorophyll-a concentration, and plant biomass accumulation. Under drought and cold stress circumstances, elevated CO_2 levels interact with melatonin, affecting plant viability. Melatonin treatment of cotton seedlings increased photosynthesis-related metrics such as net photosynthetic rate, stomatal conductance, and transpiration rate under low temperature and salt stress (Fu et al. 2023).

14.3.3.5 Interactive Effect with Other Signals

Melatonin foliar spray increased nitric oxide, NO, and hydrogen sulphide (H_2S) levels in the presence of salt stress and iron deficiency. These two chemicals, NO and H_2S, serve as the downstream signalling molecules. Iron shortage limits the absorption of iron by plants. As a result, the administration of melatonin improves the mineral nutrition of plants experiencing coupled salt stress and iron shortage through the involvement of H_2S and NO. NO and H_2S stimulate antioxidant mechanisms, which help to reduce the damaging effects of ROS. NO interacts directly with lipid radicals and ROS, increasing antioxidant enzyme activity (Kaya et al. 2020). Melatonin interacted with 24-Epibrassinolide, a type of brassinosteroid hormone, to improve tolerance to drought and salt stress in pea plants (Yusuf et al. 2024). The combination of EBL and melatonin increased photosynthetic indices, osmolyte accumulation, and antioxidant capacity.

14.3.3.6 Augmenting Melatonin Biosynthesis

O-methyl transferase 1 (*OMT1*), a gene implicated in endogenous melatonin production, was found to be upregulated during combined salinity and drought stress in wheat. *OMT1* catalyses the final but crucial step in the melatonin production pathway. *OMT1* may play a role in tolerance to coupled drought and salinity stress by boosting antioxidant defences and redox homeostasis (Shamloo-Dashtpagerdi et al. 2022). Under combined salinity and drought stress, *OMT1* gene expression was highest (approximately seven times that of non-stressed plants) in roots. The *OMT1* gene was revealed to be the most critical gene in the gene networks identified in response to drought and salinity stress. Melatonin production is regulated by hormonal signalling pathways that include abscisic acid, salicylic acid, jasmonic acid, ethylene, reactive oxygen species, and Ca^{2+}. *OMT1* expression was favourably linked with endogenous melatonin levels, antioxidant enzyme levels (namely superoxide dismutase, peroxidase, and catalase), and proline concentration. Overexpression of the *COMT1* gene in watermelon increased melatonin levels by sevenfold under a variety of stressors, including drought, cold, and salt stress (Chang et al. 2021). Overexpression of *MzSMT1* and *MzSMT9* (*Malus zumi Mats*) in *Arabidopsis* produced two to four times higher melatonin levels, which improved drought and salt stress resistance. (Zuo et al. 2014).

14.4 Conclusion

This chapter deals with the role of melatonin in combined stress tolerance. Melatonin is a natural antioxidant produced in both plants and animals. This molecule has recently gained relevance due to its protective role in stressful situations. Initial research was conducted on the role of melatonin under various stress conditions, including drought, heat, salinity, and autotoxicity. However, diverse abiotic stressors under field circumstances are always present in combination. Therefore, recent studies emphasize the role of melatonin in combined stress tolerance (Table 14.1). In this context, this chapter describes how melatonin functions during combined stress conditions, including the modulation of growth, germination characteristics, root hydraulic conductivity and root parameters, photosynthetic parameters, antioxidant capacity, and endogenous melatonin biosynthesis, as illustrated in Fig. 14.1. This may aid biotechnology or agricultural breeding strategies in increasing resistance to combination stressors in plants under field circumstances.

Table 14.1 Application of melatonin in different crops under combined stress conditions

Combined stress	Crop	Melatonin application	Role	References
Autotoxicity and soil salinization	Melon	Seed priming (16 μM MT)	Improved germination, root length, and root hydraulic conductivity	Liu et al. (2023)
Combined salt and drought stress	Wheat	Seed priming (100 μM MT)	(i) Increased root length, root vigour, and root hydraulic conductivity (ii) Increased SOD and CAT activity	Fu et al. (2024)
Combined drought and salinity stress	Wheat	Expression of *OMT1* gene, melatonin biosynthesis gene	Improved antioxidant ability, higher osmolyte concentration, and reactive oxygen species neutralization capacity	Shamloo-Dashtpagerdi et al. (2022)
Combined drought and heat stress	Tomato	Foliar spray (100 μM MT)	(i) Decreased malondialdehyde levels, electrolyte leakage (ii) Improved membrane stability, integrity, and relative tolerance index (RTI)	Annadurai et al. (2023)
Combined salinity and heat stress	Tomato	Pretreatment of plants (100 μM MT)	(i) Protection of photosynthetic machinery (ii) Decreased H_2O_2 concentration, lipid peroxidation, and protein oxidation	Martinez et al. (2018)
Salt stress (SS) and iron deficiency (ID)	Pepper plant	Foliar application (100 μM MT)	(i) Increased nitric oxide and hydrogen sulphide (ii) Improved chlorophyll content and antioxidant activities	Kaya et al. (2020)
Drought, cold, and elevated CO_2	Tomato	Seedling spray (@10 mL per plant) and irrigation (@15 mL per plant) by 1 mM MT	Enhanced net photosynthetic rate, chlorophyll-a content, and biomass accumulation	Zhou et al. (2020)
Low temperature and salinity stress	Cotton	Foliar spray (50–150 μM)	(i) Improved gas exchange parameters, SOD, and peroxidase activity (ii) Decreased lipid peroxidation	Fu et al. (2023)

MT melatonin, *OMT1* O-methyl transferase 1, *SOD* superoxide dismutase, *CAT* catalase

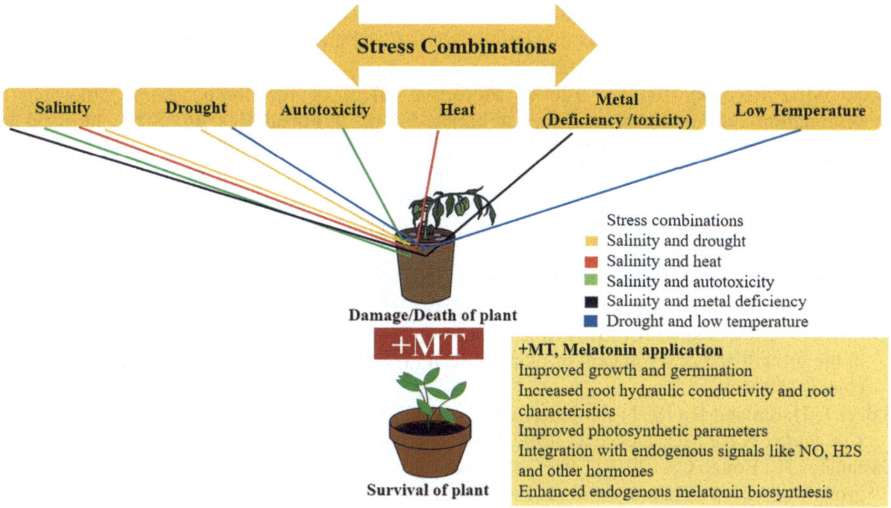

Fig. 14.1 Role of melatonin during combined stress conditions (MT, Melatonin, NO, Nitric oxide, and H_2S, Hydrogen sulphide)

References

Abdulbaki AS, Alsamadany H, Alzahrani Y, Olayinka BU (2022) Rubisco and abiotic stresses in plants: current assessment. Turk J Bot 46(6):541–552. https://doi.org/10.55730/1300-008X.2730

Ahmad S, Cui W, Kamran M, Ahmad I, Meng X, Wu X, Han Q (2021) Exogenous application of melatonin induces tolerance to salt stress by improving the photosynthetic efficiency and antioxidant defense system of maize seedling. J Plant Growth Regul 40:1270–1283. https://doi.org/10.1007/s00344-020-10187-0

Ahmad S, Wang GY, Muhammad I, Chi YX, Zeeshan M, Nasar J, Zhou XB (2022) Interactive effects of melatonin and nitrogen improve drought tolerance of maize seedlings by regulating growth and physiochemical attributes. Antioxidants 11(2):359. https://doi.org/10.3390/antiox11020359

Ahsan M, Younis A, Jamal A, Alshaharni MO, Algopishi UB, Al-Andal A, Akhtar G (2025) Melatonin induces drought stress tolerance by regulating the physiological mechanisms, antioxidant enzymes, and leaf structural modifications in *Rosa centifolia* L. Heliyon 11(1). https://doi.org/10.1016/j.heliyon.2024.e41236

Alharby HF, Fahad S (2020) Melatonin application enhances biochar efficiency for drought tolerance in maize varieties: modifications in physio-biochemical machinery. Agron J 112:2826–2847. https://doi.org/10.1002/agj2.20263

Annadurai MKK, Alagarsamy S, Karuppasami KM, Ramakrishnan S, Subramanian M, Venugopal PR, Maduraimuthu D (2023) Melatonin decreases negative effects of combined drought and high temperature stresses through enhanced antioxidant defense system in tomato leaves. Horticulturae 9(6):673. https://doi.org/10.3390/horticulturae9060673

Anwar K, Joshi R, Dhankher OP, Singla-Pareek SL, Pareek A (2021) Elucidating the response of crop plants towards individual, combined and sequentially occurring abiotic stresses. Int J Mol Sci 22:6119. https://doi.org/10.3390/ijms22116119

Arnao MB, Hernández-Ruiz J (2007) Melatonin promotes adventitious-and lateral root regeneration in etiolated hypocotyls of *Lupinus albus* L. J Pineal Res 42(2):147–152. https://doi.org/10.1111/j.1600-079X.2006.00396.x

Arnao MB, Hernández-Ruiz J (2009) Protective effect of melatonin against chlorophyll degradation during the senescence of barley leaves. J Pineal Res 46:58–63. https://doi.org/10.1111/j.1600-079X.2008.00625.x

Arnao MB, Hernández-Ruiz J (2014) Melatonin: plant growth regulator and/or biostimulator during stress? Trends Plant Sci 19:789–797. https://doi.org/10.1016/j.tplants.2014.07.006

Arnao MB, Hernández-Ruiz J (2019) Melatonin: a new plant hormone and/or a plant master regulator? Trends Plant Sci 24(1):38–48

Arnao MB, Hernández-Ruiz J (2020) Melatonin in flowering, fruit set and fruit ripening. Plant Reprod 33:77–87. https://doi.org/10.1007/s00497-020-00388-8

Asif M, Pervez A, Ahmad R (2019) Role of melatonin and plant-growth-promoting rhizobacteria in the growth and development of plants. Clean (Weinh) 47:1800459. https://doi.org/10.1002/clen.201800459

Balzer I, Hardeland R (1991) Photoperiodism and effects of indoleamines in a unicellular alga, *Gonyaulax polyedra*. Science 253(5021):795–797. https://doi.org/10.1126/science.1876838

Castañares JL, Bouzo CA (2019) Effect of exogenous melatonin on seed germination and seedling growth in melon (*Cucumis melo* L.) under salt stress. Hortic Plant J 5(2):79–87. https://doi.org/10.1016/j.hpj.2019.01.002

Chang J, Guo Y, Yan J, Zhang Z, Yuan L, Wei C, Li H (2021) The role of watermelon caffeic acid O-methyltransferase (ClCOMT1) in melatonin biosynthesis and abiotic stress tolerance. Hortic Res 8:210. https://doi.org/10.1038/s41438-021-00645-5

Chen L, Liu L, Lu B, Ma T, Jiang D, Li J, Li C (2020) Exogenous melatonin promotes seed germination and osmotic regulation under salt stress in cotton (*Gossypium hirsutum* L.). PLoS One 15(1):e0228241. https://doi.org/10.1371/journal.pone.0228241

Choudhary A, Senthil-Kumar M (2024) Drought: a context dependent damper and aggravator of plant diseases. Plant Cell Environ 47:2109–2126. https://doi.org/10.1111/pce.14863

Da Costa MV, Ramegowda Y, Ramegowda V, Karaba NN, Sreeman SM, Udayakumar M (2021) Combined drought and heat stress in rice: responses, phenotyping and strategies to improve tolerance. Rice Sci 28(3):233–242. https://doi.org/10.1016/j.rsci.2021.04.003

Duan W, Lu B, Liu L, Meng Y, Ma X, Li J, Li C (2022) Effects of exogenous melatonin on root physiology, transcriptome and metabolome of cotton seedlings under salt stress. Int J Mol Sci 23(16):9456. https://doi.org/10.3390/ijms23169456

Feng X, Wang M, Zhao Y, Han P, Dai Y (2014) Melatonin from different fruit sources, functional roles, and analytical methods. Trends Food Sci Technol 37(1):21–31. https://doi.org/10.1016/j.tifs.2014.02.001

Fu Y, Xin L, MounkailaHamani AK, Sun W, Wang H, Amin AS et al (2023) Foliar application of melatonin positively affects the physio-biochemical characteristics of cotton (*Gossypium hirsutum* L.) under the combined effects of low temperature and salinity stress. Plants 12(21):3730. https://doi.org/10.3390/plants12213730

Fu Y, Li P, Si Z, Ma S, Gao Y (2024) Seeds priming with melatonin improves root hydraulic conductivity of wheat varieties under drought, salinity, and combined stress. Int J Mol Sci 25(9):5055.https://doi.org/10.3390/ijms25095055

Hernandez-Ruiz J, Cano A, Arnao MB (2004) Melatonin: a growth-stimulating compound present in lupin tissues. Planta 220:140–144. https://doi.org/10.1007/s00425-004-1317-3

Hu W, Yang H, Tie W, Yan Y, Ding Z, Liu Y, Wu C, Wang J, Reiter RJ, Tan DX (2017) Natural variation in banana varieties highlights the role of melatonin in postharvest ripening and quality. J Agric Food Chem 65:9987–9994. https://doi.org/10.1021/acs.jafc.7b03354

Hu W, Zhang J, Yan K, Zhou Z, Zhao W, Zhang X, Yu R (2021) Beneficial effects of abscisic acid and melatonin in overcoming drought stress in cotton (*Gossypium hirsutum* L.). Physiol Plant 173(4):2041–2054. https://doi.org/10.1111/ppl.13550

Ibrahim W et al (2019) Comparative physiological analysis in the tolerance to salinity and drought individual and combination in two cotton genotypes with contrasting salt tolerance. Physiol Plant 165:155–168. https://doi.org/10.1111/ppl.12791

Jensen NB, Ottosen C-O, Zhou R (2023) Exogenous melatonin alters stomatal regulation in tomato seedlings subjected to combined heat and drought stress through mechanisms distinct from ABA signaling. Plants 12:1156. https://doi.org/10.3390/plants12051156

Jiang Y, Huang S, Ma L, Kong L, Pan S, Tang X, Mo Z (2022) Effect of exogenous melatonin application on the grain yield and antioxidant capacity in aromatic rice under combined lead–cadmium stress. Antioxidants 11(4):776. https://doi.org/10.3390/antiox11040776

Kaya C, Higgs D, Ashraf M, Alyemeni MN, Ahmad P (2020) Integrative roles of nitric oxide and hydrogen sulfide in melatonin-induced tolerance of pepper (*Capsicum annuum* L.) plants to iron deficiency and salt stress alone or in combination. Physiol Plant 168(2):256–277. https://doi.org/10.1111/ppl.12976

Khan TA, Saleem M, Fariduddin Q (2023) Melatonin influences stomatal behavior, root morphology, cell viability, photosynthetic responses, fruit yield, and fruit quality of tomato plants exposed to salt stress. J Plant Growth Regul 42(4):2408–2432. https://doi.org/10.1007/s00344-022-10713-2

Khan Z, Jan R, Asif S, Muh F, Jang Y (2024) Exogenous melatonin induces salt and drought stress tolerance in rice by promoting plant growth and defense system. Sci Rep 14:1214. https://doi.org/10.1038/s41598-024-51369-0

Kumar G, Basu S, Singla-Pareek SL, Pareek A (2022) Unraveling the contribution of OsSOS2 in conferring salinity and drought tolerance in a high-yielding rice. Physiol Plant 174:e13638. https://doi.org/10.1111/ppl.13638

Kumar R, Altaf MA, Lal MK, Tiwari RK (eds) (2023a) Melatonin in plants: a regulator for plant growth and development. Springer, Singapore

Kumar R, Sagar V, Verma VC, Kumari M, Gujjar RS, Goswami SK, Prasad PV (2023b) Drought and salinity stresses induced physio-biochemical changes in sugarcane: an overview of tolerance mechanism and mitigating approaches. Front Plant Sci 14:1225234. https://doi.org/10.3389/fpls.2023.1225234

Kumar R, Altaf MA, Lal MK, Tiwari RK (eds) (2024) Melatonin in plants: a pleiotropic molecule for abiotic stresses and pathogen infection. Springer, Singapore

Kumari M, Dubey AK, Kumar R, Kumar A (2024) Marker-assisted selection in plant breeding for stress tolerance. In: Improving stress resilience in plants. Academic Press, pp 371–387. https://doi.org/10.1016/B978-0-443-18927-2.00005-4

Kuppusamy A, Alagarswamy S, Karuppusami KM, Maduraimuthu D, Natesan S, Ramalingam K, Kanagarajan S (2023) Melatonin enhances the photosynthesis and antioxidant enzyme activities of mung bean under drought and high-temperature stress conditions. Plants 12(13):2535. https://doi.org/10.3390/plants12132535

Lal MK, Tiwari RK, Lal P, Kumar A, Kumar R (2023) Regulatory role of melatonin in post-harvest management of vegetables and fruits. In: Melatonin in plants: a regulator for plant growth and development. Springer, Singapore, pp 219–244

Li D, Wei J, Peng Z, Ma W, Yang Q, Song Z, Sun W, Yang W, Yuan L, Xu X, Chang W, Regel A, Shen J, Reiterate R, Cui X, Yu D, Chen Q (2020) Daily rhythms of phytomelatonin signaling modulate diurnal stomatal closure via regulating reactive oxygen species dynamics in Arabidopsis. J Pineal Res 68:e12640. https://doi.org/10.1111/jpi.12640

Liu K, Jing T, Wang Y, Ai X, Bi H (2022) Melatonin delays leaf senescence and improves cucumber yield by modulating chlorophyll degradation and photoinhibition of PSII and PSI. Environ Exp Bot 200:104915. https://doi.org/10.1016/j.envexpbot.2022.104915

Liu Y, Li Z, Zhong C, Zhang Y, Wang-Pruski G, Zhang Z, Wu J (2023) Alleviating effect of melatonin on melon seed germination under autotoxicity and saline-alkali combined stress. J Plant Growth Regul 42(4):2474–2485. https://doi.org/10.1007/s00344-022-10720-3

Ma X, Zhang J, Burgess P, Rossi S, Huang B (2018) Interactive effects of melatonin and cytokinin on alleviating drought-induced leaf senescence in creeping bentgrass (*Agrostis stolonifera*). Environ Exp Bot 145:1–11. https://doi.org/10.1016/j.envexpbot.2017.10.010

Martinez V, Nieves-Cordones M, Lopez-Delacalle M, Rodenas R, Mestre TC, Garcia-Sanchez F, Rivero RM (2018) Tolerance to stress combination in tomato plants: new insights in the protective role of melatonin. Molecules 23(3):535. https://doi.org/10.3390/molecules23030535

Mauriz JL, Collado PS, Veneroso C, Reiter RJ, González-Gallego J (2013) A review of the molecular aspects of melatonin's anti-inflammatory actions: recent insights and new perspectives. J Pineal Res 54(1):1–14. https://doi.org/10.1111/j.1600-079X.2012.01014.x

Medina-Santamarina J, Serrano M, Lorente-Mento JM, Garcia-Pastor ME, Zapata PJ, Valero, Guillen F (2021) Melatonin treatment of pomegranate trees increases crop yield and quality parameters at harvest and during storage. Agronomy 11(5):861

Mohamed IA, Shalby N, El-Badri AMA, Saleem MH, Khan MN, Nawaz MA, Zhou G (2020) Stomata and xylem vessels traits improved by melatonin application contribute to enhancing salt tolerance and fatty acid composition of *Brassica napus* L. plants. Agronomy 10(8):1186. https://doi.org/10.3390/agronomy10081186

Morales F, Ancín M, Fakhet D, González-Torralba J, Gámez AL, Seminario A, Aranjuelo I (2020) Photosynthetic metabolism under stressful growth conditions as a bases for crop breeding and yield improvement. Plants 9(1):88. https://doi.org/10.3390/plants9010088

Mumithrakamatchi AK, Alagarswamy S, Kuppusamy A, Djanaguiraman M, Kalarani MK, Swarnapriya R, Marimuthu S, Vellaikumar S, Kanagarajan S (2024) Melatonin imparts tolerance to combined drought and high temperature stresses in tomato through osmotic adjustment and ABA accumulation. Front Plant Sci 15:1382914. https://doi.org/10.3389/fpls.2024.1382914

Pandey S, Bhandari HS, Hardy B (2017) Economic costs of drought and rice farmers' coping mechanisms: a cross-country comparative analysis. International Rice Research Institute, Los Baños

Pelagio-Flores R, Munoz-Parra E, Ortiz-Castro R, Lopez-Bucio J (2012) Melatonin regulates *Arabidopsis* root system architecture likely acting independently of auxin signaling. J Pineal Res 53(3):279–288

Priya P, Patil M, Pandey P, Singh A, Babu V, Senthil-Kumar M (2023) Stress combinations and their interactions in plants database: a one-stop resource on combined stress responses in plants. Plant J 116:1097. https://doi.org/10.1111/tpj.16497

Rajora N, Vats S, Raturi G, Thakral V, Kaur S, Rachappanavar V, Deshmukh R (2022) Seed priming with melatonin: a promising approach to combat abiotic stress in plants. Plant Stress 4:100071. https://doi.org/10.1016/j.stress.2022.100071

Ramakrishna A, Giridhar P, Sankar KU, Ravishankar GA (2012) Melatonin and serotonin profiles in beans of Coffea species. J Pineal Res 52(4):470–476. https://doi.org/10.1111/j.1600-079X.2011.00964.x

Ranjan S, Mirchandani R, Senthil-Kumar M (2023) Abiotic stress impact on the interaction between *Macrophomina phaseolina* and crop plants. Plant Physiol Rep 29:18. https://doi.org/10.1007/s40502-023-00753-5

Reiter RJ, Sharma RN, Manucha W, Rosales-Corral S, de AlmiedaChuffa LG, Loh D, Govitrapong P (2024) Dysfunctional mitochondria in age-related neurodegeneration: utility of melatonin as an antioxidant treatment. Ageing Res Rev 101:102480. https://doi.org/10.1016/j.arr.2024.102480

Shafi A, Singh AK, Zahoor I (2021) Melatonin: role in abiotic stress resistance and tolerance. In: Plant growth regulators: signalling under stress conditions. Springer, Cham, pp 239–273. https://doi.org/10.1007/978-3-030-61153-8_12

Shamloo-Dashtpagerdi R, Aliakbari M, Lindlöf A, Tahmasebi S (2022) A systems biology study unveils the association between a melatonin biosynthesis gene, O-methyl transferase 1 (OMT1) and wheat (*Triticum aestivum* L.) combined drought and salinity stress tolerance. Planta 255(5):99. https://doi.org/10.1007/s00425-022-03885-4

Sharma A, Zheng B (2019) Melatonin mediated regulation of drought stress: physiological and molecular aspects. Plants 8:190. https://doi.org/10.3390/plants8070190

Sheikhalipour M, Mohammadi SA, Esmaielpour B, Zareei E, Kulak M, Ali S, Fotopoulos V (2022) Exogenous melatonin increases salt tolerance in bitter melon by regulating ionic balance, antioxidant system and secondary metabolism-related genes. BMC Plant Biol 22(1):380. https://doi.org/10.1186/s12870-022-03728-0

Shi H, Reiter RJ, Tan DX, Chan Z (2015) Indole acetic acid inducible positively modulates natural leaf senescence through melatonin-mediated pathway in Arabidopsis. J Pineal Res 58:26–33. https://doi.org/10.1111/jpi.12188

Shi H, Wei Y, Wang Q, Reiter RJ, He C (2016) Melatonin mediates the stabilization of DELLA proteins to repress the floral transition in Arabidopsis. J Pineal Res 60(3):373–379. https://doi.org/10.1111/jpi.12320

Siddiqui ZA, Aziz S (2024) Plant parasitic nematode-fungus interactions: recent concepts and mechanisms. Plant Physiol Rep 29(1):37–50. https://doi.org/10.1007/s40502-023-00762-4

Sultana S, Barthakur S (2023) Seed priming with melatonin improves seed germination and root system architecture in wheat (*Triticum aestivum* L.). Pharma Innov J 12:574–578

Sun Q, Zhang N, Wang J, Zhang H, Li D, Shi J, Guo YD (2015) Melatonin promotes ripening and improves quality of tomato fruit during postharvest life. J Exp Bot 66(3):657–668. https://doi.org/10.1093/jxb/eru332

Sun C, Liu L, Wang L, Li B, Jin C, Lin X (2021) Melatonin: a master regulator of plant development and stress responses. J Integr Plant Biol 63(1):126–145. https://doi.org/10.1111/jipb.12993

Tatar Ö, Brück H, Asch F (2023) Atmospheric and soil water deficit induced changes in chemical and hydraulic signals in wheat (*Triticum aestivum* L.). J Agron Crop Sci 209:242–250. https://doi.org/10.1111/jac.12620

Tian Z, Zhao M, Wang J, Yang Q, Ma Y, Yang X, Meng Y (2024) Exogenous melatonin improves germination rate in buckwheat under high temperature stress by regulating seed physiological and biochemical characteristics. PeerJ 12:e17136. https://doi.org/10.7717/peerj.17136

Tiwari RK (2021) Mechanistic insights on melatonin-mediated drought stress mitigation in plants. Physiol Plant 172:1212–1226. https://doi.org/10.1111/ppl.13307

Tiwari RK, Lal MK, Naga KC, Kumar R, Chourasia KN, Kumar D, Sharma S (2020) Emerging roles of melatonin in mitigating abiotic and biotic stresses of horticultural crops. Sci Hortic 272:109592

Tiwari RK, Lal MK, Kumar R, Mangal V, Altaf MA, Sharma S, Singh B, Kumar M (2021) Insight into melatonin-mediated response and signaling in the regulation of plant defense under biotic stress. Plant Mol Biol 109:1–15

Tiwari RK, Altaf MA, Kumar R, Lal MK (2024) Evolution of melatonin as an oxidative stress mitigator in plant. In: Melatonin in plants: a pleiotropic molecule for abiotic stresses and pathogen infection. Springer, Singapore, pp 1–20

Udhayabharathi M (2017) Physiological and metabolomic studies of drought tolerance in green gram (*Vigna radiata* L.) by exogenous melatonin. Doctoral dissertation, Tamil Nadu Agricultural University

Umapathi M, Kalarani MK, Srinivasan S, Kalaiselvi P (2022) Alleviation of cadmium phytotoxicity through melatonin modulated physiological functions, antioxidants, and metabolites in tomato (*Solanum lycopersicum* L.). Biometals 35(5):1113–1132. https://doi.org/10.1007/s10534-022-00428-y

Van Tassel DL (1997) Identification and quantification of melatonin in higher plants. University of California, Davis

Van Tassel DL, Roberts N, Lewy A, O'Neill SD (2001) Melatonin in plant organs. J Pineal Res 31(1):8–15. https://doi.org/10.1034/j.1600-079X.2001.310102.x

Wang H, Ren C, Cao L, Zhao Q, Jin X, Wang M, Zhang Y (2022) Exogenous melatonin modulates physiological response to nitrogen and improves yield in nitrogen-deficient soybean (*Glycine max* L. Merr.). Front Plant Sci 13:865758. https://doi.org/10.3389/fpls.2022.865758

Wang L, Tanveer M, Wang H, Arnao MB (2024) Melatonin as a key regulator in seed germination under abiotic stress. J Pineal Res 76(1):e12937. https://doi.org/10.1111/jpi.12937

Wolf K, Kolář J, Witters E, van Dongen W, van Onckelen H, Macháčková I (2001) Daily profile of melatonin levels in *Chenopodium rubrum* L. depends on photoperiod. J Plant Physiol 158:1491–1493. https://doi.org/10.1078/0176-1617-00561

Xie Z, Wang J, Wang W, Wang Y, Xu J, Li Z, Fu B (2021) Integrated analysis of the transcriptome and metabolome revealed the molecular mechanisms underlying the enhanced salt tolerance of rice due to the application of exogenous melatonin. Front Plant Sci 11:618680. https://doi.org/10.3389/fpls.2020.618680

Yan F et al (2021) Melatonin regulates antioxidant strategy in response to continuous salt stress in rice seedlings. Plant Physiol 165:239–250. https://doi.org/10.1016/j.plaphy.2021.05.003

Yang L, You J, Li J, Wang Y, Chan Z (2021) Melatonin promotes Arabidopsis primary root growth in an IAA-dependent manner. J Exp Bot 72(15):5599–5611. https://doi.org/10.1093/jxb/erab196

Yusuf M, Saeed T, Almenhali HA, Azzam F, Hamzah AIAH, Khan TA (2024) Melatonin improved efficiency of 24-epibrassinolide to counter the collective stress of drought and salt through osmoprotectant and antioxidant system in pea plants. Sci Hortic 323:112453. https://doi.org/10.1016/j.scienta.2023.112453

Zhang Z, Hu Q, Liu Y, Cheng P, Cheng H, Liu W, Xing X, Guan Z, Fang W, Chen S, Jiang J, Chen F (2019) Strigolactone represses the synthesis of melatonin, thereby inducing floral transition in *Arabidopsis thaliana* in an FLC-dependent manner. J Pineal Res 67:e12582. https://doi.org/10.1111/jpi.12582

Zhang M, He S, Qin B, Jin X, Wang M, Ren C, Zhang Y (2020) Exogenous melatonin reduces the inhibitory effect of osmotic stress on antioxidant properties and cell ultrastructure at germination stage of soybean. PLoS One 15(12):e0243537. https://doi.org/10.1371/journal.pone.0243537

Zhou R, Wan H, Jiang F, Li X, Yu X, Rosenqvist E, Ottosen CO (2020) The alleviation of photosynthetic damage in tomato under drought and cold stress by high CO_2 and melatonin. Int J Mol Sci 21(15):5587. https://doi.org/10.3390/ijms21155587

Zuo B, Zheng X, He P, Wang L, Lei Q, Feng C, Kong J (2014) Overexpression of MzASMT improves melatonin production and enhances drought tolerance in transgenic *Arabidopsis thaliana* plants. J Pineal Res 57(4):408–417. https://doi.org/10.1111/jpi.12180

Genomic Insights on Seed Priming Effects of Melatonin

15

Muhammad Sajid Hanif, Nawal Asif, Muhammad Tayyab, Elamin Baillo, and Hafiz Muhammad Kashif Khalil

Abstract

Melatonin, a pleiotropic molecule with dual roles as an antioxidant and signaling agent, has emerged as a transformative tool in seed priming to enhance plant resilience against abiotic stresses. This chapter synthesizes genomic, molecular, and physiological insights into melatonin's role in seed priming, emphasizing its biosynthesis pathways, cross-talk with phytohormones, and stress-alleviating mechanisms. Melatonin priming accelerates germination, improves seedling vigor, and mitigates oxidative damage by modulating transcriptional networks

M. S. Hanif (✉)
Centre for Agricultural Resources Research, Institute of Genetics and Developmental Biology, Chinese Academy of Sciences, Shijiazhuang, China

University of Chinese Academy of Sciences, Beijing, China

N. Asif
Genetics and Breeding Center for Blunt Snout Bream, Ministry of Agriculture, Shanghai, China

Key Laboratory of Freshwater Aquatic Genetic Resources, Ministry of Agriculture, Shanghai, China

National Demonstration Center for Experimental Fisheries Science Education, Shanghai Ocean University, Shanghai, China

M. Tayyab
Institute of Marine Sciences and Guangdong Provincial Key Laboratory of Marine Biotechnology, Shantou University, Shantou, China

E. Baillo
Agricultural Research Corporation (ARC), Ministry of Agriculture, Gezira, Sudan

H. M. K. Khalil
Department of Plant Breeding and Genetics/CAS-AFS, University of Agriculture, Faisalabad, Pakistan

R. K. Tiwari et al. (eds.), *Melatonin: Signal Transduction Mechanisms and Defense Networks in Plants*, https://doi.org/10.1007/978-981-96-8869-2_15

(e.g., HSFA1s, DREBs), proteomic profiles (e.g., HSPs, glycolytic enzymes), and ion homeostasis genes (e.g., NHX1, AKT1). Case studies across crops—wheat under cadmium stress, maize under salinity, and tomato under drought—demonstrate melatonin's capacity to enhance stress tolerance through ROS scavenging, osmolyte accumulation, and epigenetic regulation. However, species-specific responses and concentration-dependent efficacy underscore the need for optimized protocols. The chapter further explores evolutionary conservation in melatonin biosynthesis genes and proposes CRISPR-based engineering to amplify endogenous production. By bridging lab-based discoveries with agricultural applications, melatonin priming offers a sustainable strategy to fortify crops against climate-induced stresses, though scaling field trials and addressing ecological impacts remain critical for future adoption.

Keywords
Seed viability · Longevity · Germination · Phytomelatonin · Antioxidants

15.1 Introduction

Melatonin (N-acetyl-5-methoxytryptamine), an indoleamine derived from the amino acid tryptophan, is a pivotal signaling molecule in plants, orchestrating diverse physiological processes such as seed germination, photosynthesis, flowering, and stress adaptation (Arnao and Hernández-Ruiz 2020). Initially identified in the pineal glands of animals, melatonin was later discovered in plants in 1995 (Dubbels et al. 1995; Hattori et al. 1995), where it functions as a potent antioxidant and regulator of redox homeostasis. Its biosynthesis in plants involves a multi-step pathway where tryptophan is sequentially converted to serotonin, then to N-acetylserotonin, and finally to melatonin via the enzyme N-acetylserotonin O-methyltransferase (ASMT) (Murch and Saxena 2002). Intriguingly, Arabidopsis employs caffeic acid O-methyltransferase (COMT) as an alternative enzyme for the final methylation step, highlighting species-specific variations in biosynthesis (Byeon et al. 2015). Beyond growth regulation, melatonin mitigates abiotic stresses such as drought, salinity, and heavy metal toxicity by scavenging reactive oxygen species (ROS) and modulating stress-responsive genes (N. Zhang et al. 2015). For instance, its role in preserving chlorophyll integrity and enhancing antioxidant enzyme activity under drought stress underscores its dual function as a signaling molecule and stress protectant (Wang et al. 2013; Ye et al. 2016).

Seed priming, a pre-sowing technique, involves controlled hydration of seeds to activate metabolic processes without radical emergence, thereby enhancing germination uniformity and stress resilience (Marthandan et al. 2020). Traditional methods include hydropriming (water soaking), osmopriming (osmotic solutions), and chemical priming (phytohormones or nutrients), all aimed at accelerating imbibition, repairing DNA, and mobilizing seed reserves (Pawar and Laware 2018). Primed seeds exhibit improved vigor, faster germination, and robust seedling establishment, particularly under adverse conditions such as salinity or drought (Khan et al. 2019; Lei et al. 2021). For example, osmopriming with polyethylene glycol (PEG) enhances drought tolerance by upregulating energy metabolism and antioxidant pathways (Cui et al. 2018). This cost-effective approach minimizes crop losses in unpredictable environments, making it a cornerstone of climate-resilient agriculture.

The integration of melatonin into seed priming strategies leverages its unique biochemical properties to amplify stress tolerance. As a pleiotropic molecule, melatonin directly scavenges ROS and reactive nitrogen species (RNS), while indirectly boosting the activity of antioxidant enzymes like superoxide dismutase (SOD), catalase (CAT), and ascorbate peroxidase (APX) (Galano and Reiter 2018). Seed priming with melatonin enhances osmolyte accumulation (e.g., proline and soluble sugars), stabilizes membrane integrity, and upregulates stress-related genes, such as heat-shock factors (HSFA1s) and ion transporters (Galano and Reiter 2018; Shi et al. 2015). For instance, priming wheat seeds with melatonin under chromium stress improved α-amylase activity and reserve mobilization, mitigating metal toxicity (Lei et al. 2021). Similarly, melatonin-primed rapeseed exhibited elevated ROS scavenging and photosynthetic efficiency under drought (Khan et al. 2019). However, efficacy is concentration-dependent; excessive melatonin may inhibit germination, as observed in cabbage seeds (Posmyk et al. 2008). These findings underscore melatonin's potential as a sustainable priming agent, bridging physiological resilience and molecular adaptation to combat abiotic stresses.

15.2 Molecular and Genomic Basis of Melatonin Biosynthesis

15.2.1 Biosynthesis Pathways: Key Genes and Enzymes

Melatonin biosynthesis in plants originates from the amino acid tryptophan, proceeding through a series of enzymatic reactions. The pathway involves four key steps catalyzed by specific enzymes (Murch and Saxena 2002; Rajora et al. 2022): tryptophan decarboxylase (TDC) converts tryptophan to tryptamine; tryptamine 5-hydroxylase (TSH) hydroxylates tryptamine to form serotonin; serotonin N-acetyltransferase (SNAT) acetylates serotonin to produce N-acetylserotonin; and N-acetylserotonin O-methyltransferase (ASMT) methylates N-acetylserotonin to yield melatonin.

In *Arabidopsis thaliana*, an alternative pathway exists where caffeic acid O-methyltransferase (COMT) substitutes ASMT, performing the final methylation step (Byeon et al. 2015). While these pathways are conserved across species, the regulatory mechanisms and enzyme isoforms may vary. Environmental factors, such as light and temperature, dynamically influence the expression of biosynthesis genes (Murch et al. 2000). For instance, ASMT and COMT are stress-responsive, enabling plants to modulate melatonin levels under adverse conditions (Byeon et al. 2015). Despite progress, the full complexity of melatonin biosynthesis, including tissue-specific regulation and enzyme kinetics, remains to be elucidated.

15.2.2 Regulation Under Stress: Transcriptional and Post-transcriptional Modulation

Abiotic stressors, including drought, salinity, and heavy metals, upregulate melatonin biosynthesis by enhancing the expression of key genes. For example, under cadmium (Cd) stress, maize seedlings exhibit elevated *TDC*, *SNAT*, and *ASMT* transcript levels, correlating with increased melatonin production (Ni et al. 2018). Similarly, chromium (Cr)-exposed wheat seeds show heightened α-amylase activity and reserve mobilization when primed with melatonin, indicating stress-induced metabolic adjustments (Lei et al. 2021).

At the transcriptional level, stressors activate signaling cascades that promote the transcription of melatonin biosynthesis genes (Fig. 15.1). Proteomic studies reveal that melatonin priming upregulates stress-related proteins, such as heat shock proteins (HSPs) and antioxidant enzymes, which mitigate oxidative damage (Kołodziejczyk et al. 2016). Post-transcriptional regulation may involve enzyme stabilization or phosphorylation, though mechanistic details are less understood. Notably, melatonin biosynthesis is tightly linked to reactive oxygen species (ROS) scavenging; under stress, plants prioritize melatonin synthesis to counteract oxidative bursts, enhancing survival (Rajora et al. 2022).

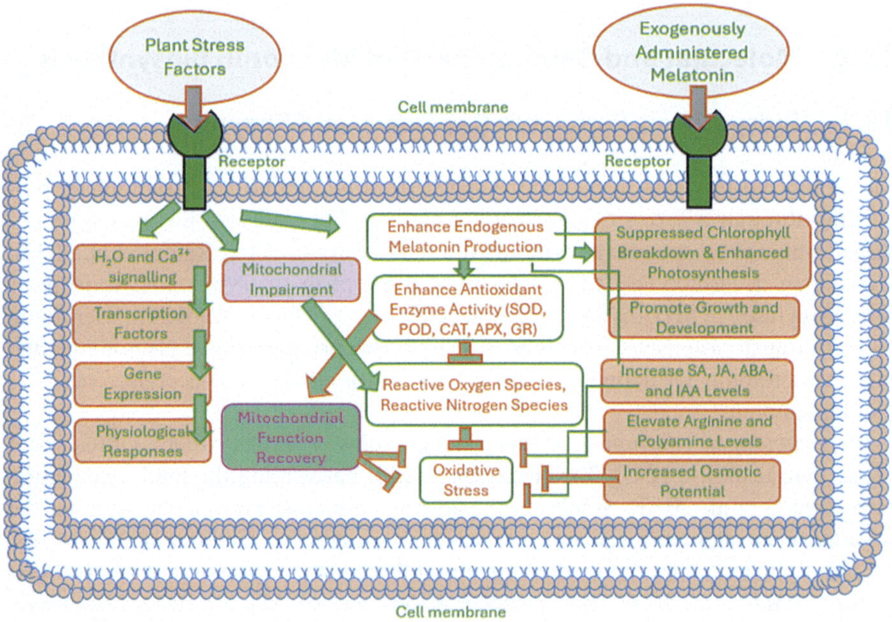

Fig. 15.1 Melatonin-mediated modulation of key physiological processes in plants. Notes: ROS—reactive oxygen species; RNS—reactive nitrogen species; SOD—superoxide dismutase; CAT—catalase; APX—ascorbate peroxidase; SA—salicylic acid; JA—jasmonic acid; IAA—indoleacetic acid

15.2.3 Cross-Talk with Phytohormones: Synergy with Auxin, Salicylic Acid, and Jasmonic Acid

Melatonin interacts intricately with phytohormone signaling pathways to coordinate stress responses. Key interactions include: Auxin (IAA): Melatonin priming enhances IAA levels in wheat under salinity, promoting root growth and ion homeostasis (Sadak 2016). This synergy likely involves melatonin-mediated activation of auxin biosynthesis genes or stabilization of IAA molecules, Salicylic Acid (SA) and Jasmonic Acid (JA): Melatonin upregulates SA and JA pathways, amplifying defense responses. For instance, in *Arabidopsis*, melatonin induces *WRKY* transcription factors and *NAC* genes, which are integral to SA/JA-mediated stress signaling (N. Zhang et al. 2015). These interactions enhance antioxidant capacity and pathogen resistance, Abscisic Acid (ABA): Though not directly a focus here, melatonin modulates ABA metabolism under drought, balancing stomatal closure and water retention (X. Li et al. 2016).

These cross-talk mechanisms enable melatonin to act as a central regulator, integrating hormonal signals to optimize growth and stress resilience (Bajwa et al. 2014; Shi et al. 2015).

15.3 Seed Priming Techniques with Melatonin

Seed priming with melatonin involves pre-treating seeds with melatonin solutions to enhance stress resilience and germination efficiency. The methodology varies depending on the crop species, stress type, and priming technique. Hydropriming, the most common delivery method, entails soaking seeds in aqueous melatonin solutions for specific durations, typically ranging from a few hours to overnight (Marthandan et al. 2020). For instance, wheat seeds exposed to cadmium stress were primed with 50–100 μM melatonin for 12 h (Ni et al. 2018), while maize seeds under salinity stress required 0.4–1.6 mM melatonin for similar durations (X. Jiang et al. 2016). Osmopriming, though less frequently studied, combines melatonin with osmotic agents like polyethylene glycol (PEG) to regulate hydration and metabolic activation (Cui et al. 2018).

Optimal melatonin concentrations are critical, as excessive doses may inhibit germination. For example, priming red cabbage seeds with 100 μM melatonin reduced growth compared to lower concentrations (Posmyk et al. 2008). Conversely, concentrations as low as 10 μM improved basil germination under salinity (Bahcesular et al. 2020). The duration of priming generally does not significantly influence outcomes, but short-term treatments (6–24 h) are preferred for practicality (Yan and Mao 2021).

15.4 Physiological Outcomes: Enhanced Germination and Stress Tolerance

Melatonin priming enhances seed germination, seedling vigor, and stress tolerance across diverse crops by modulating physiological and molecular pathways. In wheat, melatonin priming under chromium stress increased α-amylase activity, promoting starch-to-sugar conversion and reserve mobilization (Lei et al. 2021). Maize seedlings primed with melatonin exhibited improved chilling tolerance through upregulated starch metabolism and energy production (Cao et al. 2019). Similarly, tomato seeds showed enhanced root vigor and reduced oxidative damage under drought due to elevated antioxidant enzyme activity (Liu et al. 2015).

Key physiological improvements such as Melatonin boosts antioxidant enzymes (SOD, CAT, APX) and osmoprotectants (proline, sugars), mitigating oxidative damage (Khan et al. 2019). In cotton, 25 μM melatonin priming preserved chlorophyll content and PSII functionality under salinity (Y. Zhang et al. 2021). Melatonin regulates Na+/K+ ratios by upregulating vacuolar transporters (e.g., NHX1), critical for salt stress tolerance in faba bean (J. Li et al. 2019).

15.5 Case Study: Melatonin-Mediated Alleviation of Cadmium Stress in Wheat

Ni et al. (2018) demonstrates the efficacy of melatonin priming in alleviating cadmium (Cd) toxicity in wheat. Seeds primed with 50 and 100 μM melatonin showed significant improvements in seedling growth compared to non-primed seeds under 0.2 mM $CdCl_2$ stress. Melatonin reduced Cd-induced hydrogen peroxide accumulation by enhancing APX and SOD activity, thereby mitigating oxidative damage. Additionally, primed seedlings exhibited restored root elongation and biomass, attributed to improved antioxidant defense and reduced Cd uptake (Ni et al. 2018). This study underscores melatonin's role in enhancing heavy metal tolerance through dual mechanisms: direct ROS scavenging and activation of stress-responsive proteins.

15.6 Genomic and Transcriptomic Responses to Melatonin Priming

Melatonin priming induces significant molecular changes in plants, enhancing their ability to counteract abiotic stress. These responses span transcriptional, translational, and potentially epigenetic levels.

15.6.1 Stress-Responsive Gene Activation

Melatonin priming activates stress-related transcriptional networks, enabling plants to mitigate oxidative damage and maintain homeostasis. Key genes upregulated include heat shock factors (HSFA1s), which orchestrate thermotolerance by regulating heat-responsive genes (Shi et al. 2015). Under drought stress, melatonin enhances the expression of drought-responsive element-binding factors (DREBs/CBFs) and cold-regulated genes like *COR15a*, which stabilize cellular structures under dehydration (Bajwa et al. 2014; Yang et al. 2019).

In salinity stress, melatonin priming improves ion homeostasis by upregulating vacuolar Na^+/H^+ antiporters (NHX1) and K^+ transporters (AKT1), reducing cytoplasmic Na^+ toxicity (C. Li et al. 2012; J. Li et al. 2019). Additionally, genes encoding ROS-scavenging enzymes, such as superoxide dismutase (*SOD*), catalase (*CAT*), and ascorbate peroxidase (*APX*), are modulated to alleviate oxidative stress (D. Jiang et al. 2020; Y. Zhang et al. 2021). These transcriptional adjustments collectively enhance stress resilience by balancing ion uptake, osmotic regulation, and antioxidant defense.

15.6.2 Proteomic Insights

Proteomic analyses reveal that melatonin priming reprograms metabolic and stress-response pathways. For instance, in drought-stressed wheat seedlings, melatonin upregulates glycolytic enzymes (e.g., fructose-1,6-bisphosphate aldolase, glyceraldehyde-3-phosphate dehydrogenase) and energy metabolism proteins (e.g., NADP-binding proteins), enhancing ATP production under water deficit (Cui et al. 2018). Similarly, heat-shock proteins (HSPs), such as small HSP-3105 and HSP-7102, show elevated expression in melatonin-primed maize under cadmium stress, aiding protein folding and stability (Kołodziejczyk et al. 2016).

Melatonin also influences ubiquitin-mediated proteolysis, with increased expression of ubiquitin-fold modifier 1 and serine proteases, which degrade damaged proteins during stress (Cui et al. 2018). These proteomic shifts highlight melatonin's role in maintaining cellular integrity by enhancing energy production, protein repair, and turnover.

15.6.3 Epigenetic Modifications

While studies on melatonin's epigenetic role are limited, emerging evidence suggests its potential involvement in stress memory. Melatonin may influence DNA methylation and histone acetylation patterns, which regulate stress-responsive gene expression. For example, melatonin-induced activation of *HSFA1s* and *DREBs* could involve chromatin remodeling to facilitate rapid transcriptional responses upon stress recurrence (Shi et al. 2015; Yang et al. 2019). However, the precise mechanisms remain speculative, necessitating further exploration through multi-omics approaches to unravel melatonin's impact on epigenetic regulation (Deshmukh et al. 2020).

In conclusion, melatonin priming orchestrates a multi-layered molecular response, from gene activation to proteomic reorganization, enabling plants to withstand abiotic stressors. Unraveling its epigenetic contributions will deepen our understanding of long-term stress adaptation.

15.7 Crop-Specific Genomic Adaptations

Under drought stress, melatonin seed priming induces transcriptomic shifts that enhance energy metabolism and osmoprotectant synthesis. In wheat, proteomic analyses revealed upregulated expression of glycolytic enzymes such as fructose-1,6-bisphosphate aldolase, glyceraldehyde-3-phosphate dehydrogenase, and enolase, which improve energy production during germination under water deficit (Cui et al. 2018). Additionally, melatonin priming enhances the synthesis of osmoprotectants like proline and soluble sugars, aiding cellular turgor maintenance and osmotic adjustment (Khan et al. 2019). Aquaporins, particularly plasma membrane intrinsic proteins (PIPs), play a critical role in improving root hydraulic conductance. For instance, melatonin-treated maize and rice plants exhibited increased PIP expression, facilitating efficient water uptake and transport under drought conditions (Qiao et al. 2020; Tiwari et al. 2021). These genomic adaptations collectively mitigate drought-induced oxidative stress and enhance plant resilience.

Melatonin priming modulates ion homeostasis to counteract salinity stress by regulating key genes involved in Na^+ compartmentalization and K^+ retention. In cotton and wheat, melatonin upregulates vacuolar Na^+/H^+ antiporter genes (e.g., *NHX1*) and potassium transporter genes (e.g., *AKT1*), promoting Na^+ sequestration into vacuoles and maintaining optimal K^+/Na^+ ratios (X. Jiang et al. 2016; J. Li et al. 2019). This genomic regulation prevents cytoplasmic ion toxicity and stabilizes enzymatic functions. Transcriptomic studies in melatonin-primed plants also highlight enhanced expression of antioxidant enzymes such as superoxide dismutase (SOD) and ascorbate peroxidase (APX), which mitigate salinity-induced oxidative damage (Y. Zhang et al. 2021). Furthermore, melatonin preserves photosynthetic efficiency by protecting chlorophyll content and upregulating stress-responsive transcription factors like *HSFA1s*, which activate heat-shock proteins (Shi et al. 2015).

Melatonin alleviates heavy metal toxicity through genomic mechanisms involving metal chelation and phytochelatin synthesis. In tomato, melatonin priming increases phytochelatin levels, which bind cadmium (Cd) and facilitate its sequestration into vacuoles, reducing cytoplasmic toxicity (Hasan et al. 2015). Proteomic studies in maize identified melatonin-induced expression of stress-related proteins, including small heat shock proteins (HSPs) and thioredoxin, which enhance detoxification and redox homeostasis under copper stress (Kołodziejczyk et al. 2016). Additionally, melatonin reduces heavy metal uptake by modulating transporter activity; for example, in *Amaranthus*, melatonin decreased root Pb accumulation by 33% while enhancing translocation to less sensitive tissues (David et al. 2021). These adaptations are concentration-dependent, as excessive melatonin may inhibit growth, emphasizing the need for optimized priming protocols (Posmyk et al. 2008).

15.8 Comparative Genomics and Evolutionary Perspectives

The application of melatonin in enhancing abiotic stress tolerance exhibits both species-specific adaptations and evolutionary conservation across plant lineages. Comparative analyses reveal variations in molecular mechanisms and genetic pathways, reflecting the diversity of plant responses to melatonin priming.

15.8.1 Species-Specific Responses: Variations in Melatonin Efficacy Across Monocots and Dicots

Melatonin-mediated stress alleviation displays distinct patterns between monocots and dicots, likely due to differences in genetic regulation and physiological adaptations. For instance, in monocots such as wheat (*Triticum aestivum*), seed priming with melatonin under drought stress upregulated glycolytic enzymes (e.g., fructose-1,6-bisphosphate aldolase) and energy metabolism-related proteins, enhancing osmotic adjustment and recovery post-rehydration (Cui et al. 2018). Similarly, maize (*Zea mays*) seedlings under salinity stress showed improved K+/Na+ homeostasis and antioxidant enzyme activity (X. Jiang et al. 2016). In contrast, dicots like tomato (*Solanum lycopersicum*) exhibited melatonin-induced upregulation of heat-shock factors (*HSFA1s*) and polyamine biosynthesis under heat stress, mitigating ROS accumulation and preserving male fertility (Qi et al. 2018; Shi et al. 2015). Cucumber (*Cucumis sativus*), another dicot, demonstrated enhanced expression of cadmium-responsive genes (*CsHA2*, *CsHA3*) and nitrate reductase activity under heavy metal stress (Ali Shah et al. 2020). These differences highlight lineage-specific molecular strategies, where monocots often prioritize energy metabolism and ion transport, while dicots leverage transcriptional regulation and redox signaling.

Concentration-dependent effects further underscore species-specificity. For example, melatonin priming at 25 μM improved cotton (a dicot) photosynthesis under salinity, whereas higher concentrations (100–500 μM) were required to optimize germination in faba bean (*Vicia faba*, a dicot) (Dawood and El-Awadi 2015; N. Zhang et al. 2015). Conversely, excessive melatonin (100 μM) inhibited cabbage (a dicot) seedling growth, suggesting thresholds vary across taxa (Posmyk et al. 2008).

15.8.2 Evolutionary Conservation: Orthologous Genes in Melatonin Biosynthesis and Stress Adaptation

Despite species-specific responses, key components of melatonin biosynthesis and stress signaling are evolutionarily conserved. The melatonin biosynthesis pathway in plants originates from tryptophan, with enzymes like *N*-acetylserotonin methyltransferase (ASMT) catalyzing the final step. Intriguingly, Arabidopsis (*Arabidopsis thaliana*) employs caffeic acid *O*-methyltransferase (COMT) as a functional substitute for ASMT, illustrating metabolic flexibility within conserved pathways (Byeon et al. 2015). Orthologous genes involved in stress adaptation, such as vacuolar Na+/H+ antiporters (*NHX1*, *AKT1*), are upregulated by melatonin in both monocots (e.g., maize) and dicots (e.g., apple), indicating shared mechanisms for ion homeostasis (X. Jiang et al. 2016; C. Li et al. 2012).

Comparative genomics further reveals conserved transcription factors regulated by melatonin, including *WRKY*, *NAC*, and *HSFA1s*, which mediate antioxidant defense and thermotolerance across lineages (Shi et al. 2015; N. Zhang et al. 2015). For instance, *HSFA1s* activation under heat stress is conserved in Arabidopsis and tomato, while *DREB/CBF* genes enhance drought tolerance in both monocots and dicots (Bajwa et al. 2014; Yang et al. 2019). Additionally, CRISPR-based studies highlight the potential to edit orthologous melatonin biosynthesis genes (e.g., *ASMT*, *SNAT*) in crops, leveraging evolutionary conservation for trait improvement (Vats et al. 2019).

The ubiquitous presence of melatonin in phylogenetically diverse species, from bryophytes to angiosperms, underscores its ancient role in stress adaptation (Chen et al. 2003). This evolutionary persistence suggests melatonin's fundamental contribution to plant survival under fluctuating environments, making it a promising target for enhancing climate resilience in modern agriculture.

15.9 Future Directions and Applications

The potential of melatonin seed priming to enhance plant resilience against abiotic stresses has been well-documented in controlled studies. However, translating these findings into practical agricultural solutions requires addressing several research gaps and technical challenges. Below are key future directions and applications derived from current literature.

15.9.1 Multi-Omics Integration

Integrating transcriptomic, proteomic, and metabolomic approaches is critical to unravel the molecular mechanisms underlying melatonin-mediated stress tolerance (Table 15.1). While multi-omics studies have identified stress-responsive genes (e.g., *HSFA1s*, *DREBs/CBFs*) and proteins (e.g., glycolytic enzymes, antioxidant regulators) in melatonin-primed plants (Cui et al. 2018; Shi et al. 2015), the precise

Table 15.1 Summary of transcriptomic/proteomic studies on melatonin seed priming across crops under abiotic stresses

Crop	Stress type	Melatonin concentration	Key genes/proteins identified	Physiological outcomes	Reference
Wheat	Drought	100 μM	Glycolytic enzymes (fructose-1,6-bisphosphate aldolase, enolase), HSPs, ubiquitin-related proteins	Enhanced energy metabolism, autophagy, and drought resilience	Cui et al. (2018)
Arabidopsis	Heat	20 μM	*HSFA1s* (heat-shock factors)	Activation of heat-responsive genes; improved thermotolerance	Shi et al. (2015)
Maize	Copper	500 μM	Small heat-shock proteins (HSPs), thioredoxin TRp26 isoform 4	Stress reduction during germination; enhanced proteome plasticity	Kołodziejczyk et al. (2016)
Cotton	Salinity	25 μM	Photosynthesis-related genes, antioxidant enzymes (*SOD*, *CAT*)		Zhang et al. (2021)
Faba Bean	Salinity	100–500 μM	Novel proteins (unidentified)	Enhanced growth parameters, ion homeostasis (K+/Na+ balance)	Dawood and El-Awadi (2015)
Tomato	Heat	10 μM	Heat-responsive genes (unspecified)	Reduced heat-induced photoinhibition and electrolyte leakage; improved cellular protection	Xu et al. (2016)

signaling pathways remain unclear. For instance, proteomic analyses in wheat revealed melatonin-induced upregulation of energy metabolism and autophagy-related proteins under drought stress (Cui et al. 2018). Similarly, transcriptomic profiling in *Arabidopsis* highlighted melatonin's role in activating heat-shock factors (Shi et al. 2015). Future research should focus on cross-omics validation to map dynamic interactions between melatonin, stress-responsive genes, and metabolic pathways, particularly under combinatorial abiotic stresses (Liang et al. 2015).

15.9.2 CRISPR-Based Engineering

Genome editing tools, such as CRISPR/Cas9, offer promising avenues to enhance endogenous melatonin biosynthesis in crops. Key targets include enzymes like *ASMT* and *COMT*, which catalyze the final steps of melatonin synthesis (Byeon et al. 2015; Yang et al. 2019). Overexpression of *TaCOMT* in *Arabidopsis* improved drought tolerance by boosting melatonin production (Yang et al. 2019), suggesting similar strategies could be applied to staple crops. However, CRISPR applications require a deeper understanding of melatonin's genetic regulation, including tissue-specific expression patterns and interactions with phytohormones (Ansari et al. 2020; Vats et al. 2019). Additionally, editing aquaporin genes (e.g., *PIPs*) to synergize with melatonin's water uptake enhancement may further improve drought resilience (Qiao et al. 2020).

15.10 Field Applications

Scaling lab-based findings to field conditions poses significant challenges. While melatonin priming improves germination and stress tolerance in crops like wheat, maize, and tomato under controlled environments (Khan et al. 2019; Lei et al. 2021), field trials are limited. Key obstacles include optimizing melatonin concentrations for diverse crops, ensuring chemical stability under variable environmental conditions, and cost-effective application methods (Nawaz et al. 2020). For example, excessive melatonin doses can inhibit germination, as observed in cabbage (Posmyk et al. 2008). Furthermore, interactions with soil microbiota and long-term ecological impacts remain unexplored. Collaborative efforts between agronomists and biotechnologists are needed to develop standardized protocols and assess the socio-economic viability of melatonin-based practices under climate change scenarios (Marthandan et al. 2020).

15.11 Conclusion

Melatonin seed priming represents a nexus of biochemical innovation and genomic adaptation, offering a powerful yet underutilized tool to enhance crop resilience in an era of climate uncertainty. This chapter elucidates how melatonin primes seeds at molecular levels, activating stress-responsive genes (HSFA1s, WRKYs), stabilizing proteomes, and harmonizing phytohormone signaling to mitigate drought, salinity, and heavy metal toxicity. Its ability to enhance antioxidant defense, ion homeostasis, and photosynthetic efficiency underscores its pleiotropic nature. However, challenges persist: species-specific dosage requirements, limited field validation, and unresolved ecological consequences of exogenous melatonin application demand attention. Future research must prioritize multi-omics integration to decode melatonin's epigenetic influence and CRISPR-driven bioengineering to tailor biosynthesis pathways. Collaborative efforts among geneticists, agronomists, and farmers are essential to translate melatonin priming from controlled studies to real-world, climate-resilient agriculture. By harmonizing molecular insights with practical scalability, melatonin can revolutionize seed priming paradigms, ensuring food security in increasingly hostile environments.

References

Ali Shah A, Ahmed S, Yasin NA (2020) Cadmium stress consolation in melatonin supplemented *Cucumis sativus* through modulation of antioxidative defense system. Iran J Plant Physiol 10(2):3135–3154

Ansari WA, Chandanshive SU, Bhatt V, Nadaf AB, Vats S, Katara JL, Sonah H, Deshmukh R (2020) Genome editing in cereals: approaches, applications and challenges. Int J Mol Sci 21(11):4040

Arnao MB, Hernández-Ruiz J (2020) Is phytomelatonin a new plant hormone? Agronomy 10(1):95

Bahcesular B, Yildirim ED, Karaçocuk M, Kulak M, Karaman S (2020) Seed priming with melatonin effects on growth, essential oil compounds and antioxidant activity of basil (*Ocimum basilicum* L.) under salinity stress. Ind Crop Prod 146:112165

Bajwa VS, Shukla MR, Sherif SM, Murch SJ, Saxena PK (2014) Role of melatonin in alleviating cold stress in *Arabidopsis thaliana*. J Pineal Res 56(3):238–245

Byeon Y, Choi G-H, Lee HY, Back K (2015) Melatonin biosynthesis requires N-acetylserotonin methyltransferase activity of caffeic acid O-methyltransferase in rice. J Exp Bot 66(21):6917–6925

Cao Q, Li G, Cui Z, Yang F, Jiang X, Diallo L, Kong F (2019) Seed priming with melatonin improves the seed germination of waxy maize under chilling stress via promoting the antioxidant system and starch metabolism. Sci Rep 9(1):15044

Chen G, Huo Y, Tan D-X, Liang Z, Zhang W, Zhang Y (2003) Melatonin in Chinese medicinal herbs. Life Sci 73(1):19–26

Cui G, Sun F, Gao X, Xie K, Zhang C, Liu S, Xi Y (2018) Proteomic analysis of melatonin-mediated osmotic tolerance by improving energy metabolism and autophagy in wheat (*Triticum aestivum* L.). Planta 248:69–87

David OA, Jolayemi OL, Akomolafe GF, Adegoke MA (2021) Lead sequestration, immobilization and xylem cavitation in melatonin-primed *Amaranthus cruentus*. Plant Physiol Rep 26:162–171

Dawood MG, El-Awadi ME (2015) Alleviation of salinity stress on *Vicia faba* L. plants via seed priming with melatonin. Disminución del estrés salino en plantas de Vicia faba L. a través de la activación de las semillas con melatonina. Acta Biol Colomb 20:223–235

Deshmukh R, Sonah H, Belanger RR (2020) New evidence defining the evolutionary path of aquaporins regulating silicon uptake in land plants. J Exp Bot 71(21):6775–6788

Dubbels R, Reiter RJ, Klenke E, Goebel A, Schnakenberg E, Ehlers C, Schiwara HW, Schloot W (1995) Melatonin in edible plants identified by radioimmunoassay and by high performance liquid chromatography-mass spectrometry. J Pineal Res 18(1):28–31

Galano A, Reiter RJ (2018) Melatonin and its metabolites vs oxidative stress: from individual actions to collective protection. J Pineal Res 65(1):e12514

Hasan MK, Ahammed GJ, Yin L, Shi K, Xia X, Zhou Y, Yu J, Zhou J (2015) Melatonin mitigates cadmium phytotoxicity through modulation of phytochelatins biosynthesis, vacuolar sequestration, and antioxidant potential in *Solanum lycopersicum* L. Front Plant Sci 6:601

Hattori A, Migitaka H, Iigo M, Itoh M, Yamamoto K, Ohtani-Kaneko R, Hara M, Suzuki T, Reiter RJ (1995) Identification of melatonin in plants and its effects on plasma melatonin levels and binding to melatonin receptors in vertebrates. Biochem Mol Biol Int 35(3):627–634

Jiang X, Li H, Song X (2016) Seed priming with melatonin effects on seed germination and seedling growth in maize under salinity stress. Pak J Bot 48(4):1345–1352

Jiang D, Lu B, Liu L, Duan W, Chen L, Li J, Zhang K, Sun H, Zhang Y, Dong H (2020) Exogenous melatonin improves salt stress adaptation of cotton seedlings by regulating active oxygen metabolism. PeerJ 8:e10486

Khan MN, Zhang J, Luo T, Liu J, Rizwan M, Fahad S, Xu Z, Hu L (2019) Seed priming with melatonin coping drought stress in rapeseed by regulating reactive oxygen species detoxification: antioxidant defense system, osmotic adjustment, stomatal traits and chloroplast ultrastructure perseveration. Ind Crop Prod 140:111597

Kołodziejczyk I, Dzitko K, Szewczyk R, Posmyk MM (2016) Exogenous melatonin expediently modifies proteome of maize (*Zea mays* L.) embryo during seed germination. Acta Physiol Plant 38:1–18

Lei K, Sun S, Zhong K, Li S, Hu H, Sun C, Zheng Q, Tian Z, Dai T, Sun J (2021) Seed soaking with melatonin promotes seed germination under chromium stress via enhancing reserve mobilization and antioxidant metabolism in wheat. Ecotoxicol Environ Saf 220:112241

Li C, Wang P, Wei Z, Liang D, Liu C, Yin L, Jia D, Fu M, Ma F (2012) The mitigation effects of exogenous melatonin on salinity-induced stress in *Malus hupehensis*. J Pineal Res 53(3):298–306

Li X, Tan D, Jiang D, Liu F (2016) Melatonin enhances cold tolerance in drought-primed wild-type and abscisic acid-deficient mutant barley. J Pineal Res 61(3):328–339

Li J, Liu J, Zhu T, Zhao C, Li L, Chen M (2019) The role of melatonin in salt stress responses. Int J Mol Sci 20(7):1735

Liang C, Zheng G, Li W, Wang Y, Hu B, Wang H, Wu H, Qian Y, Zhu X, Tan D (2015) Melatonin delays leaf senescence and enhances salt stress tolerance in rice. J Pineal Res 59(1):91–101

Liu J, Wang W, Wang L, Sun Y (2015) Exogenous melatonin improves seedling health index and drought tolerance in tomato. Plant Growth Regul 77:317–326

Marthandan V, Geetha R, Kumutha K, Renganathan VG, Karthikeyan A, Ramalingam J (2020) Seed priming: a feasible strategy to enhance drought tolerance in crop plants. Int J Mol Sci 21(21):8258

Murch SJ, Saxena PK (2002) Melatonin: a potential regulator of plant growth and development? Vitro Cell Dev Biol Plant 38:531–536

Murch SJ, KrishnaRaj S, Saxena PK (2000) Tryptophan is a precursor for melatonin and serotonin biosynthesis in in vitro regenerated St. John's wort (*Hypericum perforatum* L. cv. Anthos) plants. Plant Cell Rep 19:698–704

Nawaz K, Chaudhary R, Sarwar A, Ahmad B, Gul A, Hano C, Abbasi BH, Anjum S (2020) Melatonin as master regulator in plant growth, development and stress alleviator for sustainable agricultural production: current status and future perspectives. Sustainability 13(1):294

Ni J, Wang Q, Shah F, Liu W, Wang D, Huang S, Fu S, Wu L (2018) Exogenous melatonin confers cadmium tolerance by counterbalancing the hydrogen peroxide homeostasis in wheat seedlings. Molecules 23:799. https://doi.org/10.3390/molecules23040799

Pawar VA, Laware SL (2018) Seed priming a critical review. Int J Sci Res Biol Sci 5(5):94–101

Posmyk MM, Kuran H, Marciniak K, Janas KM (2008) Presowing seed treatment with melatonin protects red cabbage seedlings against toxic copper ion concentrations. J Pineal Res 45(1):24–31

Qi Z-Y, Wang K-X, Yan M-Y, Kanwar MK, Li D-Y, Wijaya L, Alyemeni MN, Ahmad P, Zhou J (2018) Melatonin alleviates high temperature-induced pollen abortion in *Solanum lycopersicum*. Molecules 23(2):386

Qiao Y, Ren J, Yin L, Liu Y, Deng X, Liu P, Wang S (2020) Exogenous melatonin alleviates PEG-induced short-term water deficiency in maize by increasing hydraulic conductance. BMC Plant Biol 20:1–14

Rajora N, Vats S, Raturi G, Thakral V, Kaur S, Rachappanavar V, Kumar M, Kesarwani AK, Sonah H, Sharma TR (2022) Seed priming with melatonin: a promising approach to combat abiotic stress in plants. Plant Stress 4:100071

Sadak MS (2016) Mitigation of salinity adverse effects of on wheat by grain priming with melatonin. Int J ChemTech Res 9(2):85–97

Shi H, Tan D, Reiter RJ, Ye T, Yang F, Chan Z (2015) Melatonin induces class A1 heat-shock factors (HSFA 1s) and their possible involvement of thermotolerance in Arabidopsis. J Pineal Res 58(3):335–342

Tiwari RK, Lal MK, Kumar R, Chourasia KN, Naga KC, Kumar D, Das SK, Zinta G (2021) Mechanistic insights on melatonin-mediated drought stress mitigation in plants. Physiol Plant 172(2):1212–1226

Vats S, Kumawat S, Kumar V, Patil GB, Joshi T, Sonah H, Sharma TR, Deshmukh R (2019) Genome editing in plants: exploration of technological advancements and challenges. Cells 8(11):1386

Wang P, Sun X, Li C, Wei Z, Liang D, Ma F (2013) Long-term exogenous application of melatonin delays drought-induced leaf senescence in apple. J Pineal Res 54(3):292–302

Xu W, Cai S, Zhang Y, Wang Y, Ahammed GJ, Xia X, Shi K, Zhou Y, Yu J, Reiter RJ (2016) Melatonin enhances thermotolerance by promoting cellular protein protection in tomato plants. J Pineal Res 61(4):457–469

Yan H, Mao P (2021) Comparative time-course physiological responses and proteomic analysis of melatonin priming on promoting germination in aged oat (*Avena sativa* L.) seeds. Int J Mol Sci 22(2):811

Yang W-J, Du Y-T, Zhou Y-B, Chen J, Xu Z-S, Ma Y-Z, Chen M, Min D-H (2019) Overexpression of TaCOMT improves melatonin production and enhances drought tolerance in transgenic Arabidopsis. Int J Mol Sci 20(3):652

Ye J, Wang S, Deng X, Yin L, Xiong B, Wang X (2016) Melatonin increased maize (*Zea mays* L.) seedling drought tolerance by alleviating drought-induced photosynthetic inhibition and oxidative damage. Acta Physiol Plant 38(2):48

Zhang N, Sun Q, Zhang H, Cao Y, Weeda S, Ren S, Guo Y-D (2015) Roles of melatonin in abiotic stress resistance in plants. J Exp Bot 66(3):647–656

Zhang Y, Zhou X, Dong Y, Zhang F, He Q, Chen J, Zhu S, Zhao T (2021) Seed priming with melatonin improves salt tolerance in cotton through regulating photosynthesis, scavenging reactive oxygen species and coordinating with phytohormone signal pathways. Ind Crop Prod 169:113671

Melatonin-Induced Defense to Mitigate the Adverse Effects of Ozone Damage in Plants

16

Muhammad Tayyab, Muhammad Sajid Hanif, Nawal Asif, Tariq Dildar, and Mir Muhammad Nizamani

Abstract

Tropospheric ozone (O_3) pollution is a significant threat to plant health, leading to oxidative stress, reduced photosynthesis, and decreased crop yields. This chapter examines melatonin (N-acetyl-5-methoxytryptamine) as a powerful

M. Tayyab (✉)
Institute of Marine Sciences and Guangdong Provincial Key Laboratory of Marine Biotechnology, Shantou University, Shantou, China

M. S. Hanif
Centre for Agricultural Resources Research, Institute of Genetics and Developmental Biology, Chinese Academy of Sciences, Shijiazhuang, China

University of Chinese Academy of Sciences, Beijing, China

N. Asif
Genetics and Breeding Center for Blunt Snout Bream, Ministry of Agriculture, Shanghai, China

Key Laboratory of Freshwater Aquatic Genetic Resources, Ministry of Agriculture, Shanghai, China

National Demonstration Center for Experimental Fisheries Science Education, Shanghai Ocean University, Shanghai, China

T. Dildar
STU-UMT Joint Shellfish Research Laboratory, Shantou University, Shantou, China

Guangdong Provincial Key Laboratory of Marine Biotechnology, Shantou University, Shantou, China

International Joint Research Center for the Development and Utilization of Important Mariculture Varieties Surrounding the South China Sea Region, Shantou University, Shantou, China

M. M. Nizamani
Institute of Marine Sciences, Guangdong Provincial Key Laboratory of Marine Disaster Prediction and Prevention, Shantou University, Shantou, China

R. K. Tiwari et al. (eds.), *Melatonin: Signal Transduction Mechanisms and Defense Networks in Plants*, https://doi.org/10.1007/978-981-96-8869-2_16

phytoprotective agent against ozone-induced damage. Melatonin functions both as a direct scavenger of reactive oxygen species (ROS) and as a regulator of antioxidant defense mechanisms. It enhances enzymatic antioxidants like superoxide dismutase (SOD), catalase (CAT), and ascorbate peroxidase (APX), as well as non-enzymatic antioxidants such as ascorbate and glutathione, helping plants manage oxidative stress caused by ozone exposure. Additionally, melatonin plays a key role in regulating stress-responsive gene expression, promoting the activation of protective genes, and delaying senescence. By maintaining photosynthetic efficiency and suppressing ethylene biosynthesis, melatonin helps prolong plant vitality under stress conditions. These effects are particularly important in crops and horticultural species, where ozone exposure can lead to significant reductions in yield and quality. Case studies demonstrate the dose-dependent efficacy of melatonin in mitigating ozone stress across different plant species, showing its potential to reduce oxidative damage and enhance growth under ozone pollution. However, challenges remain in applying melatonin, such as the species-specific responses observed and the need for careful optimization of application methods. This chapter also explores practical applications, including foliar sprays and genetic engineering to boost melatonin production in plants. While promising, these methods require further research to ensure their effectiveness and ecological safety. In conclusion, melatonin holds great promise as a tool to enhance plant resilience against rising ozone levels, supporting agricultural productivity and food security in a changing climate.

Keywords

Antioxidants · Ethylene · Ozone stress · Photosynthesis · Plant resilience

16.1 Introduction

Ozone (O_3) pollution has emerged as a critical environmental stressor, significantly impairing plant health and agricultural productivity. Tropospheric ozone, a secondary pollutant formed through photochemical reactions involving nitrogen oxides (NO_x) and volatile organic compounds (VOCs) from industrial emissions, vehicular exhaust, and biogenic sources, poses severe threats to vegetation (Z. Feng et al. 2021; Y. Zhang et al. 2022). Global ozone concentrations have risen steadily due to anthropogenic activities and climate change, with projections indicating a 20–25% increase by 2050 under high-emission scenarios (Sicard et al. 2017; Wg 2013). Elevated ozone infiltrates plant tissues via stomata, triggering reactive oxygen species (ROS) overproduction, which disrupts cellular homeostasis, damages photosynthetic machinery, and accelerates leaf senescence (Apel and Hirt 2004; Grulke and Heath 2020). These effects translate into substantial economic losses, reducing yields of staple crops like wheat, rice, and soybeans by up to 15–20%, thereby threatening global food security (Fuhrer et al. 2016; Sharma and Dubey 2005).

In response to ozone stress, plants deploy multifaceted defense mechanisms, including stomatal regulation, enzymatic antioxidants (e.g., superoxide dismutase, catalase), and non-enzymatic antioxidants such as ascorbate and glutathione (Bellini and De Tullio 2019; A. A. Singh et al. 2023). However, these intrinsic defenses are often insufficient under chronic ozone exposure, necessitating exogenous interventions to enhance resilience. Melatonin (N-acetyl-5-methoxytryptamine), a phylogenetically conserved indoleamine, has recently emerged as a potent mitigator of abiotic stress in plants. Synthesized from tryptophan via serotonin through enzymatic steps catalyzed by tryptophan decarboxylase (TDC), serotonin N-acetyltransferase (SNAT), and N-acetylserotonin methyltransferase (ASMT), melatonin (MT) is ubiquitously distributed in plant tissues and regulates growth, development, and stress responses (Arnao and Hernández-Ruiz 2014; N. Zhang et al. 2015).

Melatonin's dual role as a growth regulator and stress protectant stems from its ability to scavenge ROS directly and modulate antioxidant defense systems. It enhances the activity of enzymes such as superoxide dismutase (SOD), catalase (CAT), and glutathione reductase (GR), while elevating non-enzymatic antioxidants like ascorbate and glutathione (Y. Chen et al. 2018; H. Li et al. 2017). Additionally, melatonin stabilizes membrane integrity, reduces lipid peroxidation, and regulates stress-responsive genes, thereby mitigating oxidative damage (Shi et al. 2015; J. Zhang et al. 2019). Recent studies highlight its potential in countering ozone-induced stress by suppressing ethylene biosynthesis and reinforcing the ascorbate–glutathione cycle, which preserves photosynthetic efficiency and delays senescence (C. Liu et al. 2021; Marchica et al. 2022).

The rationale for employing melatonin lies in its capacity to synergize with plant secondary metabolites, such as phenylpropanoids and flavonoids, which are critical for ozone detoxification (Korkina 2007; A. A. Singh et al. 2023). By enhancing the synthesis of protective compounds like rosmarinic acid and anthocyanins, melatonin bolsters plant antioxidant capacity under ozone stress (Pellegrini et al. 2018; Wu et al. 2021). Furthermore, melatonin's interaction with phytohormones, including jasmonic acid (JA) and salicylic acid (SA), modulates stomatal behavior and stress signaling, offering a multifaceted defense strategy (Hasan et al. 2021; Modesti et al. 2023).

Despite these advances, the application of melatonin in ozone-polluted environments remains underexplored. Field trials and transcriptomic analyses are needed to unravel gene regulatory networks and optimize melatonin-mediated defense mechanisms. This chapter synthesizes current knowledge on melatonin's role in mitigating ozone damage, emphasizing its potential to safeguard crop productivity and ecological balance in an era of escalating atmospheric pollution.

16.2 Ozone Damage in Plants: Mechanisms and Consequences

Physiological Effects Ozone (O_3) exposure induces multifaceted physiological damage in plants, primarily manifesting as visible leaf injury. Acute exposure to high O_3 concentrations causes necrotic lesions and chlorophyll loss, particularly on the upper leaf surfaces and interveinal regions of sensitive species (Jixin et al. 2018; Grulke and Heath 2020). Chronic exposure, even at lower concentrations, accelerates premature leaf senescence, characterized by reduced chlorophyll a/b ratios and total chlorophyll content (Dai et al. 2017; P. Singh et al. 2021). Stomatal dysfunction is another critical response; O_3 triggers stomatal closure via calcium ion (Ca^{2+}) signaling and phytohormonal regulation (e.g., abscisic acid [ABA], ethylene [ET]), reducing O_3 influx but simultaneously impairing gas exchange (McAdam et al. 2017; Pellegrini et al. 2013). The studies depicting the ozone-induced damage in diverse plants are enlisted in Table 16.1. This stomatal limitation compromises photosynthetic efficiency by damaging photosystem II (PSII) and disrupting pigment–protein complexes, leading to inhibited carbon assimilation (H. Wang et al. 2023; Yadav et al. 2020).

16.2.1 Biochemical and Molecular Effects

O_3-induced oxidative stress arises from the overproduction of reactive oxygen species (ROS), including hydrogen peroxide (H_2O_2) and superoxide radicals (O_2^-), which overwhelm plant antioxidant defenses (Bellini and De Tullio 2019). ROS accumulation triggers lipid peroxidation, evidenced by elevated malondialdehyde (MDA) levels, and damages cellular macromolecules such as DNA and proteins (Lie et al. 2014; Ruan et al. 2017). Antioxidant enzymes like superoxide dismutase (SOD), catalase (CAT), and ascorbate peroxidase (APX) initially increase in activity but decline under prolonged O_3 stress (Yizong and Lihua 2013). Additionally, O_3 activates programmed cell death (PCD) pathways through ethylene and salicylic acid (SA) signaling, while jasmonic acid (JA) and melatonin (MT) mitigate these effects by suppressing ROS and ethylene biosynthesis (Marchica et al. 2022; Modesti et al. 2023).

16.2.2 Long-Term Consequences

Chronic O_3 exposure results in growth retardation and yield loss, particularly in agricultural and horticultural species. Reproductive stages are severely affected, with reduced flower biomass, altered petal morphology, and diminished pollen viability, disrupting pollination and seed production (Duque et al. 2021; Yang et al. 2017). Secondary metabolism is also altered: phenolic compounds, flavonoids, and terpenoids accumulate as defense responses, but excessive diversion of resources to secondary pathways compromises primary growth (Marchica et al. 2021; A. A. Singh et al. 2023). For instance, lignin synthesis increases in poplar leaves under O_3 stress, while carotenoid levels decline in citrus, further impairing photosynthetic capacity (Cotrozzi et al. 2018; Iglesias et al. 2006). These metabolic shifts, while adaptive, reduce overall plant fitness and productivity in O_3-polluted environments (Z.-Z. Feng et al. 2020).

Table 16.1 Plant responses to ozone stress

Plant species/ variety	Experimental setup	O_3 exposure conditions	Altered secondary metabolites/ enzymes/gene expression	Plant tissue studied	Citation
Salvia officinalis	Greenhouse with fumigation chambers	120 ppb O_3 for 36 days; control with filtered air (<5 ppb)	Increased ethylene, total salicylic acid, and abscisic acid; jasmonic acid unchanged	Leaves	Marchica et al. (2022)
Glycine max "William 82"	Open-top chambers	Elevated O_3 at 100 ppb; control with <20 ppb	Enhanced lignin and phenolic compounds; reduced PAL activity; higher PPO activity	Leaves	Zhang et al. (2021)
Sida cordifolia L.	Open-top chambers	Ambient O_3 and ambient + 20 ppb	Boosted caffeic and ferulic acids; reduced PAL and PPO activity; elevated RAS activity	Leaves and roots	Marchica et al. (2021)
Tartary buckwheat "Hokkai TS"	Fumigation chamber	1000 ppb O_3 for different durations	Anthocyanins and quercetin increased; caffeic acid and rutin decreased; upregulation of PAL, CAH, ACL, CHS genes	Leaves	Marchica et al. (2020)
Cultivated & wild Brassicaceae	Controlled environment chamber	80 and 120 ppb O_3; compared to ambient	Changes in flower-associated chemical profiles	Inflorescence	Saunier and Blande (2019)
Liriodendron tulipifera	Fumigation chamber	120 ppb O_3 for 45 days; control with filtered air	PAL enzyme activity and gene expression of PAL and CHS upregulated; CAD expression decreased	Leaves	Pellegrini et al. (2018)

(continued)

Table 16.1 (continued)

Plant species/ variety	Experimental setup	O_3 exposure conditions	Altered secondary metabolites/ enzymes/gene expression	Plant tissue studied	Citation
Three wheat varieties	Open-top chambers	Ambient and ambient + 30 ppb O_3	Increases in CAD, 4CL, PAL activities; higher levels of anthocyanin, jasmonic acid, salicylic acid, total phenolics, and flavonoids	Leaves	Fatima et al. (2018)
Fraxinus excelsior (European ash)	Controlled growth chamber	150 ppb O_3 for 35 days; filtered air as control	Chlorophyll a + b, total carotenoids, and β-carotene declined; lignin and quercetrin increased; coumarins and rutin reduced	Leaves	Cotrozzi et al. (2018)
Brassicaceae, Solanaceae, Fabaceae, Ranunculaceae	Glasshouse	400 ppb O_3 for 3 min	Elevated abscisic acid levels	Leaves	McAdam et al. (2017)
Fagopyrum esculentum Moench	Closed glass chamber	391 μg m^{-3} O_3 vs ambient air	Flavone content varied by cultivar; rutin and carotenoids decreased; anthocyanins and chlorophylls unaffected	Cotyledons	Dębski et al. (2017)
Melissa officinalis L.	Perspex mini chambers	200 ppb O_3 for 3 h; filtered air control	Increases in phenols, anthocyanins, tannins, lignins, and RA; higher CAD, SKDH, PAL activities	Shoots (1 month old)	Tonelli et al. (2015)

(continued)

Table 16.1 (continued)

Plant species/ variety	Experimental setup	O_3 exposure conditions	Altered secondary metabolites/ enzymes/gene expression	Plant tissue studied	Citation
Salvia officinalis	Fumigation chamber	Avg. 120 ± 13 ppb O_3 for 90 days; filtered air	Enhanced gallic, catechinic, rosmarinic, and caffeic acids; decreased chlorophyll a/b, lutein, β-carotene	Leaves	Pellegrini et al. (2015)
Melissa officinalis L.	Fumigation chamber	80 ppb O_3 for 5 h; O_3-free filtered air	PAL and RAS activities increased; gene expression of PAL, 4CL, TAT, and RAS upregulated	Leaves	Döring et al. (2014)

16.3 Melatonin in Plant Defense: An Overview

Melatonin (N-acetyl-5-methoxytryptamine) is synthesized in plants through multiple enzymatic pathways originating from the amino acid tryptophan. The primary route involves the conversion of tryptophan to tryptamine by tryptophan decarboxylase (TDC) in the cytoplasm. Tryptamine is subsequently hydroxylated by tryptamine-5-hydroxylase (T5H) in the endoplasmic reticulum to produce serotonin. Serotonin then undergoes N-acetylation via serotonin-N-acetyltransferase (SNAT), forming N-acetylserotonin, which is finally methylated by acetylserotonin methyltransferase (ASMT) or caffeic acid O-methyltransferase (COMT) to yield melatonin (Back et al. 2016; Sun et al. 2021). Alternative pathways exist, including the direct methylation of serotonin to 5-methoxytryptamine followed by N-acetylation in chloroplasts (Pathway II) or hydroxylation of tryptophan to 5-hydroxytryptophan prior to decarboxylation (Pathway III/IV) (Arnao and Hernández-Ruiz 2018). Melatonin biosynthesis occurs in mitochondria, chloroplasts, and cytoplasm, with tissue-specific variations. Roots and leaves are primary production sites, but melatonin is translocated to meristems, flowers, and fruits, where it accumulates in higher concentrations in young reproductive tissues and declines during senescence (Van Tassel et al. 2001; Y. Wang et al. 2017). For instance, 2-day-old Pharbitis nil seedlings exhibit elevated melatonin levels compared to mature tissues. Environmental stressors such as drought, salinity, and cold dynamically regulate the expression of biosynthetic genes (e.g., TDC, SNAT, ASMT), enhancing melatonin production as part of stress adaptation (Ahn et al. 2021). Melatonin influences plant developmental processes, including seed germination, root elongation, and senescence delay, partly through auxin-like activity (Arnao and Hernández-Ruiz 2018). It shares a biosynthetic precursor (tryptophan) with indole-3-acetic acid (IAA), enabling synergistic regulation of growth. For example, low melatonin concentrations (0.1 μM) stimulate lateral root formation in Brassica juncea, while higher doses (100 μM) inhibit growth (Q. Chen et al. 2009). Melatonin functions as a potent scavenger of reactive oxygen species (ROS) and reactive nitrogen species (RNS), surpassing the efficacy of glutathione and vitamin (Sun et al. 2021). It directly neutralizes free radicals and enhances the activity of antioxidant enzymes, including superoxide dismutase (SOD), catalase (CAT), peroxidase (POX), and glutathione reductase (GSSG-R) (C. Li et al. 2015). By maintaining redox homeostasis, melatonin mitigates oxidative damage to lipids, proteins, and nucleic acids under abiotic stress (Shi et al. 2015). Additionally, it stabilizes photosynthetic machinery by protecting chloroplast ultrastructure and reducing electron leakage in mitochondria (Reiter et al. 2007; N. Zhang et al. 2013). Melatonin interacts with stress-responsive signaling cascades, including mitogen-activated protein kinase (MAPK) pathways and calcium (Ca^{2+}) signaling. For instance, under drought stress, melatonin upregulates Asmap1 and Aspk11 (MAPK genes) and transcription factors involved in antioxidant defense (Shi et al. 2015). It also modulates Ca^{2+} flux to regulate stomatal closure and osmotic balance (Zhou et al. 2016). Cross-talk with phytohormones such as abscisic acid (ABA), jasmonic acid (JA), and gibberellins (GA) further amplifies stress tolerance. Melatonin suppresses ABA biosynthesis genes (e.g., NCED3) while activating ABA catabolic genes (e.g., CYP707A1/2), thereby alleviating drought-induced stomatal closure (C. Li et al. 2015). Similarly, it enhances methyl jasmonate (MeJA) accumulation under cold stress, reinforcing defense signaling (H. Li et al. 2021). In summary, melatonin serves as a multifunctional molecule in plants, integrating biosynthesis plasticity, ROS scavenging, and strategic modulation of stress-signaling networks to enhance resilience against abiotic challenges.

16.4 Mechanisms of Melatonin-Induced Defense Against Ozone Stress

Melatonin mitigates ozone-induced oxidative stress by enhancing both enzymatic and non-enzymatic antioxidant systems. Under ozone stress, reactive oxygen species (ROS) such as superoxide anions (O_2) and hydrogen peroxide (H_2O_2) accumulate, damaging cellular components. Melatonin directly scavenges ROS through its electron-rich aromatic structure, which donates electrons to neutralize free radicals (Debnath et al. 2019; N. Zhang et al. 2015) Additionally, exogenous melatonin upregulates key ROS-scavenging enzymes, including superoxide dismutase (SOD), catalase (CAT), ascorbate peroxidase (APX), and glutathione peroxidase (GPX)

(Y. Chen et al. 2018; Turk et al. 2014). For instance, in maize exposed to salinity, melatonin increased SOD and CAT activity, reducing oxidative damage (Y. Chen et al. 2018). Similarly, APX and GPX activities were elevated in tomato and wheat under combined stress conditions, maintaining redox homeostasis (Martinez et al. 2018; Turk et al. 2014). Melatonin also enhances the ascorbate–glutathione (AsA–GSH) cycle, boosting levels of reduced glutathione (GSH) and ascorbate (AsA) while lowering oxidized forms (GSSG, DHA), thereby preserving cellular integrity (H. Li et al. 2017; Nawaz et al. 2018).

Melatonin modulates stress-responsive gene expression to counteract ozone-induced damage. While specific genes like PR1 or PAL are not explicitly discussed in the review, melatonin is shown to regulate transcription factors such as NAC, WRKY1, and MYB, which are critical for stress adaptation (Gao et al. 2019). For example, in naked oat under salinity, melatonin upregulated NAC and DREB1, enhancing stress tolerance (Gao et al. 2019). Furthermore, melatonin suppresses senescence-associated genes (e.g., SAG12, ORE1) by delaying chlorophyll degradation and lipid peroxidation, as observed in apple leaves under drought (P. Wang et al. 2013). These regulatory effects stabilize cellular functions under oxidative stress.

Melatonin interacts with hormonal signaling pathways, including salicylic acid (SA) and ethylene, to mitigate ozone stress. It synergizes with SA to enhance pathogen defense and abiotic stress responses (Arnao and Hernández-Ruiz 2018). Notably, melatonin suppresses ethylene biosynthesis, a hormone linked to senescence and stress sensitivity. By inhibiting ethylene production, melatonin delays leaf senescence and maintains photosynthetic efficiency, as demonstrated in drought-stressed apple plants (P. Wang et al. 2013). This cross-talk ensures a balanced hormonal response, favoring stress resilience over senescence. Melatonin enhances enzymatic detoxification systems, including glutathione S-transferases (GSTs), which conjugate toxic metabolites to GSH for safe compartmentalization. In tomato under combined salinity and heat stress, melatonin increased GST activity, reducing oxidative damage (Martinez et al. 2018). Additionally, melatonin preserves membrane integrity by lowering lipid peroxidation and malondialdehyde (MDA) levels, as seen in bermudagrass exposed to heavy metals (Xie et al. 2018). This dual action boosting detoxification enzymes and protecting membranes limits ozone-induced cellular damage. Ozone stress disrupts chloroplast ultrastructure and inhibits Rubisco activity (Fig. 16.1). Melatonin safeguards photosynthetic machinery by maintaining chlorophyll content and reducing ROS accumulation in chloroplasts (Huang et al. 2019). In maize under drought, melatonin-treated seedlings exhibited higher photosynthetic efficiency due to preserved chloroplast integrity and reduced H O levels (Huang et al. 2019). Similarly, in cucumber, melatonin mitigated cold-induced photoinhibition by enhancing antioxidant capacity in chloroplasts (H. Zhang et al. 2014). These mechanisms collectively sustain carbon fixation and energy production under stress.

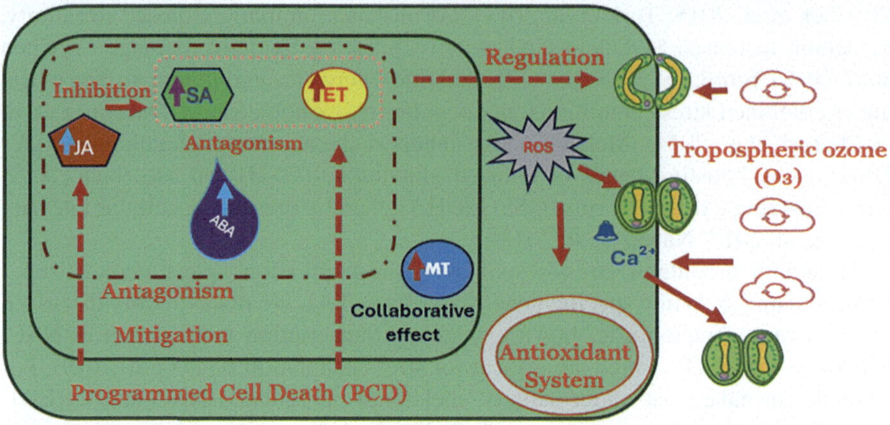

Fig. 16.1 Conceptual diagram illustrating ozone (O_3)-induced regulation of phytohormones and associated plant defense mechanisms. SA, salicylic acid; ET, ethylene; JA, jasmonic acid; MT, melatonin; ABA, abscisic acid; ROS, reactive oxygen species. The initial defense strategy involves calcium signaling and hormone-mediated stomatal regulation in response to O_3 exposure. Elevated ROS levels induce oxidative stress, activating antioxidant defenses composed of enzymatic and non-enzymatic components. ROS further modulate hormone signaling pathways, leading to additional layers of defense activation within the plant. (Adapted from Aono et al. (2005), Aslam et al. (2021), Gilroy and Breen (2022), Hasan et al. (2021), Kangasjärvi et al. (2005), C. Liu et al. (2021), McAdam et al. (2017), Modesti et al. (2023), and Vainonen and Kangasjärvi (2015))

16.5 Case Studies: Melatonin Application in Ozone-Stressed Plants

Melatonin has demonstrated significant protective effects in crop plants exposed to ozone (O_3) stress. In Phaseolus vulgaris (common bean), foliar application of melatonin at 1 mM (Mel_L) mitigated O_3-induced damage by preserving photosynthetic performance, reducing visible foliar injury, and enhancing antioxidant enzyme activities such as glutathione transferase. Additionally, melatonin increased osmoprotectants like proline and abscisic acid, while reducing malondialdehyde (MDA) levels, indicative of reduced membrane damage (Vougeleka et al. 2024). Similarly, in Nicotiana tabacum (tobacco), higher melatonin concentrations (3 mM, Mel_H) were required to alleviate O_3 stress. Mel_H maintained photosynthetic efficiency, minimized membrane damage, and reduced oxidative stress markers, though ethylenediurea (EDU) showed superior efficacy (Vougeleka et al. 2024). These findings highlight melatonin's dose-dependent and species-specific protective roles in crops.

16.5.1 Model Plants

While *Arabidopsis thaliana* studies were absent in the provided data, insights from transgenic tobacco (Nicotiana tabacum) offer mechanistic clarity. Overexpression of the melatonin synthesis gene VvASMT1 (from grapevine) in tobacco reduced endogenous ethylene biosynthesis under O_3 stress, enhancing antioxidant capacity and alleviating oxidative damage (C. Liu et al. 2021). Conversely, overexpression of the ethylene biosynthesis gene VvACO2 in grape leaves exacerbated O_3 sensitivity, underscoring melatonin's role in suppressing ethylene pathways to confer stress tolerance (C. Liu et al. 2021). These results suggest conserved melatonin–ethylene crosstalk mechanisms across species.

16.5.2 Horticultural Species

In horticultural plants, melatonin improved resilience to O_3 through multifaceted mechanisms. In grapevine (Vitis vinifera), melatonin pretreatment inhibited ethylene biosynthesis genes (e.g., ACO2), preserved photosynthetic function, and enhanced antioxidant enzyme activities, thereby reducing leaf injury (C. Liu et al. 2021). Similarly, in Malus crabapple "Hong Jiu," exogenous melatonin (0.5–12.5 μM) and methyl jasmonate (MJ) pretreatment reduced MDA accumulation, boosted antioxidant enzymes (SOD, POD, CAT), and elevated non-enzymatic antioxidants (e.g., ascorbic acid), improving the Air Pollution Tolerance Index (APTI) (Qiu et al. 2019). These studies emphasize melatonin's role in delaying senescence and maintaining fruit quality under oxidative stress.

16.6 Experimental Methods

Application methods and dosage optimization critically influence melatonin's efficacy. Foliar spraying was predominantly used, as seen in bean and tobacco studies, where direct leaf application ensured rapid uptake (Vougeleka et al. 2024). Dosage varied by species: 1 mM sufficed for bean, while tobacco required 3 mM. Timing also mattered; pretreatment before O_3 exposure was essential for maximizing protection, as demonstrated in Malus crabapple, where melatonin applied 24 h prior to stress significantly enhanced resilience (Qiu et al. 2019). Root application was not explored in the provided studies, suggesting foliar methods are preferred for targeted delivery.

16.7 Practical Applications and Challenges

Melatonin has been applied in fruit production through diverse methods, including foliar sprays, seed priming, and postharvest immersion, with concentrations tailored to species-specific requirements (S. Liu et al. 2020; Medina-Santamarina et al. 2021). For instance, foliar application of 5 ppm melatonin reduced fruit drop in pomegranate by enhancing flower retention (Usanmaz et al. 2022), while immersion in 500 μM melatonin delayed ripening in mango and banana (Cao et al. 2018; S. Liu et al. 2020). Precision agriculture technologies, such as drone-assisted spraying, could optimize melatonin delivery by enabling targeted dosing and minimizing waste, particularly in large-scale orchards. Such technologies may also facilitate real-time monitoring of stress conditions, allowing for timely melatonin applications to mitigate abiotic stressors like drought or salinity (Tiwari et al. 2021). Genetic engineering offers promising strategies to enhance endogenous melatonin biosynthesis in fruit crops. Overexpression of key enzymes in the melatonin pathway, such as serotonin-N acetyltransferase (SNAT) and N-acetylserotonin methyltransferase (ASMT), has been shown to elevate melatonin levels, thereby improving stress resilience (Mir et al. 2020; S. K. Sharma et al. 2023). For example, transgenic apple rootstocks overexpressing MdASMT1 exhibited increased drought tolerance by modulating abscisic acid (ABA) metabolism and maintaining water balance (C. Li et al. 2015). Similarly, CRISPR/Cas9-mediated editing of TDC (tryptophan decarboxylase) in kiwifruit enhanced seedling thermotolerance through upregulated heat shock protein (HSP) expression (Xia et al. 2021). These biotechnological interventions not only reduce reliance on exogenous melatonin applications but also provide a sustainable pathway to develop stress-resilient cultivars. Despite its potential, several challenges hinder the widespread adoption of melatonin in horticulture. Cost-effectiveness remains a concern, as optimal concentrations vary widely across species (e.g., 0.1 μM for grapevines vs. 500 μM for mango), complicating standardized protocols (Gökbayrak et al. 2020; S. Liu et al. 2020). Scalability is limited by the need for precise environmental monitoring and controlled application systems, particularly in resource-constrained regions. Species-specific variability in melatonin uptake and metabolism further complicates universal guidelines; for instance, melatonin delays ripening in climacteric fruits like banana but accelerates it in non-climacteric grapes (Verde et al. 2019; Xu et al. 2018). Additionally, ecological impacts of exogenous melatonin runoff into soil and water systems are poorly understood, with potential risks to non-target organisms and microbial communities (Suresh et al. 2023). Addressing these challenges requires interdisciplinary research to optimize application protocols, develop cost-effective formulations, and assess long-term ecological consequences.

16.8 Conclusion

Ozone pollution continues to be a formidable challenge for global agriculture, significantly impairing photosynthesis, accelerating plant senescence, and ultimately reducing crop yields. This chapter highlights the crucial role of melatonin as a dual-function phytoprotectant in mitigating ozone-induced damage. By acting as a potent antioxidant, melatonin directly scavenges reactive oxygen species (ROS), which are overproduced during ozone exposure, while also enhancing the plant's inherent antioxidant defense systems. Through the upregulation of key enzymatic (SOD, CAT, APX) and non-enzymatic (ascorbate, glutathione) antioxidants, melatonin helps plants maintain cellular integrity under oxidative stress. Additionally, melatonin stabilizes the photosynthetic machinery, thus preserving the plant's ability to produce energy, and suppresses ethylene, a hormone that accelerates senescence and exacerbates stress. Case studies across various crops, such as *Phaseolus vulgaris* (common bean), *Nicotiana tabacum* (tobacco), *Vitis vinifera* (grapevine), and *Malus crabapple* (crabapple), underline melatonin's dose-dependent efficacy. These studies show that melatonin can mitigate the adverse effects of ozone pollution in a range of plant species, although the optimal dosage and application methods can vary. This variability underscores the challenge of achieving species-specific optimization, a key hurdle in effectively implementing melatonin-based treatments across diverse agricultural settings. The practical application of melatonin, whether through foliar sprays, genetic engineering, or precision agriculture techniques, holds significant promise for bolstering plant resilience in regions affected by high ozone concentrations. However, further research is needed to address critical gaps in field validation, ensuring both the ecological safety and cost scalability of these approaches. As ozone pollution intensifies, the integration of melatonin into agricultural practices could provide an essential tool for safeguarding food security. To fully realize melatonin's potential, future research should focus on interdisciplinary collaborations to better understand its molecular mechanisms, refine application protocols, and develop strategies to incorporate it into climate-resilient farming systems. As atmospheric ozone levels continue to rise due to climate change, leveraging melatonin's protective properties will be crucial in sustaining crop productivity, preserving ecological balance, and ensuring a food-secure future in the face of escalating environmental challenges.

References

Ahn H-R, Kim Y-J, Lim Y-J, Duan S, Eom S-H, Jung K-H (2021) Key genes in the melatonin biosynthesis pathway with circadian rhythm are associated with various abiotic stresses. Plants 10(1):129

Aono M, Kanna M, Ogawa D, Murata Y, Rakwal R, Agrawal GK, Tamogami S, Iwahashi H, Kubo A, Tamaoki M (2005) Regulation of ascorbate contents by jasmonate-mediated signaling pathway in Arabidopsis during ozone exposure. Phyton 45:201–207

Apel K, Hirt H (2004) Reactive oxygen species: metabolism, oxidative stress, and signal transduction. Annu Rev Plant Biol 55(1):373–399

Arnao MB, Hernández-Ruiz J (2014) Melatonin: plant growth regulator and/or biostimulator during stress? Trends Plant Sci 19(12):789–797

Arnao MB, Hernández-Ruiz J (2018) Melatonin and its relationship to plant hormones. Ann Bot 121(2):195–207

Aslam S, Gul N, Mir MA, Asgher M, Al-Sulami N, Abulfaraj AA, Qari S (2021) Role of jasmonates, calcium, and glutathione in plants to combat abiotic stresses through precise signaling cascade. Front Plant Sci 12:668029

Back K, Tan D, Reiter RJ (2016) Melatonin biosynthesis in plants: multiple pathways catalyze tryptophan to melatonin in the cytoplasm or chloroplasts. J Pineal Res 61(4):426–437

Bellini E, De Tullio MC (2019) Ascorbic acid and ozone: novel perspectives to explain an elusive relationship. Plants 8(5):122

Cao S, Shao J, Shi L, Xu L, Shen Z, Chen W, Yang Z (2018) Melatonin increases chilling tolerance in postharvest peach fruit by alleviating oxidative damage. Sci Rep 8(1):806

Chen Q, Qi W, Reiter RJ, Wei W, Wang B (2009) Exogenously applied melatonin stimulates root growth and raises endogenous indoleacetic acid in roots of etiolated seedlings of *Brassica juncea*. J Plant Physiol 166(3):324–328

Chen Y, Mao J, Sun L, Huang B, Ding C, Gu Y, Liao J, Hu C, Zhang Z, Yuan S (2018) Exogenous melatonin enhances salt stress tolerance in maize seedlings by improving antioxidant and photosynthetic capacity. Physiol Plant 164(3):349–363

Cotrozzi L, Campanella A, Pellegrini E, Lorenzini G, Nali C, Paoletti E (2018) Phenylpropanoids are key players in the antioxidant defense to ozone of European ash, *Fraxinus excelsior*. Environ Sci Pollut Res 25:8137–8147

Dai L, Li P, Shang B, Liu S, Yang A, Wang Y, Feng Z (2017) Differential responses of peach (*Prunus persica*) seedlings to elevated ozone are related with leaf mass per area, antioxidant enzymes activity rather than stomatal conductance. Environ Pollut 227:380–388

Debnath B, Islam W, Li M, Sun Y, Lu X, Mitra S, Hussain M, Liu S, Qiu D (2019) Melatonin mediates enhancement of stress tolerance in plants. Int J Mol Sci 20(5):1040

Dębski H, Wiczkowski W, Szawara-Nowak D, Bączek N, Piechota M, Horbowicz M (2017) The effect of tropospheric ozone on flavonoids and pigments content in common buckwheat cotyledons. Ecol Chem Eng 24(3):457

Döring AS, Pellegrini E, Della Batola M, Nali C, Lorenzini G, Petersen M (2014) How do background ozone concentrations affect the biosynthesis of rosmarinic acid in *Melissa officinalis*? J Plant Physiol 171(5):35–41

Duque L, Poelman EH, Steffan-Dewenter I (2021) Effects of ozone stress on flowering phenology, plant-pollinator interactions and plant reproductive success. Environ Pollut 272:115953

Fatima A, Singh AA, Mukherjee A, Agrawal M, Agrawal SB (2018) Variability in defence mechanism operating in three wheat cultivars having different levels of sensitivity against elevated ozone. Environ Exp Bot 155:66–78

Feng Z-Z, Yuan X-Y, Li P, Shang B, Ping Q, Hu T-J, Liu S (2020) Progress in the effects of elevated ground-level ozone on terrestrial ecosystems. Chin J Plant Ecol 44(5):526

Feng Z, Agathokleous E, Yue X, Oksanen E, Paoletti E, Sase H, Gandin A, Koike T, Calatayud V, Yuan X (2021) Emerging challenges of ozone impacts on Asian plants: actions are needed to protect ecosystem health. Ecosyst Health Sustain 7(1):1911602

Fuhrer J, Val Martin M, Mills G, Heald CL, Harmens H, Hayes F, Sharps K, Bender J, Ashmore MR (2016) Current and future ozone risks to global terrestrial biodiversity and ecosystem processes. Ecol Evol 6(24):8785–8799

Gao W, Feng Z, Bai Q, He J, Wang Y (2019) Melatonin-mediated regulation of growth and antioxidant capacity in salt-tolerant naked oat under salt stress. Int J Mol Sci 20(5):1176

Gilroy E, Breen S (2022) Interplay between phytohormone signalling pathways in plant defence–other than salicylic acid and jasmonic acid. Essays Biochem 66(5):657–671

Gökbayrak Z, Engin H, Kiraz H (2020) Effects of melatonin and IAA on adventitious root formation in rootstock 5BB and cv. cabernet sauvignon (*Vitis vinifera* L.). Kahramanmaraş Sütçü İmam Üniv Tarım Doğa Derg 23(4):835–841

Grulke NE, Heath RL (2020) Ozone effects on plants in natural ecosystems. Plant Biol 22:12–37

Hasan MM, Rahman MA, Skalicky M, Alabdallah NM, Waseem M, Jahan MS, Ahammed GJ, El-Mogy MM, El-Yazied AA, Ibrahim MFM (2021) Ozone induced stomatal regulations, MAPK and phytohormone signaling in plants. Int J Mol Sci 22(12):6304

Huang B, Chen Y-E, Zhao Y-Q, Ding C-B, Liao J-Q, Hu C, Zhou L-J, Zhang Z-W, Yuan S, Yuan M (2019) Exogenous melatonin alleviates oxidative damages and protects photosystem II in maize seedlings under drought stress. Front Plant Sci 10:677

Iglesias DJ, Calatayud Á, Barreno E, Primo-Millo E, Talon M (2006) Responses of citrus plants to ozone: leaf biochemistry, antioxidant mechanisms and lipid peroxidation. Plant Physiol Biochem 44(2–3):125–131

Jixin C, Siyuan Y, Zhan C, He S (2018) A review on plant indicating of atmospheric ozone pollution. World For Res 31:21–27

Kangasjärvi J, Jaspers P, Kollist H (2005) Signalling and cell death in ozone-exposed plants. Plant Cell Environ 28(8):1021–1036

Korkina LG (2007) Phenylpropanoids as naturally occurring antioxidants: from plant defense to human health. Cell Mol Biol 53(1):15–25

Li C, Tan D-X, Liang D, Chang C, Jia D, Ma F (2015) Melatonin mediates the regulation of ABA metabolism, free-radical scavenging, and stomatal behaviour in two Malus species under drought stress. J Exp Bot 66(3):669–680

Li H, Chang J, Chen H, Wang Z, Gu X, Wei C, Zhang Y, Ma J, Yang J, Zhang X (2017) Exogenous melatonin confers salt stress tolerance to watermelon by improving photosynthesis and redox homeostasis. Front Plant Sci 8:295

Li H, Guo Y, Lan Z, Xu K, Chang J, Ahammed GJ, Ma J, Wei C, Zhang X (2021) Methyl jasmonate mediates melatonin-induced cold tolerance of grafted watermelon plants. Hortic Res 8:57

Lie GW, Ye LH, Xue L (2014) Effects of ozone stress on major plant physiological functions. Acta Ecol Sin 34(2):294–306

Liu S, Huang H, Huber DJ, Pan Y, Shi X, Zhang Z (2020) Delay of ripening and softening in "Guifei" mango fruit by postharvest application of melatonin. Postharvest Biol Technol 163:111136

Liu C, Kang H, Wang Y, Yao Y, Gao Z, Du Y (2021) Melatonin relieves ozone stress in grape leaves by inhibiting ethylene biosynthesis. Front Plant Sci 12:702874

Marchica A, Cotrozzi L, Detti R, Lorenzini G, Pellegrini E, Petersen M, Nali C (2020) The biosynthesis of phenolic compounds is an integrated defence mechanism to prevent ozone injury in *Salvia officinalis*. Antioxidants 9(12):1274

Marchica A, Ascrizzi R, Flamini G, Cotrozzi L, Tonelli M, Lorenzini G, Nali C, Pellegrini E (2021) Ozone as eustress for enhancing secondary metabolites and bioactive properties in *Salvia officinalis*. Ind Crop Prod 170:113730

Marchica A, Cotrozzi L, Lorenzini G, Nali C, Pellegrini E (2022) Antioxidants and phytohormones act in coordination to regulate sage response to long term ozone exposure. Plants 11(7):904

Martinez V, Nieves-Cordones M, Lopez-Delacalle M, Rodenas R, Mestre TC, Garcia-Sanchez F, Rubio F, Nortes PA, Mittler R, Rivero RM (2018) Tolerance to stress combination in tomato plants: new insights in the protective role of melatonin. Molecules 23(3):535

McAdam EL, Brodribb TJ, McAdam SAM (2017) Does ozone increase ABA levels by non-enzymatic synthesis causing stomata to close? Plant Cell Environ 40(5):741–747

Medina-Santamarina J, Serrano M, Lorente-Mento JM, García-Pastor ME, Zapata PJ, Valero D, Guillén F (2021) Melatonin treatment of pomegranate trees increases crop yield and quality parameters at harvest and during storage. Agronomy 11(5):861

Mir AR, Faizan M, Bajguz A, Sami F, Siddiqui H, Hayat S (2020) Occurrence and biosynthesis of melatonin and its exogenous effect on plants. Acta Soc Bot Pol 89(2):8922

Modesti M, Marchica A, Pisuttu C, Risoli S, Pellegrini E, Bellincontro A, Mencarelli F, Tonutti P, Nali C (2023) Ozone-induced biochemical and molecular changes in *Vitis vinifera* leaves and responses to *Botrytis cinerea* infections. Antioxidants 12(2):343

Nawaz MA, Jiao Y, Chen C, Shireen F, Zheng Z, Imtiaz M, Bie Z, Huang Y (2018) Melatonin pretreatment improves vanadium stress tolerance of watermelon seedlings by reducing vanadium concentration in the leaves and regulating melatonin biosynthesis and antioxidant-related gene expression. J Plant Physiol 220:115–127

Pellegrini E, Trivellini A, Campanella A, Francini A, Lorenzini G, Nali C, Vernieri P (2013) Signaling molecules and cell death in *Melissa officinalis* plants exposed to ozone. Plant Cell Rep 32:1965–1980

Pellegrini E, Francini A, Lorenzini G, Nali C (2015) Ecophysiological and antioxidant traits of *Salvia officinalis* under ozone stress. Environ Sci Pollut Res 22:13083–13093

Pellegrini E, Campanella A, Cotrozzi L, Tonelli M, Nali C, Lorenzini G (2018) What about the detoxification mechanisms underlying ozone sensitivity in *Liriodendron tulipifera*? Environ Sci Pollut Res 25:8148–8160

Qiu Y, An K, Sun J, Chen X, Gong X, Ma L, Wu S, Jiang S, Zhang Z, Wang Y (2019) Investigating the effect of methyl jasmonate and melatonin on resistance of Malus crabapple "Hong Jiu" to ozone stress. Environ Sci Pollut Res 26:27761–27768

Reiter RJ, Tan D, Manchester LC, Simopoulos AP, Maldonado MD, Flores LJ, Terron MP (2007) Melatonin in edible plants (phytomelatonin): identification, concentrations, bioavailability and proposed functions. World Rev Nutr Diet 97:211–230

Ruan YN, Xu S, Guo L, Zhu MZ, Wang C, Li SY, Wang HY (2017) Effects of elevated ozone concentrations on reactive oxygen metabolism and related gene expression in *Ginkgo biloba* leaves. Ying Yong Sheng Tai Xue Bao = J Appl Ecol 28(11):3479–3486

Saunier A, Blande JD (2019) The effect of elevated ozone on floral chemistry of Brassicaceae species. Environ Pollut 255:113257

Sharma P, Dubey RS (2005) Drought induces oxidative stress and enhances the activities of antioxidant enzymes in growing rice seedlings. Plant Growth Regul 46:209–221

Sharma SK, Sharma I, Bhushan B (2023) A review of melatonin metabolism with reference to plants. Int J Adv Res Biol Sci 10(9):1–10

Shi H, Jiang C, Ye T, Tan D-X, Reiter RJ, Zhang H, Liu R, Chan Z (2015) Comparative physiological, metabolomic, and transcriptomic analyses reveal mechanisms of improved abiotic stress resistance in bermudagrass [*Cynodon dactylon* (L). Pers.] by exogenous melatonin. J Exp Bot 66(3):681–694

Sicard P, Anav A, De Marco A, Paoletti E (2017) Projected global ground-level ozone impacts on vegetation under different emission and climate scenarios. Atmos Chem Phys 17(19):12177–12196

Singh P, Kannaujia R, Narayan S, Tewari A, Shirke PA, Pandey V (2021) Impact of chronic elevated ozone exposure on photosynthetic traits and anti-oxidative defense responses of *Leucaena leucocephala* (Lam.) de wit tree under field conditions. Sci Total Environ 782:146907

Singh AA, Ghosh A, Agrawal M, Agrawal SB (2023) Secondary metabolites responses of plants exposed to ozone: an update. Environ Sci Pollut Res 30(38):88281–88312

Sun C, Liu L, Wang L, Li B, Jin C, Lin X (2021) Melatonin: a master regulator of plant development and stress responses. J Integr Plant Biol 63(1):126–145

Suresh TR, Pradhan J, Kumar S, Singh H, Prasad K, Jaiswal A, Kumari G (2023) Melatonin discovery and divergent biosynthetic pathways in plants. In: Melatonin in plants: a regulator for plant growth and development. Springer, Singapore, pp 1–18

Tiwari RK, Lal MK, Kumar R, Chourasia KN, Naga KC, Kumar D, Das SK, Zinta G (2021) Mechanistic insights on melatonin-mediated drought stress mitigation in plants. Physiol Plant 172(2):1212–1226

Tonelli M, Pellegrini E, D'Angiolillo F, Petersen M, Nali C, Pistelli L, Lorenzini G (2015) Ozone-elicited secondary metabolites in shoot cultures of *Melissa officinalis* L. Plant Cell Tissue Organ Cult 120:617–629

Turk H, Erdal S, Genisel M, Atici O, Demir Y, Yanmis D (2014) The regulatory effect of melatonin on physiological, biochemical and molecular parameters in cold-stressed wheat seedlings. Plant Growth Regul 74:139–152

Usanmaz S, Aşkın MA, Kahramanoğlu İ (2022) Effects of foliar melatonin application on the flowering and fruit set of pomegranate (*Punica granatum* L. Wonderful). Acta Hortic 1349:63–72

Vainonen JP, Kangasjärvi J (2015) Plant signalling in acute ozone exposure. Plant Cell Environ 38(2):240–252

Van Tassel DL, Roberts N, Lewy A, O'Neill SD (2001) Melatonin in plant organs. J Pineal Res 31(1):8–15

Verde A, Míguez JM, Gallardo M (2019) Melatonin and related bioactive compounds in commercialized date palm fruits (*Phoenix dactylifera* L.): correlation with some antioxidant parameters. Eur Food Res Technol 245:51–59

Vougeleka V, Risoli S, Saitanis C, Agathokleous E, Ntatsi G, Lorenzini G, Nali C, Pellegrini E, Pisuttu C (2024) Exogenous application of melatonin protects bean and tobacco plants against ozone damage by improving antioxidant enzyme activities, enhancing photosynthetic performance, and preventing membrane damage. Environ Pollut 343:123180

Wang P, Sun X, Li C, Wei Z, Liang D, Ma F (2013) Long-term exogenous application of melatonin delays drought-induced leaf senescence in apple. J Pineal Res 54(3):292–302

Wang Y, Wang Q, Liu M, Bo C, Wang X, Ma Q, Cheng B, Cai R (2017) Overexpression of a maize MYB48 gene confers drought tolerance in transgenic arabidopsis plants. J Plant Biol 60:612–621

Wang H, Li M, Yang Y, Sun P, Zhou S, Kang Y, Xu Y, Yuan X, Feng Z, Jin W (2023) Physiological and molecular responses of different rose (*Rosa hybrida* L.) cultivars to elevated ozone levels. Plant Direct 7(7):e513

Wg I (2013) The physical science basis. In: Contribution of Working Group I to the Fifth assessment report of the Intergovernmental Panel on Climate Change. Cambridge University Press, New York, p 1535

Wu S, Wang Y, Zhang J, Gong X, Zhang Z, Sun J, Chen X, Wang Y (2021) Exogenous melatonin improves physiological characteristics and promotes growth of strawberry seedlings under cadmium stress. Hortic Plant J 7(1):13–22

Xia H, Zhou Y, Deng H, Lin L, Deng Q, Wang J, Lv X, Zhang X, Liang D (2021) Melatonin improves heat tolerance in *Actinidia deliciosa* via carotenoid biosynthesis and heat shock proteins expression. Physiol Plant 172(3):1582–1593

Xie C, Xiong X, Huang Z, Sun L, Ma J, Cai S, Yu F, Zhong W, Chen S, Li X (2018) Exogenous melatonin improves lead tolerance of bermudagrass through modulation of the antioxidant defense system. Int J Phytoremediation 20(14):1408–1417

Xu L, Yue Q, Xiang G, Bian F, Yao Y (2018) Melatonin promotes ripening of grape berry via increasing the levels of ABA, H2O2, and particularly ethylene. Hortic Res 5:41

Yadav DS, Mishra AK, Rai R, Chaudhary N, Mukherjee A, Agrawal SB, Agrawal M (2020) Responses of an old and a modern Indian wheat cultivar to future O3 level: physiological, yield and grain quality parameters. Environ Pollut 259:113939

Yang N, Wang X, Zheng F, Chen Y (2017) The impact of elevated ozone on the ornamental features of two flowering plants (*Tagetes erecta* Linn. and Petunia hybrida Vilm.). Int J Environ Pollut 61(1):29–45

Yizong H, Lihua S (2013) Antioxidant mechanism of plants under ozone stress. Asian J Ecotoxicol 4:456–464

Zhang N, Zhao B, Zhang H, Weeda S, Yang C, Yang Z, Ren S, Guo Y (2013) Melatonin promotes water-stress tolerance, lateral root formation, and seed germination in cucumber (*Cucumis sativus* L.). J Pineal Res 54(1):15–23

Zhang H, Zhang NA, Yang R, Wang L, Sun Q, Li D, Cao Y, Weeda S, Zhao B, Ren S (2014) Melatonin promotes seed germination under high salinity by regulating antioxidant systems, ABA and GA 4 interaction in cucumber (*Cucumis sativus* L.). J Pineal Res 57(3):269–279

Zhang N, Sun Q, Zhang H, Cao Y, Weeda S, Ren S, Guo Y-D (2015) Roles of melatonin in abiotic stress resistance in plants. J Exp Bot 66(3):647–656

Zhang J, Gao F, Jia H, Hu J, Feng Z (2019) Molecular response of poplar to single and combined ozone and drought. Sci Total Environ 655:1364–1375

Zhang X, Zhang X, Wang T, Li C (2021) Metabolic response of soybean leaves induced by short-term exposure of ozone. Ecotoxicol Environ Saf 213:112033

Zhang Y, Han Z, Li X, Zhang H, Yuan X, Feng Z, Wang P, Mu Z, Song W, Blake DR (2022) Plants and related carbon cycling under elevated ground-level ozone: a mini review. Appl Geochem 144:105400

Zhou X, Zhao H, Cao K, Hu L, Du T, Baluška F, Zou Z (2016) Beneficial roles of melatonin on redox regulation of photosynthetic electron transport and synthesis of D1 protein in tomato seedlings under salt stress. Front Plant Sci 7:1823

Metabolomics-Mediated Profiling of Melatonin-Induced Metabolic Diversity and Function in Plant

17

Iram Batool, Ahsan Ayyaz, Fakhir Hannan, Muhammad Ahsan Farooq, and Weijun Zhou

Abstract

Sugar metabolism in plants is a complex network of interconnected pathways that regulate the production, conversion, storage, and transport of carbohydrates, which serve as the primary source of energy for plant growth and development. The key processes in sugar metabolism include photosynthesis, the process by which plants convert light energy into chemical energy, producing glucose and other sugars. Glycolysis is the breakdown of glucose into pyruvate, producing ATP (adenosine triphosphate) as an energy source for cellular processes. The conversion of glucose into starch for storage in plant tissues and the subsequent breakdown of starch into sugar when energy is required. Sucrose metabolism conversion of glucose into sucrose, the primary transport sugar in plants, which is used to distribute energy to various tissues. Each of these pathways plays a vital role in plant energy management. However, under abiotic stress conditions, these processes can become compromised, leading to reduced energy availability and growth. Therefore, understanding the regulation of sugar metabolism under stress is critical for improving crop performance and resilience. One of the key functions of melatonin in plant metabolism is its regulation of photosynthesis, which directly affects sugar production. Photosynthesis is essential for converting light energy into chemical energy stored in the form of sugars, and any impairment in this process can lead to reduced sugar accumulation and energy deficit in plants. Starch is synthesized in chloroplasts during periods of high

I. Batool · A. Ayyaz · F. Hannan · W. Zhou (✉)
Institute of Crop Science, Ministry of Agriculture and Rural Affairs Key Laboratory of Spectroscopy Sensing, Zhejiang University, Hangzhou, China
e-mail: wjzhou@zju.edu.cn

M. A. Farooq
Zhejiang Provincial Key Laboratory for Water Environment and Marine Biological Resources Protection, College of Life and Environmental Science, Wenzhou University, Wenzhou, China

R. K. Tiwari et al. (eds.), *Melatonin: Signal Transduction Mechanisms and Defense Networks in Plants*, https://doi.org/10.1007/978-981-96-8869-2_17

photosynthetic activity and stored in vacuoles or amyloplasts for later use. Melatonin serves as a key regulator of starch metabolism, dynamically modulating both starch biosynthesis and degradation to maintain energy homeostasis in plants under changing climatic conditions.

Keywords
Genomics · Photosynthesis · Ionic balance · Redox networks · Antioxidant

17.1 Introduction

Melatonin (N-acetyl-5-methoxytryptamine) is a small indoleamine molecule that was originally identified in animals, where it plays a central role in regulating circadian rhythms and sleep patterns (Zhao et al. 2015). However, in recent years, its presence and biological significance in plants have been discovered, offering insights into its complex functions across various physiological and metabolic processes. Melatonin has been identified as a potent antioxidant, growth regulator, and stress modulator in plants. It plays a significant role in mitigating oxidative damage induced by environmental stressors, such as drought, salinity, extreme temperatures, and heavy metal toxicity (Korkmaz et al. 2014). Beyond its role in stress mitigation, melatonin has been implicated in regulating sugar metabolism in plants, which is crucial for energy homeostasis and growth under normal and stressed conditions. Sugar metabolism, which includes processes such as photosynthesis, glycolysis, starch synthesis, and sucrose metabolism is essential for providing energy to plants. This is particularly critical during adverse environmental conditions, when energy production and storage must be tightly regulated (Thalmann and Santelia 2017). This study explores the current understanding of the role of melatonin in sugar metabolism in plants, with particular emphasis on its effects under abiotic stress conditions. We delve into the mechanisms by which melatonin influences sugar production, transport, storage, and breakdown, and how these processes help plants cope with stresses such as water deficit, high salinity, and temperature fluctuations. Additionally, we discuss how melatonin interacts with other plant hormones to optimize metabolic processes under stress and its potential application in improving crop yield and stress resilience.

17.1.1 Sugar Metabolism in Plants

Sugar metabolism in plants is a complex network of interconnected pathways that regulate the production, conversion, storage, and transport of carbohydrates, which serve as the primary source of energy for plant growth and development (Rolland et al. 2002). The key processes in sugar metabolism include photosynthesis by which plants convert light energy into chemical energy, producing glucose and other sugars. Glycolysis is the metabolic pathway that breaks down glucose into pyruvate, generating ATP (adenosine triphosphate) as an energy source for cellular processes (Hossain and Uddin 2011). Starch synthesis and breakdown involve the conversion of glucose into starch for storage in plant tissues, followed by the mobilization of starch back into sugars when energy is needed. Sucrose metabolism, on the other hand, entails the conversion of glucose into sucrose the primary transport sugar in plants which facilitates energy distribution to various tissues (Gupta and Kaur 2005). Each of these pathways plays a vital role in plant energy management. However, under abiotic stress conditions, these processes can become compromised, leading to reduced energy availability and growth. Therefore, understanding the regulation of sugar metabolism under stress is critical for improving crop performance and resilience.

17.1.2 Role of Melatonin in Photosynthesis

One of the key functions of melatonin in plant metabolism is its regulation of photosynthesis, which directly affects sugar production. Photosynthesis is essential for converting light energy into chemical energy stored in the form of sugars, and any impairment in this process can lead to reduced sugar accumulation and energy deficit in plants. Several studies have highlighted that melatonin has a positive influence on photosynthetic efficiency in both normal and stressed conditions (Borek and Ratajczak 2010). Melatonin has been shown to increase chlorophyll content in plant leaves, which enhances the plants' ability to capture light energy for photosynthesis. By improving the light-harvesting capacity of leaves, melatonin helps increase the rate of photosynthesis even under suboptimal environmental conditions (McMullan et al. 2013). In plants subjected to water stress or high salinity, melatonin treatment has been found to maintain higher chlorophyll levels, thus mitigating the reduction in photosynthetic efficiency. Melatonin also enhances the activity of photosystem II (PSII), a crucial component of the photosynthetic apparatus. Experimental evidence from *Arabidopsis* and other models shows that melatonin mitigates stress-induced photosynthetic inhibition by optimizing PSII electron transport chain efficiency (Kostopoulou et al. 2015). By improving the functioning of the photosynthetic machinery, melatonin contributes to enhanced energy production, which is vital for maintaining plant growth and stress tolerance. Melatonin impact extends beyond chlorophyll content and photosystem activity; it also affects carbon fixation in the Calvin cycle. Ribulose-1,5-bisphosphate carboxylase-oxygenase (RuBisCO), the enzyme responsible for carbon fixation, is upregulated in the presence of melatonin, leading to increased carbon assimilation (Wang et al. 2013). This enhanced carbon fixation is critical for maintaining sugar production under conditions of water deficit and other stresses that typically reduce photosynthetic activity.

17.1.3 Role of Melatonin in Starch Metabolism

Starch is a vital energy reserve in plants, and its metabolism is crucial for maintaining sugar homeostasis, especially during periods of stress (Fig. 17.1). Starch synthesis occur in chloroplasts during periods of high photosynthetic activity and stored in vacuoles (Koch 1996). Melatonin plays a pivotal role in regulating starch synthesis and breakdown, thereby ensuring that plants have access to a reliable energy source when needed. Melatonin influences starch biosynthesis by regulating the expression of genes encoding key enzymes involved in the process, such as starch synthase and starch branching enzymes. These enzymes catalyze the formation of starch from glucose. Under normal conditions, melatonin promotes starch accumulation in leaves and storage organs. During periods of stress, this process becomes even more crucial as plants must rely on stored carbohydrates for survival. In response to energy demands, plants mobilize stored starch by converting it back into sugar. Melatonin has been shown to enhance the activity of enzymes such as starch phosphorylase and amylopectin debranching enzymes, which break down starch into simple sugars like glucose and fructose. This breakdown of starch ensures that plants can access energy even during periods of low photosynthesis and nutrient stress impair their ability to produce sugars (Zhao et al. 2015).

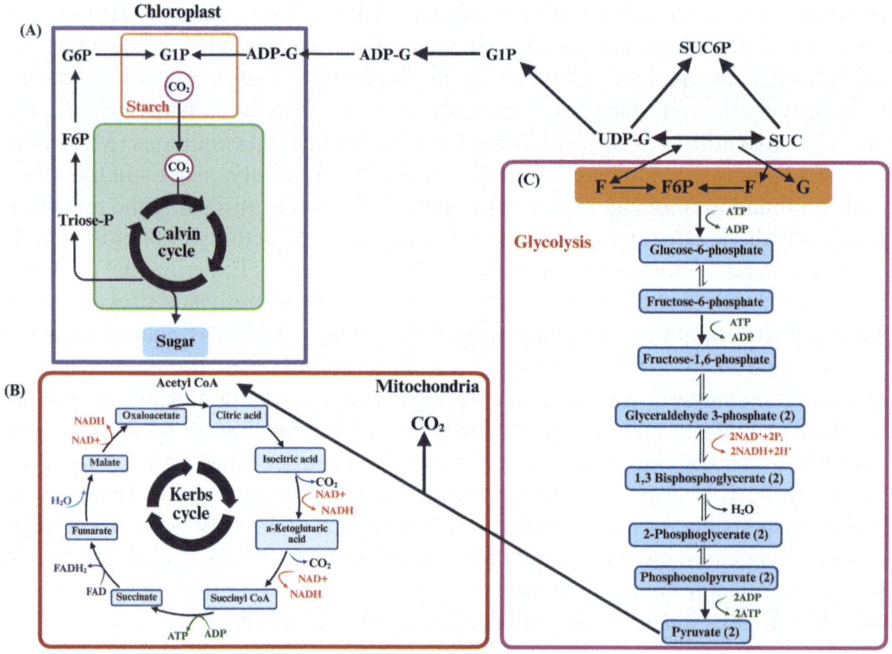

Fig. 17.1 Breakdown of glucose into pyruvate, generating ATP and NADH, which are essential for cellular functions. Light reactions of photosynthesis, improving the generation of ATP and NADPH, which are required for the Calvin cycle and conversion of acetyl-CoA into energy-rich molecules like ATP, NADH, and $FADH_2$ (**A, B and C**), respectively

17.1.4 Melatonin and Sucrose Metabolism

Sucrose is the primary sugar transported in plants, and its metabolism plays a vital role in distributing energy throughout plant tissues (McMullan et al. 2013). Melatonin regulates sucrose synthesis and hydrolysis, ensuring the proper balance of sucrose levels during stress. Melatonin regulates enzymes involved in both sucrose synthesis and its breakdown into glucose and fructose (Lazár et al. 2013). Sucrose synthase (SuSy) is one such enzyme that is modulated by melatonin. Studies have shown that melatonin promotes sucrose hydrolysis, especially under drought or salt stress, leading to increased glucose and fructose availability for energy production and osmotic regulation (Szafrańska et al. 2016). Melatonin also influences invertase, an enzyme that catalyzes the conversion of sucrose into glucose and fructose. During stress conditions, this activity is upregulated by melatonin, providing an immediate source of energy for metabolic processes. By enhancing invertase activity, melatonin ensures that plants maintain energy production during stress, supporting their survival and growth. Glycolysis is the process by which glucose is broken down into pyruvate, generating ATP and NADH, which are essential for cellular functions (Fig. 17.1). Melatonin-mediated regulation of glycolytic enzymes plays a key role in maintaining energy production during abiotic stress. Enzymes such as hexokinase, phosphofructokinase, and pyruvate kinase are all influenced by melatonin, helping plants generate energy even under water deficit, salinity, or temperature extremes. In particular, melatonin has been found to enhance glycolysis under salt stress, promoting the breakdown of stored sugars and facilitating energy production. This is essential for maintaining cellular metabolism, especially when external sources of energy (such as photosynthesis) are compromised.

17.2 Physiological and Metabolomic Analyses Reveal Mechanisms of Improved Abiotic Stress Resistance in Crop Plants in Response to Exogenous Melatonin

Abiotic stresses, including drought, salinity, extreme temperatures, and heavy metals, pose significant challenges to crop productivity worldwide (Kolář and Macháčkova 2005). These stresses disrupt plant growth, metabolism, and development, leading to significant yield losses (Fig. 17.2). As global climate change exacerbates the frequency and intensity of these stresses, improving crop resilience to abiotic stress has become a priority for agricultural research (Arnao and Hernández-Ruiz 2006). To address this challenge, plant scientists are exploring various strategies, one of the most promising being the use of exogenous melatonin. Melatonin, a small indoleamine, is best known for its regulatory roles in circadian rhythms and antioxidative processes in animals. In plants, melatonin has emerged as a key regulator of physiological processes, particularly under abiotic stress conditions. Exogenous melatonin has been shown to enhance plant growth, maintain photosynthetic efficiency, regulate water balance, improve ion homeostasis, and alleviate oxidative damage under various stresses. While the physiological and biochemical

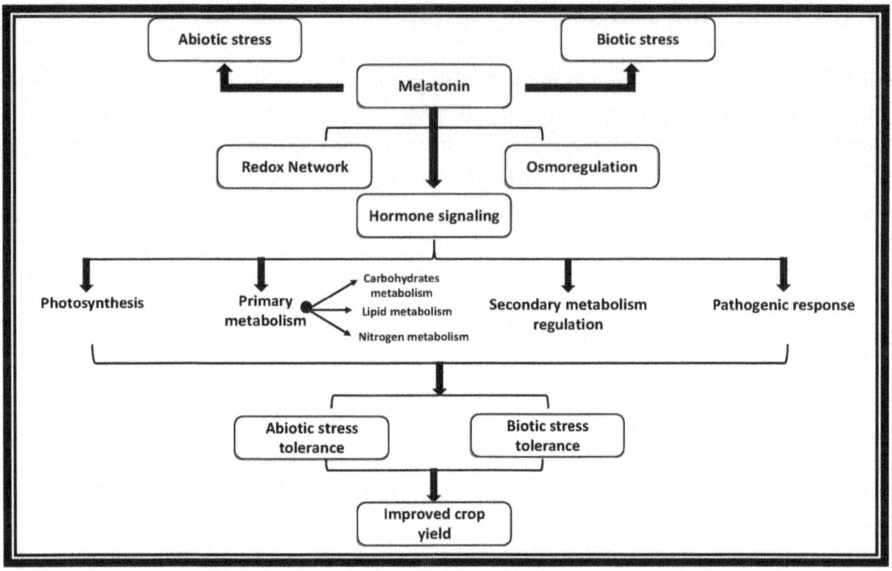

Fig. 17.2 Role of melatonin in regulation of plant primary and secondary metabolism under abiotic and biotic stressful conditions in improving plant growth and productivity

effects of melatonin in plants are well-documented, recent studies have increasingly focused on the molecular mechanisms that underlie melatonin-induced stress resistance, particularly through metabolomic, transcriptomic, and proteomic analyses.

17.2.1 Metabolomic Changes Induced by Exogenous Melatonin

Metabolomic analysis provides a comprehensive overview of the small molecules involved in metabolic processes that are affected by external treatments. Exogenous melatonin alters a range of metabolites in plants, particularly those involved in primary metabolism (such as sugars, amino acids, and organic acids) and secondary metabolism (such as phenolics and flavonoids). These metabolic changes play a crucial role in melatonin-mediated stress resistance. Carbohydrates, particularly sucrose, glucose, and starch, are vital for energy production, osmotic regulation, and stress tolerance in plants. Several studies have shown that exogenous melatonin treatment enhances the accumulation of soluble sugars and starch in plants under abiotic stress (Krasensky and Jonak 2012). Melatonin regulates sucrose synthase and invertase activity, leading to increased sucrose hydrolysis, which provides energy during stress (Tan et al. 2012). In addition, melatonin influences the starch–sucrose interconversion process by promoting starch biosynthesis in leaves and its breakdown in roots, thus optimizing carbohydrate distribution across plant tissues. Amino acids, such as proline and glutamine, play a key role in osmotic adjustment and protecting cellular structures during stress. Exogenous melatonin treatment has been shown to increase proline accumulation, which acts as an osmoprotectant and helps stabilize proteins and membranes under drought and salt stress. Additionally, melatonin regulates the nitrogen metabolism pathway, enhancing the synthesis of key amino acids and contributing to better nitrogen use efficiency under stress. Secondary metabolites such as phenolic compounds, flavonoids, and alkaloids are important for plant defense and stress tolerance (Fig. 17.2). Melatonin has been shown to enhance the accumulation of these compounds in response to abiotic stresses. Flavonoids, for instance, have antioxidative properties that help protect plants from ROS-induced damage. Melatonin treatment induces the expression of genes involved in phenylpropanoid metabolism, increasing the synthesis of phenolic acids and flavonoids, which contribute to enhanced oxidative stress resistance plants (Posmyk et al. 2008; Zhao et al. 2011).

17.2.2 Exogenous Melatonin Counteracts NaCl-Induced Damage by Regulating Carbohydrate Metabolism in Crop Plants

The increasing salinization of soils due to natural and anthropogenic activities is a significant global concern, especially for agricultural systems. High levels of sodium chloride (NaCl) in soil result in osmotic stress, ion toxicity, and oxidative damage, severely affecting the growth and productivity of crop plants (Munns and Gilliham 2015). Among various abiotic stressors, salt stress represents a major environmental challenge for crop production worldwide, impacting plant physiological and biochemical functions, particularly in carbohydrate metabolism (Basyuni et al. 2019). The disruption of photosynthesis, starch synthesis, sucrose metabolism, and cellular respiration under salt stress limits plant growth and yield potential (Butcher et al. 2016). Recent studies have highlighted the potential of melatonin, a small indoleamine molecule, as a phytohormone involved in plant growth regulation, particularly under abiotic stress conditions. Melatonin has been shown to play a crucial role in regulating antioxidant defense, ion homeostasis, and osmotic regulation in plants subjected to various stresses, including salinity (Zhang et al. 2014; Siddiqui et al. 2020). Among the diverse mechanisms by which melatonin enhances plant tolerance to NaCl-induced stress, its ability to regulate carbohydrate metabolism has garnered increasing attention. Previously, melatonin has been reported to enhance resistance against salinity stress in various crops, including rice (Byeon and Back 2014), cucumber (Zhang et al. 2014), apple (Wang et al. 2013), and soybean (Wei et al. 2015). Under salt stress, melatonin helps maintain cellular integrity and osmotic balance, thereby mitigating the detrimental effects of NaCl (Li et al. 2012; Kamiab 2020). Its actions are largely mediated through its antioxidant properties, regulation of ion homeostasis, and modulation of stress-related genes (Bahcesular et al. 2020). Recent studies have also underscored its role in enhancing carbohydrate metabolism, which plays a central role in plant energy management under stress. Salt stress causes an imbalance in plant metabolism, with significant effects on photosynthesis, starch, and sucrose metabolism. High concentrations of Na^+ in plant tissues disrupt the chloroplast structure, leading to reduced photosynthetic efficiency and starch synthesis (Tan et al. 2012). The disruption of these metabolic processes directly affects carbohydrate allocation, which is essential for plant growth and development under stress.

The process of photosynthesis involves the conversion of light energy into chemical energy, with the production of glucose and other sugars. Salt stress, however, impairs this process by reducing chlorophyll content, decreasing CO_2 assimilation, and ultimately lowering glucose production (Yang and Lu 2005). Additionally, high salinity induces osmotic stress, which leads to stomatal closure, limiting water availability for photosynthesis and thus reducing carbohydrate production.

Starch metabolism is also significantly impacted by salt stress. Starch is synthesized in chloroplasts from glucose produced during photosynthesis and stored in vacuoles. Under stressful conditions, plants often shift from energy production to energy storage, and salt stress impairs both starch biosynthesis and its subsequent mobilization (Munns and Gilliham 2015). This results in reduced starch storage, further exacerbating the energy deficit in plants. Sucrose is the primary transport sugar in plants, acting as a carrier of energy and carbon. The synthesis of sucrose occurs in the cytosol from glucose and fructose, while its breakdown is catalyzed by enzymes like invertase and sucrose synthase. Under salt stress, sucrose synthesis is inhibited due to reduced glucose availability and the disruption of enzyme activities (Ye et al. 2019). Furthermore, excess Na^+ accumulation leads to osmotic imbalance, hindering sucrose transport and its distribution across plant tissues. Sucrose hydrolysis into glucose and fructose helps plants generate energy under stress conditions. Melatonin has been shown to influence this process by regulating the expression of genes involved in sucrose synthesis and hydrolysis, enhancing energy production and osmotic regulation during salt stress.

Exogenous melatonin has been shown to mitigate the negative effects of NaCl stress on photosynthesis. By increasing chlorophyll content and photosystem II activity, melatonin enhances light absorption and electron transport, thereby improving CO_2 assimilation (Zhang et al. 2016). These effects help counteract the reductions in carbohydrate production that typically occur under salt stress. Moreover, melatonin promotes the activation of RuBisCO and phosphoenolpyruvate carboxylase (PEPC), two key enzymes involved in carbon fixation, which helps to maintain glucose production even in saline conditions (Zhang et al. 2020). Melatonin also modulates the light reactions of photosynthesis, improving the generation of ATP and NADPH, which are required for the Calvin cycle. These processes ultimately enhance the overall production of sugars, supporting plant growth and metabolic function under NaCl-induced stress. Melatonin plays a significant role in regulating starch metabolism under salt stress. By influencing the activity of starch synthase, starch phosphorylase, and amylase, melatonin enhances the synthesis and mobilization of starch. Under normal conditions, starch is stored in vacuoles in the form of energy reserve. However, during stress conditions, plants rely more on starch breakdown to meet their energy needs. Melatonin modulates this process by activating enzymes that convert starch into glucose and fructose, ensuring a steady energy supply (Zheng et al. 2017).

In addition to promoting starch synthesis, melatonin facilitates starch degradation, which helps release glucose for energy production when photosynthesis is impaired. The increased availability of glucose during salt stress aids in maintaining cellular metabolism and osmotic regulation. Melatonin influences the expression of genes involved in sucrose biosynthesis and hydrolysis, ensuring that sucrose is synthesized and mobilized efficiently during salt stress. Studies have shown that exogenous melatonin treatment increases the activity of invertase and sucrose synthase, enzymes responsible for sucrose breakdown into glucose and fructose (Wei et al. 2015). By enhancing these enzymatic activities, melatonin helps maintain glucose and fructose levels, which are essential for energy production and osmotic balance under salt stress. Furthermore, melatonin regulates sucrose transporters to ensure that sucrose is efficiently distributed across plant tissues. This regulation helps improve the plants ability to allocate sugars to sink tissues such as roots and fruits, ensuring growth and development even under stressful conditions (Yang et al. 2020).

17.2.3 Melatonin Regulates Photosynthesis and Metabolic Pathways of Starch Starvation in Drought Stress

Drought stress is one of the most severe abiotic stress factors affecting plant growth and productivity. It disrupts cellular water balance, impairs photosynthesis, and leads to metabolic changes that ultimately reduce crop yield (França et al. 2000; Wu et al. 2008). Under drought conditions, plants must adapt to limited water availability, often by altering physiological and biochemical processes. One critical metabolic adjustment in drought stress involves starch metabolism, as starch serves as an energy reserve that can be mobilized to sustain cellular activities during periods of water deficit. Melatonin, a bioactive molecule known for its antioxidant properties, has emerged as an essential regulator of plant responses to various abiotic stresses, including drought (Kanwar et al. 2018). Recent studies suggest that melatonin can influence key processes such as photosynthesis, starch metabolism, and gene expression, enhancing plant survival during drought (Wang et al. 2009). Therefore, this study explores the role of melatonin in regulating functional components of photosynthesis, gene expression, and starch metabolism, specifically under conditions of drought-induced starch starvation.

17.3 Melatonin and Its Role in Plant Stress Responses

Melatonin is a pleiotropic molecule; in plants, melatonin is synthesized from tryptophan through the action of a series of enzymes and has been identified in various tissues such as leaves, stems, flowers, and roots. While its antioxidant properties are widely acknowledged, melatonin also plays a key role in regulating a variety of biological processes including growth, development, and responses to environmental stressors such as drought, salinity, and temperature extremes (Liang et al. 2019). Melatonin mechanisms of action include direct antioxidant activity, modulation of gene expression, regulation of ion homeostasis, and involvement in various signaling pathways. It has been shown to alleviate the damaging effects of oxidative stress, a common result of drought conditions, by scavenging reactive oxygen species (ROS) and enhancing the plants' antioxidant defense system. Recent studies have suggested that melatonin plays a crucial role in improving drought resistance through its effects on photosynthesis, starch metabolism, and gene expression related to stress responses.

17.3.1 Drought Stress and the Impact on Photosynthesis and Starch Metabolism

Photosynthesis is a critical process for plants, as it provides the energy required for growth and development. Drought stress significantly impairs photosynthesis by reducing stomatal conductance, leading to decreased carbon dioxide (CO_2) uptake, and by generating ROS that damage the photosynthetic apparatus. Stomatal closure, while conserving water, limits CO_2 diffusion into the leaf, causing a reduction in the rate of photosynthesis. In addition, drought-induced oxidative stress can lead to the damage of the photosystem II (PSII), which is the primary site for light-dependent reactions in photosynthesis (Zhang et al. 2013). Starch serves as the primary carbohydrate storage in plants, acting as an energy reservoir that can be mobilized during periods of insufficient photosynthetic activity, such as during drought stress. Under normal conditions, starch synthesis and degradation are tightly regulated by enzymes like ADP-glucose pyrophosphorylase (AGPase), starch synthase, and amylase. During drought stress, when photosynthetic activity is compromised, the plant relies on stored starch to meet its energy needs. The mobilization of starch into soluble sugars is crucial for maintaining cellular functions during water deficit (Liang et al. 2015). However, prolonged drought can result in starch depletion, leading to a condition known as "starch starvation." This condition exacerbates the detrimental effects of drought, as the plant becomes unable to generate sufficient energy and metabolites. Thus, the regulation of starch metabolism, especially during drought-induced starch starvation, becomes a crucial aspect of drought resistance. Melatonin has been shown to influence both starch biosynthesis and mobilization, aiding plants in coping with energy deficits during drought stress.

17.3.2 Melatonin Regulation of Photosynthesis in Drought Stress

Melatonin has been found to improve the efficiency of photosynthesis under drought conditions by enhancing the stability of photosynthetic components (Fig. 17.3). A key component affected by drought is Photosystem II (PSII), which is particularly vulnerable to oxidative damage. Melatonin antioxidant activity helps protect PSII from oxidative damage by scavenging ROS. Research has shown that melatonin treatment results in higher chlorophyll content, reduced ROS production, and better performance of PSII under drought stress (Bota et al. 2004). This leads to improved

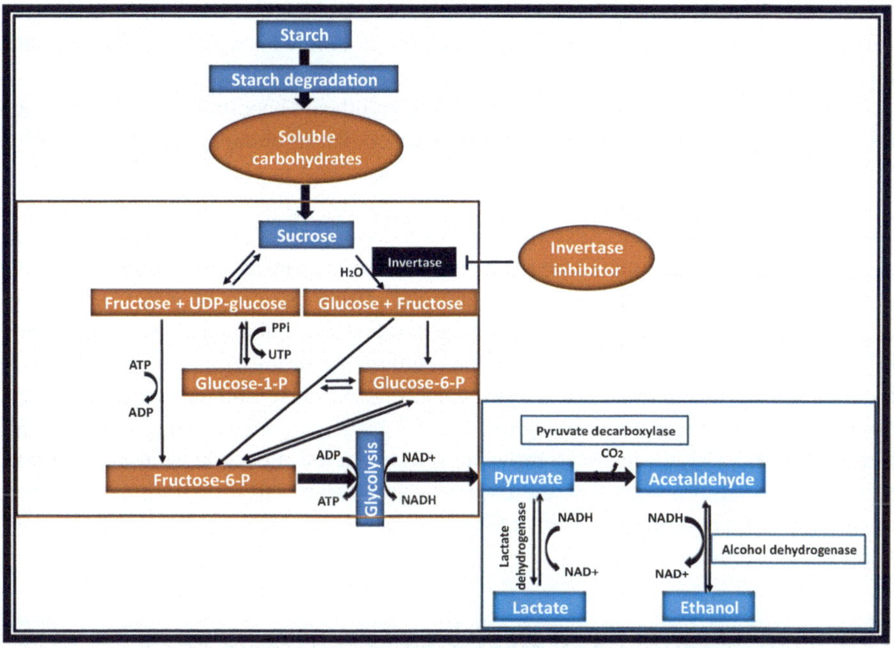

Fig. 17.3 Melatonin regulate the mobilization of stored starch by modulating the activity of enzymes such as amylase and isoamylase, which break down starch into soluble sugars

photosynthetic capacity, even in water-limited conditions. In addition to its protective effects on the photosystem, melatonin has been shown to influence the electron transport chain involved in photosynthesis. By regulating the activity of key enzymes in the light reaction, melatonin ensures that plants maintain adequate ATP production and NADPH availability for carbon fixation, despite the reduced water availability (Foyer and Noctor 2000). Stomatal closure under drought stress is a common adaptive response aimed at minimizing water loss. However, excessive stomatal closure limits CO_2 uptake, thus impairing photosynthesis. Melatonin has been shown to regulate stomatal movement by modulating the expression of genes involved in guard cell signaling. Studies have indicated that melatonin treatment can reduce excessive stomatal closure, thereby maintaining a balance between water conservation and CO_2 uptake, improving photosynthesis under drought conditions (Fu and Huang 2001). This modulation of stomatal behavior allows plants to optimize their photosynthetic efficiency while conserving water. Melatonin has also been reported to promote the synthesis of chlorophyll, the main pigment involved in capturing light for photosynthesis. Chlorophyll degradation is often accelerated under drought stress, leading to a decrease in the efficiency of light absorption. By enhancing chlorophyll biosynthesis and preventing premature chlorophyll degradation, melatonin ensures that plants maintain high photosynthetic efficiency, even under stress conditions. Gene expression modulation is central to how plants respond to drought stress, as it enables them to adapt to changing environmental conditions. Melatonin has been shown to regulate the expression of a variety of genes involved in stress tolerance, antioxidant defense, water regulation, and metabolism.

During drought, the regulation of starch biosynthesis becomes critical for maintaining an energy reserve. Melatonin has been shown to enhance starch accumulation by regulating the expression of enzymes involved in starch biosynthesis. For example, melatonin promotes the activity of ADP-glucose pyrophosphorylase (AGPase), the key enzyme in starch synthesis, by upregulating its gene expression. As a result, melatonin-treated plants accumulate more starch, providing them with a larger energy reserve that can be mobilized during periods of water deficit. As drought progresses, plants may experience starch starvation due to insufficient photosynthesis. In response, melatonin helps regulate the mobilization of stored starch by modulating the activity of enzymes such as amylase and isoamylase, which break down starch into soluble sugars (Fig. 17.3). The mobilized sugars provide an alternative energy source for cellular activities, including respiration and protein synthesis. Melatonin enhances the expression of these starch-degrading enzymes, ensuring that plants can effectively utilize their starch reserves during drought stress (Zhao et al. 2015).

17.4 Exogenous Supplementation of Melatonin Alters Representative Organic Acids and Enzymes of Respiratory Cycle as Well as Sugar Metabolism During Heavy Metal Stress in Plants

Heavy metal toxicity is a major environmental concern for crop production worldwide, significantly affecting plant growth, development, and yield. Metals like cadmium (Cd), lead (Pb), arsenic (As), and mercury (Hg) accumulate in the soil and are taken up by plants, leading to various physiological and biochemical disturbances (Samanta et al. 2020). These disturbances include oxidative stress, inhibition of essential enzymatic activities, disruption of metabolic pathways, and alterations in respiration and sugar metabolism. Under heavy metal stress, plants undergo changes in cellular metabolism, including reduced photosynthetic efficiency, altered carbohydrate allocation, and imbalanced energy production through respiration (Abbas et al. 2018; Khan and Gupta 2018). In recent years, melatonin, a molecule known for its antioxidant properties and role in regulating growth, development, and stress responses in plants, has been recognized for its potential to alleviate the detrimental effects of heavy metal stress (Hasan et al. 2019). Melatonin, a plant hormone-like compound, is produced endogenously but can also be applied exogenously to improve plant resilience against abiotic stresses. Its protective effects in plants are often attributed to its ability to regulate reactive oxygen species (ROS) and modulate metabolic pathways, including respiration and sugar metabolism. This study aims to explore the role of melatonin supplementation in regulating organic acids, respiratory enzymes, and sugar metabolism in crop plants under heavy metal stress. By examining how melatonin influences key metabolic pathways, including the tricarboxylic acid (TCA) cycle, glycolysis, and the pentose phosphate pathway, as well as the activity of key enzymes involved in these processes, to mitigate the adverse effects of heavy metals.

Under heavy metal stress, the respiratory cycle in plants is often impaired, leading to altered energy production and metabolic dysregulation. Organic acids, which are intermediates in the respiratory cycle, play a pivotal role in cellular energy metabolism. The respiratory cycle consists of two primary phases: glycolysis and the tricarboxylic acid (TCA) cycle. Both cycles produce ATP and reduced cofactors (NADH and $FADH_2$) necessary for cellular activities (Arnao and Hernandez-Ruiz 2018). The TCA cycle (also known as the citric acid cycle or Krebs cycle) involves the conversion of acetyl-CoA into energy-rich molecules like ATP, NADH, and $FADH_2$. Key organic acids such as citric acid, malic acid, and oxaloacetic acid are intermediates in the TCA cycle. These organic acids not only serve as energy carriers but also play a role in the regulation of oxidative stress. Under heavy metal stress, the activity of TCA cycle enzymes may be impaired due to oxidative damage, leading to a reduction in energy production (Das et al. 2019). Melatonin has been shown to restore TCA cycle activity by enhancing the expression and activity of enzymes such as citrate synthase, isocitrate dehydrogenase, and malate dehydrogenase (Ahammed et al. 2020). By enhancing TCA cycle function, melatonin helps to mitigate the negative effects of heavy metal stress on energy production and cellular metabolism. Glycolysis, the breakdown of glucose into pyruvate, is another crucial metabolic pathway that generates energy in the form of ATP. Pyruvate is subsequently transported to the mitochondria, where it is converted into acetyl-CoA and enters the TCA cycle. Heavy metals can inhibit glycolysis by disrupting key enzymes such as hexokinase, phosphofructokinase, and pyruvate kinase, leading to reduced ATP production. Melatonin supplementation has been shown to enhance glycolytic activity under heavy metal stress by upregulating the expression of glycolytic enzymes. Additionally, melatonin modulates the levels of key organic acids (e.g., pyruvate and lactate), thereby sustaining energy production pathways even under heavy metal stress (Sun et al. 2020). Exogenous melatonin application can alleviate the disruption of sugar metabolism under heavy metal stress by modulating the expression and activity of key enzymes involved in both the synthesis and breakdown of sugars. Melatonin has been shown to increase the activity of enzymes like SPS, invertase, and hexokinase, which enhance the conversion of sucrose to glucose and fructose, providing readily available sugars for cellular metabolism in red cabbage under toxic copper. Melatonin may regulate the flux through the pentose phosphate pathway, increasing NADPH production, which is crucial for maintaining cellular redox balance and supporting the synthesis of stress-related molecules, such as antioxidants and osmo-protectants. Melatonin has also been shown to promote the accumulation of soluble sugars, such as glucose, fructose, and sucrose, which act as osmotic regulators, helping to maintain cell turgor under heavy metal stress (Sun et al. 2020). These sugars can also protect proteins, lipids, and other macromolecules from damage caused by ROS, further enhancing stress tolerance.

17.5 Comparative Metabolomic Analysis Highlights the Involvement of Sugars in Melatonin-Mediated Innate Immunity Against Bacterial Pathogen in Crop Plants

Crop plants face numerous biotic and abiotic stressors, with bacterial infections being one of the most significant challenges to agricultural productivity worldwide. Bacterial pathogens can cause a wide range of diseases in crops, resulting in significant yield losses. The plant innate immune system plays a critical role in defending against such pathogens through a series of complex, multifaceted responses that include physical barriers, biochemical changes, and activation of immune signaling pathways (Lee et al. 2014).

Melatonin, a bioactive molecule traditionally associated with circadian rhythm regulation, has recently emerged as a key regulator of plant defense mechanisms, including those involved in innate immunity against bacterial pathogens. Recent studies have provided insights into how melatonin mediates plant immunity, with a particular focus on the modulation of various metabolic pathways in *Arabidopsis* susceptibility to an avirulent pathogen (Lee et al. 2015). One area that has garnered attention is the role of sugars, both as energy sources and as signaling molecules, in the melatonin-mediated immune response. Sugars are critical for supporting immune-related processes such as cell wall reinforcement, production of antimicrobial compounds, and the activation of defense signaling pathways. Metabolomic analyses have begun to reveal how melatonin supplementation alters sugar metabolism and how this may influence the plants response to bacterial pathogens.

Melatonin supplementation has been shown to induce significant changes in sugar metabolism, particularly in response to bacterial pathogen attack. These changes support the plants' immune responses and help to mitigate the detrimental effects of bacterial infections. Glucose: Glucose is one of the primary energy sources for plants and plays a central role in immune responses. Melatonin-treated plants show higher glucose levels during bacterial infections, which is crucial for energy-dependent defense mechanisms such as ROS production and the synthesis of phytoalexins. Elevated glucose levels may also support the rapid synthesis of cell wall components, which act as physical barriers to pathogen invasion. Sucrose, the main transport sugar in plants, plays a key role in regulating immune responses by activating sugar-signaling pathways (Zhao et al. 2015). Sucrose not only provides energy but also acts as a signaling molecule that can activate defense genes, particularly those involved in the synthesis of PR proteins. Fructose, another sugar involved in energy metabolism, has been found to increase in plants treated with melatonin under pathogen stress. Fructose may play a role in enhancing cellular responses to pathogen attack, such as the production of secondary metabolites and the activation of stress-related signaling pathways.

17.6 Conclusions

Melatonin is a key regulator of sugar metabolism in plants, particularly under abiotic stress conditions. Its role in enhancing photosynthetic efficiency, regulating starch and sucrose metabolism, and promoting glycolysis is critical for maintaining energy balance and supporting growth under stress. Melatonin synergizes with key plant hormones (ABA, cytokinins, and auxin) to optimize sugar metabolism and allocation, enhancing plant survival under stressful conditions. Comparative physiological, metabolomic, and transcriptomic analyses have revealed that melatonin mediates a complex network of responses involving antioxidant defense, ion homeostasis, stress signaling, and photosynthesis regulation. By enhancing energy metabolism, promoting osmotic balance, and regulating stress-responsive genes, melatonin significantly improves plant tolerance to drought, salinity, and other abiotic stresses. Exogenous melatonin supplementation plays a critical role in mitigating the detrimental effects of heavy metal stress in crop plants by modulating key metabolic pathways, including respiration and sugar metabolism. Melatonin enhances the activity of enzymes involved in the TCA cycle, glycolysis, and sugar metabolism, ensuring that plants maintain energy production and metabolic homeostasis under stress. Additionally, melatonin's antioxidant properties help reduce oxidative damage, while its ability to regulate sugar metabolism provides plants with essential energy reserves during periods of heavy metal exposure. Recent studies have highlighted the important role of sugars in melatonin-mediated innate immunity against bacterial pathogens in crop plants. Comparative metabolomic analyses provide valuable insights into how melatonin regulates sugar metabolism, enhancing the plants' ability to defend against bacterial infection. Melatonin affects the levels of key sugars such as glucose, sucrose, and fructose, which provide the energy necessary for immune responses and support the synthesis of cell wall components and antimicrobial compounds. While the mechanisms by which melatonin regulates sugar metabolism and immunity are still being unraveled, the available evidence suggests that melatonin supplementation could be a promising strategy for improving plant resistance to bacterial pathogens. Future research should focus on identifying the specific enzymes and signaling pathways involved in sugar regulation under melatonin treatment and pathogen stress, as well as investigating the potential for using melatonin as a practical tool for enhancing crop immunity in agricultural system.

References

Abbas G, Murtaza B, Bibi I, Shahid M, Niazi NK, Khan MI, Amjad M, Hussain M, Natasha N (2018) Arsenic uptake, toxicity, detoxification, and speciation in plants: physiological, biochemical, and molecular aspects. Int J Environ Res Public Health 15:59

Ahammed GJ, Wu M, Wang Y, Yan Y, Mao Q, Ren J, Ma R, Liu A, Chen S (2020) Melatonin alleviates iron stress by improving iron homeostasis, antioxidant defense and secondary metabolism in cucumber. Sci Hortic 265:109205

Arnao MB, Hernández-Ruiz J (2006) The physiological function of melatonin in plants. Plant Signal Behav 1:89–95
Arnao MB, Hernandez-Ruiz J (2018) Melatonin and its relationship to plant hormones. Ann Bot 121:195–207
Bahcesular B, Yildirim ED, Karaçocuk M et al (2020) Seed priming with melatonin effects on growth, essential oil compounds and antioxidant activity of basil (*Ocimum basilicum* L.) under salinity stress. Ind Crops Prod 146:112165
Basyuni M, Wasilah M, Hasibuan PAZ et al (2019) Salinity and subsequent freshwater influences on the growth, biomass, and polyisoprenoids distribution of *Rhizophora apiculata* seedlings. Biodiversitas 20:288–295
Borek S, Ratajczak L (2010) Storage lipids as a source of carbon skeletons for asparagine synthesis in germinating seeds of yellow lupine (*Lupinus luteus* L.). J Plant Physiol 167:717–724
Bota J, Medrano H, Flexas J (2004) Is photosynthesis limited by decreased Rubisco activity and RuBP content under progressive water stress? New Phytol 162:671–681
Butcher K, Wick AF, Desutter T et al (2016) Soil salinity: a threat to global food security. Agron J 108:2189–2200
Byeon Y, Back K (2014) Melatonin synthesis in rice seedlings in vivo is enhanced at high temperatures and under dark conditions due to increased serotonin N-acetyltransferase and N-acetylserotonin methyltransferase activities. J Pineal Res 56:189–195
Das P, Manna I, Sil P, Bandyopadhyay M, Biswas AK (2019) Exogenous silico alters organic acid production and enzymatic activity of TCA cycle in two NaC stressed indica rice cultivars. Plant Physiol Biochem 136:76–91
Foyer C, Noctor G (2000) Oxygen processing in photosynthesis: regulation and signaling. New Phytol 146:359–388
França MGC, Thi ATP, Pimentel C, Rossiello ROP, Zuily-Fodil Y, Laffray D (2000) Differences in growth and water relations among *Phaseolus vulgaris* cultivars in response to induced drought stress. Environ Exp Bot 43:227–237
Fu J, Huang B (2001) Involvement of antioxidants and lipid peroxidation in the adaptation of two cool-season grasses to localized drought stress. Environ Exp Bot 45:105–114
Gupta AK, Kaur N (2005) Sugar signalling and gene expression in relation to carbohydrate metabolism under abiotic stresses in plants. J Biosci 30:761–776
Hasan MK, Ahammed GJ, Sun S, Li M, Yin H, Zhou J (2019) Melatonin inhibits cadmium translocation and enhances plant tolerance by regulating sulfur uptake and assimilation in *Solanum lycopersicum* L. J Agric Food Chem 67:10563–10576
Hossain MA, Uddin SN (2011) Mechanisms of waterlogging tolerance in wheat: morphological and metabolic adaptations under hypoxia or anoxia. Aust J Crop Sci 5:1094–1101
Kamiab F (2020) Exogenous melatonin mitigates the salinity damages and improves the growth of pistachio under salinity stress. J Plant Nutr 43:1468–1484
Kanwar MK, Yu J, Zhou J (2018) Phytomelatonin: recent advances and future prospects. J Pineal Res 65:e12526
Khan E, Gupta M (2018) Arsenic–silicon priming of rice (*Oryza sativa* L.) seeds influence mineral nutrient uptake and biochemical responses through modulation of Lsi-1, Lsi-2, Lsi-6 and nutrient transporter genes. Sci Rep 8:10301
Koch KE (1996) Carbohydrate modulated gene expression in plants. Annu Rev Plant Physiol Plant Mol Biol 47:509–540
Kolář J, Macháčkova I (2005) Melatonin in higher plants: occurrence and possible functions. J Pineal Res 39:333–341
Korkmaz A, Değer Ö, Cuci Y (2014) Profiling the melatonin content in organs of the pepper plant during different growth stages. Sci Hortic 172:242–247
Kostopoulou Z, Therios I, Roumeliotis E et al (2015) Melatonin combined with ascorbic acid provides salt adaptation in *Citrus aurantium* L. seedlings. Plant Physiol Biochem 86:155–165
Krasensky J, Jonak C (2012) Drought, salt, and temperature stress induced metabolic rearrangements and regulatory networks. J Exp Bot 63:1593–1608

Lazár D, Murch SJ, Beilby MJ, Khazaaly S (2013) Exogenous melatonin affects photosynthesis in Characeae Chara australis. Plant Signal Behav 8:E23279

Lee HY, Byeon Y, Back K (2014) Melatonin as a signal molecule triggering defense responses against pathogen attack in Arabidopsis and tobacco. J Pineal Res 57:262–268

Lee HY, Byeon Y, Tan DX, Reiter RJ, Back K (2015) Arabidopsis serotonin N-acetyltransferase knockout mutant plants exhibit decreased melatonin and salicylic acid levels resulting in susceptibility to an avirulent pathogen. J Pineal Res 58:291–299

Li C, Wang P, Wei Z et al (2012) The mitigation effects of exogenous melatonin on salinity-induced stress in *Malus hupehensis*. J Pineal Res 53:298–306

Liang C, Zheng G, Li W, Wang Y, Hu B, Wang H et al (2015) Melatonin delays leaf senescence and enhances salt stress tolerance in rice. J Pineal Res 59:91–101

Liang D, Ni Z, Xia H, Xie Y, Lv X, Wang J et al (2019) Exogenousmelatonin promotes biomass accumulation and photosynthesis of kiwifruit seedlings under drought stress. Sci Hortic 246:34–43

McMullan CJ, Schernhammer ES, Rimm EB et al (2013) Melatonin secretion and the incidence of type 2 diabetes. JAMA 309:1388–1396

Munns R, Gilliham M (2015) Salinity tolerance of crops—what is the cost? New Phytol 208:668–673

Posmyk MM, Kuran H, Marciniak K, Janas KM (2008) Presowing seed treatment with melatonin protects red cabbage seedlings against toxic copper ion concentrations. J Pineal Res 45:24–31

Rolland F, Moore B, Sheen J (2002) Sugar sensing and signaling in plants. Plant Cell 14:185–205

Samanta S, Banerjee A, Roychoudhury A (2020) Melatonin application differentially modulates the enzymes associated with antioxidative machinery and ascorbateglutathione cycle during arsenate exposure in indica rice varieties. Plant Biol. https://doi.org/10.1111/plb.13181

Siddiqui MH, Alamri S, Alsubaie QD, Ali HM (2020) Melatonin and gibberellic acid promote growth and chlorophyll biosynthesis by regulating antioxidant and methylglyoxal detoxification system in tomato seedlings under salinity. J Plant Growth Regul 39:1488

Sun C, Lv T, Huang L, Liu X, Jin C, Lin X (2020) Melatonin ameliorates aluminum toxicity through enhancing aluminum exclusion and reestablishing redox homeostasis in roots of wheat. J Pineal Res. https://doi.org/10.1111/jpi.12642e12642

Szafrańska K, Reiter RJ, Posmyk MM (2016) Melatonin application to *Pisum sativum* L. seeds positively influences the function of the photosynthetic apparatus in growing seedlings during paraquat-induced oxidative stress. Front Plant Sci 7:1663

Tan DX, Hardeland R, Manchester LC, Korkmaz A, Ma S, Rosales-Corral S, Reiter RJ (2012) Functional roles of melatonin in plants, and perspectives in nutritional and agricultural science. J Exp Bot 63:577–597

Thalmann M, Santelia D (2017) Starch as a determinant of plant fitness under abiotic stress. New Phytol 214:943–951

Wang WB, Kim YH, Lee HS, Kim KY, Deng XP, Kwak SS (2009) Analysis of antioxidant enzyme activity during germination of alfalfa under salt and drought stresses. Plant Physiol Biochem 47:570–577

Wang P, Sun X, Li C et al (2013) Long-term exogenous application of melatonin delays drought-induced leaf senescence in apple. J Pineal Res 54:292–302

Wei W, Li QT, Chu YN et al (2015) Melatonin enhances plant growth and abiotic stress tolerance in soybean plants. J Exp Bot 66:695–707

Wu F, Bao W, Li F, Wu N (2008) Effects of drought stress and N supply on the growth, biomass partitioning and water-use efficiency of Sophora davidii seedlings. Environ Exp Bot 63:248–255

Yang X, Lu C (2005) Photosynthesis is improved by exogenous glycinebetaine in salt-stressed maize plants. Physiol Plant 124:343–352

Yang H, Dai L, Wei Y et al (2020) Melatonin enhances salt stress tolerance in rubber tree (*Hevea brasiliensis*) seedlings. Ind Crop Prod. https://doi.org/10.1016/j.indcrop.2019.111990

Ye L, Zhao X, Bao E et al (2019) Effects of arbuscular mycorrhizal fungi on watermelon growth, elemental uptake, antioxidant, and photosystem II activities and stress-response gene expressions under salinity-alkalinity stresses. Front Plant Sci 10:863

Zhang N, Zhao B, Zhang HJ, Weeda S, Yang C, Yang ZC et al (2013) Melatonin promoteswater-stress tolerance, lateral root formation, and seed germination in cucumber (*Cucumis sativus* L.). J Pineal Res 54:15–23

Zhang HJ, Zhang N, Yang RC et al (2014) Melatonin promotes seed germination under high salinity by regulating antioxidant systems, ABA and GA4 interaction in cucumber (*Cucumis sativus* L.). J Pineal Res 57:269–279

Zhang P, Senge M, Dai Y (2016) Effects of salinity stress on growth, yield, fruit quality and water use efficiency of tomato under hydroponics system. Rev Agric Sci 4:46–55

Zhang T, Shi Z, Zhang X et al (2020) Alleviating effects of exogenous melatonin on salt stress in cucumber. Sci Hortic. https://doi.org/10.1016/j.scienta.2019.109070

Zhao Y, Qi LW, Wang WM, Saxena PK, Liu CZ (2011) Melatonin improves the survival of cryopreserved callus of *Rhodiola crenulata*. J Pineal Res 50:83–88

Zhao H, Su T, Huo L et al (2015) Unveiling the mechanism of melatonin impacts on maize seedling growth: sugar metabolism as a case. J Pineal Res 59:255–266

Zheng X, Tan DX, Allan AC et al (2017) Chloroplastic biosynthesis of melatonin and its involvement in protection of plants from salt stress. Sci Rep. https://doi.org/10.1038/srep41236

The manufacturer's authorised representative in the EU is Springer Nature Customer Service Centre GmbH, Europaplatz 3, 69115 Heidelberg, Germany. If you have any concerns regarding our products, please contact ProductSafety@springernature.com

Printed and bound by CPI Group (UK) Ltd, Croydon, CR0 4YY
01/10/2026
02225915-0011